A look back at Earth by Voyager I in February 1990 from above the lunar highlands. Photograph ASP-14-2383, courtesy of Lyndon B. Johnson Space Center, NASA, Houston. Photograph provided by R. E. Stevenson, Del Mar

E. Uchupi K.O. Emery

Morphology of the Rocky Members of the Solar System

With a Foreword by Robert S. Dietz

With 150 Figures

Springer-Verlag

Berlin Heidelberg New York
London Paris Tokyo
Hong Kong Barcelona
Budapest

Professor Dr. Elazar Uchupi
Professor Dr. Kenneth O. Emery

Woods Hole Oceanographic Institution (W.H.O.I.)
Department of Geology and Geophysics
Woods Hole, MA 02543
USA

ISBN 978-3-642-87552-6   ISBN 978-3-642-87550-2 (eBook)
DOI 10.1007/978-3-642-87550-2

Library of Congress Cataloging-in-Publication Data. Uchupi, Elazar. Morphology of the rocky members of the solar system / E. Uchupi, K.O. Emery. p. cm. Includes bibliographical references and index. ISBN-13: 978-3-642-87552-6
1. Planets—Geology. 2. Satellites. 3. Satellites—Outer planets. 4. Planetology. I. Emery, K.O. (Kenneth Orris), 1914- . II. Title. QB603.G46U23 1993 559.9'2—dc20 93-30159

© Springer-Verlag Berlin Heidelberg 1993
Softcover reprint of the hardcover 1st edition 1993

Typesetting: Macmillan India Ltd., Bangalore 25

32/3145/SPS–5 4 3 2 1 0 – Printed on acid-free paper

Begin at the Beginning,
and Go On
Till You Come to the End : Then Stop.

The King in ALICE IN WONDERLAND

Foreword

These words are written on the 500th anniversary of Columbus' discovery of the New World. Surely the deep-space exploration of other worlds in our Solar System over the past few decades is an event of similar magnitude. Man has traveled far enough to see Spaceship Earth suspended alone in black space. And he has voyaged even farther to marvel at the crescent Earth rising over the Moon's cratered terrain. Instrumented spacecraft have toured the entire Solar System even beyond the ninth planet Pluto. This work of science *Morphology of the Rocky Members of the Solar System* is an inquiry about our extended home. As with the Darwinian and Copernican paradigms, the nature of our planetary system, as the extended world around us, has great significance for those who ponder the human condition. The deep-space views of our Planet Ocean with its sweeping clouds, and moving oceans and creeping continents must rank as the greatest photograph ever taken. Viewing Spaceship Earth hanging in the vast void is an almost frightening experience. We are so alone! It is easy to understand why so many are attracted to a simpler account of origins, like the allegorical tale of creation written in heroic style (but eschewing math, maps, figures, tables, references, and evidence) in the first eleven chapters of Genesis.

This treatise examines the morphology of the six rocky planets and their 27 satellites from a broad perspective. Aspects examined are their internal constitution and volcanic/tectonic activity (endogenic processes); their surface modification by wind, water, ice, mass-wasting, etc. (exogenic processes); and their modification by impact cratering (an exotic process). Remarkably, asteroid/cometary cratering was regarded as a nonprocess on Earth until recently but now these star scars (astroblemes) appear to dominate surface morphology on terrestrial (rocky) planets.

The landing of man on the Moon during the 1960s is best known. But the grand tour of the outer planets (Voyagers 1 and 2), the Mars missions, and the 23 encounters with Venus topped off with its mapping by Magellan are events of far greater scientific value. There are those who decry these expensive feats as a waste of money, but what a comment it would be on the human race if, although we had the technology to do so, we did not try! Mark Twain once chided scientists for spinning so many speculations from a paucity of data. But in reading this synthesis, one is impressed by the ingenuity of scientists in understanding the planets although we have certain ground truth only from lunar rocks. The usual behest is "Let's go and see" but better advice would seem to be "Let's think about it." Some personal comments on the Moon by astronauts ("Wow, fantastic!") are not really meaningful. In the bibliography of about 1000 references, interpretations from instrumental results overwhelm personal observations.

The Earthlike planets and their satellites have orbited the Sun since its nuclear furnace burst into flame 4.55 Ga (billion years ago). This book should appeal to those having a nontrivial interest in Earth history, astronomy, or the grand scheme of things. Crammed with facts and figures, it is not easy reading but nevertheless its perusal is richly rewarding. The interpretations are constrained by the current data set which is expanding rapidly. Some conclusions may not stand the test of time as it is the nature of science to be tentative, falsifiable, testable, and corrigible, so we can expect a better and more exact understanding as we approach the twenty-first century.

Before 1950 historical geology concerned mainly the Phanerozoic Era beginning in the Cambrian with a clearly legible fossil record. Recently, the veil has been lifted from the Precambrian with radiometric dating providing chronological control. Even so, our knowledge hits a blank wall about 3.0 Ga ago. Fortunately, an earlier readable history appears to be preserved on the Moon, Mercury, and especially Mars although the Hadean Era (4.55 to 4.0 Ga ago) is once again wiped out by accretionary bombardment. Yet, the large variety of planetary bodies gives hope that initial events such as core formation eventually can be worked out. Planets are heat engines, so the generation and the ridding of internal heat are basic to their evolution. It is understandable that small bodies, whose surface areas are large relative to their volume, radiate heat readily and so remain as solid coreless rocks in space. But why does not Venus, with a planetary radius like that of Earth, exhibit plate tectonics? Many mysteries remain to be explained.

Why is Earth so special? The presence of an ocean and an atmosphere is a partial answer. Earth uniquely has three volumes or "spheres" – hydrosphere, atmosphere, and lithosphere (liquid, gaseous, and solid states of matter). Geometrically, volumes intersect–forming plains; thus there is the First Surface (land/air), the Second Surface (air/water), and the Third Surface (water/ocean floor). It is in the nature of things that energy exchanges take place primarily across surfaces; hence their importance in the scheme of things. Only Earth has a hydrosphere which is a substance needed for the existence of life. Our planet is a remarkable place. There may be other similar planets in our universe, but even in a galaxy of ten billion stars the number of Earth-like planets must be a small number. Can we ever get there? Probably not, for if so, they would already be here. The question remains: where is everybody?

Nearly three decades ago we learned how our world works – by plate tectonics and uniquely so. Among Solar System members, there is a carapace of eight major lithospheric plates and many lesser ones which drift and rotate a few centimeters per year interacting along their boundaries by subducting, shearing, or spreading apart. Mountains, volcanoes, earthquakes, and even the continents and ocean basins are the result. Hypsographic bimodality is unique to Earth and is the hallmark of a highly evolved planet where basalt is extruded as a partial melt of mantle peridotite and then is remelted to create the sialic rocks of the granitoid continental plateaus. A convecting body like the Sun is understandable and so is a solid rock in space which has gravitationally collapsed into a sphere like the Moon. But a planet with surficial plates drifting only as fast as one's fingernails grow is nonintuitive. This must be why we were so slow in understanding our own planet.

The Moon was the first celestial body to be a source of wonder. To ancient Greeks, it was a mirror reflecting a map of the Mediterranean region, but in 1609 Galileo's telescope discerned the presence of mountains and maria ('seas'). Before 1965 there was a consensus that the lunar craters were volcanic and the Apollo project was biased toward the search for such plutonian features. But it quickly became evident that lunar highlands or terrae are saturated with impact craters. Even the maria are giant waterless impact basins dating back to the terminal bombardment nearly 4 billion years ago. The Moon is so dry that basalts erupted then are less weathered than Hawaiian flows extruded yesterday. With the drift of the continents over the eons Mother Earth has changed her expression slowly, but the Man-in-the-Moon virtually forever has retained a fixed stare. Mercury, like the Moon, is saturated with craters and became quiescent 3 billion years ago. At the University of Illinois in the late 1930s, this writer proposed writing a doctoral thesis about the lunar surface, but it was rejected as not a suitable subject for geological study. Not in their wildest dreams did anyone then anticipate that 30 years hence man would walk on the lunar surface. Now space travel is taken for granted.

Mars, the red planet, has long been the subject of unconstrained speculation and science fiction. Sadly, Percival Lowell's ideas about canals and intelligent life were proved totally wrong when viewed in the stark reality of the Mariner and Viking spacecraft missions. Mars is now dry, cold, and inhospitable to life although there may once have been shallow lakes whose deposits may one day reveal fossil evidence of extinct primitive life. Mars is unique in having polar ice caps that strongly wax and wane yearly. Its two small satellites, Phobos and Deimos, appear to be captured asteroids. Although it lacks surficial water we do know that Mars has ice and a thin atmosphere sufficient to generate dust storms which obscure the surface. It also has a dense core and has evolved a basaltic crust but apparently lacks granitic highlands. The Tharsis region is a great bulge with volcanic edifices and capped by Olympus Mons, the highest mountain in the Solar System. Extending radially away from Tharsis is Valles Marineris, a giant gash of uncertain origin. Like nearly all extraterrestrial bodies, Mars is a museum of archaic landforms; major geological activity ceased more than three eons ago. Although no surface samples have been returned from Mars, certain meteorites called SNCs (eucrites or shocked basaltic rocks) probably are martian samples blasted off its surface by asteroidal impact and then orbited into space for millions of years before colliding with Earth. The Mars Orbiter launched in 1992 will mark a new return to Mars and promises a wealth of new information provided by new sophisticated instrumentation. Will man ever visit this most Earthlike of planets? Very likely it is in the offing although decades away. It is even remotely conceivable that we can modify the martian environment or perhaps ourselves make Mars a second habitable oasis in space. Presently, Mars, with its complex landforms and varied history, remains a wonderful planet for scientific contemplation.

Radar imaging has pierced the thick cloud veil of Venus and poured out an avalanche of new data from the Magellan mission – more than from all previous space flights combined. Revealed is an arid searing (500 °C) terrain where lead would melt and which apparently was resurfaced catastrophically some 500 million years ago. Volcanic constructs dominate the scene, but there are also nearly 1000 impact craters. The term astrobleme used on Earth to describe the eroded or buried remains of an impact crater is quite unnecessary on Venus as nearly all such features are pristine with little degradation. In contrast to Mars with an atmospheric pressure only 1% that of Earth, venusian air is 100 times as thick as Earth's and is dominantly carbon dioxide. Thus, the surface has a shield against impacting bodies equivalent in effect to 1 km of water. The ice-ball comets probably break up high in the atmosphere and only the larger asteroids get through. Mead, the largest impact crater, is 275 km across. Volcanoes abound but it is doubtful that any are active, for in spite of their fresh appearance they are mostly tens and even a few hundred million years old. The overall generation of lava is probably only about 10% of the rate on Earth. As a wholly different kind of world, Venus holds fascination but is more like Hell than Heaven.

The Copernican revolution was a blow to man's ego. Jerusalem was not the center of the universe nor was the Earth for it revolved around the Sun. The Sun, in turn, was an ordinary main sequence star in the vast Milky Way, an ordinary galaxy among billions in the universe. But comparative planetology reveals that the Earth is nevertheless remarkably unique. Life-friendly habitats capable of housing a biosphere are a one in a billion chance. Not only were conditions right on Earth for biopoiesis (the origin of life) but the environment was stable for eons permitting the evolution from dawn species, to photosynthesis, to the eucaryotic cell, to multicellular organisms, to vertebrates, and to sentient man. Surely, the greatest disappointment of our generation is not finding even primitive life on Mars or elsewhere. There seems little chance that we ever can communicate with intelligent life elsewhere because, if it were possible, they would have contacted us already.

The surface of a sphere, the perfect geometric form, is finite but boundless. All points on a sphere are geometrically equal but hardly so for Earth's ecology. Most of our globe is so inhospitable that higher life is confined to the green oases. We regard our environment or biosphere as benign, but oxygen, a product of life, is a highly corrosive poison, and water is the greatest of all solvents. Through evolution life has adapted to this milieu although magmatic rocks rapidly weather to sand and clay. We live under stress which causes time's arrow of evolution to fly faster sacrificing individuals for the good of the species along the way. Land is more stressful than the sea, requiring locomotion and coping with desiccation, gravity, freezing, etc. Evolution proceeds faster on land so that the mammals which now dominate the ocean returned to the sea after evolving on land. As the Earth winds down, life winds up but a quartz crystal from the Precambrian is the same yesterday, today, and forever.

It is said that we must leave home to understand the place from whence we came. The bizarre collection of neighboring planetary bodies emphasizes the unbelievable uniqueness of Earth with its vast blue planet-encircling ocean of liquid water from which rise the tan island continents. A hypothetical Man-in-the-Moon (I suppose he should be called a lunatic) would see no sign of life on Earth except for coral reefs. He would wonder at the shallow continental shelves and whether rising sea level would drown the continents. But land life has existed since the Silurian and so it never will be completely drowned because plate tectonics provide a feedback mechanism which relates the freeboard of continents to the depth of the oceans. Earth's surface temperatures (permitting water to occur in all three states of matter) and atmospheric composition always have been conducive to the origin and evolution of life. Although some other stars may have stable planetary systems, other habitable abodes such as our must be rare in the extreme.

The new chapters in today's standard textbooks of geology are plate tectonics, man's role as a geological agent (environmental geology), and planetology. Clearly, we must accept the new paradigm of astrocatastrophism – the role of impacting asteroids and comets in shaping the external morphology of the planetary bodies. This is most evident on the Moon and Mercury but also is clear on Mars and Venus. Only about 150 astroblemes (ancient, deeply eroded impact scars) are currently identified on Earth as this record has been largely obliterated by tectonism, erosion, and sedimentation, making discovery difficult. It took a 10-year search to find the buried 200-km-diameter Chixulub crater in Yucatan that was apparently the source of the worldwide iridium-rich fallout layer at the Cretaceous-Tertiary boundary and its associated mass extinctions. Precambrian examples of giant impact scars are the Sudbury Basin in Canada and the Vredefort Ring in South Africa. Although plate tectonics remains the Earth's game plan, cosmic cannonballs are wild cards punctuating the geological record and even perturbing the evolution of life.

Comparative planetology promises to tell us much about Earth. Size is, of course, a critical factor and planets must dissipate heat from radioactive decay, gravitational collapse, and other sources. Comparative planetology is a new discipline awaiting the attention of insightful scientists and this synthesis is such an effort. The early pages of Earth history are tattered and torn and the opening chapters are entirely missing. In the vast abyss of time the understanding of our beginnings must be sorted out with reference to the entire suite of rocky celestial bodies. Archaic realms like Mercury, Moon, and Mars hold clues which one day will raise the curtain on Earth's early days. Truly remarkable discoveries lie ahead. Be assured, however, Earth for all its failings will remain everyone's favorite planet and water will remain the fluid of choice.

A final personal note. It was a pleasure and an honor to be asked to write a brief Foreword for this scholarly work. I first met K. O. Emery in 1935 at the University of Illinois. In the midst of the Great Depression, he had arrived by train (in an empty box car) from California while I had hitch-hiked out from New Jersey. We both then fell under the tutelage of Francis P. Shepard, the founder of marine geology. Together, we completed our bachelor's, master's, and doctoral degrees migrating back and forth between Illinois and the Scripps Institution of Oceanography in California. Elazar Uchupi can be best described as Emery's protégé, collaborating initially as Emery's student during his tenure as a professor of geology at the University of Southern California. In 1962, they moved as a team to the Woods Hole Oceanographic Institution. Probably their most notable joint opus to date has been their 1984 two-volume treatise on The Geology of the Atlantic Ocean. In the present book, the usual authorship has been reversed. In my view, both Emery and Uchupi are geniuses in the Edisonian sense – 10% inspiration and 90% perspiration. Workaholics is the operative word, the type of people one admires but does not envy. As dedicated naturalists, they are relentlessly driven by intellectual curiosity, a desire to know everything. Emery once told me that he writes

scientific papers so that he can forget and get on to new things. Alas!
Unlike elephants which sometimes, I am told, forget Emery and Uchupi
never seem to. They seem to be the first persons since Leibnitz to know
everything – at least in their chosen fields of science.

Tempe, Arizona   ROBERT S. DIETZ
September 1993   Professor of Geology Emeritus
                                               Arizona State University

Preface

In many ways the exploration of extraterrestrial space, especially within the Solar System parallels previous exploration of the lands and then of the oceans of the Earth's surface. In all three examples the active part of the explorations required prior developments or adaptations of technology, and the continuity and sustenance of the explorations depended on perceived economic or political benefits. Explorations of the lands began with increased human populations followed by gathering into small cities in Mesopotamia by 5000 B.C. and in Egypt, development of agriculture, masonry, weaving, mining, metal-working, and the creation of a leisure class that invented writing and numerical notation, all by about 3000 B.C. Surplus products led to trade, land transport, coinage, a form of banking, and iron weaponry by 2000 B.C. Earlier religions and laws became more formalized as did alphabets and literature by 1500 B.C. Migrations of peoples occurred in earliest times because of changes in climates and political needs, increasing in numbers and distances throughout this period and afterward, but migrations were limited eventually by the inability to traverse the ocean. Some thinkers turned their attention to estimating the size of the Earth (Eratosthenes, about 225 B.C.) mapping it, and developing the concept that the other planets and the Sun orbited in an Earth-centered system (Ptolemy, A.D. 170). Others began to learn its composition (India–earth, air, fire, and water) about 1500 B.C.; later, various Arabs and then Europeans and Americans had discovered 9 of the 103 presently known chemical elements by A.D. 0, 12 by A.D. 1500, and 84 by A.D. 1900.

Oceanic exploration could occur only after ocean-worthy ships had been developed for both trade and war, as early as 2650 B.C. in the Mediterranean Sea (Egyptians, Cretans, Phoenicians, Greeks). By 1200 B.C. the Phoenicians reached Britain to trade for tin, and by 600 B.C. a Phoenician fleet had circumnavigated Africa clockwise from the Red Sea under contract with the Egyptian King Necho II, but there were no immediate continuations because economic or political benefits were not evident. About 600 B.C. the Carthagenian Hanno sailed down the west coast of Africa, and Pytheas of Massilia sailed to north of Britain in 220 B.C. By about A.D. 700 Viking ship building had improved so that long voyages in the Atlantic Ocean resulted in discovery of Helgoland in A.D. 698, Iceland by the Irish in A.D. 800

and by Vikings in A.D. 862, Greenland in A.D. 900 by Eric the Red, and Newfoundland in A.D. 1000 by Leif Ericson. A whole new epoch began with the Portuguese much later, chiefly through systematic training by Prince Henry the Navigator, Joao Diaz to Cape Bojador in A.D. 1434; to the Gulf of Guinea in 1470, Diogo Cam to the Congo River in A.D. 1482, and Vasco daGama to India in 1498. Crossing of the Atlantic Ocean was accomplished with Spanish funds by Cristoforo Columbo in 1492–1500, Amerigo Vespucci in 1499, and with Portuguese funds by John Cabot to Labrador in 1497, and Pedro Alvarez Cabral to Brazil in 1500. In 1500 also came Juan de la Cosa's map of the New World. In 1522 one of Ferdinand Magellan's five ships circumnavigated the globe carrying only 18 survivors of the original 239 men who began the 3-year voyage. These navigations were for economic purposes – to avoid the Arab transit control of spices, silks, and other luxury items from India and farther east by sailing around Africa or finding a direct passage across the Atlantic and Pacific Oceans. Required were not only good ships but reasonably accurate navigational instruments and knowledge: the astrolabe by Arabs in A.D. 850, mariner's compass in A.D. 1125, general concepts of navigation provided by Prince Henry about A.D. 1440. After 1520 there were many exploratory voyages by Europeans to claim new regions and their resources or to plunder the products gained by other Europeans (Francis Drake, 1572, and others), but a by-product was the filling of gaps in geographical knowledge. The results led to a more complete understanding of the geography of the whole Earth than was possible from land habitats alone in the Middle East, Asia, northern Africa, and Europe, as illustrated by new maps of the world (Olaus Magnus, 1539; Gararardus Mercator, 1595; and globes by Martin Behaim, 1492; and others). Moreover, the exploration of ocean space provided access to new areas for subsequent land exploration and habitation – thus 'The New World' of North America, South America, and many smaller areas and islands unknown to Europe before 1500.

The third realm of space to be explored is beyond the Earth – outer space. Curiously, this realm was perhaps the first to be given much thought by early humans because cursory optical examination was easy – viewing the light from sun, moon, planets, and stars. By about 4500 B.C. Egypt had produced a calendar of 360 days with 12 months of 30 days, perhaps mainly as an aid to agriculture and trade. It was adjusted slightly many times in later centuries by other cultures. After about 3000 B.C. astronomy (including astrology) was systematized in Egypt, Babylonia, India, and China. Even Cheop's pyramid of about 2500 B.C. in Egypt records important astronomical alignments. By 500 B.C. China had measured the angular height of the Sun with respect to the Earth's axis of rotation and China, Egypt, and Greece had invented sundials. At least China and Greece were able to predict solar eclipses;

Babylonia had established the length of the lunar months; and the Mayas in Central America had developed their own calendar and knowledge about planetary movements. By A.D. 963 the Arab Al Sûfi published his Book of Fixed Stars and discovered the Andromeda Nebula; and at Toledo a table of the positions of stars came in A.D. 1080; Nicolaus Copernicus stated in 1512 without the approval of the Pope that the Earth and other planets revolve around the Sun; Tycho Brahe built a 6-m quadrant and a 1.6-m celestial globe and plotted the positions of 777 stars; Galileo Galilei in 1610 discovered the revolution of the larger moons of Jupiter around their host and the phases of their illumination by the Sun; Johannes Kepler empirically obtained his laws of planetary motion in 1618; Johann Hevel described the lunar surface in 1647 and Giovanni Riccoli named many lunar features in 1651; Giovanni Cassini determined the rotations of Jupiter, Mars, and Venus in 1665; Isaac Newton developed the understanding of gravitational laws for the Solar System in 1666; Edmund Halley in 1705 correctly predicted the return of Halley's comet for 1758.

Subsidiary but necessary advances ensued: nautical almanac in 1767; J. J. Leland catalogued 47390 stars in 1802; William Herschel discovered binary stars in 1802; Christian J. Doppler discovered a red shift of stars in 1842; Lion Foucault measured the speed of light in 1862; G. V. Schiaparelli discovered the synchronous rotations and revolutions of Mercury and Venus in 1889; Viktor F. Hess – cosmic radiation in 1912; Max Wolf described the Milky Way and asteroids by photography in 1920; Henry Draper catalogued the spectra of 225000 stars in 1924; James H. Jeans and T. C. Chamberlin proposed the origin of the Solar System by a passing star in 1926 and 1928, respectively; Edwin P. Hubble measured the red shift of nebulae and advocated the expansion of the universe in 1929–the Big Bang; C. W. Tombough discovered Pluto in 1930; a United States' rocket missile traveled 5000 km/h and reached 125 km altitude in 1942. These and very many other observations and inferences about the Solar System were made from the Earth's surface by scientists of many nations, but mostly by Europeans and later Americans between 1600 and 1950. The invention of the computer in 1942, the transistor in 1947, and more powerful rocket missiles contributed greatly. Finally, the Soviet Union and the United States opened the way to closer examination via their Sputnik I in 1957 and Explorer I in 1958.

It is clear that in many ways the slow and gradual prelude and development of travel into outer space parallels the earlier travel on land and on the ocean at the Earth's surface. All had early beginnings but had to await the growth of broad interest and adaptation of technologies largely initiated for other purposes. Land travel required beasts of burden, wheeled vehicles, and weapons. Ocean travel required ocean-worthy ships, navigational instruments, and seamanship, but it led to a new phase of land travel. Early interest in extraterrestrial space had continued by visual, photographic, radar, and other means of observation, and finally rocketry brought humans to the nearest of the heavenly bodies – the Moon. If the trend continues to parallel the exploration of land and ocean, it may lead to longer and more distant space travel and perhaps to at least limited habitation such as may be needed for economic and political advantages. Because space travel lagged land travel by about 5000 years and ocean travel by about 1000 years, the future extent of knowledge about planets and stars cannot yet be fully visualized, nor can the future extent of space travel and habitation. Another 5000 years is needed!

Knowledge of the compositions and development of the Solar System and of its individual members may now be increasing at a rate that will double in less than a decade. This increase in knowledge permits the formulation of more rational hypotheses regarding the origin of the Earth than were previously possible. In fact, we have been able to learn about some aspects of its geohistory and oceans that cannot yet be learned from examination of the Earth alone. This same rapid increase in knowledge and the development of new technologies may well cause the methods and conclusions that are summarized here to become obsolete soon. Nevertheless, our efforts at synthesis may enhance and speed the future arrival at a much better understanding of how the Earth came into being, when and how the Earth may reach its end, and whether or where other Earth-like habitable planets may exist.

Our approach in this book was to describe each of the planets of the Solar System, their satellites, and groups (meteorites, asteroids, and comets) in turn, making comparisons with other bodies and trying to arrive at reasons for the differences that exist. The subject is so large that we are still at an early stage in the sequence of observation, interpretation, understanding, and utilization—farther in some aspects of the sequence than in others. Therefore, we hope that our attempt here will serve to interest others in carrying the effort farther in the near future.

Woods Hole,   E. UCHUPI
September 1993   K. O. EMERY

Acknowledgments

We wish to express our appreciation to the following individuals whose help made the task of writing this book much easier: D. Helpern of the Jet Propulsion Laboratory, Pasadena, CA; R. E. Stevenson, Del Mar, CA; Lisa Vazquez, National Aeronautics and Space Administration, Lyndon B. Johnson Space Center, Houston, TX; Carolyn Y. Ng, P. D. Lowman, Jr., and J. R. Heirtzler of the Goddard Space Flight Center, National Aeronautics and Space Administration, Greenbelt, MD; Joyce M. Miller, Assistant Director of Operations, Ocean Mapping Development Center, Graduate School, University of Rhode Island, Kingston, RI; Nancy Soderberg, D. W. Twichell, W. C. Schwab, and W. P. Dillon of the Marine Branch, US Geological Survey, Woods Hole, MA; S. C. Solomon formerly at Massachusetts Institute of Technology and now at Carnegie Institution of Washington, Washington, D.C., and Suzanne E. Smrekar of Massachusetts Institute of Technology, Cambridge, MA; and Pam Foster, R. A. Goldsmith, Colleen Hurter, Marie Johnson, S. R. Gegg, W. N. Lange, G. M. Purdy, Susan S. Putnam, N. Shimizu, and Joy J. Travis of Woods Hole Oceanographic Institution, Woods Hole, MA.

Our greatest debt is to the many space technicians and scientists whose efforts produced the imagery, data, and interpretations of this information that we have built upon. Their contributions are acknowledged in the form of many text references to their publications and in the bibliography – nearly 1000 citations, of which half date from 1984 or later. Their work also provided the basis for many of our figures and tables, as acknowledged in the captions. The success of the scientists, in turn, depended on the abilities of the engineers who designed and built the satellites and sensors that made the images and recorded the data. The countries that sponsored this space research have justly been proud of the results and of the people who have been at the forefront of the exploration. One method of expressing this pride has been the issuance of postage stamps depicting many of the satellites, some of the images that were obtained, and some of the explorers. We, therefore, selected many of the postage stamps that show satellites and their images of the Universe and of members of the Solar System and include them in this book as unnumbered illustrations at the end of appropriate chapters.

Contents

1 Introduction

In the vastness of space, nine planets, 27 moons having diameters larger than 100 km, and thousands of smaller bodies orbit around a star that is not much different from many other stars in the Universe. The third planet from the Sun appears quite blue from space as a result of its extensive hydrosphere, an envelope that made possible the creation of life. Streaks of white depict the circulation of its atmospheric cover. Since inception, its biosphere has deeply influenced the Earth's development, so much so that it can be conceived as being the prime agent of the Earth's morphological evolution. The geohistory of Earth includes the effects of another agent – massive lateral tectonic transport of crustal plates induced by convection circulation within the planet's interior – a tectonic regime that has led to changes in its surface morphology, its climate, and even its biosphere as the latter adapted to ever-changing habitats. In this inquiry we look at Earth's rocky companions in the Solar System to determine how similar or different was their development from that of the Earth, then we attempt to ascertain the causes of these differences. Such a goal has been made possible by recent observations of the planets via unmanned satellites and actual landings by man and vehicles; these observations have yielded many publications during the past two decades. As our own geological field experience has been restricted to our planet, we have made extensive use of these publications.

We entered this inquiry through general curiosity about the early history of the Earth, the nature of its companions in the Solar System, and the reasons why there are so many differences between them. Previous general syntheses are by Head and Solomon (1981), Hartmann (1983), Carr et al. (1984), Greeley (1984), and others will follow as surely as more facts from future landings will produce more knowledge. This book is an updating of knowledge of the Solar System and of implications for the evolution of the Earth for the benefit of geologists. The synthesis of our findings was prepared hopefully to arouse the curiosity of other geologists and perhaps to serve as a new general platform for others to build upon without having to make similar massive reviews of the recent literature like the one that was undertaken by us. In addition, it may conceivably help increase the science literacy of students beyond the general level within typical textbooks (Freeman 1993).

The way that the reality of the history of the Earth and the rest of the universe was envisioned during the past and today is influenced by the mores of human culture at these different times. During past millenia physical reality was considered a manifestation of the spiritual, and we looked to a Creator. During those early times the physical and spiritual realms were conceived as being inseparable pieces of the same reality. Today, we are more interested in physical aspects and look to science for explanations. Most modern thinkers, particularly in the western world, see the physical and spiritual realms as quite distinct from one another, a philosophy that has produced considerable disagreement between the followers of these two realms – a conflict between those who look at the fabric of the Universe and see the culmination of a series of random events, and those who look at the same Universe and see a planned grand design.

Humans have long been so familiar with the topography of the particular part of the Earth's surface where they lived that exposure to different topography in other regions often surprises distant travelers. The early investigations of geomorphology of the land during the nineteenth century by Lyell (1850) in many editions of his *Principles of Geology* and of the ocean floor by Maury (1855) in editions of his *The Physical Geography of the Sea* led to further investigations that continue to the present. During the early twentieth century only a few authors (T. C. Chamberlin, William M. Davis, Walther Penck, Eduard Suess, and J. H. F. Umbgrove, for example) concentrated their efforts on the Earth as a whole and dealt with the broader aspects of geomorphology. Most studies were and are restricted to small regions of the Earth and have been conducted to levels of progressively increasing detail.

Prior to 400 years ago, knowledge of the members of the Solar System beyond the Earth and its Moon was restricted to brightness, color, and movements that differ greatly from those of stars. Eclipses (the earliest was reported in China prior to 2000 B.C.) usually were considered as omens (*omina*, ominous) for subsequent events, and the interest in precise prediction of lunar and planetary movements came from astrologers, beginning long before the codification of astrology about 1000 B.C. in cultures of both Asia and Europe (Needham 1959, vol. III, pp. 392–408). Similarly, the need for a simple but accurate earthly calendar to date past events, to coordinate economies (mainly agricultural), and to make future appointments led very early to improved and independent observations of timings of the Sun, Moon, planets, and stars by priesthoods in many diverse cultures of the Earth (Wilson 1937).

Later came the utilization of celestial bodies for coastal and then distant-ocean navigation (Taylor and Richey 1969). Ephemerides, almanacs, tables of lunar conjunctions for Mercury, Venus, Mars, Jupiter, and Saturn, and tables for lunar eclipses and astrology were written in cuneiform characters at least as early as 747 B.C. (Sachs 1955). The Romans, probably long before Julius Caesar's calendar reform of 45 B.C., named the days of the week for the Sun, Moon, and the five visible planets (most of which already had been named for Roman gods): Solis, Lunae, Martis, Mercurii, Jovis, Veneris, and Saturni. These day names were continued in Latin-derived French, Italian, Spanish, Portuguese, and Romanian, but the Saxons substituted some of their own gods that had similar attributes (Tiw for Mars, Wotan for Mercury, Thor for Jove, and Frigg for Venus) and from these were derived the names used in German and English (the latter: Sunday, Monday, Tuesday, Wednesday, Thursday, Friday, and Saturday). The Japanese names for the days of the week were derived from the Sun, Moon, and the five planets that already had been named after the ancient 'elements' of fire, water, wood, metal, and mud. Russians accept the European names for the planets but number some weekdays in consecutive order. The seven-branched palm tree denoted on some coins of Israel (especially on those minted during the revolt against the Romans led by Bar Kochba during A.D. 132–135) may have symbolized both the same seven heavenly bodies visible to the unaided eye and the days of the week, but the days of the week are numbered consecutively. They end with Shabatt (the seventh day and one of rest after six days of Creation – according to their religion, this practice also is followed on printed calendars used by western societies).

The bodies of the Solar System (houses) and their movements viewed from Earth against the 12 constellations (signs of the Zodiac) or against subdivisions within each sign, form the basis for astrology. The relative positions of these bodies at the date of birth of an individual, or at the initiation of organizations and groups, including nations, was supposed to permit predictions of later events to affect them, and positions at given later dates were supposed to indicate propitious times for actions or for useful answers to questions – basically a system of divination. Because of the intricate and changing relationships of the seven houses to the 12 signs, astrology probably was the ancestor to astronomy because of the need for increased precision of naked-eye celestial observations. Astrology has had a long and complex history marked by considerable variations in interpretation as it evolved from Babylonian through Assyrian, Persian, Egyptian, Greek, Indian, Arabic, and Chinese, to Western schools. When the heliocentric solar system displaced the Earth-centered concept during the seventeenth and eighteenth centuries and when additional planets were discovered, astrology became scientifically untenable and marked by increasing fraudulence against gullible populations. Nevertheless, the astrological attributes of persons born under the signs of the seven 'planets' known to the ancients have become ingrained in our literature: Mercury – swift; Venus – female, feminine; Mars – male, martial; Jupiter, Jove – jovial; Saturn – saturnine, morose; Moon – lunacy, looney; Sun – bright? The outer planets (Uranus, Neptune, and Pluto) have no such attributions, because they were discovered only in 1781, 1846, and 1930, respectively. The seven earlier recognized bodies that move differently from the stars (planets from *planetes*, Greek for wanderers) also provided the names of the days of the week supported by the seven-day length of the phases of the Moon and the biblical seven-day Creation and rest – all contributing to the power of the number seven.

A bronze calendar-calculator was recovered from a depth of about 55 m off Anticythera, a small island between Greece and Crete (Grant 1990, pp. 66–69). When set in motion on its axle it gave the movements of the Moon and Sun and the rising and setting of constellations and planets (Mercury, Venus, Mars, Jupiter, Saturn, and possibly others as well). It was made in Rhodes about 87 B.C., repaired twice, the last time in 80 B.C. when it was reset before being lost at sea. Apparently, it was built to help in the compilation of an astronomical calendar. The complexity of this device challenges our general belief about the technological backwardness of Greek and Roman worlds.

In the New World the Mesoamerican Mayan astronomers and chronologists created a sophisticated calendrical system in a solar calendar of 365-plus days which they organized into 18 months of 20 days plus a period of 5 extra days. Two other Mayan calendars included a lunar one based on 29 and a fraction days between two successive new moons, and a Venus calendar that integrated five 584-day orbits of Venus with eight 365-day solar years. They also used a 260-day calendar, the tzokin, that is a pure invention as it does not correspond with any recognized astronomical phenomena. In the Mayan Dresden Codex, painted in the thirteenth century A.D. is a table of the movements of Venus, an eclipse table, a zodiac, possible references to other planets, and cosmological statements about the relationships of time, space, and the gods (Henderson 1981, pp. 92, 144; Willey 1966, pp. 132–135). Probably some of the many other codexes that were destroyed by the Spanish invaders as works of the devil contained other interesting astronomical information.

About 330 B.C. Aristotle in his Book III *Physica* considered the general mechanics of movement in the solar system, describing his concept of concentric spherical shells of transparent crystalline ether, each sphere turning independently. The inner four of these shells were the terrestrial ones, the lithosphere, hydrosphere, atmosphere, and pyrosphere (the elements of the ancients – earth, water, air, and fire – the fire here being from meteorite trails and auroras; Siscoe 1991). These four shells were thought to be enclosed within the lunar shell that in turn was overlain by a series of other outer shells separately containing the planets and the Sun. Beyond the shell with the Sun was another that contained all the stars. Differential rotations of the celestial shells provided the different rates and directions of their contained bodies. A mechanical improvement over Aristotle's spherical shells was made by Ptolemy's Earth-centered Solar System described in his *Almagest* of A.D. 145, made largely to account for apparent reversals of movements of the planets along their orbital paths as viewed from Earth. These concepts by both Aristotle and Ptolemy were attempts to explain relative movements without knowledge of gravity (being just descriptive – essentially symptoms and not causes).

Only much later (seventeenth century A.D.) could the morphology of members nearest to Earth be studied from telescopic observations and photographs. The past decade has yielded enormous amounts of new information from orbiters (digitized and transmitted by radio back to Earth) and even from the landing of probes; other more direct and comprehensive observations were made by six pairs of astronauts who landed on the Moon. These new sources provide knowledge of the morphology far beyond that which was available only a few decades ago. Most of the new imagery from orbiters is very broad (regional in scope), while some other images are highly detailed (the surface beneath and near the legs of landers and photographs from Moon vehicles and astronauts); however, less is available on an intermediate scale of tens of kilometers. This means that most global-wide knowledge is restricted to first-order morphological units (mountains, hills, plateaus, and plains) with only incidental information being yielded about second- and third-order features (escarpments, valleys, peaks, terraces, volcanic flows, landslide scars, and dunes). The first-order morphological features of whole planets or moons exhibit many differences from those of the Earth because of differences in major structure of the bodies (thickness and composition of crust and mantle, presence or absence of a core, a convection system, and geometry of the systems), and in eroding, weathering, and transporting agents (presence or absence of hydrosphere and atmosphere). Thus, a comparison of the morphology of the various members of the Solar System must illustrate major variations produced by unlike conditions of origin, crustal emplacement, and subsequent alteration of the surfaces. Such variety may illustrate stages in the evolution of the Earth during its past history – an investigation of some morphologies that no longer are produced on Earth and the influence of those morphologies upon later development of the Earth's surface.

2 Origin of the Solar System

2.1 General Properties of Galaxies, Stars, and Sun

Our Universe is believed by many astronomers and geologists to have begun 10 to 20 Ga ago with an outrush of material during the 'Big Bang.' Other cosmologists question the concept of the Big Bang, stating that it reflects search for a creation and a beginning (Burbidge 1992; Powell 1992); they proposed that creation is continuous and occurs in a series of little bangs. Was the Big Bang the beginning of time? If not, what came before it (Hawking 1988)? Should one visualize the Cosmos as having a beginning without a beginning, and an end without an end, as Merritt (1932, p. 30) described *Khalk'rus*? The present concensus is that the expansion of the Universe after the big bang appears to have gone through two stages: first, a period of rapid acceleration and inflation, when all distorting influences diminished quickly, and second, a phase when the Universe developed its present highly symmetrical state of expansion (Barrow 1991, pp. 49–50; Halliwell 1991). Stars, galaxies, and clusters of galaxies aggregated within this outrushing material when the Universe was about 1 Ga old (Riordan and Schramm 1991, pp. 3, 21; Powell 1992). Measurements on the broadest scale indicate nearly uniform radiation within the Universe, but recently the Cosmic Radiation Explorer (a satellite of the National Aeronautics and Space Administration – NASA) has mapped small variations in temperature attributed to the irregularities in radiation produced by the Big Bang. Within the past decade more detailed mapping has shown a nonuniform distribution of galaxies separated by huge voids, sheets of galaxies (such as the Great Wall, the largest coherent structure known within the Universe), and large foam-like distributions of galaxies (Spergel and Turok 1992). More than 90% of the material in the Universe consists of dark unseen matter generated during the inflation phase and required by density considerations. This dark matter is in a form that is difficult to detect, possibly as black holes, brown dwarfs, or planetary blobs the size of Jupiter or smaller (Riordan and Schramm 1991, pp. 61–62).

Studies of spiral galaxies reveal that young stars are concentrated in the spiral arms, showing that some process or condition enhances star formation there (Cameron 1978). The structure and dynamics of galaxies indicate that interstellar material revolves around the center of the galaxy. As this cloud of material passes from an interarm area into an arm of the galaxy (a gravitational trough), it decelerates and slowly becomes compressed to form young stars (Reeves 1978b).

Once formed, the stars continue in their circumgalactic path. Transit through the arm may require about 100 Ma, thus the more massive stars must explode before leaving the arms.

The Milky Way is a spiral galaxy having a bright, central, lens-shaped nucleus containing millions of stars, about 100 000 light-years wide and 10 000 light-years thick (1 light-year = 9.46×10^{12} km). It is slowly rotating, with the stars in the galaxy's arms orbiting its center about once every several hundred million years. The Milky Way is one of a small cluster of 17 galaxies known as the Local Group. It is one of about 100 000 million galaxies within the field of vision of a modern astronomical telescope; each galaxy contains about 100 000 million stars. The nearest galaxies to the cluster containing the Milky Way are the Magellanic Clouds, two irregular galaxies in the Southern Hemisphere sky, and the Andromeda Galaxy in the Northern Hemisphere sky, each about 2 million light years away. There are three types of galaxies: those whose radiation observed on Earth comes from the constituent stars; a second type for which a portion of the radiation originates in interstellar gases illuminated by hot stars, and a third type whose nucleus is brighter than the associated stars. Quasars are examples of this last type, known as active galaxies (Courvoisier and Robson 1991). Quasars are believed to be powered by gravitational energy associated with massive and dense black holes.

Many stars within a galaxy are double stars whose orbits around each other tend to evolve into accretion disks because interactions between particles tend to cancel random motions and thereby produce a flat rotating disk (Cannizzo and Kaitchuck 1992). Probably a nuclear disk of such origin produced our planetary system, all of whose members orbit the Sun in the same direction and in nearly the same plane. Our Sun, an average-sized yellow star near the inner edge of one of the spiral arms (Short 1975, p. 15; Hawking 1988, p. 37), belongs to the GO spectral class of stars within the Hertzsprung-Russell diagram of luminosity versus temperature for the main sequence of star distributions (Cloud 1978, fig. 6). Such stars generate their energy by converting hydrogen to helium. It is these small stars, rather than the larger and hotter ones that are considered more likely to have planetary systems (Short 1975, p. 16); they have life spans of about 10 Ga. Stars having solar masses of 2 may last 1 Ga, those with solar masses of 4 may last about 0.1 Ga, and those with solar masses of 10 may last only a few million years before exploding into supernovas (Hartmann 1983, p. 89). To the high temperatures of these supernovas are attributed the production of heavy elements. As the Solar System contains considerable amounts of the heavy elements, concentrated in the planets, our Sun must be a second- or third-generation star with its ancestors having ended as supernovas, and their debris being incorporated within later reincarnations (Hawking 1988, p. 120). We can expect that the Sun will have used up its hydrogen fuel supply in about 5 Ga and will swell to a red giant and die, leaving the dead planets to continue orbiting but as lifeless bodies.

The diameter of the Sun is 1 392 000 km, about 109 times that of the Earth and nearly four times the distance from the Earth to the Moon; it has a mass of 1.99×10^{30} g (330 000 times that of Earth) and a density of 1.41 g/cm^3. Its rotational period at the equator is 25.4 days, thus it has a very small amount of angular momentum relative to its planets. Its movement at the present position in its orbit is toward the constellation Lyra at 19.4 km/s.

2.2 The Solar System

The nine planets that orbit the Sun can be grouped into two series. The four nearest ones, Mercury, Venus, Earth (the largest of the group), and Mars, are the inner, terrestrial, or rocky planets. Their interiors consist primarily of relatively dense silicates with or without nickel-iron cores. The next group of four planets, Jupiter, Saturn, Uranus, and Neptune, are known as the major, giant, Jovian, or gaseous planets because of their much larger size and their lower densities that result from a predominance of lighter elements and compounds (Stevenson 1978). Jupiter and possibly Saturn have a primary hydrogen-helium composition, but Uranus and Neptune consist of material derived from combinations of carbon, nitrogen, and oxygen (Stevenson 1978; Williams 1978; Hartmann 1983, p. 294). Pluto, the outermost planet, is about the size of the Moon and has a satellite, Charon, that is about 52% of the diameter of Pluto (Lunine et al. 1989). Except for Mercury and Pluto, these planets have orbits around the Sun that can best be described mathematically as an ellipse and all planets revolve around the Sun in a counter-clockwise or prograde direction. Except for Venus, Uranus, and Pluto that rotate clockwise or retrograde, each planet and the Sun spins counterclockwise around its axis. Except for Neptune and Pluto the average distance of each planet from the Sun is governed by a regular numerical position called Bode's Law:

	M	V	E	M	A	J	S	U	N	P
	0	3	6	12	24	48	96	192	384	768
+ 4	4	7	10	16	28	52	100	196	388	772
÷ 10	0.4	0.7	1.0	1.6	2.8	5.2	10.0	19.6	38.8	77.2

This sequence is quite close to the actual distance in astronomical units (1 AU is the mean distance between the Earth and Sun: 1.4×10^8 km) of each planet from the Sun: Mercury (0.4), Venus (0.72), Earth (1.0), Mars (1.5), Asteroids (2.8), Jupiter (5.2), Saturn (9.5), Uranus (19.2), Neptune (30.0), and Pluto (39.5). Another important distance measure for planets and satellites is the Roche Limit (about 2.5 times the radius of the central body), inside of which an orbiting body is likely to be fragmented by tidal forces.

Any hypothesis put forward to explain the origin of the Solar System is constrained by the following facts, as summarized by Short (1975, p. 45) and Hartmann (1983, p. 123):

1. Except for Neptune and Pluto the planets lie at regular distances from the Sun following Bode's law.

2. All planets move counterclockwise around the Sun within a narrow zone about the ecliptic plane; these orbits are nearly circular ellipses.

3. Except for Venus, Uranus, and Pluto all planets and the Sun rotate in a counterclockwise or prograde direction, with axial obliquities of less than 29°.

4. Most satellites, except for some of those belonging to Jupiter and Saturn, revolve counterclockwise around their host planets.

5. The Sun possesses less than 2% of the angular momentum but nearly all the mass of the Solar System.

6. The Sun turns on its axis in almost the same plane as the planets, and it rotates in the same direction as most planets. Most planets and asteroids rotate with similar periods of about 5 to 10 hours, except where tidal forces have slowed rotation to coincide with revolution.

7. Major planet-satellite systems resemble minature solar systems.

8. The rocky planets are small, dense, slow in rotation, have few satellites and are closer to the Sun. The gaseous planets are large, less dense, rotate faster, have many satellites, and are farther from the Sun.

9. Some meteorites contain inclusions of minerals having unique isotopic abundances formed at higher temperatures than occurred for other meteoritic materials; meteorites also differ in detailed chemical and geological properties from known terrestrial and lunar rocks.

2.3 Development of Thought About the Origin

Any reasonable discussion of the shape and distribution of surface morphologies of members of the Solar System must consider the sources, compositions, and ages of crustal materials. All these properties are interrelated with concepts of the origin of the Solar System, as factors that controlled possible causes and subsequent modifications of the morphologies. Primitive animistic legends of the origin necessarily ignored questions of composition, age, and source, because methods for investigating these questions had not yet been developed. The ancient Egyptian, who expressed creation as an event comparable to the annual Nile flood cycle, for example, believed that the first ground emerged from the primeval waters. From this ground supposedly arose the life-giving force of the Sun from which the rest of creation originated (Quirke and Spencer 1992, p. 60). Nevertheless, these animistic origins (emergence from an embryonic state, cosmic egg, world parents, divers that brought Earth from the sea floor) apparently were recognized later as unrealistic, and recourse was to higher powers (gods). Assyrio-Babylonian texts describe an epic, the *Enuma elish*, recorded about 2500 B.C. that recounts Marduk's six-day creation of the Earth from the body of the evil monster Tiamat. This six-day origin was rewritten, but without the conflict of good and evil forces, within the Hebrew account (Genesis) probably during the Babylonian Exile (586 to about 538 B.C.), and later it was fully accepted by many Christians and Moslems (Koran, suras VII, XI, XXV, XLI, L). By spoken command Yahweh created the Earth, its plants and animals, and the Sun, Moon, and stars in six days. The beginning of this creation is dated on current (1992) Israeli coins as 5752 years ago (3760 B.C.), and it was recomputed from biblical chronology by Archbishop James Ussher in A.D. 1654 to start at 9.00 a.m. on 26 October 4004 B.C. Both dates are vastly unreasonable according to both geology and isotopic geochronology, most of which evolved only during the present century. The biblical account has internal inconsistencies as well, because it says that the Sun, Moon, and stars were created on the fourth day after the Earth was formed, yet there were stated to be evenings and mornings before the Sun was present. Moreover, plants supposedly were created and lived before the Sun's presence permitted photosynthesis.

Long continued belief in the early concept of secondary position for the Sun relative to Earth is supported in later books of the Bible (*Joshua 10:12–14; Isaiah 38:8*, written about 1430 and 710 B.C., respectively) that described the halting of of Sun in its course around the Earth to permit completion and winning of a battle by the Jews and as a sign of Yahweh's aid. In *Joshua 10* also the Moon was halted. Alternately, a sudden stop of Earth's rotation to halt the relative motion of Sun/Moon would have caused absolutely devasting tectonic and biological disturbances on the Earth's surface and also would have halted the perceived relative movement of the stars. A similar halt of the Star of Bethlehem is implied by the widely known biblical acount (*Matthew 2:9*) of the birth of Christ and its attendance by the Magi. A better explanation may be that the Magi were eastern astrologers who were attracted by an alignment of Jupiter (the king's planet), Saturn (the shield of Palestine), Mars, and the constellation Pisces (epochal event). This alignment may have been interpreted by them as meaning that a cosmic ruler or king was to appear then in Palestine. The alignment occurs every 805 years including 6 B.C., according to Prof. Maier of Western Michigan University (US News and World Report, 21 Dec. 1992, p. 85) and confirmed by astronomical computations of A. C. Emery of Memphis, TN.

About 1400 years after Ptolemy's *Almagest* the development of thought in western Europe produced the concept of a heliocentric solar system controlled by gravity. This concept was led by Copernicus in A.D. 1543 and later supported by Tycho Brahe's precise observations in about 1600 of changing planetary positions, Kepler's laws of elliptical orbits, and Galileo's telescopic observations in about 1610. Galileo's observation of moons orbiting Jupiter vastly complicated both Aristotle's idea of planets revolving around the Earth within concentric crystalline shells, and Ptolemy's complex system of hinged and moving radial arms that connected the Moon, planets, and Sun to the Earth. Nevertheless, the Catholic church so firmly opposed a heliocentric system rather than an Earth-centered one that it rejected the former, perhaps partly as an attempted counter to the religious Reformation and freeing of thought that was initiated and symbolized by Martin Luther's nailing of his Ninety-five Theses on the door of the church at Wittenberg, Germany during 1517. The views of Copernicus, Kepler, and Galileo were confirmed by Newton's *Principia* in 1687, and they permitted the later calculation of precise orbits and interplanetary landings for satellites whose imagery, measurements, and sampling are the basis for most of this book. The results of many astronomical, geological, and geochemical studies made during the past few decades support the concept that the Earth and other members of the Solar System were formed simultaneously and about 4.55 billion years (4.55 Ga) ago. This date corresponds with the general date of 4.0 by B.P. inferred by Jeffreys (1992, p. 268) from the rate of recession of the Moon from the Earth caused by tidal friction. The much younger age of the Solar System conceived by ancient thinkers (which perhaps reflects only the general date of the beginning of written records) obviously would put serious constraints on chemical and structural modifications of the Earth – thereby accounting for the simplicity of early views of Earth processes and natural history that led to a perceived need for all-powerful deities.

2.4 Modern Concepts About the Origin

2.4.1 Sun, Planets, and Moons

The theories that have been proposed for the origin of the Solar System can be classified in two groups: monistic or evolutionary concepts by which the Solar System was formed as a single unit, and dualistic or catastrophic concepts by which the system was formed through the interaction of the Sun with a foreign body. The earliest monistic theory was the vortex theory of Descartes who in 1644, suggested that the gas in the Universe developed a series of vortices with each vortex producing a star surrounded by a body of gas having minor vortices; the planetary system condensed from these swirling eddies (see Mather and Mason 1939, p. 14; Short 1975, p. 46; Hartmann 1983, p. 97; Brush 1986). The second monistic models were those of Kant in 1755 and Laplace in 1796, both of whom proposed that the Solar System formed from a rotating cloud of gas and dust (Mather and Mason 1939, pp. 88–89; 143–150; Struve 1949; Short 1975, p. 45). According to Kant, contraction of the nebula and formation of the Solar System was caused by gravity, whereas Laplace proposed that it was due to condensation. As described by Laplace, the outer rim of the rotating cloud would tend to shed concentric rings of material, with each ring later condensing into a planet. Whereas Laplace postulated that in an early phase the cloud was a uniform gaseous disk, Alfven (1978a) proposed that the planets were formed from segments of the cloud having different chemical compositions. This banded structure in the cloud was supposed to have been caused by selective ionization of groups of elements during the infall of neutral gas toward the Sun as it formed (see also Arrhenius 1978). As pointed out by Moore (1933, p. 8), there are numerous difficulties with the nebular concept as first proposed by Laplace, the most obvious being that the Sun should be rotating faster than it does. In addition, material thrown off the nebula at its equator would be continuous, not in the form of separate rings. Even if such rings had been formed, they could not have condensed into planets, and even if the planets developed from the rings they would have a backward (clockwise) rotation rather than the existing forward (counterclockwise) rotation. The nebular concept also does not explain why the maximum momentum is within the outer planets rather than on the Sun, why the moons of Mars rotate faster than the planet, why some of the satellites including the Moon revolve in orbits that do not coincide with the plane of the host planet's equator, and why some of the moons of Jupiter and Saturn revolve in a retrograde direction.

The first dualistic theory was proposed in 1745 by Buffon, who suggested that the Solar System was produced from materials ejected from the Sun when it collided with a comet (Mather and Mason 1939, pp. 56–64). Later studies demonstrated that comets have too little mass to have affected the Sun upon impact (Short 1975, p. 46). The shortcomings of the Kant/Laplacian nebular hypothesis also led the geologist Chamberlin and the astronomer Moulton to propose, in 1901, the second dualistic theory (Chamberlin and Salisbury 1930; Struve 1949; Page and Page 1966, pp. 222–230; and Reeves 1978a). Their planetesimal hypothesis postulated that during an encounter between two stars a filament was withdrawn from one (the Sun). Orbital aspects imparted during extraction of the filament were thought to account for the orbital characteristics of the Solar System. In time, the filament condensed to form planetesimals that grew into the planets through collision and accumulation. In 1916 this concept was modified by Jeans (1916) and Jeffreys (1916, 1918) to propose that the Sun underwent major distortion in shape as a second star passed within grazing distance (Page and Page 1966, p. 224). In 1939 Lyttleton proposed that the Sun originally had a companion star and that a third star induced the Sun's companion to escape, leaving a filament of gas revolving around the Sun (Shapley 1937; Lyttleton 1940; Page and Page 1966, p. 226). A variation of this concept, proposed by Woolfson (1978), was that a diffuse protostar that passed near the Sun lost material that was captured by the Sun. All these hypotheses called for what must have been a very rare event, a close encounter of a star with the domain of the Solar System. Eddington (see Barton 1939; Page and Page 1966, p. 211) estimated that such an event may occur for only one out of a hundred million stars. In addition, Spitzer (1939, 1941; see also Page and Page 1966, pp. 217–220) demonstrated that any matter torn out of the Sun in such a manner would simply explode in a matter of hours and could not condense into planets. Others also showed that a filament of gas pulled from the Sun would dissipate rapidly, because gravity could not hold it, and thus the gas could not evolve into planets (see Page and Page 1966, p. 237). Hypotheses based on the extraction of materials from the Sun by passing bodies are no longer credible.

Although the nebular hypothesis was initially dismissed, it forms the basis for the modern view of the origin of the Solar System (Cameron 1978; Prentice 1978; Reeves 1978a). The concensus among scientists today is that formation of our Solar System was not a unique event in the history of the Universe, and that there is a high probability that similar planetary systems have been generated throughout our galaxy and within millions of other galaxies as well. However, this does not mean that each planetary system is graced with the uniqueness of Earth, a uniqueness that permitted the creation and evolution of life. At present, cosmologists hold several views regarding the origin of the Solar System. Two nebular models have been proposed for the origin. In one model the Sun was formed first from one cloud, and the planets came from a second cloud captured by the Sun. In the second model the Sun and planets were created from the same source, a vast rotating nebula. Another hypothesis assumes that when the Sun formed it left no residual gas (Reeves 1978a). This young star had a strong magnetic field that attracted and captured a dense cloud of particles of silicates and metals mixed with hydrogen, helium, and other gases. An exploding supernova may have produced this cloud. The planets were formed from this gradually added cloud. Still another model assumes that the Sun originally was part of a binary system and that for some reason the sister star disintegrated and part of the gas that it produced was captured by the Sun. Goudas et al. (1978) proposed that the Sun was not born alone, and as a result of a close encounter with at least two of its star companions it escaped. As it escaped, severe tidal action caused it to become partially fragmented. Probably it is this escaping star and some of the fragments that left with it that evolved into our Solar System. This escape supposedly occurred about 4.6 Ga ago. The present concensus is that both Sun and planets were formed from the same cloud injected with material from a neighboring supernova. The Sun was created by gravitational contraction and central condensation of the cloud, and the planets were created from a cocoon nebula that enveloped the protosun. The mass of the nebula that formed the planets is believed to have been about 0.15 solar masses with only 0.03 solar masses being retained within the planets, the remainder having been blown away by stellar wind (Hartmann 1983, p. 124).

Radiogenic dates from meteorites, Moon rocks, and reconstruction of the original chemistry of parental material for a 3.59-Ga old gneiss from Greenland indicate that our Solar System is about 4.55 Ga old (Patterson 1956; Kirsten 1978; Hartmann 1983, p. 121). Observations and deductions show that stars like our Sun form in clusters from nebular concentrations of interstellar material of atoms, molecules, and dust grains. The heavier elements (Ca, Al, Fe, Mg, and C) are concentrated on the grains and depleted in the gas. This material, which has approximately the same composition as our Sun, is the parent of stars and planetary systems (Hartmann 1983, p. 90). As the Sun became denser and hotter, its luminosity increased linearly with time; the early Sun at about 4.5 Ga was 25 to 30% less luminous than today (Kasting 1987).

Molecules and dust composed of mineral grains present within ordinary rocks occur in dense cool environments, where collisions leading to planetesimal aggregation are enhanced. These cool nebulas would tend to evolve near stars newly formed by gravitational collapse (Hartmann 1983, p. 93). Under some conditions of rotation and turbulence the central condensation of the nebula could divide into binary or multiple star systems (Larson 1978) orbiting around each other, although some of these bodies would be too small to be true stars. Statistical analyses suggest that 50 to 80% of all star systems contain two or more stars. The more massive neighbors would tend to explode as supernovas only a few million years after cluster formation, but others may gradually spiral together to merge into a single body (Cameron 1978). Thus the suggestion was made by Kuiper (1955) that planets are analogs of these stars. They are simply low-mass companions whose distribution is controlled by angular momentum, turbulence, and mass distribution in the collapsing protostellar cloud. Each newly formed star is surrounded by a dense cloud of gas and dust that is tens or hundreds of astronomical units in diameter. This envelope of dust and gas is known as a cocoon nebula; an example is Star R in the constellation Monoceros (Hartmann 1983, p. 98). Radiation pressure or solar wind would cause the cocoon nebula to shed as much as 0.4 of its solar mass, exposing the star at the T-Tauri stage.

As described by Hartmann (1983, p. 104), the T-Tauri stars display fast rotations, have variable luminosity, are accompanied by variable nebulas and strong magnetic fields, and occur in regions having abundant dust clouds and young stars. The dust at this stage is still too thin to allow planet formation. Nebulas associated with some of this type of star have expansion velocities of 70 to 200 km/s, possibly indicative of the final outward dispersion of the cocoon nebulas by stellar winds. Most of the T-Tauri stars, however, appear to be disk-shaped systems with outflow in some regions and inflow in others. This outflowing stellar wind also carries off angular momentum, thereby supposedly reducing the central star's angular momentum and slowing it to the rotational rate observed in the Sun (Kuhi 1966, 1978). Another factor that may have slowed rotation of the Sun is the magnetic field created during its partial loss of nebula (Schwartz and Schubart 1969; Alfven 1978b). In the viscous accretion model where large effective viscosity is produced by turbulence associated with convective instability at right angles to the midplane of the nebula, viscous stresses transport the angular momentum in the direction of the planets and the nebular mass toward the protosun (see Boss 1990, and references therein). In the three-dimensional model discussed by Boss, angular momentum is transferred to the nebular gas by gravitation torques. In time, much of this transferred angular momentum was carried out of the system when the gas was blown away by the solar wind.

Gas near the surface of T-Tauri stars contains more concentrated lithium than is possessed by the Sun, the interstellar medium, or gases in the outer reaches of the nebulas, but concentrations are comparable with those in planetary material (Hartmann 1983, p. 106). This lithium may be the result of acceleration of atoms by solar flares and disturbances of solar magnetic fields of the T-Tauri stars that causes them to collide with other atoms in the inner nebula and splits them to form the lighter lithium atoms. Thus the chemistry of the rocky planets was affected by early stellar conditions. Another observation critical to our understanding of the origin of the Solar System is the presence of xenon[129] (formed from the decay of iodine[129], with its half-life of 0.017 Ga) and magnesium[26] (formed from the decay of aluminum[26], with its half-life of 0.72 Ma) in many 4.6-Ga old meteorites (Hartmann 1983, p. 120). These unstable radioactive minerals are believed to have been created by nuclear reactions in supernovas; thus the protosun's nebula may have been contaminated by a nearby supernova that exploded a few million years before the Solar System formed (Cameron and Truran 1977; Black 1978). Possibly, the radioactive material accumulated first in presolar interstellar grains derived from various supernovas and later it was added to the protosolar nebula (Clayton 1977, 1978). The distribution of these short-lived radioisotopes in meteorites and the Earth suggest that the interval between the explosion of the supernova and the planet-forming processes was very short, with most meteorites having been formed within 20 Ma and the planets having developed within 100 Ma after the explosion. These variations in isotopic composition among the planets and meteorites indicate that the composition of the nebula that produced the Sun was not uniform and was injected with supernova-derived material.

The solar nebula was disk shaped and rotated in the plane of the Sun's equator. Collisions between finely dispersed material tended to cancel out noncircular motions until a system of parallel circular coplanar orbits was established. Once the nebula became disk shaped and rotated in approximate hydrostatic equilibrium (inward directed gravity forces balancing gas pressure and outward directed centrifugal forces), its rate of contraction decreased and it cooled by infrared radiation from temperatures that exceeded 2000 K during contraction. As the nebula cooled and condensation temperatures were reached, microscopic dust grains formed. The mineral grains were in equilibrium with the surrounding gas and other minerals at a particular temperature and pressure. The earliest crystals to form were the refractory compounds (tungsten, osmium, zirconium, and oxides of calcium, titanium, and aluminum). Materials that condensed at temperatures of 1625 to 1125 K occur in carbonaceous chondrites where they were trapped when the meteorite formed later (Hartmann 1983, p. 131). As the temperature continued to drop, iron and nickel condensed to form an alloy; magnesium silicates condensed as rocky particles at temperatures of 1400 to 1300 K, and these were followed by sodium and potassium silicates (feldspars). At still lower temperatures the grains reacted with the gas, and by the time temperatures reached 500 K the dust consisted of olivine, feldspars, and oxidized iron minerals. This mineral assemblage is preserved in 4.6-Ga old meteorites. With even lower temperatures, carbon, in the form of graphite, carbon-rich silicate minerals, and complex organic molecules, began to appear. From 500 to 200 K water vapor in the cloud reacted with some of the dust grains to form hydrated minerals (serpentine, tremolite, and talc). In regions of the cloud having temperatures of about 200 K many more minerals were hydrated, and ice crystals began to form. On the outer parts of the cloud where temperatures were even lower, ammonia and methane formed ice – some of which became mixed with water to form hydrated ices. As the grains aggregated into bodies having meter dimensions and the nebular cloud dissipated, sunlight was able to sublimate any unshielded ice in the inner region, and water-ice became restricted to the outer reaches of the Solar System beyond the Asteroid Belt. Thus, the composition of each planet is governed by its formation temperature which, in turn, is a function of its heliocentric distance (Goettel and Barshay 1978). The gross composition of the planets can be explained as though, at some stage of cooling, the nebular gas was blown away by solar wind, and each planet retained residual condensates at temperatures corresponding to its distance from the Sun. Mercury is in the refractory-metal-rich part of the system, Venus and Earth are in the lower density silica-rich part, Mars has a still lower density and is rich in oxidized-iron compounds, and the gaseous planets have very low densities and are rich in ices (Hartmann 1983, p. 134).

A scenario described by Williams (1978) and McCrea (1978) is somewhat different. They proposed the initial existence of a number of protoplanets identical in mass and having a composition like that of Jupiter. The protoplanets nearer the Sun (where only iron and silicates are condensable) formed solids that settled gravitationally into the centers of the planets. After formation of these planets solar winds and/or solar tides caused major loss of the gaseous outer material, leaving behind the inner iron and silicates. Farther from the Sun, ammonia and methane are condensable, producing massive icy cores at Jupiter and Saturn. The original protoplanets may have been larger than the present Jupiter, decreasing to their present dimensions through loss of some hydrogen and helium. The nature of the Uranus/Neptune type of planet resulted from the energy released during formation of the icy cores; this energy was sufficient to evaporate the residual outer layers of hydrogen and helium. One of the most interesting effects of the solar wind is its concentration of ammonia (NH_3) and methane (CH_4) at a distance from the Sun corresponding to the positions of the outermost planets and their satellites and of asteroids and comets, where they reacted to form cyanide (HCN) compounds. Reflectance spectra show that the cyanide is present on low-albedo bodies. There is a strong possibility, but not a certainty, that the low albedo is the result of conversion of the hydrogen cyanide in the presence of water into simple carbon-bearing compounds, perhaps primitive proteins as black films on the hard surfaces of the bodies.

Condensation probably took two forms: one was the homogeneous chemical-equilibrium model where compounds existing at any particular time were able to reach chemical equilibrium with the surrounding media, and the other was the inhomogenous model where the dust aggregated into bodies centimeters to kilometers in dimension, preventing it from reacting with later materials. In the first model metallic cores would have required melting and drainage of nickel-iron to their centers, whereas in the second model metallic cores and silicate mantles would have been formed by sequential condensation. The first (homogeneous accretion) model appears to be compatible with the Earth's internal structure (Condie 1989, pp. 2–29).

The next problem to be addressed is the means by which the dust grains become aggregated into planets. In the first stage, collisions and gas drag would have established nearly circular orbits. According to Harris (1978), the basic processes that controlled formation of the planets were gravitational instability, conversion of radial gradients of orbital motion into random motion between planetesimals, collisions leading to damping of random motions, and aggregation and/or fragmentation of the planetesimals. The earliest stages of grain accretion were influenced by the nebular gas in which they were immersed and by perturbations from Jupiter, which had formed prior to nebular dissipation (Heppenheimer 1978). As particles passed near each other, gravitational forces proportional to the particle masses caused minor perturbations in orbits, making them eccentric and causing collisions at about 0.5 km/s, one-tenth the speed of modern asteroid collisions. Drag effects on bodies of 0.1- to 10-m diameters would have been even less, 0.01 to 1 cm/s. Under these conditions tiny dust grains would have tended to adhere to form clumps held together by electrostatic forces (Hartmann 1983, p. 135). The first planetary matter probably resembled the fluffy clumps formed by aggregation of snowflakes. That the Solar System planets grew from interactions between small particles is suggested by their nearly planar circular orbits, and their characteristic prograde rotations. As the aggregates grew larger, they continued to perturb each other during near-collisions, and as they grew their motions departed from gas motions and their approaches increased in speed. In time these velocities became equal to the escape velocities from the larger bodies but not from the largest in the swarm. These larger bodies, which are abundant enough and large enough to have gravitational effects, tend to act as 'gravitational spoons' stirring the smaller planetesimals in the nebula (Hartmann 1983, p. 135).

When a small object hits a larger one at a speed so low that it does not fracture but bounces away, its rebound speed may be so low that it falls back and adds to the mass of the larger body. By this process the larger bodies would grow even larger. According to Hartmann (1983, p. 136), the smallest bodies in the swarm may not be able to grow in this manner, because they would be hit at speeds greater than their escape velocities and their rebounds would be too fast for accretion to occur. This process of growth by random collision is known as collision accretion. It is in this manner that dusty grains first formed aggregates, and later bodies reached several kilometers in diameter. Planetesimal growth also may have been enhanced by gravitational collapse of a dust layer to form bodies a few kilometers in diameter. This layer of kilometer-scale bodies in turn became gravitationally unstable, aggregating into a second generation of about 100-km-scale planetesimals. Greenberg (1989) described two scenarios for planetary growth: one where growth is the result of accretion of small planetesimals, and the other by independent formation of large embryos that accreted one another in a violent phase. An example of the latter is the Earth/Moon system where the Moon was formed as a result of a giant impact with protoearth.

Hartmann (1983, p. 138) described the accretion process as having four successive stages: (1) chemical condensation of grains (micrometer scale), (2) electrostatic clumping of grains (submillimeter scale), (3) accretion of clumps by collision (meter scale?), and (4) gravitational collapse of grains and clumps (kilometer scale?). Continued growth among the rocky planets and asteroids was by collision accretion. Model studies by Greenberg et al. (1978a, b) indicate that for a variety of starting velocities, a swarm of 1-km bodies could produce a few 500- to 1000-km-diameter 'planet embryos' in about 10 000 years. Similar studies by Wetherill (1985) using three-dimensional Monte Carlo simulations demonstrated that rocky planet accretion in the absence of gas drag produced a distribution comparable with that of the present Solar System. After 1.8 Ma 18% of the mass had accumulated in the final planets, by 9 Ma 66% of the mass was in the final planets, and by 250 Ma almost all the bodies had accreted into planets with masses and orbits resembling those of the present rocky planets. Weissman (1989) believed that accretion of planetary atmospheres occurred during the first Ga of the accretionary history and that subsequent impacts have had little effect on adding, modifying, and/or eroding the existing atmospheres. Another result of such impacts during growth of the rocky planets would have been partial melting that led to core formation by gravitational segregation; thus, formation of the iron core would have been contemporaneous with growth. These collisions also may have been responsible for removal of the primordial terrestrial atmosphere. In time the planets reached nearly their present mass, with each being scarred by craters dating back to 4 Ga, a record of the cleanup of the interplanetary bodies. The effect of these cleanups and subsequent impacts was to produce planets having prograde rotations of 5- to 20-h periods and rotational axes nearly perpendicular to the ecliptic (Harris 1977). The whole process of planetary evolution described above may be exemplified by the ring systems of Jupiter, Saturn, and Uranus. The particles in the Jupiter rings appear to consist of rocky dust or dirty water, those of Saturn – frozen water, and those of Uranus may be rocky or carbonaceous. These rings may have been formed from the nebula that produced the planets, but they never accreted and were left stranded, or they may have resulted from the fragmentation of planetesimals or satellites by tidal stresses within the Roche Limit or by collision of a planetesimal with an inner satellite or with the planet itself.

The satellite (from *satelles*, Latin for attendant) systems in the Solar System do not appear to be just simple small images of the host planets (Hartmann 1983, pp. 143–151). To begin with, the past history of orbits cannot be traced by contemporary celestial mechanics beyond 1 Ma ago. Secondly, some of the irregularities in satellite orbits are associated with resonance between pairs of satellites (Saturn's Enceladus 1 : 2 resonance with Dione, Mimas 1 : 2 resonance with Tethys, and Titan in a 3 : 4 resonance with Hyperion – all are geometries that came into being during tidal evolution of satellite orbits). Thirdly, tidal effects tend to reduce the inclination of orbits so that present geometries may not reflect the original conditions. And fourthly, planets may have lost their satellites or have captured them in geologically recent time, some of the satellites may have fragmented after formation, and there even may have been satellite exchange between planets. For example, the early satellites of Mercury and Venus may have been removed by tidal friction (Harris 1978). Tidal transfer of angular momentum from the Earth to the prograde Moon has forced the Moon outward, but retrograde satellites are drawn toward the host planets; thus Triton may become disrupted by coming within Neptune's Roche Limit or it may crash into the planet in about 100 Ma to 1 Ga, thereby altering the planet's inclination. The association of a retrograde satellite and a prograde planet probably is the result of capture of a planetesimal that approached too near the planet. Examples of such captures may be Phobos and Deimos of Mars, the smaller outer satellites (J6 to J13) of Jupiter, Phoebe – the outermost retrograde satellite of Saturn, and most of the outer asteroids. Hartmann (1983, p. 146) suggested that all these satellites originated in the region of Jupiter or Saturn and later were captured when they were perturbed into planet-crossing orbits. Such captures also may have been enhanced by the massive extended atmospheres possessed by the early planets, by close approach to an existing satellite, or by collision with a satellite followed by reaggregation of the debris to form a new satellite. Although the question of how the four prograde and

four retrograde outer satellites of Jupiter arrived at their present positions remains
to be answered, the most reasonable explanation appears to be that the satellites
are fragments produced at their present orbits. The large regularly spaced satellites
with zero inclination circular orbits may be accretion products of nebulas around
the host planets. Examples of such satellites are Io, Europa, Ganymede, and
Callisto around Jupiter.

2.4.2 The Meteoritic Complex

2.4.2.1 Meteorites

Nebular debris that was neither accreted onto planets and satellites nor ejected
from the Solar System has survived in the form of meteorites, asteroids, and comets,
known collectively as the meteoritic complex. The planetesimals and their fragments that interact with the Earth and survive passage through the atmosphere are
called meteorites. According to Short (1975, p. 31), about 500 tons of rock debris
ranging from dust to rare large meteorites enter the Earth's atmosphere each day.
Most are known as micrometeorites, and only about 500 objects larger than 10 cm
in diameter survive passage through the atmosphere each year, with more than
two-thirds of them dropping into the ocean. However, early in the history of the
Solar System the newly formed planets were more efficient in sweeping up
neighboring debris, and bombardment of the planets was much more intense than
it is now. At that time, planetesimals in orbits similar to those of the planets had
half-lives of about 15 to 20 Ma, whereas the remaining scattered debris has half-
lives of 20 to 100 Ma (Hartmann 1983, p. 163). Apollo data from different sites on
the Moon indicate that the impact rate about 4 Ga ago was about a thousand times
greater than at present. Present cratering on the other rocky planets and the Moon
relative to that on the Earth ranges from 0.5/1.3/3.3 (minimum/most likely/max-
imum values) on Mercury to 0.5/0.7/1.4 on Venus, 0.5/0.7/0.9 on the Moon, and
0.6/1.3/2.6 on Mars (Hartmann 1983, p. 164, table 6.1).

Meteorites, many of which represent records of chemical and physical processes that occurred in the nebula from which the Solar System was formed, can be
classified in three groups: stony, stony-iron, and iron (Larimer 1978). The stony
group constitutes about 94% of the meteorites and in turn is subdivided into
chondrites and achondrites. The chondrites, which comprise 80 percent of all the
meteorites landing on Earth, are seed-size glassy grains. They may represent melted
droplets formed early in the history of planetesimal collisions (Kerridge 1978), and
they consist of olivine, pyroxene, iron, and iron sulfide. Herndron (1978) believed
that the presence of iron in other than metallic form implies that the chondrites
were formed after hydrogen became depleted by a factor of hundreds to thousands
relative to solar matter. The mechanism that caused the hydrogen loss has yet to be
resolved.

By composition, chondrite meteorites consist of three groups: ordinary chondrites (the most abundant), enstatite chondrites containing high concentrations of
that mineral, and carbonaceous chondrites. The first two groups formed at higher

temperatures and are somewhat more depleted in volatile compounds than are carbonaceous chondrites. Carbonaceous chondrites (or black stony meteorites) have undergone so little heating that they retain carbon compounds, other volatile elements, and water bound within minerals. They are not the only bodies in the Solar System that consist partly or wholly of carbonaceous material; a similar type of material also occurs in the satellites of Mars (Phobos and Deimos), comets, interplanetary dust, and perhaps some Jovian satellites (Wilkening 1978; Cronin et al. 1988). Most of the iron in carbonaceous meteorites has the oxide form as either FeO in silicates or Fe_2O_3 in magnetite (Grossman et al. 1979). Based on their carbon (in the form of complex organic polymers), water, volatiles, and mineral compositions, the carbonaceous chondrites can be divided into four classes: CI, CM, CO, and CV, with CI and CM having the higher carbon and volatile contents and representing the chemically most primitive meteorites (McSween 1979; Wolfteich 1989). Types CO and CV are more modified, retaining about 1% water and having less carbon and more chrondules (Short 1975, p. 39; Hartmann 1983, p. 169). Thus the CI and CM bodies that have whole rock ages of 4.6 Ga have escaped alteration since the nebula evolved into the Sun and its orbiting bodies. Both the CI and CM chondrites have albedos lower than 6%, and their reflection spectra are characterized by steep absorption in wavelengths less than 0.6 μm because of the Fe^{2+}–Fe^{3+} charge transfer band in the silicates (Feierberg et al. 1985), and a water-absorption band near 3 μm (Gaffey and McCord 1978; Feierberg et al. 1981).

The CI types, with their close chemical similarity to the Sun, are very rare, only five falls having been recorded to-date (Wolfteich 1989, and references therein). The CI chondrites are hydrated phyllosilicates, lacking chondrules or inclusions, and commonly they are brecciated, with the fragments being cemented together by hydrous magnesium-sulfate and calcium-sulfate veins deposited from aqueous solutions (Zolensky and McSween 1988). Other characteristic features of CI carbonaceous meteorites are the presence of magnetite, carbon as carbonate or organic carbons (Kerridge et al. 1979; Sears and Dodd 1988), and a mineral suite formed by secondary hydrous alterations at low temperatures (Kerridge and Bunch 1979). The CM types contain inclusions and chondrules, are brecciated, and display evidence of hydrous alteration and of shock effects due to hypervelocity impact (Kerridge and Bunch 1979; McSween 1979). Thus both CI and CM meteorites originated from a planetesimal regolith that underwent meteoritic impact and aqueous processes. The aqueous alteration and deposition of magnetite in the CI meteorites did not occur during the nebular condensation phase (Kerridge et al. 1979; Zolensky and McSween 1988), but were the result of alterations in the near-surface regions of planetesimals (Bunch and Chang 1980). Water and other volatiles were released from preexisting phases by heat produced either from the decay of short-lived radionuclides or from repeated impacts (Zolensky and McSween 1988). A planetesimal having a 200-km diameter is capable of retaining water for about 200 Ma before evaporation into spacèa period long enough to account for the alteration seen in meteorites (Kerridge and Bunch 1979).

The achondrites resemble volcanic lavas on Earth and have been heated, melted, and/or shattered to such a degree as to destroy the chondrule structures. They comprise only 8% of all meteorites recorded on Earth. This group consists of eucrites, which usually are brecciated as though by impact and display bubbly textures like lava flows on Earth; they consist of about equal amounts of anorthite and pigeonite (Hartmann 1983, p. 173; Wolfteich 1988), and the very rare SNC meteorites (four Shergottites composed mainly of medium-grained basalt; three Nakhlites composed mainly of augite, and one Chassignite which is a dunite; Wolfteich 1988). Whereas the other meteorites have formation ages of 4.4 to 4.6 Ga, the SNC meteorites have average ages of 1.31 Ga and exhibit properties that indicate origin on a planet.

Stony-iron meteorites consist of two types. In one type the mesosiderite silicates consist of plagioclase and pyroxene minerals; they resemble surface lavas or crustal silicate rocks in contact with metal. In the second type, the pallasite silicates (principally olivine) resemble mantle material and reflect an environment where mantle rocks were in contact with a planetary iron core or at least with surfaces of iron blobs near the core. Iron meteorites are classified according to their nickel content, crystal structure, and associated parameters (Hartmann 1983, pp. 173–174).

Still another kind of meteorites are tektites. These small glass objects, the largest of which have masses of 200 to 300 g, are rich in silica (about 73%), and they tend to occur in restricted areas of the Earth probably related to a nearby impact crater. Their high silica content indicates that tektites came from the Earth rather than the Moon, whose igneous rocks are silica poor. Wind-tunnel experiments show that the tektites either entered or reentered the atmosphere at high speeds, causing melting and reshaping of the clast. Entry trajectories calculated for tektites from Australia led Chapman and Larson (1963) to speculate that the tektites originated on the Moon, but other workers have proposed that they represent fragments of fused silicate blasted out when a stony meteorite struck the Earth, then flying outward into space, solidifying, and falling back to Earth (Hartmann 1983, p. 178).

Cosmic Ray Exposure (CRE) ages measured from radioactive and stable isotopes created by nuclear reactions with cosmic rays that penetrate approximately 1 m indicate that iron meteorites have been exposed in space for about 100 Ma, stony ones for 10 Ma, SNC meteorites for 2.6 to 0.5 Ma, and tektites for less than 300 years and possibly only for minutes. Although some first-generation meteorites date back to the early phases of Solar System formation, CRE ages tend to be concentrated between ages of 20 and 4.5 Ma ago (Hartmann 1983, pp. 176–177). Consequently, the major collisions that created some meteorites must have occurred during the recent history of the Solar System. The lower exposure ages for stony meteorites may be the result of a greater capability for fragmentation and loss for stony rather than iron meteorites, and the low exposure times of tektites suggest that they originate in the Earth-Moon region.

CRE ages indicate that meteorites were emplaced in Earth-intercepting orbits soon after they were fragmented to their present sizes. These ages and the large amount of included inert gases also show that the meteorites originated from our general part of the Solar System (Chapman 1983). Orbits calculated from photographic records of several meteorites as they entered the atmosphere reveal that they came from the Asteroid Belt (Feierberg and Drake 1980; Wetherill and Chapman 1988). Such an origin also is suggested by similarities in reflectance spectra of the eucrite meteorites, the large asteroid Vesta (Feierberg and Drake 1980; Feierberg et al. 1980; McSween 1987), and the low albedos and spectral characteristics of class C asteroids that comprise 75% of the planetesimals in the main Asteroid Belt. As described by Hartmann (1983, p. 180), collisions in the Asteroid Belt eject fragments into orbits perturbed by Jupiter resonances. These perturbations in turn send meteorites into orbits that intercept Earth. An example of such fragments are Earth-approaching Apollo asteroids a few kilometers in diameter.

The young crystallization age of SNC meteorites indicates that these bodies must have come from a planet large enough to have undergone magmatic activity about 1.3 Ga ago. Chemical incompatibility eliminates Mercury, the Moon, and the Jovian satellite Io as sources. Venus' dense atmosphere and large gravitational field, which inhibit ejection of debris, also eliminate Venus as a potential source. This leaves Mars as the only reasonable source. A martian source for these young crystallized fragments is supported by magnetic measurements on five SNC meteorites, indicating that the rocks crystallized in a very small magnetic field (Cisowoski 1986; Collinson 1986). The oldest martian lava flows on the periphery of Olympus Mons and Tharsis are 1.1 to 1.6 Ga old (based on crater abundance in the Tharsis region; Wood and Ashwal 1981), and these values are comparable with the ages of the SNC meteorites. Recently, Mouginis-Mark et al. (1992) identified 25 impact craters in the Tharsis region as potential sources for the SNC meteorites. In addition, reflection spectra of the martian dark regions contain absorption bands caused by augite clinopyroxenes (Wood and Ashwal 1981), and the composition of the soil where the Viking spacecraft landed matches the composition of Shergottite (Baird and Clark 1981). The lower Ca/Si ratio of the martian soil relative to the SNC meteorites may be due to partial removal of calcium from the soil as carbonate (Warren 1987). Gases in impact melt glasses from a Shergottite meteorite found in Antarctica closely resemble the composition of the martian atmosphere analyzed by Viking (Bogard and Johnson 1983; Becker and Pepin 1984; Bogard et al. 1984; Wiens et al. 1986). All these observations suggest that SNC meteorites may have originated in Mars. Ejection from Mars is believed to be the result of oblique impact by a meteorite large enough to cause the resulting ejecta to reach escape velocities of 5 km/s (Nyquist 1983, 1984; O'Keefe and Ahrens 1986). Some meteorites also may have originated on the Moon. That such an origin is possible is suggested by the discovery of a golf-ball-sized unshocked meteorite in Antarctica during 1981; it closely resembles some Apollo samples from the Moon (Kerr 1983). Thus, as for the debris that possibly originated on Mars, some sort of mechanism allowed this body to reach a lunar escape velocity of at least 2.4 km/s.

2.4.2.2 Asteroids

The Asteroid Belt between Mars and Jupiter consists of at least 3445 named and numbered bodies (Kowal 1988) and about 1 million others, with mean distances of most asteroids from the Sun being between 2.1 and 3.3 AU. The total mass of asteroids is about 3×10^{20} g, about 4% of the mass of the Moon. Counting all those that are larger than 1 km, the total may be about 1 million (Binzel et al. 1991). Within the Belt are hundreds of asteroids having diameters larger than 100 km, 12 with diameters larger than 250 km, and three exceeding 500 km in diameter. Ceres, the largest asteroid, has a diameter of 1020 km; this is followed by Vesta at 549 km and Pallas at 538 km, all in the asteroid midbelt. Hygiea with a diameter of 443 km and Davida at 341 km are at the outer edge of the Belt (Hartmann 1983, pp. 192–193; Table 1). Asteroids tend to be irregular in shape, whereas moons larger than 1000 km are more spherical; Hartmann (1983, p. 201) ascribed this difference to a greater pressure in the central parts of the larger bodies than the strength of the material. Although Ceres contains nearly half the total mass of all asteroids, the asteroid size distribution appears to be compatible with the concept of fragmentation of a swarm of fewer, but larger, bodies. The excessive mass of Ceres may have resulted from accretionary growth within the Belt. A hump in the cumulative curve of asteroid size distribution at a diameter of 60 km may reflect an initial Gaussian size distribution modified by numerous later collisions and fragmentations (Hartmann and Hartmann 1968). Other workers (Chapman 1977; Davis et al. 1979) proposed that the hump instead may indicate differences in strengths of the bodies in the Asteroid Belt. Chapman and Davis (1975) proposed that the initial belt was more massive than at present and that it ground itself to its present textural distribution, with the smallest pieces being displaced out of the belt by the Poynting-Robertson light effect (the interaction of light with centimeter-scale particles), which causes the planetesimals to spiral inward toward the Sun, and by forces of solar radiation that cause the bodies to move outward.

The asteroids have orbits that are more inclined than those of the planets (as much as 30° from the ecliptic versus mainly less than 3° for the planets), and their eccentricities are 0.1 to 0.3, compared with planetary eccentricities less than 0.1. Because of its large mass Jupiter has a strong perturbing influence on these bodies. Asteroids with periods of one-fourth, one-half, one-third, or two-thirds of Jupiter's period undergo repeated resonance of perturbation that causes rapid changes in their orbital characteristics, kicking them out of that commensurate orbit and creating narrow unpopulated ranges of semimajor axes – the Kirkwood Gap (Hartmann 1983, p. 191). Under some circumstances, however, commensurabilities tend to stabilize the asteroids rather than to change orbital elements. Sites of stable asteroid orbits occur at the Lagrangian points of Jupiter's orbit (60° ahead and behind the planet). These asteroids, the Trojan Asteroids, oscillate as much as 20° back and forth from these two points. Analogs of the Trojan Asteroids are planetesimals at the Lagrangian points in the orbits of the Saturn satellites, Tethys and Dione. The Mars-crossing asteroids are located on the inner fringe of the Asteroid Belt, and they cross inside Mars' orbit. Some of these bodies, the Amor Asteroids, come within 1.3 to 1.0 AU of the Sun and thus approach Earth's orbit. The Apollo Group, consisting of 30 known asteroids with diameters larger than 1 km, and possibly as many as 700 ± 300 others cross Earth's orbit. The Apollo and Amor asteroids have half-lives that are only a few percent of the age of the

Solar System; therefore they must be replaced or replenished periodically from the Asteroid Belt. The Aten Asteroids, having semimajor orbital axes less than the Earth's, lie mainly inside Earth's orbit, but they may cross outside near their aphelion.

Asteroids also occur in the outer Solar System. One, Hidalgo, has an orbit more like a comet than an asteroid. The 310- to 400-km diameter Chiro, which passes between Saturn's satellites every 10 000 years, has a shorter orbit. It may move into the inner Solar System on an orbit comparable to an Apollo satellite or a short-lived comet (Hartmann 1983, p. 196). As described by Everhart (1979), these outer Solar System bodies display a variety of orbits that change from one type to another because of perturbations by the giant planets. The asteroids tend to cluster into groups within similar orbital elements known as the Hyramaya families. Although as many as 100 of these families have been proposed, only about a dozen are well established. Of the 34 Mars-crossing asteroids, 20 can be identified within four families. Each family is believed to have been formed by the collision and fragmentation of a large asteroid. The spectra of some families display a uniform composition, but other families have mixed spectra. In the former, the family was formed by breakup of a single parent of uniform composition, whereas in the latter the family either was formed from an inhomogeneous parent or did not originate from collision. One family, which includes Flora, appears to have been derived from a binary asteroid with only the satellite being broken by the collision and with Flora surviving intact.

Remote sensing methods, including polarimetry, radar, spectroscopy, amount of polarization, and infrared reflectance spectrophotometry, indicate that asteroids have compositions similar to those of meteorites. They also display zonal variations (Gradie and Tedesco 1982). Rare high-albedo E type (possibly related to enstatite chondrites) and the S type (possibly related to common chondrites) occur mainly at the inner edge of the Asteroid Belt. The most abundant C-type asteroids have low albedos of 2 to 7%; such low values have been attributed to the presence of graphite and other carbon-based minerals (Feierberg et al. 1985). Comparison of 0.1- to 1.1-μm spectra of 20 C asteroids and low albedo meteorites indicate that most C asteroids are similar to hydrous CI and CO chondrites (Gaffey and McCord 1978). Their low albedo values also suggest that this type of asteroid contains large amounts of ice disguised by opaque carbonaceous material (Hartmann 1983, pp. 199–200). Spectral data from Ceres appears to indicate the presence of chemically bound water and montmorillonite; this large asteroid probably also has water frost or ice on its surface (Lebofsky 1978, 1980; Lebofsky et al. 1981). Other kinds of asteroids that are present on the outer fringes of the Belt are D types. These are as dark as C types, but they are much redder and probably contain lower temperature condensates than those of C types. Thus, the asteroids on the outer edge of the Asteroid Belt are a combination of organic matter, carbonaceous rocky material, ice, and representatives of the inner edge of the Belt of stony volatile-poor bodies. These latter bodies may have originated in the inner Solar System and were thrown outward by Earth to be stored in the Asteroid Belt.

The brightness displayed by many asteroids varies with time, a variation that is ascribed either to end-over-end tumbling of an elongate body, or to the asteroids having dark and light markings perhaps hemispheric in pattern. Asteroid surfaces tend to be peppered with craters, reflecting numerous collisions; their lack of rough relief is due to subsequent small-scale collisions that subdued the topographic grain. Images of Gaspra, an S-asteroid with a 7-h counterclockwise rotation, as recorded by the Galileo spacecraft en route to Jupiter, show that the surface of this $19 \times 12 \times 11$-km asteroid, a fragment of a larger piece, is disfigured by ridges, grooves (?), and small craters superimposed on the ridges. The cratering age of the surface is about 200 Ma, and the ridges may have been produced by collisions with smaller bodies in the Asteroid Belt (Simarski 1991; Belton et al. 1992). Another feature of Gaspra includes linear features 200 to 400 m wide and as much as 2 km long. They may be similar to grooves on Phobos and caused by catastrophic impacts (Belton et al. 1992; Veverka et al. 1993). Distribution of color on the surface of the asteroid, brighter materials on the highs and darker on the lows, suggests that some of the regolith has migrated downslope exposing the mafic materials that comprise Gaspra. Small satellites, similar in composition and environment to Mars' satellites Phobos and Deimos, may display different impact morphologies, indicating that collisions occurred at different times, or that the rate of degradation by surface processes is different, as on Phobos and Deimos. Although asteroids rotate quite rapidly, with periods as short as 2 h, centrifugal forces do not exceed gravity to allow loose material to be thrown off. Thus, many asteroids have dusty surfaces, as verified by polarimetric data (Hartmann 1983, pp. 201, 210). On asteroids that are 10 to 11 km in diameter the regolith created by repeated collisions may be only a few centimeters thick, whereas megaregoliths that are a few kilometers thick may accumulate on larger asteroids.

Some asteroid surfaces consist of pristine primitive material, whereas others (4 Vesta, for example) have a basaltic achondritic surface resembling lava. Some asteroids also appear to have satellites. Pallas may have an 85-km diameter satellite located about 2.7 Pallas radii away, and the 153-km 9 Metis may have a 60-km satellite (Wang et al. 1981). Such configurations must be short lived, only about 100 000 years (Binzel and van Flandern 1979). Satellites outside the synchronous point would migrate outward until they become independent asteroids, and those inside the synchronous point would move inward and come to rest on the primaries (Hartmann 1983, p. 207). The largest Trojan, the 150×300 km 624 Hektor, appears to consist of two elongate ellipsoidal bodies in near contact (Weidenschilling 1980). This arrangement, in turn, tends to increase the rate of spin of the host; for example, 1580 Betulia with a diameter of 7 km and a 3-km bump on its side has a period of 6.1 h (Todesco et al. 1978). If minor satellites are common and have short lives, there must be a mechanism to create new ones or reseparate old ones. Major collisions between comparably sized asteroids and which result in chaotic brecciation and release of pairs of coorbiting fragments may be a possible mechanism (Davis et al. 1979; Hartmann 1979).

Asteroids previously were thought to be fragments of a shattered planet that occupied the orbit defined by Bode's Law at 2.8 AU from the Sun. Now, according to Binzel et al. (1991), they are believed to be materials for a planet that failed to form. The large number of asteroids, more than 1 million, confined within identifiable belts implies that crater-forming collisions occur, but that the average spacing between asteroids (several million kilometers) means that collisions are significant only over geological time scales. These collisions produce additional asteroids, so that the origin of asteroids must be a combination of protoplanetary materials and of their fragments from subsequent collisons.

2.4.2.3 Comets

Interest in comets (the name is derived from a Greek word meaning 'long-haired one') goes back before the beginning of systematic astronomy during the third millennium B.C. in the Tigris-Euphrates Valley (Bailey et al. 1990, pp. 7–36). For example, a megalith marking at Trapain Law, Midlothian, Scotland dated about 2000 B.C. shows a circular feature with a tail that may represent a comet. Later, during the Han Dynasty in 168 B.C. a Chinese astronomer classified comets into 27 types on the basis of their tail morphologies.

Surrounding the Solar System 50 000 AU distant from the Sun, is the Oort Cloud (named for the Dutch astronomer, Jan Oort, who proposed it in 1950). It consists of a swarm of several trillion comets having highly eccentric orbits inclined at high angles to the ecliptic (Short 1975, p. 29; Weissman 1990). Bursts of gas from the surface of the comets cause sporadic departures from elliptical orbits and acceleration or deceleration of a comet. The comets move at speeds of about 0.1 km/s relative to the Sun and are perturbed by the nearer stars; these perturbations occasionally send one into the Solar System. For example, the comet Schwassmann-Wachmann 1 which is in a nearly circular orbit between Jupiter and Saturn (Hartmann et al. 1982) and at least seven other comets spend periods of a few months to a few years near Jupiter. The explosion in the Tunguska region of Siberia on 30 June 1908 is believed to have been caused by an encounter of a comet with Earth (Kirova 1964; Turco et al. 1982). That such an explosion was due to collision with Earth of a 20- to 60-m diameter piece of a comet is suggested by the absence of a crater or of substantial meteoritic fragments. Another famous visitor to the inner Solar System is Halley's Comet that arrives every 76 or 77 years. According to Strouhal (1992, p. 240), a fragment of a document from the reign of Thutmose III (about 1479–1425 B.C.) may possibly be the earliest report of this comet (James 1988, p. 6). The sudden appearance of this bright object has been viewed by countless civilizations, as far back as the Babylonians, as an omen of bad tidings – the destruction of Jerusalem in A.D. 70, the defeat of Attila the Hun in 451, the Norman invasion of England in 1066 (its visit is shown on the Bayeux Tapestry), and the possibility that the Turks would invade Europe after they captured Constantinople in 1453 (Hartmann 1983, p. 215). During 1066, the comet was seen in western Europe for seven days during April. In the same year, William, Duke of Normandy, crossed the English Channel on 27 September, landing in Pevensy Bay on 28 September with an army of 12 000 including 1500 mounted

troops and several hundred sailors. The army was transported across the English
Channel in 50 fighting ships (one of which, the Mora, was presented to William by
his wife Matilda) and 200 larger vessels. This army engaged the Anglo Saxon force
of 6300 to 7500 under the leadership of King Harold Godwineson, on 14 October
at the battle of Hastings, six months after the appearance of Halley's Comet
(Linkletter 1966, pp. 1, 205–219).

Orbital studies indicate that shooting stars, meteor showers, and meteoritic
swarms are debris detached from comets. Samples of this fine-grained debris
recovered on Earth consist of aggregates of micrometer- and submicrometer-
diameter grains resembling CI-type carbonaceous chondrite debris (Wilkening
1978). Nickel-bearing iron sulfide and crystals of olivine and pyroxene (Brownlee
et al. 1977), grains having a magnetic composition mainly of iron (Perseid Shower),
and others having a glassy rather than a crystalline structure also have been
reported (Gemenid Shower). These particles are believed to have originated from
icy comet nuclei (Hartmann 1983, p. 228). The large volume of dust particles
dispersed by comets throughout interspace is responsible for the Zodiacal Light, a
faint reflection of sunlight off dust grains in the ecliptic plane as well as for the
Gegenschein Glow at the antisolar point.

When the comets approach within 1 or 2 AU of the Sun, interactions between
their constituents and solar radiation pressure produce a glowing tail of diffuse
particles streaming more or less radially away from the Sun over distances of 75 to
150 million km (Short 1975, p. 29). There are two general types of comet tails.
Type I is essentially straight with an emission spectrum of ionized gas. Type II is
broad, diffuse, gently curved, and consists of grains of dust that scatter sunlight.
Some comets display both types of tails (Hartmann 1983, p. 219). The 15 000- to
1 000 000-km-diameter head of a comet consists of gases and dispersed particles in
the form of a very low density cloud or coma with a diffuse outer boundary. The
core of the coma is less than one to a few tens of kilometers in diameter with a total
mass of less than 10^{20} g; it possibly consists of silicates and complex hydrocarbons
embedded in a water-ice and frozen gas matrix of ammonia, carbon dioxide,
cyanogen (hydrogen cyanide), and other gases (Short 1975, p. 29; Delsemme 1978)
including clathrate hydrates (Blake et al. 1991). Interceptions with Halley's Comet
by the spacecrafts Giotto from Europe, Vega 1 and 2 from the former USSR, Suisei
and Sakigake from Japan, and the International Cometary Explorer from the
United States in March 1986 provided a wealth of information on the physical
structure and chemistry of this comet (Bailey et al. 1990, pp. 425–467). Its body
dimension is about $16 \times 8 \times 7.5$ km with a geometric albedo of a few percent. Its
surface temperature ranged from about 300 K to more than 400 K, a range
indicating that the surface of the nucleus is very dark, porous, insulating, and
thermally decoupled from the sublimating ices below. The dust associated with the
comet has a composition comparable to CI carbonaceous chrondrites, and of the
many ices detected on Halley, water is the most common – comprising 80% of the
volatiles – with carbon monoxide and carbon dioxide ices constituting 5 to 15%
and about 3%, respectively. Methane forms about 1 to 2% and ammonia about
0.2%. Clathrate hydrates in the cometary ice are believed to form by rearrange-

ments in the solid state during the warming of vapor-deposited amorphous ices in a vacuum (Blake et al. 1991). Interaction of dust and ice in the nucleus causes irregular outbursts of brightness. As a result of the rapid evaporation and dispersion of their nuclear materials, comets that repeatedly pass near the Sun have life spans of only a few thousand years.

Several theories have been proposed for the origin of comets. Some cosmologists believe that comets represent planetesimals of primeval matter formed from the original gas/dust cloud during the early history of the Solar System. They supposedly were kicked out of the system by perturbations in the cloud; thus, exclusive of hydrogen, the comets are the best surviving examples of the original composition of the cloud. Some authors have proposed that comets are being created continuously from gases and dust in the Solar System and in interstellar space as the Solar System moves through the galaxy. The presence of hydrogen cyanide and methyl cyanide, common components of interstellar dust in Comet Kohoutek, which passed around the Sun during late 1973 and early 1974, may be a verification that these bodies originate far beyond the confines of the Solar System (Short 1975, p. 29). Recent studies indicate that comet formation was due to condensation in a molecular cloud environment prior to formation of the Sun (if primordial) or of other parent stars (if captured; Bailey et al. 1990, p. 466).

2.4.3 Companion Star and the Tenth Planet

Evolution of the Earth's biosphere may have been influenced by galactic processes (Bailey et al. 1990, pp. 389–423). For example, from a study of the fossil record of the past 600 Ma Raup and Serkoski (1984) suggested that mass extinctions of flora and fauna on the Earth occurred regularly every 26 Ma, at least during the past 250 Ma. A similar analysis by Rampino and Stothers (1984) yielded a periodicity between 31 and 33 rather than 26 Ma, and an analysis by Fisher and Arthur (1977) gave 32 Ma. Such periodicity cannot be explained by occasional bombardment of the Earth by two or more nearly simultaneous impacts by meteorites or large fragments of a comet nucleus, as occurred during extinction at the Cretaceous/Tertiary boundary (Alvarez and Asaro 1990). Hatfield and Camp (1970) proposed that the extinctions were the result of the periodicity of the Sun's orbit about the galactic center and its movement perpendicular to the galactic plane. Davis et al. (1984) and Whitmore and Jackson (1984) suggested that the Sun has a small faint companion, which supposedly is in a distant orbit and has a period of 26 million years; they named it Nemesis. This possible companion star spends most of its time outside the Oort Cloud, but every 26 Ma, as it passes through perihelion it cuts a path through the cloud. During this passage Nemesis perturbs the orbits of many comets in the cloud, sending a shower of many of these objects into the inner Solar System to bombard the Earth. The extinctions may be a cumulative effect of comet impacts that affect the Earth's climate. Such a periodic bombardment of Earth appears to be supported by age-constrained impact craters. Using 16 of the 103 craters ranging in age from 0.1 to 1970 ± 100 Ma described by Grieve in 1982 (in

1987 Grieve described an additional 17 craters giving a total of 120 structures), Alvarez and Muller (1984) arrived at a periodicity of 28.4 Ma. Rampino and Stothers (1984) and Schwartz and James (1984) suggested another mechanism for the extinctions. They noted that the Sun tends to oscillate above and below the plane of the Milky Way with a half-period of 31 and 33 Ma. Possibly as the Solar System passes through the galactic plane it encounters giant molecular clouds that might perturb the Oort Cloud and send a shower of comets to the Earth. Whitmore and Matese (1985) proposed instead that the comet shower was caused by a tenth planet, Planet X. This planet supposedly circulates beyond Pluto within a hypothetical ring of comets, the so-called Kuiper belt. They proposed that gravitational perturbation from the other planets would cause the highly inclined orbit to precess into the flattened disk of the comets every 26 Ma, producing a comet shower.

The reality of this whole scenario of periodic bombardment of the Earth by comets and associated mass extinction has been questioned (Weissman 1990). The analytical methods of Raup and Serkoski (1984), for example, are uncertain because they tabulated some local rather than global, disappearances and also identified some changes in the names of flora and fauna as extinctions. As there are no accurate dates for the times of each extinction, the dating is based on linear interpolations between dated levels. The periodicity that one obtains also reflects the geological time scale that one uses, thus, times of mass extinctions vary from 26 to 33 Ma and some even disappear (Weissman 1990). An impact periodicity established on the basis of sampling only 16% of a crater population, also is suspect. In addition, Weissman believed that only two of the 103 craters described by Grieve in 1982 may have been formed by comets, and to-date no evidence comparable with the iridium layer at the Cretaceous/Tertiary boundary has been found associated with the supposed older extinctions. If the region of the Earth has been bombarded by a comet shower every 26–33 Ma this record should be preserved on the Moon, where geologic processes are much slower than on Earth. If there has been a comet shower every 26 Ma, the number of craters on the Moon would be eight times as many as those found (Weissman 1990). Moreover, no firm evidence of a companion star has been discovered to date. Binary systems that have the indicated separation distance of the Sun and Nemesis are rare in the Universe, and also the orbital characteristics proposed for Nemesis would be quite unstable. Perturbations by passing stars and a weak gravitational bond to the Sun would enhance its chances of escape from the Solar System. Although Nemesis could stay bound for 3.2 Ga, it probably would leave the system in about 600 Ma (Weissman 1990). As the Solar System dates from about 4.55 Ga, if the companion star tends to escape in about 600 Ma, then Nemesis must have been added to the Solar System rather recently, possibly by capture of a random passing star, an unlikely scenario. Stellar perturbations also would cause the orbital period of Nemesis to vary periodically, causing the star motion to depart from an initial 26-Ma period. Weissman (1990) suggested an alternative model in which Nemesis formerly was much closer to the Sun and has slowly migrated outward because of stellar perturbations. If this is so, its period previously was much shorter, and comet showers and impacts on Earth then must have been more frequent; however, data from the Moon where the cratering record extends to 3.0 Ga, suggest the possiblity that the cratering rate was lower during the past (Weissman 1990). Weissman also believed that comet showers caused by encounters with giant molecular clouds during Solar System oscillations are unlikely. These clouds are spread 150 to 250 light years above and below the galactic plane, the same distance by which the Solar System oscillates above and below the plane. As the Sun is only about 50% less likely to encounter the cloud at the end points of the oscillations than it is at its crossing of the galactic plane, there would not be a periodic shower of comets. The possibility that the shower is due to a tenth planet also is suspect. Such a body would tend to perturb the comets also when it approaches and leaves the ecliptic plane, not just when it was exactly within it, thus it would tend to smear out any periodicity.

3 Morphology of Planets and Satellites

In the broadest sense, we consider that the morphologies of rocky members of the Solar System (Table 1, listed in order of distance from their host) can be grouped into three categories: endogenic, exogenic, and exotic. Endogenic provinces are ones that were produced by internal forces that caused plate movements and interplate and intraplate magmatic/tectonic activities. Exogenic provinces are ones that owe their origin to external processes such as weathering, erosion, transportation, and deposition from earlier rock surfaces; therefore, they require the presence of a hydrosphere (or other liquid) and an atmosphere (Table 2, listed in order of decreasing diameter). Exotic provinces are ones created by bombardment of the planets or satellites by planetesimals and comets; this bombardment has produced large regions of terrae (impact craters), planitias (flat plains of debris ejected from impact sites), and impact melts (country rock melted by the kinetic energy of impact on rocky planets and moons).

The morphologic features are produced by geological processes that cannot be observed closely because they are seasonal (as on Earth), because they occur infrequently (as meteorite or asteroid impacts), or because they occurred during past eons (as on most planets and satellites), after which atmospheric or crustal conditions changed. Thus, identification of the origin of large to small morphologic features that are formed only seasonally or rarely or were formed during the distant past, largely depends upon the transfer of knowledge about morphologic features on the Earth to small-scale or large-scale images of features on other planets and satellites; such images are obtained by radar scans, altimeter surveys, photographs, spectroscopy, and polarimetry. Great support for distant visual interpretations is provided by information from samples, and from chemical analyses of such samples obtained by landers and especially by human observers (as on the Moon). Such support finally settled the long controversy about whether the craters on the Moon were made by volcanoes or by meteorite impacts.

Morphology of Planets and Satellites

Table 1. Data for planets and their major satellites. (After Smith et al. 1979a, b, 1986, 1989; Morrison 1982; Hartmann 1983. pp. 472–475; Burns 1986a; Morrison et al. 1986; Greeley 1987, p. 16; Lunine et al. 1989; Moore 1990, pp. 115–120; Pollack et al. 1991; Stern 1993)

Object	Revolution period (Earth days)	Rotation period	Equator diameter (km)	Density (g/cm³)	Gravity[a]	Escape velocity (m/s)	Distance from host (10³ km)
Mercury	88	59	4879	5.43	0.38	4300	
Venus	225	243	12104	5.27	0.9	10400	
Earth	365	1	12756	5.52	1.0	11200	
Moon	27	27	3476	3.34	0.16	2380	384.4
Mars	780	1	6794	3.93	0.38	5000	
Phobos	0.3	0.3	27×21×19[b]	~1.90	0.3–0.5	16	9.4
Deimos	1.3	1.3	15×12×11[b]	~1.50	0.3	6	23.5
Asteroids							
Ceres	1684	0.4	1020	?		~650	
Vesta	1326	0.2	549	~2.90		~350	
Pallas	1687	0.3	538	?		~340	
Hygeia	2031	~0.7	443	?		~280	
Davida	2071	0.4	341	?		~220	
> 1000 smaller ones							
Jupiter	4332	0.4	143884	1.33	2.64	60200	
Io	1.8	1.8	3630	3.55		2560	422
Europa	3.6	3.6	3138	3.04		2100	671
Ganymede	7.1	7.1	5262	1.93		2780	1070
Callisto	16.7	16.7	4800	1.83		2430	1883
12 smaller moons							

Saturn	10760	0.4	120536	0.71	1.16	32300	
Mimas	1	1	394	1.2		100	186
Enceladus	1	1	502	1.20		200	238
Tethys	2	2	1060	1.21		400	295
Dione	3	3	1120	1.43		900	377
Rhea	5	5	1530	1.24		600	527
Titan	16	16?	5120	1.88		2470	1222
Hyperion	21	?	350 × 235 × 200[b]	?		200	1481
Iapetus	79	79	1437	1.16		700	3561
Phoebe	550	9.4	220	?		100	12952
8 smaller moons							
Uranus	30685	0.7	51118	1.20	1.17	22500	
Miranda	1	?	472	1.35		500	130
Ariel	3	?	1158	1.66		1200	191
Umbriel	4	?	1190	1.51		1200	266
Titania	9	?	1610	1.68		1600	436
Oberon	13	?	1550	1.58		1500	583
10 smaller moons							
Neptune	60193	0.7	50538	1.64	1.2	23900	
Triton	6	?	2705	2.08		~2100	354
Nereid	360	?	340	?		~600	552
6 small ones							
Pluto	90472	6.4	2300	2.1	low	~1270	
Charon	5.4	6.4	1200	1.3		~600	19

[a] Gravity of Earth = 1.
[b] Dimensions of irregular satellites.

Table 2. Atmospheres of rocky planets and larger moons. (After Hartmann 1983, pp. 390–391, 431, 472–475; Moore 1990, 115–120)

Object	Surface temperature (°K)			Pressure (m bar)	Atmospheric gases present
	Max.	Min.	Calc.		
Mercury	700	100	452	< 0.3	None
Venus	731	731	731	90000	CO_2, N_2, H_2O
Earth	310	260	281	1000	N_2, O_2, Ar
Moon	280	100	280	0	None
Mars	280	150		10	CO_2, N_2, Ar
Phobos				0	None
Deimos				0	None
Asteroids					
Ceres			171	0	None
Vesta			175	0	None
Pallas				0	None
Hygeia				0	None
Davida				0	None
Jupiter	150		120	High	H_2, He, CH_4
Ganymede	145	85	107	~ 0	
Callisto	153	79	122	0	None
Io	135		106	~ 0	$?SO_4$
Europa	125	85	103	0	None
Saturn	160		88	High	H_2, He, CH_4
Titan	136		97	1500	N_2, CH_4, ?Ar
Rhea			70	0	None
Iapetus			85	0	None
Dione			70	0	None
Tethys			70	0	None
Enceladus			70	0	None?
Mimas			70	0	None
Hyperion			70	0	None
Phoebe			70	0	None
Uranus	110		59	High	H_2, He, CH_4
Titania			59	0	None
Oberon			59	0	None
Umbriel			59	0	None
Ariel			59	0	None
Miranda			59	0	None
Neptune	110		48	High	H_2, He, CH_4
Triton	38		42	0.1	$?CH_4$
Neried				0	None
Pluto			37	0.3–0.9	N_2, CH_4
Charon			37		None?

4 Earth

4.1 Early History, Structure, and Hydrosphere

Accretion of the Earth and its companions from chondritic planetesimals appears to have been completed by about 4.55 Ga, 100 Ma after formation of the solar nebula, and according to Condie (1989, p. 29) 1 Ma before the Sun went through a T-Tauri stage and the resulting strong solar winds blew away the most volatile elements from the inner Solar System and the early atmospheres of the inner planets. In the three-dimensional model described by Boss (1990) compressional heating during nebula formation caused the gross depletion of volatiles on Earth relative to their solar abundances. The inner nebula may have experienced temperatures of 1500 K that were regulated by vaporization of iron grains. Remote sensing data indicate that planetesimals of the inner Asteroid Belt underwent post-accretionary heating during the first few million years of Solar System history, a heating that led to extensive melting and magmatic differentiation (Gaffey 1990). This heating, which may have been caused by electrical induction during the T-Tauri stage or by short-lived radioisotope activity, also may have affected the planetesimals in the inner Solar System whose accretion produced the rocky planets. Such planetesimal differentiation must have influenced the accretionary and post-accretionary geohistory of the rocky planets. Accretions from volatile-depleted planetesimals differentiated into metallic cores and silicate mantles and allowed the Earth's core to form during accretion (Taylor and Norman 1990). Some investigators suggested that formation of the Earth may have been homogeneous, involving simultaneous condensation and accretion of compounds from a hot nebula as it cooled (Condie 1989, pp. 26–29). Others proposed an inhomogeneous model with an iron core and with the silicate mantle condensing sequentially, and still others believed that giant impacts influenced the evolution of the Earth, including the ejection of a primitive atmosphere, a source for melting the Earth, and a component for the mantle and core (Ahrens 1990; Melosh 1990; Newson and Sims 1991). Ahrens and O'Keefe (1989) also stated that possibly the core settled out at the same time that accretion occurred; thus there were some deviations from homogeneity during accretion of the planet. In a scenerio described by Benz and Cameron (1990), impact of the Earth by a giant planetesimal caused iron from the core of the impactor to penetrate the mantle and settle atop the Earth's core where it was heated to several ten thousand degrees. Most of the mantle of the impactor and a considerable volume of the Earth's mantle were ejected, but later some fell back onto the proto-earth. There was some heating of all parts of the interior of the proto-earth, and the surface layers reached temperatures of 16 000 K and were vaporized. It is this intense surface heating coupled with the presence of an orbiting disk of rock vapors and magmas that Benz and Cameron (1990) believed caused ejection of the early terrestrial atmosphere.

The Earth's crust above the Mohorovičić Discontinuity has two crustal divisions: oceanic and continental. Continental crust comprises 79% of the total crust by volume and is less than 20 to more than 70 km thick (Condie 1989, pp. 74, 130). It is layered, its upper part is silicic and its lower part mafic, and its average composition is comparable to andesite. Oceanic crust is mafic (like the lower continental crust), is layered, and is about 5 km thick except in zones of plate convergence where it may reach 12 km. The Earth has a magnetic field whose magnetic pole is at an angle of 11.5° with the rotation axis. Magnetic reversals occur on the average about every 30 000 years. Apparently, as the poles shift (north to south or vice versa during reversals) they follow the same path across the Earth. As reported by Kerr (1991b), this persistence implies that the solid mantle is somehow involved in the reversals because the upper core is fluid and thus cannot retain the track trends of previous reversals. Magnetic anomalies on continental crust mimic near-surface rocks and structural discontinuities, and on oceanic crust they are linear and parallel to mid-ocean ridges. Broad Bouguer gravity anomalies on continental crust reflect inhomogeneities in the lower crust or upper mantle, while narrower ones indicate near-surface structural features. In oceanic regions the anomalies range from -30 to $+250$ mgal on oceanic islands, are negative over trenches, range from $+250$ to $+350$ mgal in basins, from $+200$ to $+250$ mgal on ridges, and from 0 to $+200$ mgal over marginal and inland sea basins (Condie 1989, pp. 102–103). The calculated heat flow is 55 mW/m^2 on continents and 95 mW/m^2 over oceanic basins. The total heat loss from the Earth is about 42×10^{12} W with 12×10^{12} W from the continents and 30×10^{12} W from the oceans (Sclater et al. 1980); approximately 25% of this loss is via hydrothermal circulation in mid-ocean ridges (Davies 1980).

Distributions of seismic wave velocities permit the mantle to be divided into an upper level, a transitional zone, and a lower level. The upper mantle extends from the Mohorovičić discontinuity down to 400 km and includes the lower part of the 50- to 200-km thick lithosphere and the upper part of the asthenosphere which extends from the base of the lithosphere to a depth of about 700 km. At the top of the asthenosphere is a 50- to 100-km thick low-velocity zone. The transitional zone between depths of 400 and 1000 km is characterized by three sharp depthward increases in velocity. The lower mantle from 1000 to 2900 km contains a constant increase in velocity and density to just above the core, where the gradients of both flatten. The base of the lower mantle, known as the D″ region, is patchy, laterally heterogeneous, and probably stratified, with a possible discontinuity 280 km above the core. This D″ layer may receive molten iron from the outer core by capillary action to form aggregations of iron minerals (Powell 1991). Possibly, the layer contains the remainder of the original mantle or of old subducted plates. These mantle impurities have become mixed with core material to form a distinct layer on the surface of the core. Plumes rise from this part of the lower mantle (Kerr 1991a) and carry traces of core material to the Earth's surface, explaining the presence of ^{3}He in volcanic gases (Powell 1991). The large spherical head of a new plume can cause uplift, flood-basalt volcanism, and possibly regional-scale metamorphism or crustal melting and crustal extension. The narrow tails that follow the heads create the hot-spot tracks. Examples of volcanic activity associated with a plume include the flood basalts of the Karroo province of southern Africa, the Deccan Traps of western India, the North Atlantic Tertiary Igneous Province, and the Late Archean granite-greenstone terrane of the Yilgarn Block of western Australia. Examples of plume tail activity include the Yellowstone plume and the Hawaii-Emperor volcanic chain (Hill et al. 1992). Major seismic discontinuities within the mantle occur at 400 and 670 km and minor ones are at 550 and 1050 km. The upper mantle consists mainly of olivine associated with garnet (mainly below shields), pyroxene (in a thin layer beneath oceanic and tectonic regions), and plagioclase and amphiboles beneath oceanic regions (pyrolite model; Ringwood 1977). Olivine in the lower upper mantle is transformed to spinel at 400-km depth, forming most of the transitional zone. The 670-km discontinuity may reflect the transition from spinel ($\pm$ garnet) to perovskite and periclase structures (Condie 1989, p. 63). Recent analyses of deep mantle xenoliths transported by an eruptive kimberlite at Jagersfontein, South Africa (Haggerty and Sautter 1990) indicate that the mantle from about 400 to 670 km is a mixture of eclogite and lherzolite, with the seismic transition at 400 km due mostly to transition from olivine to beta-spinel (Sautter et al. 1991). The lower mantle is believed to consist of magnesium-iron silicates, strontium plumbate, ilmenite, plus some perovskite and periclase (Emery and Uchupi 1984, p. 187, and references therein).

The results of seismic tomography indicate that large inhomogeneities exist in the mantle and that they extend below the 670-km discontinuity to the mantle-core boundary. Geochemical isotopic data of oceanic basalts show the presence of a minimum of four and perhaps six or more long-lived reservoirs of end-member mantle (Condie 1989, p. 243). Two models have been proposed to explain the origin and survival of these mantle heterogeneities. In the layered model there are two separate convecting cells: the one in the upper mantle above the 670-km discontinuity is believed to have been depleted by extraction of continental crust from the original mantle, while another in the lower mantle is either pristine or less depleted. Tomographic studies show, however, that subducted slabs penetrate below the 670-km discontinuity, indicating that the convection system that drives plate tectonics involves the entire mantle (Kerr 1991a). This observation casts doubt on the validity of the two-layered model. A second model, the plum-pudding model has a convection system that involves the whole mantle; the mantle is depleted (pudding) and is embedded with blobs (plums). These plums range from a few meters to more than 10 km in diameter and may persist for billions of years. The core-mantle interface has a relief of no more than 5 km.

Shock-wave experimental data indicate that the core, extending from a depth of 2900 km to the center of the Earth (6378 km), consists of an iron and nickel alloy containing light elements. Some workers believed that the core formed from iron melts in the mantle and that such melts sank gravitationally to the center of the Earth within 20 ± 10 Ma after planetary accretion (Newson and Sims 1991). Stevenson (1990) questioned the gravitational origin of the core, thinking that the metallic-iron rich liquid would not be able to percolate through a mostly solid silicate mantle prior to macrosegregation and diapiric descent. He proposed instead that the core either was formed from iron supplied to the deep mantle by Rayleigh-Taylor instabilities of a silicate-iron suspensate in the upper mantle, and from the deep mantle to the core by Darcy flow, or the mantle was completely or nearly completely molten and the iron droplets settled to the core or into ponds that could descend to the core. Arculus et al. (1990) advocated that metal production near the Earth's surface may have involved molten silicate with additional separation of liquid metal from, and in equilibrium with, a solid silicate at high pressures in the mantle. Newson (1990) stated that siderophile element depletion from the mantle is more compatible for origin of a core formed by multiple stages of accretion and core formation. The light elements may have been incorporated into the core when it formed, or they may represent pieces of silicate mantle that periodically were added to the core, thus the core is in a state of continuing change, as summarized by Powell (1991). The outer part of the core does not transmit shear seismic waves, suggesting that it is in a liquid stage. Convection currents driven by latent heat from crystallization of iron on the surface of the inner core allow the outer core to become well mixed. The Earth's magnetic field is due to the dynamo action caused by these currents that flow at rates of several kilometers per year, a million times faster than convection in the mantle. The inner core, with a radius of 1220–1230 km, is a solid near its melting point or is partly molten (Condie 1989, pp. 67–72). Seismic oscillations suggest that its iron crystals may be aligned – an orientation caused by extremely slow convection currents flowing at a rate of only a few centimeters per year.

None of the other rocky planets has a crust comparable with that of the Earth, a crust that is related to plate tectonics and the presence of a hydrosphere degassed from the mantle (Condie 1989, p. 401). The geohistory of Earth is one of crustal mobility where the continental/oceanic basin geometry is continuously changing on a geological time scale. An excellent example of such mobility is Antarctica which, during the Proterozoic was juxtaposed against western North America and now is separated by thousands of kilometers (Dalziel 1991; Hoffman 1991; Moores 1991; Stump 1992). Whereas Rubey (1951) proposed that the hydrosphere accumulated slowly over geological time, most growth of both the atmosphere and the hydrosphere probably occurred within 100 Ma after accretion, with xenon isotopic data demonstrating that the Earth was degassed during the first 50 Ma. Nesbit (1987, p. 327) described continents as a mixture of granitoid melts generated above subduction zones that depend on the presence of deep oceans for their creation and survival. Thus, the critical factors that made possible the formation of continents on Earth appear to be the presence of water and plate tectonics. Continental plateaus have been eroded and in part transported back into the mantle. Geological data clearly demonstrate that by about 3.5 Ga ago, about 1 Ga after accretion, there existed on Earth well-defined continents and oceans, that continental crust was roughly the same thickness as today, and that the hypsometric curve (see Sect. 4.3) had roughly the same shape as now (Nesbit 1987, pp. 146–149). The available evidence also suggests that mid-ocean ridges always have been sub-aqueous. The oldest well-documented crust is in the Acosta Gneisses of the Slave Craton of northwestern Canada where it has an age of 3.96 Ga, in Enderby Land on Antarctica where it has an age of 3.9 Ga, and in the Amitsoq terrain of southwestern Greenland where it has an age of 3.8 Ga (Black et al. 1986; Moorbath et al. 1986; Bowring et al. 1990). Isotopic data indicate that the Acosta Gneisses were derived from a terrane that had experienced a complex history prior to 3.9 Ga ago. Another very old crust occurs in Mozambique where it is bordered by a 3.5-Ga ophiolite belt (de Wit et al. 1987). The presence of 4.0 to 4.2 Ga detrital zircons in early Archean sediments of Australia (Froude et al. 1983) suggests that small continental islands may have come into existence by that time.

How were the continents formed? The large amount of heat available on the accreting Earth and the presence of a dense blanketing atmosphere may have caused extensive melting and formation of a magma ocean, as has been proposed also for the Moon. Meteoritic impacts may have enhanced the melting (Condie 1989, pp. 337–340). The early Archean Earth may have resembled the ponded magma in Kilauea crater in Hawaii (Duffield 1972), with the crust overlying the 100- to 1000-km deep oceanic magma being fragmented by active spreading ridges, subduction zones including ones where two plates were subducted at the same time, and the presence of transform faults. These plate tectonic movements accommodated the large heat loss and the vigorous convection in the mantle. This model of a magma ocean is a point of contention among geologists, with some arguing that fractionations which should have occurred during a magma-ocean stage have not been documented (McFarlane and Drake 1990; Ringwood 1990) and others arguing that a magma ocean would have been too unstable for crystal fractionation to occur (Tonks and Melosh 1990). Sasaki (1990) also advocated a magma ocean on the accreting Earth – due to the presence of a solar type (H_2O-He) atmosphere that melted the Earth's surface when the Earth exceeded 0.2 times its present mass. Reduction of the surface melt by hydrogen supposedly added a large amount of water vapor to the atmosphere. As discussed by Davies (1990), such a magma ocean is needed only for an insolated surface, as postulated by Sasaki, or a late giant planetesimal impact. As the planet cooled, it may have acquired a global tectonic grid of orthogonal or near-orthogonal fracture patterns that may control the orientation of younger structures. Such a grid has been observed on the Moon, Mercury, and Mars (Bryan 1986, and references therein). Topographic trends on Venus are quite different from those of the other rocky planets, with morphologic features trending at less than 45° to the equator (Sharpton and Head 1986).

Xenon isotopic data demonstrate that the Earth was degassed during the first 50 Ma, and that the early crust was the floor of the early oceans. As at present, the crust was generated along mid-ocean belts and probably was komatiitic in composition. Heat loss was much greater then than now, so this primitive crust must have formed at rates 4 to 6 times faster than present crusts. If the excessive heat was lost by increasing the length of the mid-ocean ridge, its production may have been comparable with present rates. Recycling of the crust took place at subduction sinks – recycling that may have been helped by meteoritic bombardments which were important at that time (Condie 1989, p. 342). During Archean time the plates probably were much smaller, the depth of the ocean floor was much less (higher heat flow), and if the amount of water was the same as at present the surfaces of the continental fragments were more deeply submerged. A near absence of continental deposits within the Archean succession appears to verify this. The continents were produced at subduction sinks where the komatiitic crust underwent small degrees of melting to produce felsic magmas that underplated the mafic and komatiitic-arc systems; true granites did not appear in the geological record until 3.0 Ga ago and did not become important until 2.6 Ga ago. These granites were produced by partial melting or fractional crystallization of tonalites (Condie 1989, p. 344). Evolution of the early crust is the history of four rock types (komatiites, basalts, tonalites, and granites) with komatiites (high temperature Mg-rich basalts) being formed first and granites last. As described by Condie, early Archean komatiites and basalts were produced at mid-ocean ridges and basalts at subduction sinks. They were hydrated by reaction with ocean waters and partly melted at sinks to produce tonalites, which in turn partly melted or underwent fractional crystallization to produce granites. As summarized by Meissner and Mooney (1991), continental crustal formation from accreted island arcs is related to three processes: (1) delamination where the lower mafic crust and uppermost mantle of island arcs are recycled into the mantle; (2) partial melting and crustal differentiation of juvenile crust with the silicic/intermediate melts rising in the crust and the more mafic residue forming a new Mohorovičić discontinuity; or (3) intrusion of a large volume of alkali magma. Probably the second process is the most common one.

An alternate origin for continental crust has been proposed by Lowman (1989), who believed that the crust of the rocky planets (Moon, Mercury, Venus, and Mars) was the product of early global non-plate tectonic differentiation. In this model sea-floor spreading on Earth was initiated later as a result of meteoritic bombardment. Lowman proposed that massive andesitic overplating during degassing of the Earth created a global crust of heavily cratered interlayered andesites, minor basalts, and sediments represented today by the intensely deformed granulites of the lower continental crust. A heavy bombardment 4.0 Ga ago disrupted this crust and the impact basins became loci for mantle upwelling, basaltic magmatism, and sea-floor spreading. Basaltic overplating, underplating, and repeated tectonic deformation thickened the early continental crust. Partial melting of the underplated basalt generated tonalitic magmas that intruded the continental crust and added to its thickness. The granitoids and gray gneisses on Archean shields and granulites of the lower crust are remnants of this basaltic intrusion and of the primordial continental crust. The greenstone belts and dike swarms in the late Archean and Proterozoic were emplaced during a second basaltic magmatic differentiation. As the continental crust increased in thickness and rigidity it produced a change of tectonic regime to its present style 2.5 Ga ago.

This style is one of deformation, metamorphism, anatexis (melting of preexisting rocks), and changes in the geometry of continental blocks. Lowman believed that continental crustal growth since the Archean has been one of thickening as a direct or indirect result of the creation of basalt in the subcontinental tectosphere.

Continental crustal growth was by a combination of magma additions (by underplating and overplating), thrusting and stacking of crustal rocks, continental and island-arc collisions, and accretion of sediment wedges. Crustal growth during the early Archean prior to 3.0 Ga was slow; it increased during the late Archean 3.0 to 2.7 Ga ago, and increased again in the late Proterozoic 2.0 to 1.9 Ga ago. The North American continent, for example, was assembled by continental collisions largely between 1.95 and 1.75 Ga ago. What is the significance of this apparent crustal growth with time? Is it possible that the distribution is not real, that the older crust was much more extensive in the past and since then it has been recycled? Is it possible that the absence of older crust is a reflection of an absence of plate tectonics prior to about 2.7 Ga ago? As pointed out by Davies (1990), in regard to the crustal evolution of the early Earth we are left with mainly questions and few answers. Possibly the Archean continental crust (Figs. 1 and 2) was much

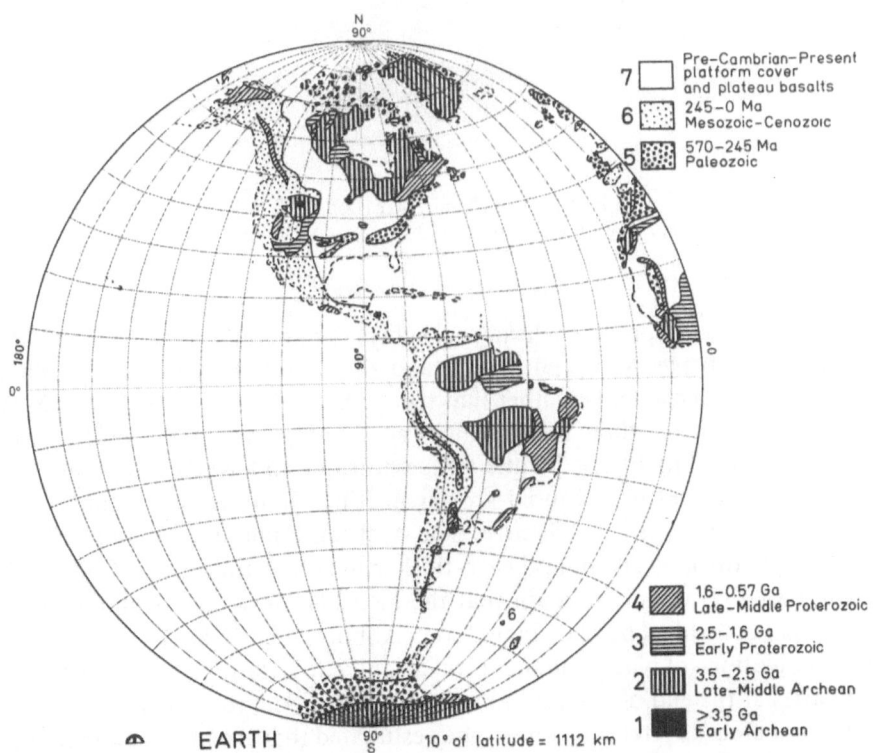

Fig. 1. Crustal provinces of the Earth, western hemisphere (Long. 0° is on the right, Long. 90°W in the *middle* and Long. 180°W on the *left*.) (After Porada 1979, fig. 1; Emery and Uchupi 1984, Plates XIA and XIB; Condie 1989, plate 1; Dalziel 1991, fig. 1)

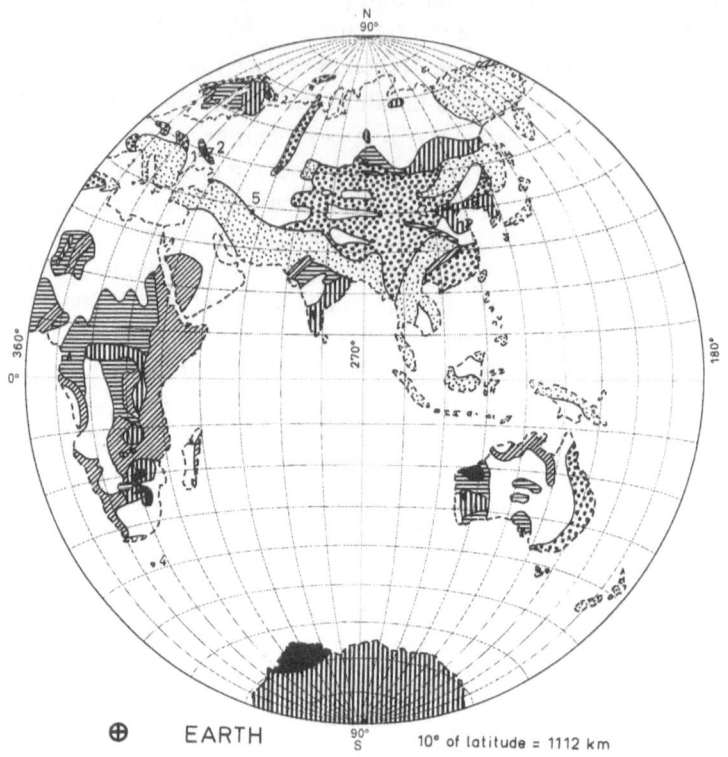

Fig. 2. Crustal provinces of the Earth, eastern hemisphere (Long. 180°W is on the right, Long. 270°W in the middle, and Long. 360°W on the left). (Condie 1989, plate 1, fig. 10.12)

thicker than now. Exposed Archean rocks display mineral assemblages character-istic of depths between 30 and 50 km, and they are underlain by crust that is 30–40 km thick. This implies either that the Archean crust was much thicker than now (60–80 km) or that subsequent underplating has been able to keep pace with uplift. The higher mantle temperature in mid-ocean ridges during the Archean also should have caused greater melting, more magma formation, and consequently a thicker oceanic crust. The Archean komatiitic oceanic crust may have been as thick as 20 km and even thicker in subduction zones (Condie 1989, p. 346). Based on the observation that the distribution of continental areas is to some extent controlled by plate boundaries, Abbott and Menke (1990) used Monte Carlo simulations of plate boundaries on a sphere to determine the length of plate boundaries at the end of Archean time 2.4 Ga ago. From the size distribution and normalized areas of Archean cratons they estimated that the plate boundary at the end of the Archean was about 1.7 to 2.7 times the present length, with a preferred mean of 2.2 times. This result supposedly favors a model in which sea-floor spreading rates were about 16% slower than now and oceanic crust was more than 20 km thick. These speculations, however, tend to be contradicted by data from the

3.5-Ga old Jamestown ophiolite complex in Swaziland of southern Africa (de Wit et al. 1987), which indicate that the oceanic crust, at least at that time and in that area, was only about 3 km thick.

The geohistory of the Earth is one of extensive translations where megacontinents broke up, ocean basins formed, and ocean basins in turn were consumed to form other megacontinents. Because of the persistence of these events in the mantle, high-resolution seismic tomographic models of the upper mantle have made possible reconstruction of these plate-tectonic events (Anderson et al. 1992). Spreading rates were not uniform throughout the Earth's history. For example, between 120 and 80 Ma ago there was a 50 to 75% increase in formation of ocean crust due to a 'superplume' that originated near the core/mantle boundary. The South Pacific Superswell under Tahiti is the remnant of this plume (Larson 1991). Spreading rates, except for a secondary peak near the Cretaceous-Tertiary boundary, decreased slowly afterward to reach the present, nearly steady state. Associated with this scenario of continental construction and destruction by mantle convection (Gurnis 1988, 1990) were changes in sea level (Kominz and Bond 1991) and climatic changes including extensive continental glaciation 2800, 2400–2300, 900–600, 450, 350–250, and 15–0 Ma ago. Changes in the Earth's orbital characteristics, sun-spot activity, deep ocean circulation, as well as changes in the biosphere may have enhanced these refrigeration events (Hayes et al. 1976; Condie 1989, p. 437; Reid 1993; Wright 1993). Continental uplift, which altered atmospheric circulation, also may have been one of the causes of the Quaternary glaciation (Ruddiman and Kutzbach 1990), but others have suggested that the Cenozoic uplift was the result rather than the cause of the climatic change – a change that enhanced weathering and erosion and led to isostatic rebound and rejuvenation of relief (Molnar and England 1990).

4.2 Atmosphere and Biosphere

In addition to its plate tectonics and hydrosphere, other features that make the Earth unique are its atmosphere and biosphere. The present atmosphere consists chiefly of nitrogen, oxygen, and small amounts of such gases as argon and carbon dioxide. The low abundance of noble gases in this atmosphere, when compared with solar type gas, indicates that this is not a primitive atmosphere representing residual gases left after accretion, degassing of the planet, or extraterrestrial sources. Sasaki (1990) believed that the composition of the primitive atmosphere was a solar type consisting of H_2O-He with accretionary processes enhancing its water and volatile content. The presence of water modified the opacity and molecular weight of the atmosphere. Others postulated that this primitive atmosphere was either dominantly of CH_4 with smaller amounts of NH_3, H_2O, He, and H_2O and was reducing (Oparin 1953), or of CO_2, CO, H_2O, and N_2 (Abelson 1966). Because CH and NH_3 are destroyed by photolysis, and ammonia is destroyed by ultraviolet radiation and is highly soluble in water, the Abelson model may be more realistic than the Oparin one (Condie 1989, p. 387). This primitive atmosphere is believed to have been dissipated early in the history of the Earth by solar winds during the T-Tauri phase of the Sun (Newson and Sims 1991). The present atmosphere thus must have formed by degassing of the planet. Extensive degassing occurred probably during the first 50 Ma after accretion, with some degassing continuing for at least a few hundred million more years. The degassing on the Earth was especially large because of exposure of much of the lithosphere and even of mantle caused by a glancing impact of a large planetesimal on the Earth. The material that was removed formed about 80% (the remainder was contributed by the impactor) of the Moon, whose volume is about 2% that of the present Earth. By 4 Ga ago an atmosphere rich in CO_2 and H_2O was in place (Kasting 1987).

This concept of the planets being formed first and then the atmosphere and oceans by endogenic processes has been challenged during the past two decades. Some workers (see Ahrens and O'Keefe 1989, and references therein) have postulated that proto-atmospheres developed during accretion of the planets. Ahrens and O'Keefe stated that impact devolatilization of the Earth and Venus occurred when the planets reached about 0.12 of their present radii and complete impact degassing occurred when the planets reached 0.3 their present size. The same authors also postulated that because of its lower mass, planetesimal devolatilization did not begin on Mars until the planet radius was 0.3 of its present size, and complete devolatilization may not have been attained even when it reached its present size. In the course of accretion of Earth and Venus a mass of water equivalent to several terrestrial oceans may have been ejected into space by cratering. Ahrens and O'Keefe suggested that this erosion of coaccreting atmospheres was more effective on Mars and that Mars has lost 1 to 100 times the total water inventory originally accreted onto the planet.

As described by Sasaki (1900), the molten surface of a magma ocean before it solidified (according to Davies 1990, a magma ocean generated by a giant planetesimal impact probably cooled to a mush in about 1000 years) reduced the degassed water to hydrogen. This conversion may have oxidized iron to iron oxide and changed the surface of the Earth from a reducing to an oxidizing environment. When the impact energy flux decreased at the end of accretion, the water in the atmosphere condensed and a proto-ocean and drainage system was established. The opacity of this CO_2 and H_2O atmosphere also produced a greenhouse effect, preventing the Earth from freezing at the time when the Sun was 25–30% less luminous than it is today. That the Earth never experienced such low temperatures is indicated by 3.8-Ga old sediments that reveal the presence of oceans and running water (Condie 1989, p. 391). Absence of glaciation when luminosity was low is believed to have been caused by a high content of atmospheric CO_2 prior to 2.8 Ga ago, but glaciation on Gondwana during the Late Ordovician was at the time that atmospheric CO_2 was high because of low rates of continental weathering and high volcanicity (see Kastings 1992, and references therein) and low reception of solar radiation caused by Gondwana's position near the South Pole (Crowley and Baum 1991). Thus, as Kasting pointed out, what goes around comes around; previously low solar luminosity was seen as a problem and a high carbon dioxide content was looked upon as the solution, and later a high content of carbon dioxide is seen as the problem and low luminosity is perceived as the solution.

Tajika and Matsui (1990) stated that if there was no continental growth the atmospheric CO_2 would not be reduced and there would be an increase in temperature with increasing solar luminosity. Yet, during the early Archean the continental masses were small and weathering of silicates on land must have been minor. Either the early Archean continental masses were larger and have been extensively recycled or some other processes were responsible for preventing a runaway greenhouse effect. During the so-called aquaplanet stage, when continents were non-existent or only minor islands were present, the carbon cycle must have been dominated by submarine processes. Probably large-scale hydrothermal circulation and associated weathering of the ocean floor and biological activity on the komatiitic mid-ocean ridges maintained the carbon dioxide in the atmosphere at a reasonable level. As the continents evolved, weathering of silicates on land reduced the atmospheric CO_2 content to a minor but important component. It is these processes that prevented a greenhouse effect and dehydration of the planet (Nesbit 1987, p. 335; Walker 1985, 1990a). This decrease in carbon dioxide and subsequent temperature decrease may in part have been responsible for the Huronian episode of continental glaciation during the early Proterozoic.

Addition of oxygen to the atmosphere is due to photosynthesis. Its appearance seems to have had three stages: (1) when free oxygen was not present; (2) when small amounts were present in the atmosphere and surface ocean water, and (3) when oxygen invaded the atmosphere-ocean system (Condie 1989, p. 436). Sediment data indicate that stage 2 began about 2.4 Ga ago (appearance of red beds, increase of sulfates, and the near disappearance of detrital uraninite-pyrite deposits). Stage 3 began 2.0–1.8 Ga ago with the near disappearance of banded iron deposits and the appearance of eucaryotic organisms. Life may have formed on Earth about 4.0 Ga ago in volcanic centers and submarine hydrothermal vents where a supply of organic compounds and mineral catalysts was present. Life in such environments would have been protected from the massive planetesimal bombardment that the Earth incurred prior to 3.8 Ga ago. To-date the oldest remains of life have come from 3.8-Ga old sedimentary rocks from Isua, Greenland, and from 3.5-Ga-old Archean strata in Australia and South Africa (Schidlowski et al. 1975; Schopf 1976; Cloud 1978, p. 128; Schopf et al. 1983; de Wit et al. 1987; Walker 1990b; also see discussion by Horgan 1991 on the origin of life). As pointed out by Walker, the presence of life documents a terrestrial environment that was neither too hot nor too cold and contained water. The first life forms probably were anaerobic fermenting heterotrophs. Aerobic photoautotrophs capable of living in the presence of free oxygen had appeared by 2.8 Ga ago, and organisms dependent on oxygen had appeared by 2.0 Ga ago and become widespread by 1.7 to 1.5 Ga ago. Eukaryotes appeared by 2.1 Ga ago (Han and Runnegar 1992) and had become abundant and diversified by 1.2–1.0 Ga ago (Cloud 1978, p. 134; Condie 1989, p. 425). By 1.0 Ga ago the ozone screen was in place in the upper atmosphere, and life was protected from most ultraviolet radiation. Knoll (1991) ascribed the proliferation of the Ediacarian fauna at the end of the Proterozoic to a pronounced increase in atmospheric oxygen. This change was tectonically driven, with hydrothermal and volcanic activities enhancing anomalously high rates of burial of organic matter. That burial in turn promoted an increase in the oxidizing potential of the atmosphere and hydrosphere and allowed the microscopic metazoans to evolve into macroscopic forms and eventually to dominate the animal world. From these small beginnings life proliferated, with metazoans appearing 700 Ma ago, organisms with calcareous and siliceous skeletons 500 Ma ago, vertebrates 450 Ma ago, vascular plants 425 Ma ago, mammals 225 Ma ago, flowering plants 150 Ma ago, and man 3 Ma ago (Condie 1989, p. 437). This biosphere acting in concert with inorganic chemical processes, weathering on the continents, and hydrothermal activity in mid-ocean ridges has controlled the Earth's surface environment since its inception (Nesbit 1987, pp. 337–339). If Lovelock (1979) is correct in supposing that life controls all aspects of the Earth including its atmosphere, hydrosphere, sedimentation, and albedo, that have allowed water to persist on the Earth for billions of years, then (as Nisbet pointed out) one can conclude that the biosphere must be the ultimate architect of Earth's morphology.

4.3 Hypsometry

The morphology of the Earth, in terms of origin and history of development of endogenic and exogenic provinces was mapped and discussed by Uchupi and Emery (1991a). Far less detailed treatment for the Earth is needed here to compare with the morphologies of other rocky planets and moons, because much less is known about these latter. Our earlier treatment showed the importance of crustal plate movements in generating morphological provinces on Earth. These movements are attributed to cellular circulation within the underlying mantle. The earliest stage for crustal plates on the Earth followed the accumulation of low-density crustal material in a large area (a megacontinent) surrounded by denser crustal material also derived from the molten magma of the mantle. As discussed by Nance et al. (1986), Bercovici et al. (1989), and Gurnis (1988, 1990), such accumulations of thick, light material tended to insulate and reduce heat loss from the mantle, leading to doming and lateral flow of the mantle that dragged the overlying crust and eventually rifted it. Others believe that breakup is the result of high angular momentum possessed by supercontinents and that such angular momentum produces long-lived stresses (Murphy and Nance 1992, and references therein). A belt of denser crust soon began to form within the newly opening rift at either side of a central midrift ridge. As the rifted parts of the megacontinent continued to drift apart, the opposite sides of these parts overrode earlier denser crusts in subduction belts bordering the continents. Thus, the rifting side of a continent became the site of a belt of endogenic production of irregular dense crust by igneous activity and the subducting side became a belt of endogenic production of irregular light crust by more acidic volcanism accompanied by faulting and folding. Continued subduction removed all the older dense crust and produced a new megacontinent of light crust that eventually rifted to initiate a new cycle of production of dense crust. The subduction also caused addition to the continents of allochthonous terranes (former microcontinents, island arcs, oceanic ridges, and seamounts). This cycle of rifting, drifting, subduction, and regrowth of megacontinents has repeated itself many times ever since the Earth formed a low-density continental crust soon after its origin. Murphy and Nance (1992) indicated that this supercontinental cycle has an average period of about 500 Ma; other studies suggest that the cycle may be much shorter. For example, the Wilson cycle in western Africa began with the rifting of the West Craton 800 Ma ago and terminated 650 to 700 Ma ago with continental collision – a duration of

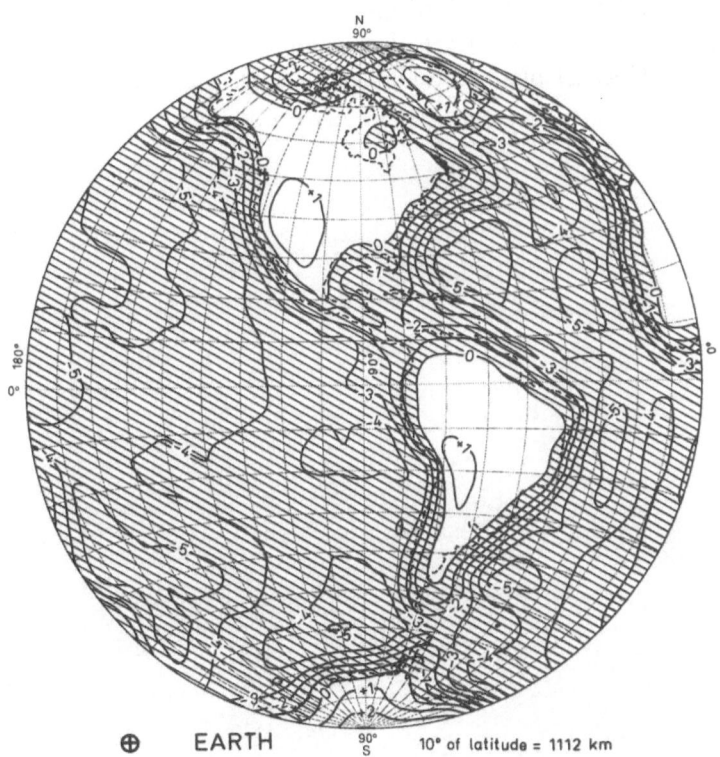

Fig. 3. Generalized 1-km interval contours for the Earth, western hemisphere. Contours are based on a digital tape from the NGDC degraded by R. A. Goldsmith (Woods Hole Oceanographic Institution) to average elevations (including ice sheets) within areas having a radius of 5 degrees and plotted for contouring at 5° centers. This method, of course, causes the zero contour to deviate from the true shoreline (*dashed line*). The degradation was intended to make contours of the Earth's surface to be somewhat comparable in detail with those of other planets and moons obtained by harmonic analysis on widely-spaced graticules. *Diagonally lined pattern* denotes areas below computed sea level. Map is a Lambert azimuthal equal area projection (Wilhelms 1987, plates); the same is used on all other global maps in this book

100–150 Ma (Bertrand and Caby, 1978). According to Uchupi and Emery (1991b), in the Paleozoic and Meso-Cenozoic Wilson cycles in the North Atlantic and western Tethys lasted from 450/250 to 210 Ma, and in the Norwegian-Greenland Sea it began 350 Ma ago and has not yet come to completion.

Development of a hydrosphere or ocean allowed the areas of lighter and higher crust to be known as continents and the areas of lower and denser crust as ocean basins (Figs. 3 and 4). On Earth the presence of a hydrosphere accentuates the pattern of lighter and denser crusts (as continents and ocean basins), even though the hydrosphere has exhibited changing levels during the past, with the highest levels occurring during times of most active rifting and the lowest levels occurring during times of accumulated megacontinents. As a result, the most precise hypsometric curve of the Earth contains two frequency levels of relatively flat topography (Table 3, Fig. 5) such that the present mean elevation of the land is 0.84 km above sea level and the mean depth of the deep-ocean floor is 3.80 km below sea level. The difference between the two levels has changed during the past, episodically according to stages of plate tectonics and secularly because of erosion on the

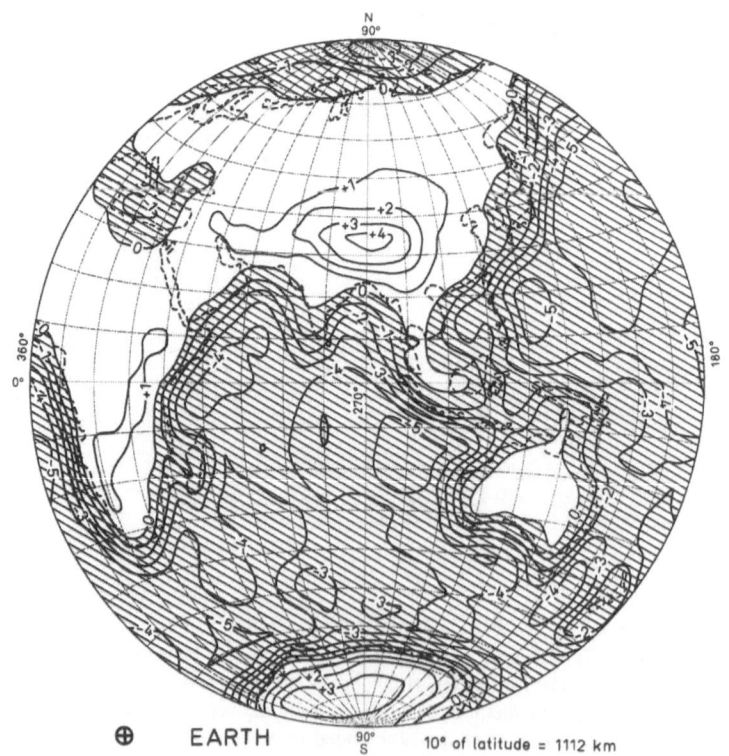

Fig. 4. Generalized 1-km interval contours for the Earth, eastern hemisphere. *Symbols* and sources as for Fig. 3

Table 3. Hypsometry of Earth, Moon, Mercury, Venus and Mars

Elev. (km)	Earth[a] (cum.%)	Earth[b] (cum.%)	Earth[c] (cum.%)	Moon[d] (cum.%)	Mercury[e] (cum.%)	Venus[f] (cum.%)	Mars[g] (cum.%)	Mars[h] (cum.%)
								0.0
> +25								0.1
+20							0.0	0.3
+15							0.6	0.5
+10						0.0	1.3	2.1
+9	0.0					0.0	2.6	3.2
+8	0.0				0.0	0.1	4.0	4.8
+7	0.0			0.0	0.4	0.1	5.6	6.5
+6	0.2	0.0		0.1	5.3	0.1	9.6	10.2
+5	0.6	0.1		2.1	10.1	0.6	15.0	17.6
+4	1.8	0.8	0.0	5.8	17.2	2.3	30.4	38.7
+3	3.7	2.1	0.0	13.2	31.3	4.9	46.5	50.6
+2	8.1	5.4	2.5	26.0	65.0	12.7	62.0	60.7
+1	30.0	25.0	12.5	46.9	93.5	39.1	73.9	72.2
+0	38.0	33.0	26.1	68.1	98.9	93.3	84.5	87.8
−1	41.5	39.7	41.0	90.2	100.0	100.0	93.3	94.5
−2	46.0	48.1	56.6	99.7			99.0	99.4
−3	61.2	67.8	79.4	100.0			100.0	100.0
−4	83.0	91.6	98.1				100.0	
−5	98.5	100.0	100.0					
−6	100.0							
−7	100.0							
−8	100.0							
−9	100.0							
−10								

[a] Kossina's (1921, p. 53) calculations based on topographic contours.
[b] NGDC, average topography within 5° circles, 5° graticule; Figs. 3 and 4
[c] NGDC, average topography within 14.2° circles, 20° graticule.
[d] Bills and Ferrari (1977), degree-12 harmonics of contours based on rader altimetry from satellites and photogrammetry from satellites and Earth; Figs. 15 and 16
[e] Clark et al. (1988, fig. 26), topography from Earth-based radar.
[f] Masursky et al. (1980), contours, derived from radar altimetry from satellites.
[g] US Geological Survey (1976), contours of Figs. 28 and 29 from data of Figs. 26 and 27 plus photogrammetry.
[h] Bills and Ferrari (1978), contours of Figs. 26 and 27 from degree-16 harmonics of data from Earth-based radar and other materials.

continents and deposition of the erosional products on the ocean floor. This curve, by Kossinna, was compiled from contour maps of continents and ocean basins, and the part for all ocean basins was supported by results obtained by Murray (1888) and Murray and Hjort (1912, p. 132) and closely confirmed by Menard and Smith

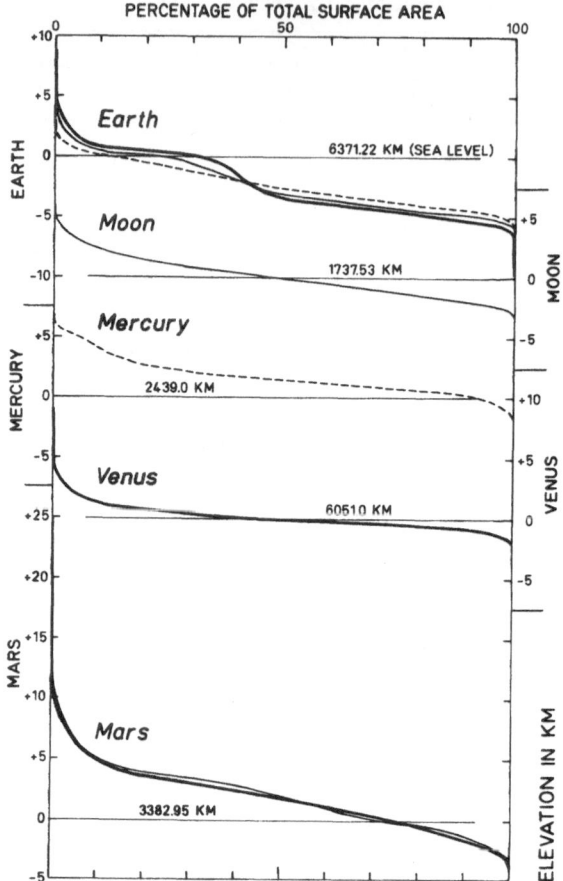

Fig. 5. Hypsometric curves for the Earth, Moon, Mercury, Venus, and Mars.

Earth – *wide continuous* line fom Kossinna (1921) obtained by his measurement of the areas between elevation contours at 1-km intervals plus contours at + 0.5, + 0.2, − 0.2 and − 0.5 km;

    – *narrow line* from 2° digital tape from NGDC (probably TGP-0013-MT of mid-1960s) degraded to average elevations centered within 5° circles at 5° spacings of latitude and longitude, plotted on a Lambert azimuthal equal-area projection from which areas between 1-km contours were measured by weighing;

    – *dashed line* from 2° digital tape from NGDC degraded to provide average elevations within 14.2° circles centered at 20° spacings of latitude and longitude.

Moon – From contours obtained by harmonic analysis to degree-12 (Bills and Ferrari 1977) relative to mean sphere of 1737.53-km radius, transferred to Lambert azimuthal equal-area projection and measured by weighing of areas between contours.

Mercury – Cumulative curve derived from frequency curve of elevations relative to 2439.0-km reference sphere. From Earth-based radar data averaged by 10° longitudinal bins (Clark et al. 1988).

Venus – Derived from radar altimetry data measured aboard Pioneer and Venera 15 (Masursky et al. 1980; Anonymous 1989; Ford and Pettingill 1992 corrected the mean radius to 6051.84 km).

Mars – *wide continuous line* from contours obtained by harmonic analysis to degree-16 (Bills and Ferrari 1978) relative to 6.1-mbar surface, transferred to Lambert azimuthal equal-area projection, from which areas between contours were measured by weighing.

    – *narrow line* from contours by US Geological Survey (1976) transferred to Lambert azimuthal equal-area projection, from which areas between contours were measured by weighing

(1966, fig. 8). All of these authors measured the areas between contours, but the contours were based upon vastly increasing numbers of elevation data. The other hypsometric curves for the Earth in Table 3 and Fig. 5 were derived from global averages obtained from the National Geophysical Data Center (NGDC) digital tapes plotted at 5° and 20° spacings, respectively, on the Earth's surface and then contoured. Clearly, the curves derived from global averages are less detailed and smoother than the one from Kossinna, and they fail to show the flattenings atop continental crust and oceanic crust.

4.4 Morphology

4.4.1 Endogenic

At present, land occupies about 28% of the Earth's surface of 510×10^6 km^2, but endogenic provinces comprise about 65% of the land area and 49% of the ocean floor area (Figs. 6 and 7; Table 4). Lands plus ocean floor shallower than 1 km constitute 192×10^6 km^2 or 38% of the Earth's surface (Menard and Smith 1966). The percentages of endogenic areas probably were larger on land during times of major consolidation of megacontinents and larger on ocean floors during times of major rifting (the opposite is true for percentage areas of exogenic provinces), but this is a much more complex matter for the entire Earth's surface than for only a single continent or ocean basin. Although most of the source materials for exogenic provinces on Earth must be provided by endogenic processes, both endogenic and

Table 4. Areas of morphological provinces (10^6 km^2)

	Endogenic	Exogenic	Exotic Cratered	Exotic Impact Melts	Total
Earth (total)	276	234	0.07	< 1	510 km^2
(land)	98	51	0.07	< 1	149 km^2
(ocean)	178	234	0	0	361 km^2
Moon (total)	6	0	32	< 1	38 km^2
Mercury (48%)	27	0	9	< 1	36 km^2
Venus (89%)	176	< 1	0.4	< 1	198 km^2
Mars (total)	27	54	64	< 1	145 km^2
Earth	54	46	< 1	< 1	100%
Moon	16	0	84	< 1	100%
Mercury	76	0	24	< 1	100%
Venus	99	< 1	< 1	< 1	100%
Mars	19	37	44	< 1	100%

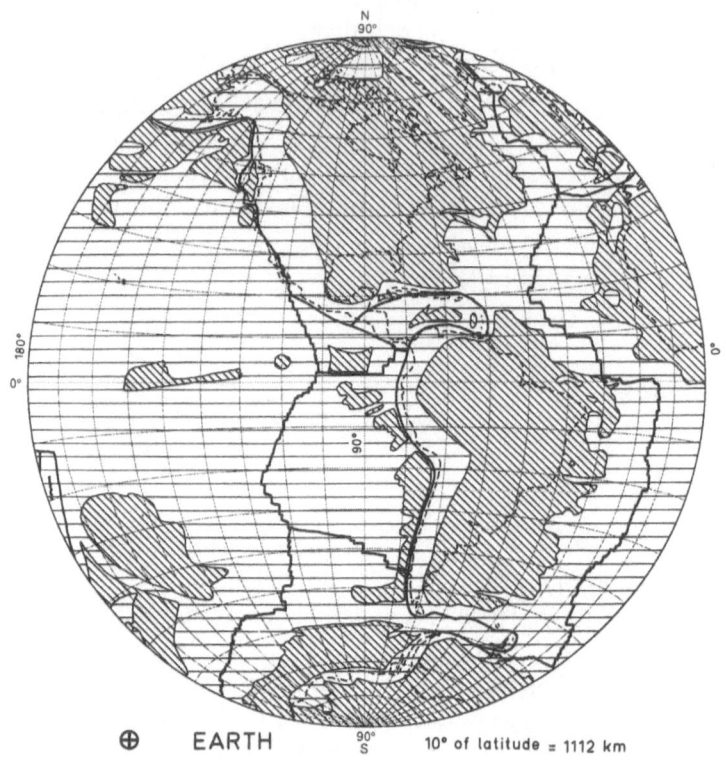

Fig. 6. Morphological provinces of Earth – western hemisphere. *Dashed line* present coastlines; *wide jagged line* crustal plate boundaries; *horizontal shading* areas have endogenic surface morphology; *diagonal shading* areas have exogenic morphology. After Uchupi and Emery (1991a; Figs. 1 and 2) on a Lambert equal-area projection

exogenic processes operate simultaneously. Mountain building and erosion are concurrent, but the mountain building occurs at faster rates and more continuously on the ocean floor than on land. The longer interval between episodes of endogenesis on continents than on the ocean floor probably is the result of the damping effect caused by the greater thickness of crusts on continents. In fact, most of the tectonism on continents in the subduction phase has been transmitted from ocean floors.

On Earth the fundamental cause of most endogenic morphology is plate tectonics and associated midplate stresses due to ridge push, slab pull, and collisional resistance (Zoback 1992), and mantle plumes (Richardson 1992), unlike on all other rocky members of the Solar System. This exclusiveness is because Earth contains a hydrosphere (oceans and other sites of abundant water) which makes the lithosphere, asthenosphere, and mantle less viscous than those layers on

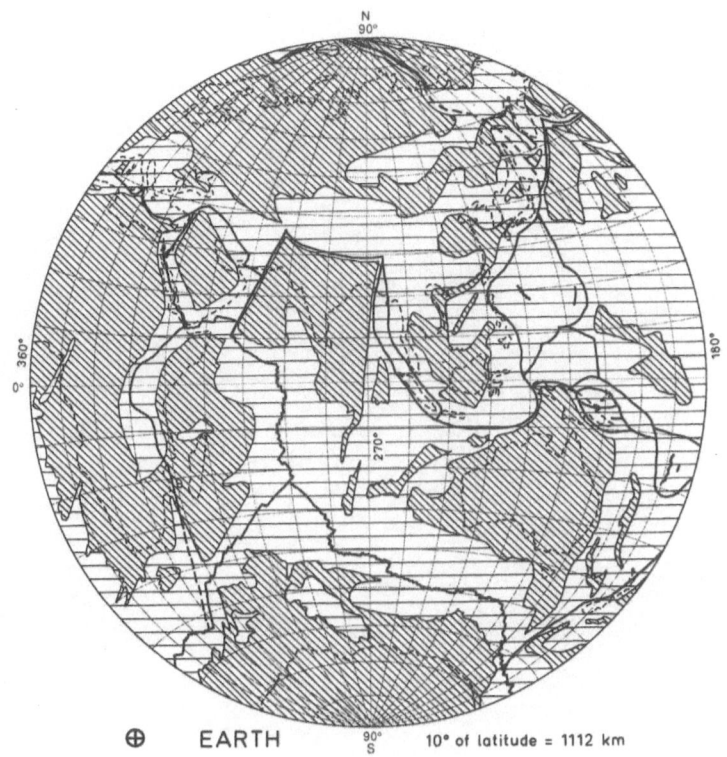

Fig. 7. Morphological provinces of Earth–eastern hemisphere. *Symbols* and source as for Fig. 6

other planets. The low viscosity of the mantle rocks allows the cellular motion of the mantle to help dissipate by convection the heat from an original high temperature of Earth's accretion plus subsequent heat from interior radioactivity. The cellular movement of mantle cells drags with it the overlying lithospheric plates through the intermediate mobile layers of asthenosphere. Where the adjacent mantle cells rise and diverge, the overlying lithosphere becomes rifted and the rifted pieces drift farther apart. Where adjacent mantle cells converge and sink, they force the overlying lithospheric plates also to converge. Where mantle cells of opposite direction of movement lie along the same axis (separated by a transverse belt of no movement), the overlying lithospheric plates have a laterally divergent or translation movement – essentially separated by a major strike-slip fault (Fig. 8).

The surface morphologies produced by these plate movements have major dimensions on Earth. Those of divergence start with a rift valley across a large plate, commonly that of a megacontinent that is a descendent from several previous episodes of divergence and convergence. Continued divergence causes the rifted pieces of the plate to drift apart, and rising magma along the widening rift valley

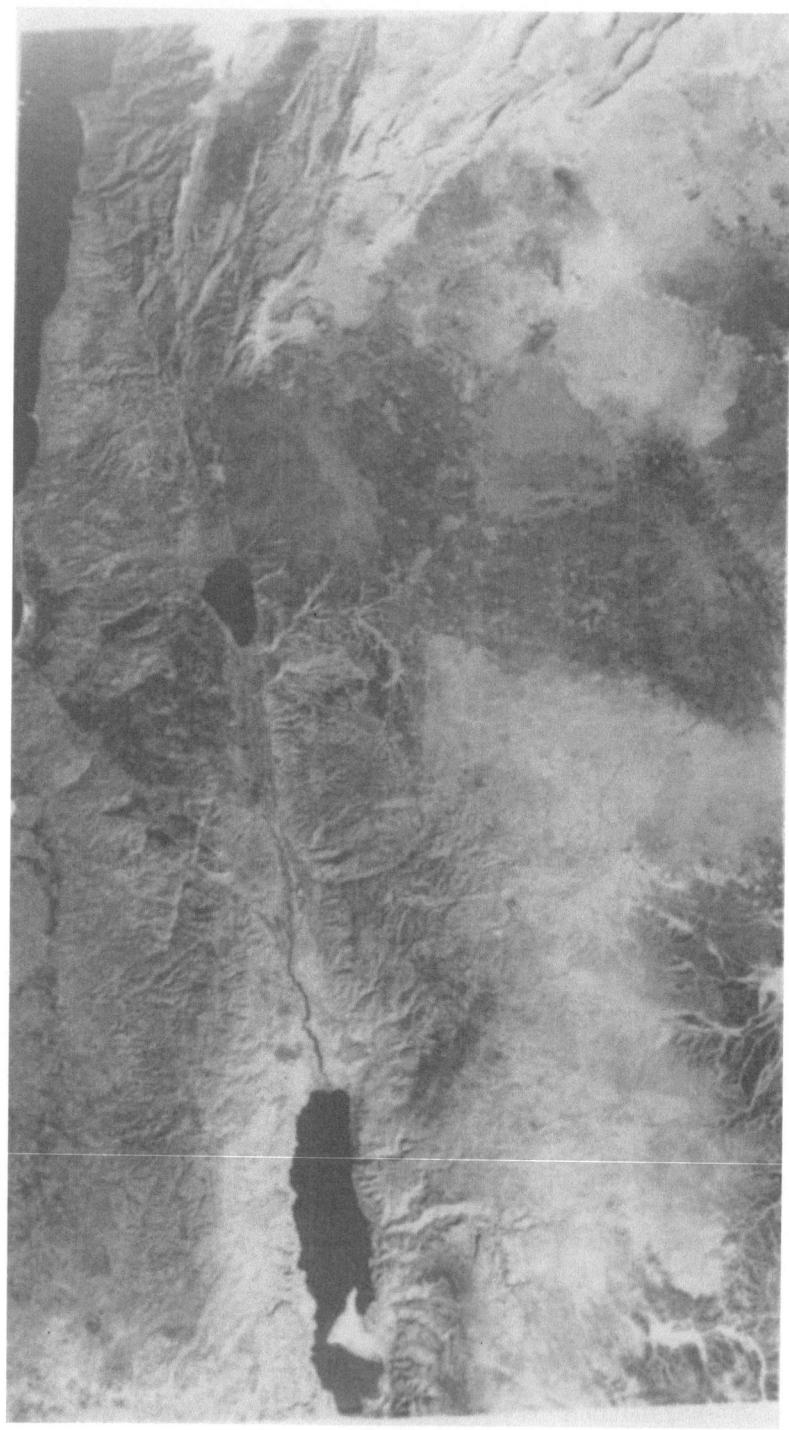

Fig. 8. LANDSAT view of the Dead Sea left-lateral shear along which at least 105 km of motion has occurred during the past 22 Ma. Along the belt of shear are the Dead Sea (*bottom*), the Jordan River, and the Sea of Galilee. Photograph (Everett et al. 1986, p. 113) centered at Lat. 32°30′N, Long. 35°45′E

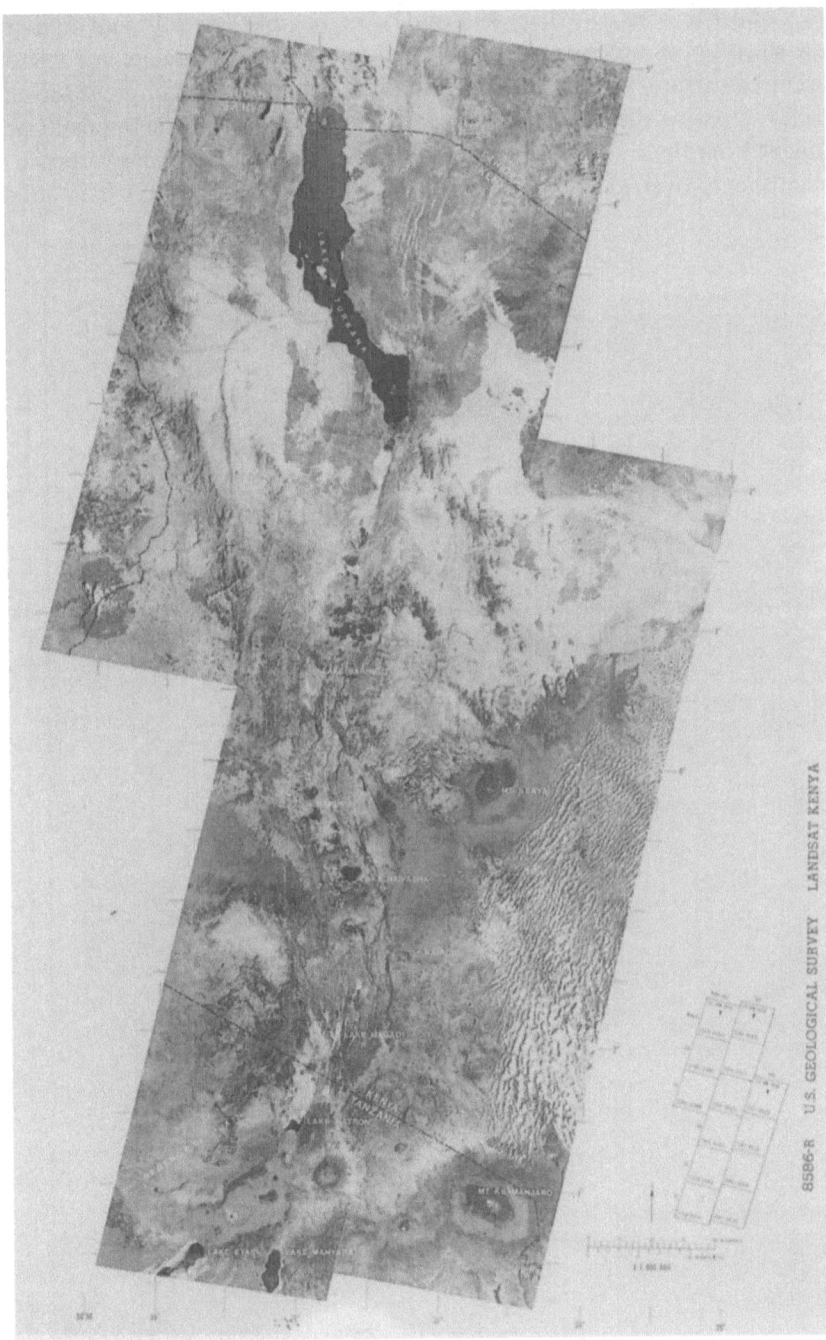

Fig. 9. A middle segment of the 4000-km-long East African Rift System that extends from the Zambezi River to the Afar Triangle. The rift cuts through Precambrian rocks and is filled with Miocene and younger volcanic and sedimentary deposits. View is centered about Lat. 0°00′; Long. 36°00′E. LANDSAT photograph 8586-R from the US Geological Survey

fills the increasing gap between sialic continents with denser simatic oceanic crust (Figs. 9 and 10). Thus, the divergent plates become bordered by long half grabens whose steep slopes are lowered by listric faults and by a sediment apron from the adjacent continent. The continuing active rift belt is the site of great seismic and volcanic activity – the great Mid-Ocean Ridge (Fig. 11). On the opposite side of a continent from the rift, both continental fragments are sites of convergence where the continents override oceanic crust, or where the oceanic crust is subducted

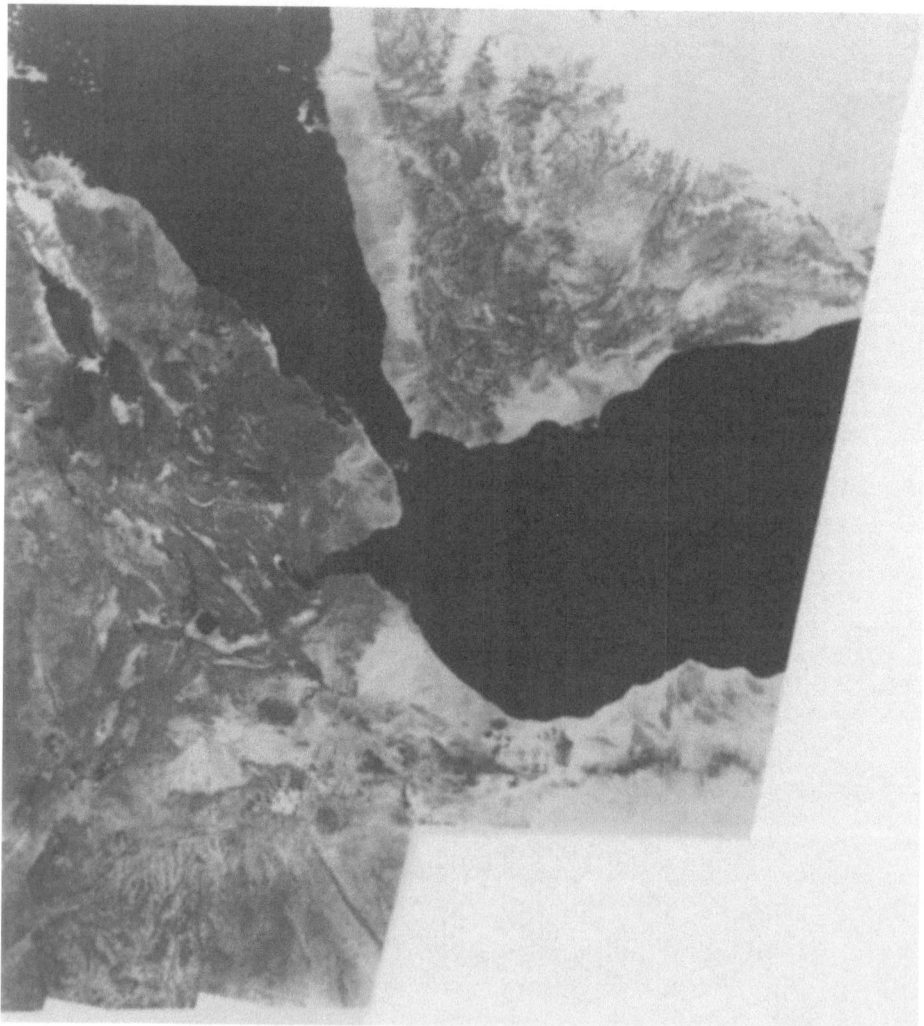

Fig. 10. The Afar Triangle at the Red Sea and Gulf of Aden triple junction. The triangle is in transition from continental rifting to sea-floor spreading. LANDSAT photograph centered at Lat. 12°00′N, Long. 43°30′E. (Everett et al. 1986, p. 108)

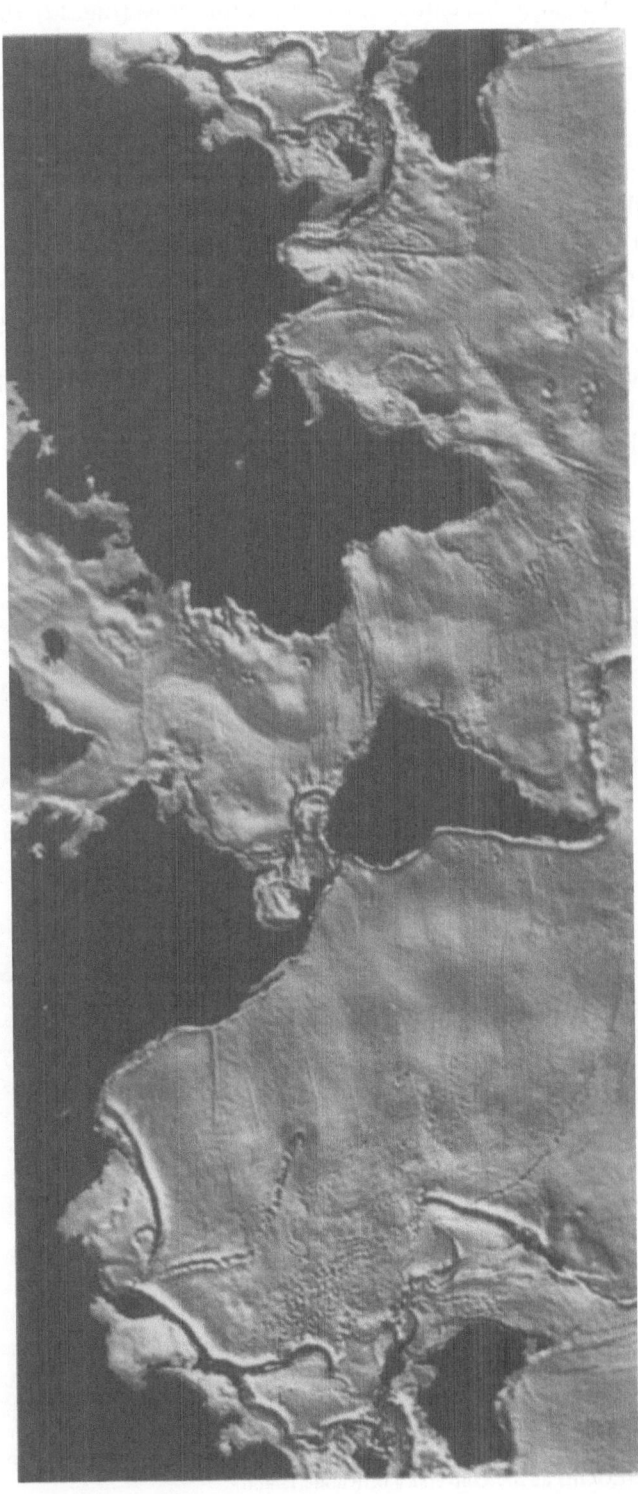

Fig. 11. Topography of the World ocean based on GEOS-S and SEASAT satellite altimetry data. The mid-ocean ridge is especially well shown in the Atlantic Ocean, whereas deep-ocean trenches (belts of plate convergence) are best shown in the Pacific and Indian oceans; fracture zones and hot-spot chains of volcanoes are evident in all oceans. Photograph courtesy of C. Koblinsky, NASA Goddard Space Flight Center, Greenbelt. (Emery and Aubrey 1991, p. 27)

beneath the continental crust or beneath a different oceanic crust (Fig. 12). In their early stages these belts of convergence are marked by long deep oceanic trenches and by lines of volcanoes along the edges of the continental or oceanic crusts that are being underthrust; later when continental crusts meet, the junction becomes a system of folded mountains accompanied by intrusions and extrusions of igneous rocks and by regional metamorphism. The collision of India and Eurasia about 55 Ma ago (Klootwijk et al. 1992) produced the Himalaya Mountains, uplifted the Tibetan Plateau, and created a terrain that is nearly half that of the conterminous United States with an average elevation of 5 km (Fig. 13). According to Harrison et al. (1992), the Earth has not experienced a catastrophic event of this magnitude for longer than 1 Ga. Of all the theories that have been suggested for the elevation of a terrain of these dimensions, including underthrusting of India, delayed continental underplating, and continental injection, Harrison et al. favor a model involving continental extrusion, development of a crustal-scale thrust ramp, and

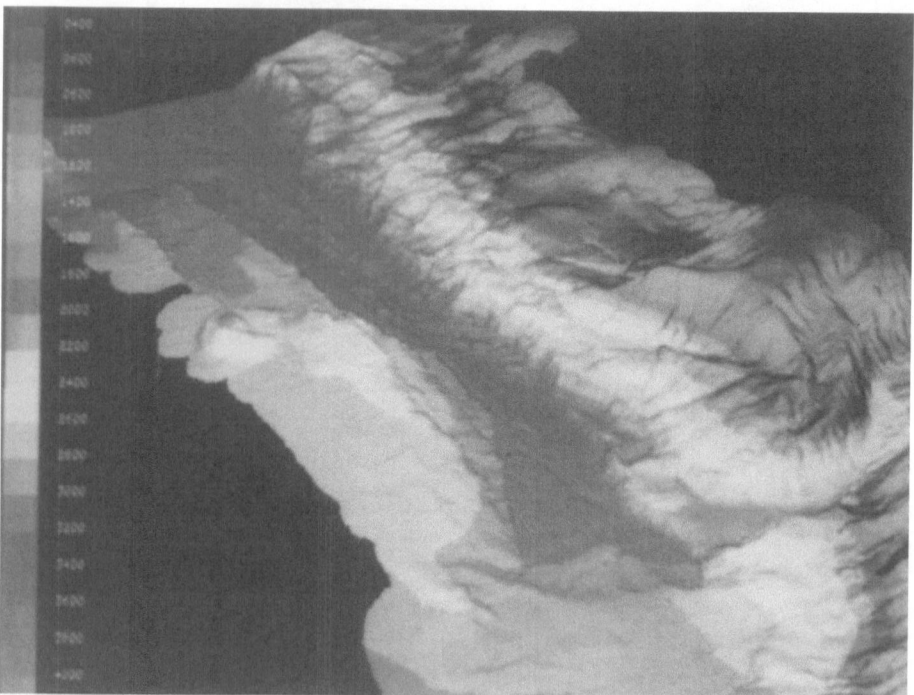

Fig. 12. Relief diagram (centered at about Lat. 49°S, Long. 75°W) of the western South American convergent margin along which a spreading ridge (the Chile Ridge) is being subducted. Note small volcanoes along the axis of the trench, which trends north-south. Relief map was computer-generated from SEABEAM swath-mapping-system data obtained aboard R/V Conrad of the Lamont-Doherty Geological Observatory by S. Cande. Photograph courtesy of Joyce Miller of Northeast Consortium Oceanographic Research (NECOR) at the University of Rhode Island

Fig. 13. The Himalaya Mountains, a product of the collision of India and Eurasia during the past 55 Ma. Photograph S17-120-009 from the Space Shuttle facing southerly from about Lat. 30°N, Long 85°E, courtesy of NASA, Lyndon B. Johnson Space Center, Houston

lithospheric delamination. Associated with the Himalayan collision zone is a system of strike-slip faults and rift basins, such as Qaidam Basin, formed by eastward migration of China away from the zone of collision (Fig. 14; Molnar and Tapponnier 1975). Lastly, the continental margins of translation or lateral relative movement become major strike-slip faults having high continental crust on one side and low oceanic crust on the other side. All these features produced by plate movements have become known to geologists only during the past three decades as major morphologies of Earth, even though they control many of the sites of another endogenic morphology – those produced by igneous extrusion onto the surface of the continental and oceanic crusts.

A major characteristic of endogenic morphological provinces is the addition of new material from below the Earth's surface. This material is added in a narrow belt along areas of plate divergence, mid-ocean ridges, areas of convergence, and areas of mantle plume upwelling. Most of the irregular topography of the oceanic crust is in the form of ridges of volcanic rock that mark former and present positions of spreading belts correlatable with magnetic-reversal patterns in showing the widening by divergence of ocean basins during the past (Figs. 6, 7, 15, and 16). Emplacement of the new rock is in the form of parallel dikes and volcanoes.

Fig. 14. Image of northwestern Qaidam Basin of Asia looking northward from about Lat. 39°N, Long. 91°E between Altyn Shan at the north and the Kuniun Mountains at the south. The basin was formed by movements along east-northeast-trending left-lateral faults and is filled with Tertiary continental deposits. The basin fill was folded and compressional events affected sediments as young as Quaternary (Everett et al. 1986). The surface is mantled by alluvial fans and evaporite deposits distal from the mountain front, and they appear to be wind-scoured. LANDSAT photograph courtesy of P. D. Lowman of Goddard Space Flight Center, NASA, Greenbelt

The morphology of the approximately 90 000-km Mid-Ocean Ridge system varies along its length, a variation caused by changes in spreading rates that are functions of the rate at which magma is supplied to the system. The East Pacific Rise has spreading rates of 6 to 17 cm/year, a crest with only a few hundred meters of relief and a width of 5 to 20 km (Fig. 17). In contrast, the Mid-Atlantic Ridge has a spreading rate of about 3 cm/year and is dominated by a rift valley several

Fig. 15. Some characteristics of endogenic morphological provinces of the Earth – western hemisphere. Projection and meridians as for Fig. 6. *diagonally lined areas* oceanic crust (Vogt et al. 1981; Larson et al. 1982); *blank areas* continental crust; *short wide lines* position of crustal spreading belts of active plate divergence (Clark 1981; Vogt et al. 1981; Cande et al. 1989); *open circles* position of active hot-spot swells (Crough 1983); and *dots* position of volcanoes active during the Quaternary (Cande et al. 1989)

kilometers deep and 20 to 30 km wide (Fig. 18; Macdonald and Fox 1990). Ridge systems are disrupted by discontinuities that can be grouped into four types according to shape, size, and longevity (Macdonald et al. 1988). On fast-spreading ridges first-order discontinuities consist of transform faults, while second-, third-, and fourth-order discontinuities are large overlapping spreading belts, small overlapping spreading belts, and slight deviations in axial linearity, respectively. Slow-spreading ridges also are disrupted by transform faults and second-, third-, and fourth-order discontinuities such as bends or jogs in the rift valley, gaps between chains of volcanoes, and smaller gaps within chains of volcanoes. First-order discontinuities tend to segment the ridge system at intervals of 300 to 500 km and they last at least several million years, whereas second-order features that partition the Mid-Ocean Ridge at scale lengths of 50–300 km last only 0.5 to 3.0 Ma. Third-order discontinuities at length scales of 30 to 100 km and fourth-order features at length scales of 10 to 50 km are shorter lived. The first- and

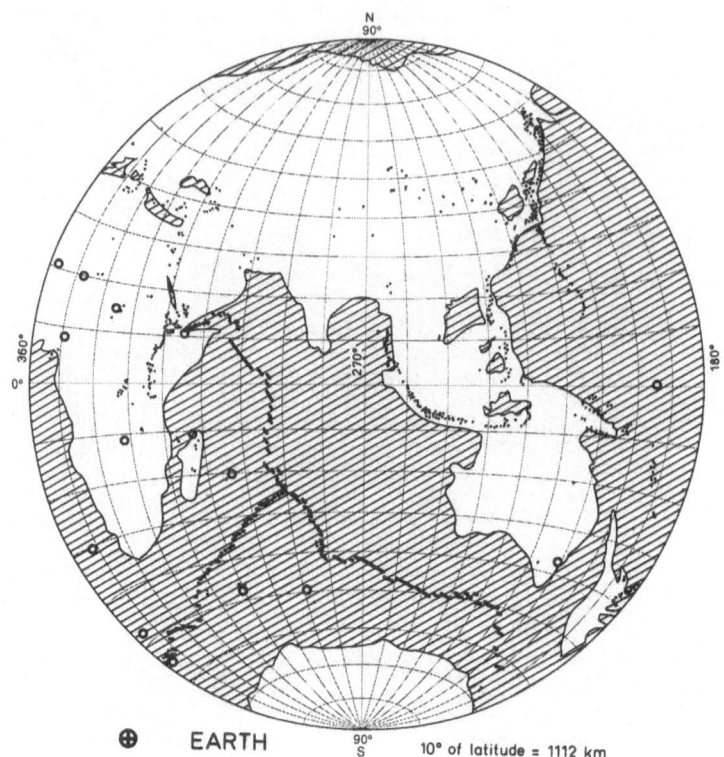

Fig. 16. Some characteristics of endogenic morphological provinces of the Earth, eastern hemisphere. *Symbols* as for Fig. 15

second-order breaks image a sustained long wave pattern of convective upwelling and melt segregation in the upper mantle, whereas the third- and fourth-order breaks are caused by local perturbations in melt delivery. Macdonald et al. (1988) proposed that this segment system consists of lozenge-shaped (in plan view) igneous units with each unit having distinct physical and chemical properties with sizes that are controlled by the scale of the parent magma body. Closely associated with a slow-spreading Mid-Atlantic Ridge neovolcanic belt are volcanoes and seamounts. Smith and Cann (1990), for example, identified 481 seamounts on the Mid-Atlantic Ridge at Lat. 24°–30°N. This style of construction, volcanic pile-ups, and hummocky plains, is missing from the fast-spreading East Pacific Rise – a difference related to the size and activity frequency of magma chambers along the ridge axis (Smith and Cann 1992).

Ancient deceased volcanoes occur on land but occur more especially on the ocean floor, where they are represented by seamounts and guyots (flat-topped seamounts). Most of the dead oceanic volcanoes had formed near earlier positions of mid-ocean spreading ridges along belts of plate divergence. Similar belts on land mark the early stages of rifting such as the Great Basin, Rio Grande Rift, and the

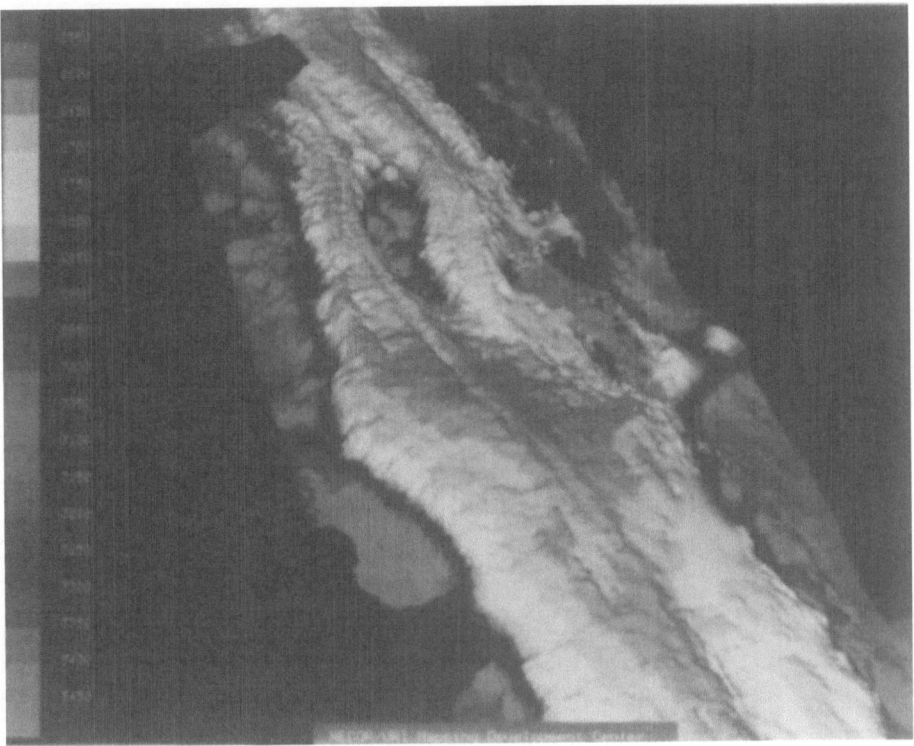

Fig. 17. Relief diagram of the East Pacific Rise at Lat. 9°N, a fast-spreading mid-ocean ridge system. Features of interest are the overlapping spreading axes and associated basin (*depression near top*) and the small relief of the median valley (whose axis trends north-south). Vertical exaggeration of the relief diagram is × 7. Bathymetry used to generate this diagram was compiled by Stacey A. Tigh and P. J. Fox of the University of Rhode Island. Photograph courtesy of Joyce Miller of NECOR at the University of Rhode Island

East African Rift all of which may define future oceanic spreading belts and future oceans.

Most of the presently active volcanoes on land and in the oceanic basins occur in belts of plate convergence, and they are especially prominent along the volcanic 'Rim of Fire' around the Pacific Ocean (Fig. 19). The most common volcanic rocks along these subduction belts are basaltic andesites (immature oceanic arcs) or andesites (more mature oceanic arcs and continental convergent margins). The magmas associated with arcs tend to have imprints of various processes including partial melting of a mantle source, fractional crystallization, contamination by subducted sediments and continental crust, derivation from continental crust, and magma mixing. Andesites and basalts on arcs are emplaced as flows in both subaerial and submarine environments, but the more felsic magmas tend to erupt as plinyan eruptions in the form of ash and dust. The cores of these arc systems consist of batholiths comprised of numerous plutons ranging in composition from

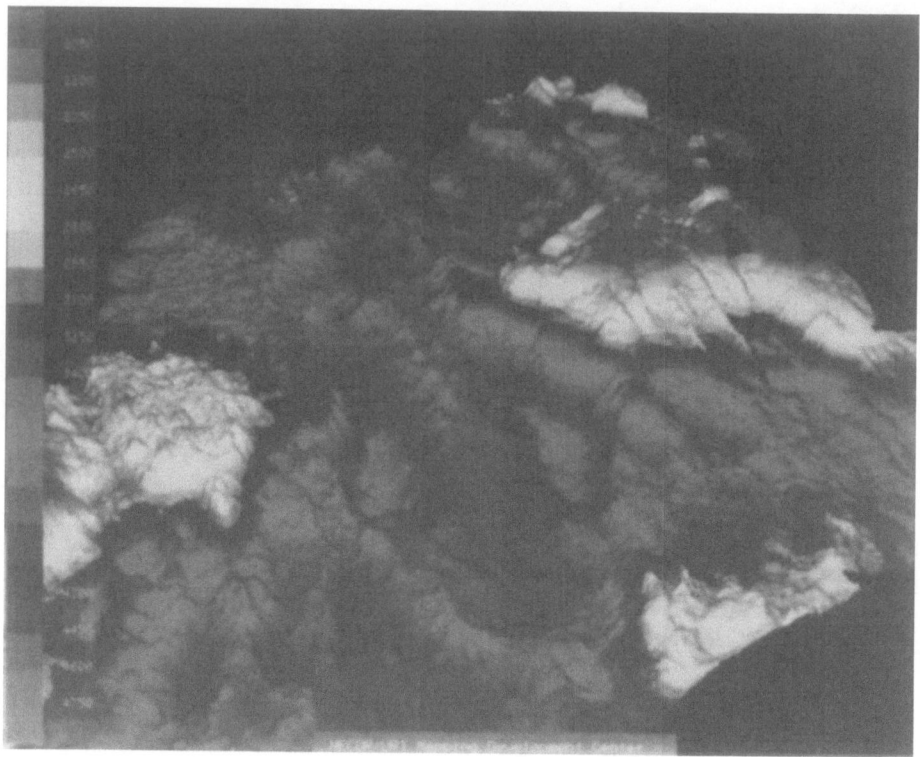

Fig. 18. Relief diagram of the Mid-Atlantic Ridge, a slow-spreading ridge system. Note the ruggedness of the median valley in contrast to the valley on the East Pacific Rise (Fig. 17). The bend of the valley at the north is a second-order discontinuity in the trend of the ridge. This diagram is based upon data collected by J. S. Semper of the University of Rhode Island and G. M. Purdy of Woods Hole Oceanographic Institution. Photograph courtesy of Joyce Miller of NECOR

diorite to granite. The Sierra Nevada in California is an example of such an arc core. Andesites or basaltic andesites that originate in magma reservoirs at depths of 50 to 100 km tend to reach the surface of the arcs because they are quite fluid, whereas the felsic magmas that are more viscous tend to crystallize before reaching the surface (Condie 1989, pp. 218–220, 229). Also associated with subduction are faults and folds that augment volcanism in producing coast-parallel mountains along belts of plate convergence. Subduction of aseismic ridges and accretion of microcontinents have added to the structural complexities of subduction belts. For example, western North America is a complex of collected terrains docked onto the North American craton during the Mesozoic (Fig. 20). These terrains converged on the North American continent obliquely, leading to large-scale latitudinal transport and formation of marginal seas (Wilson et al. 1991, and references therein; see also Uchupi and Emery 1991a).

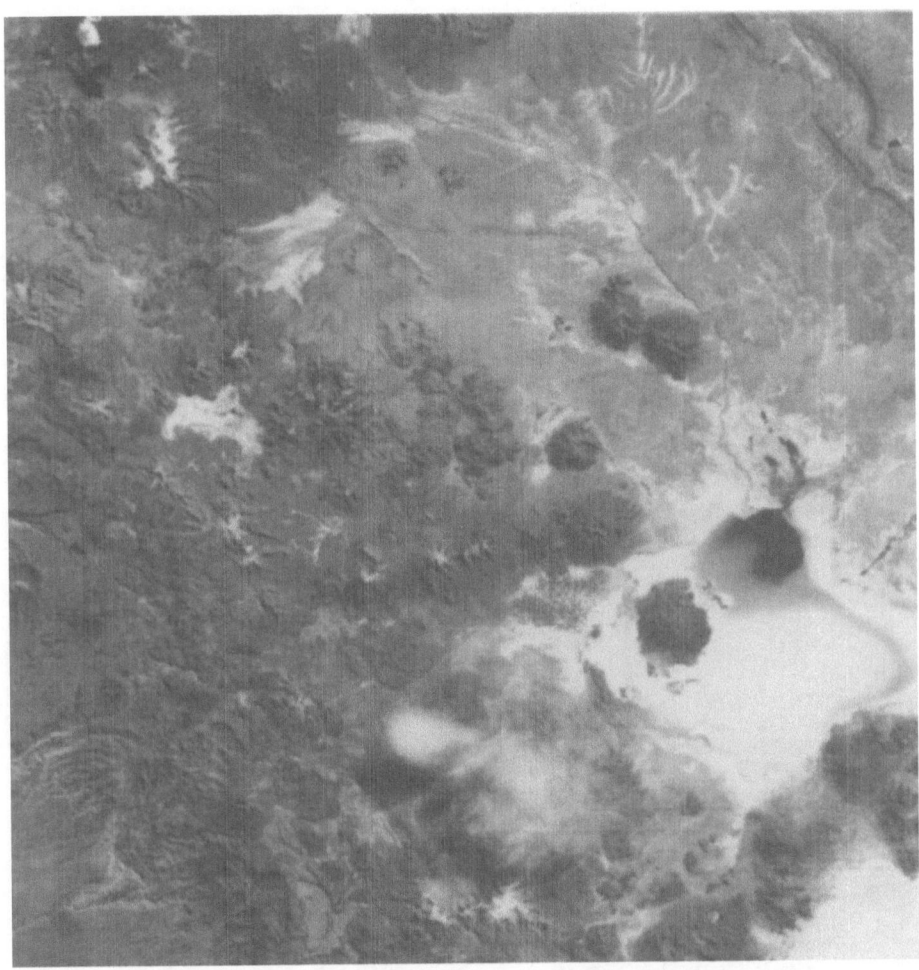

Fig. 19. Volcanic terrain in eastern Chile (Lat. 19°S, Long. 6°W) along the convergent margin of western South America. These stratocones were emplaced from Late Tertiary to Holocene (Everett et al. 1986) and are part of the Pacific 'Rim of Fire.' LANDSAT photograph courtesy of P. D. Lowman of the Goddard Space Flight Center, NASA, Greenbelt

Another source of new rock is the lower mantle, from which magma rises to form hot spots – broad areas shallower than normal oceanic regions and which are formed mainly by doming of previous oceanic lithosphere though surmounted by volcanoes. These hot spots are nearly stationary at depth, but their intraplate swells and volcanoes formed lines or traces as the crustal plates that contain them migrated across the plumes rising from hot-spot sources. The sources remain active for tens of millions of years to the entire age of an ocean basin, as long as 200 Ma. As shown in Figs. 11, 15, and 16 and in maps by Duncan (1991), most of the active

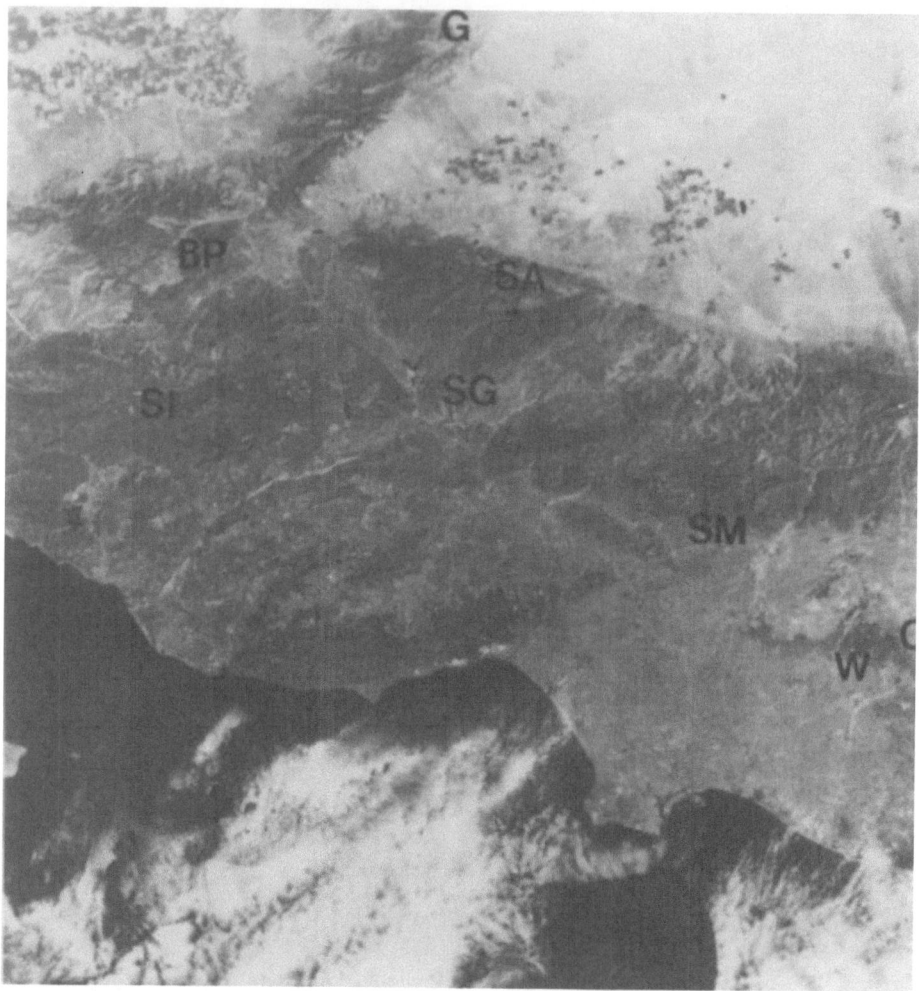

Fig. 20. Endogenic terrain of southern California created by the accretion of allochthonous terrains onto the North American craton during the Mesozoic. These terrains underwent extensive large-scale horizontal transport along transcurrent faults, a transport that still continues. **BP** Big Pine Fault; **C** Chino Fault; **G** Garlock Fault displaying left-lateral shift; **SA** San Andreas Fault displaying right-lateral shift; **SG** San Gabriel Fault with right-lateral shift; **SM** Sierra Madre Fault; and **W** Whittier Fault. LANDSAT photograph centered at Lat. 34°20′N, Long. 118°30′W, courtesy of P. D. Lowman of the Goddard Space Flight Center, NASA, Greenbelt

hot spots lie above oceanic crust, but some are within areas of continental crust, especially in Africa. The concentration of present hot-spot activity in Africa probably portends further rifting within that continent, a stationary part of the megacontinent of Pangaea that underwent extensive rifting beginning about 200 Ma ago to form the present distribution pattern of continents and ocean basins. Many hot spots have trails (Crough 1983) that began within continents or near continental margins, indicating their probable role in initiating rifting of megacontinents to form the presently existing ocean basins. These hot-spot trails provide supporting evidence for the present directions of plate migration and for past changes in these directions (Hawaiian-Emperor Seamount Chain). As a consequence of plate motion, volcanic edifices cannot reach the dimensions of volcanic structures on single-plate planets, such as Mars, or be as intense as on Io, where volcanic activity is driven not by radioactive heating but by continuous flexing of Io by Jupiter's gravitational field (Schneider and Spencer 1992, and references therein). Heat generated in this manner on Io is radiated into space via volcanoes or hot spots that are several hundred degrees hotter than surrounding terrain.

Regardless of the depth of origin, the magma that produces volcanoes and plains forms an endogenic morphology that is definite and recognizable; in contrast, the magma that does not reach the surface but produces intrusive igneous bodies cannot be recognized by morphology until it is unroofed by erosion. Therefore, the volume ratios for intrusive and extrusive materials are far better known for the Earth than for other members of the Solar System. Greeley and Schneid (1991) estimated the ratios as 8.5 for the Earth and applied this ratio to Venus, the Moon, and Mars to estimate these bodies' volume of intrusive rocks from measures of area and inferred thicknesses of extrusive materials (Table 5). Considering the masses of these bodies (Greeley and Schneid 1991), or their surface areas (as in Table 5), it appears clear that the volumes of extrusive morphology per 10^6 km^2 per year since the last major epoch of resurfacing are approximately proportional to the size of the body. The fastest production of extrusive (and thus also of intrusive) materials occurs on the Earth followed by Venus, Mars, and the Moon.

4.4.2 Exogenic

Even while it was being formed, the tectonic endogenic morphology of the Earth began to be modified by weathering and erosion, with transportation and redeposition of the products forming exogenic provinces. These exogenic processes have

Table 5. Estimated annual production of extrusive and intrusive materials on Venus, Earth, Mars and Moon (Greeley and Schneid 1991)

Body	Surface area (10^6 km^2)	Time span (10^9 yr)	Extrusive production (km^3/yr)	(/10^6 km^2/yr)	Total Magma production (km^3/yr)	(/10^6 km^2/yr)
Earth	510	0.18	3.9	0.0076	30	0.059
Venus	460	1.0	2.0	0.0043	19	0.041
Mars	145	3.9	0.018	0.00012	0.17	0.0012
Moon	38	3.85	0.0024	0.000063	0.025	0.00066

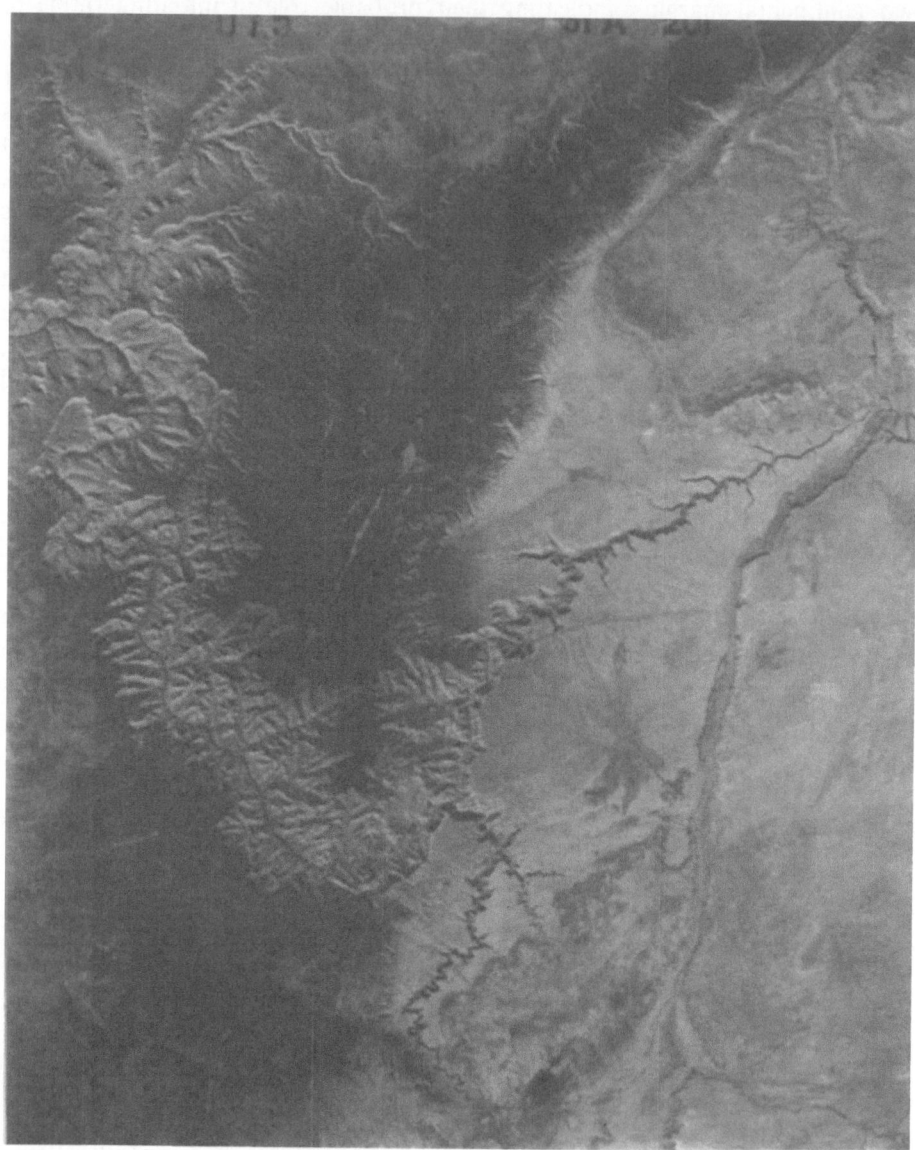

Fig. 21. Exogenic morphology on Earth – the Grand Canyon of Arizona cutting through generally flat-lying sediments of the Phanerozoic atop Precambrian igneous and metamorphic rocks. Space Shuttle Hand-Held Earth Observation Photograph (top is toward northeast) at about Lat. 36°N, Long. 112°W. Challenger flight STS-61A, altitude 290 km. Photograph courtesy of R. E. Stevenson, Del Mar

created a vast mosaic of present terrains ranging from deep eroded Precambrian cratons and eroded mountain chains to glacial-fluvial features in high latitudes (including the present polar regions), vast networks of fluvial systems such as valleys, deltas in most climates, alluvial fans (bajadas and playas in desert climates), lakes, marshes and swamps, various types of eolian and karst topographies in continental settings, and terrigenous sediment wedges, carbonate platforms, atolls, and vast expanses of terrigenous and pelagic (carbonate and silicate) blankets on the ocean floor (Figs. 21–32). Authigenic processes and deposition of such materials as phosphorite, manganese oxide, and siliceous strata further enrich the oceanic realm. Similar exogenic terrains formed by the same processes, but not

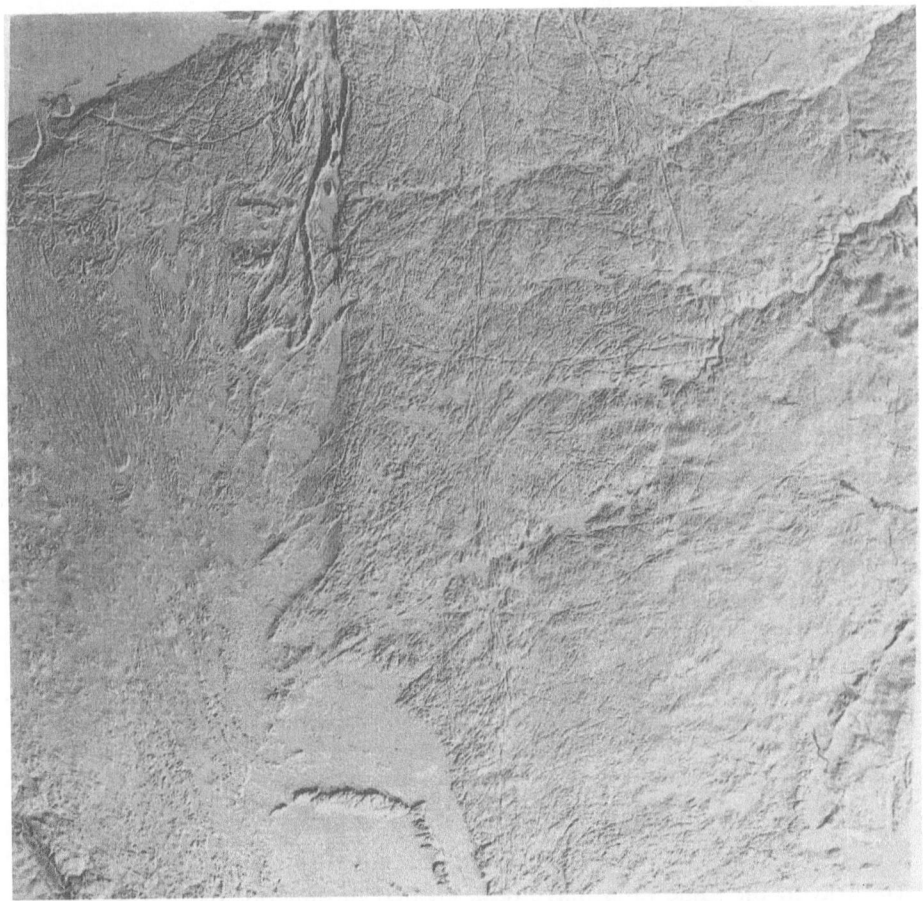

Fig. 22. Deeply eroded Precambrian Canadian Shield with the Bear Province on the *left* and Slave Province on the *right*. The provinces may be separated by a suture along the low escarpment. This low-relief terrain may be considered an old-age stage of geomorphic evolution. LANDSAT photograph at Lat. 67°N, Long. 110°W (Everett et al. 1986, p. 67)

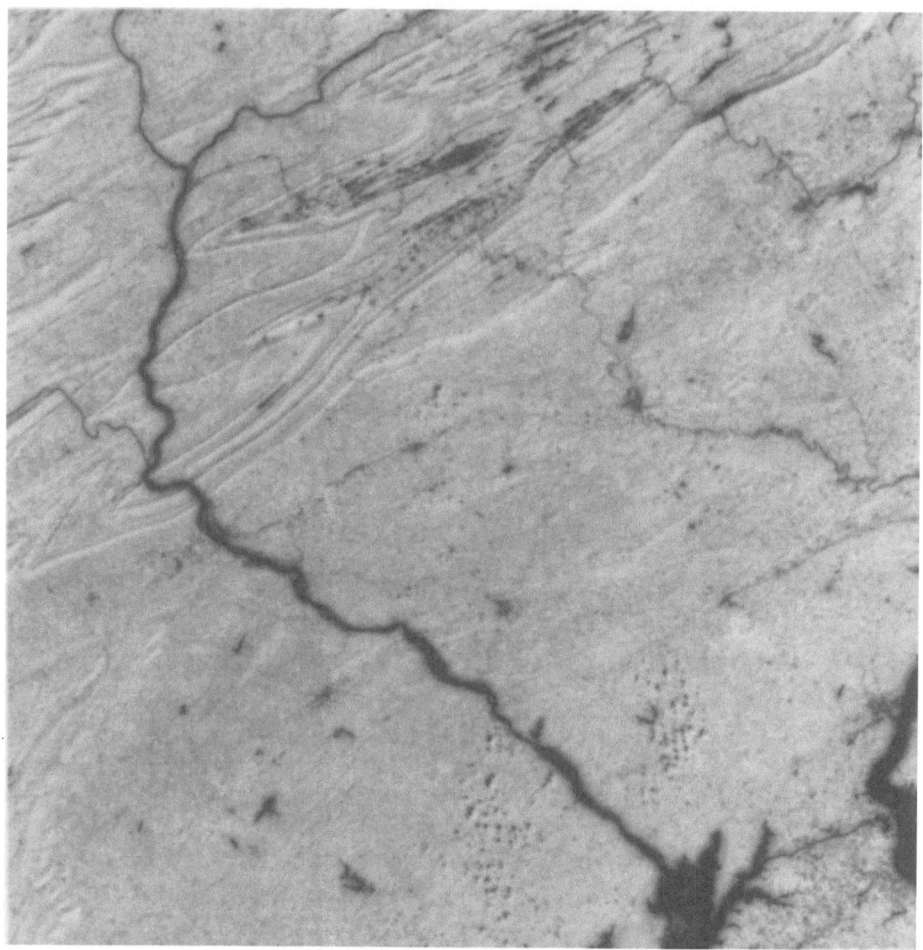

Fig. 23. LANDSAT image of the deeply eroded Valley and Ridge terrain of the eastern United States at about Lat. 40°N, Long. 76°W, a shelf and slope sedimentary sequence deformed during closing of the Iapetus Ocean at the end of the Paleozoic. East of the folded belt is a complex of allochthonous terrains that were docked onto the North American craton during closure of the Paleozoic ocean. These terrains have been eroded to low relief. Toward the southeast the allochthonous terrains are overlain by Triassic rift structures, formed during the opening of the present Atlantic Ocean, and by undeformed Mesozoic-Cenozoic strata that reflect the present geological regime of the region. The river that extends across the photograph is the Susquehanna emptying into Chesapeake Bay. Northeast of the bay is the Delaware River. Photograph courtesy of P. D. Lowman of Goddard Space Center, NASA, Greenbelt

necessarily at the same rates, have been created throughout the Earth's history. These features are not the final products of external processes because they may undergo additional modification within the exogenic realm.

Reduction of topographic relief is not restricted to exogenic processes, because endogenic processes such as massive flood basalts and slope failures triggered

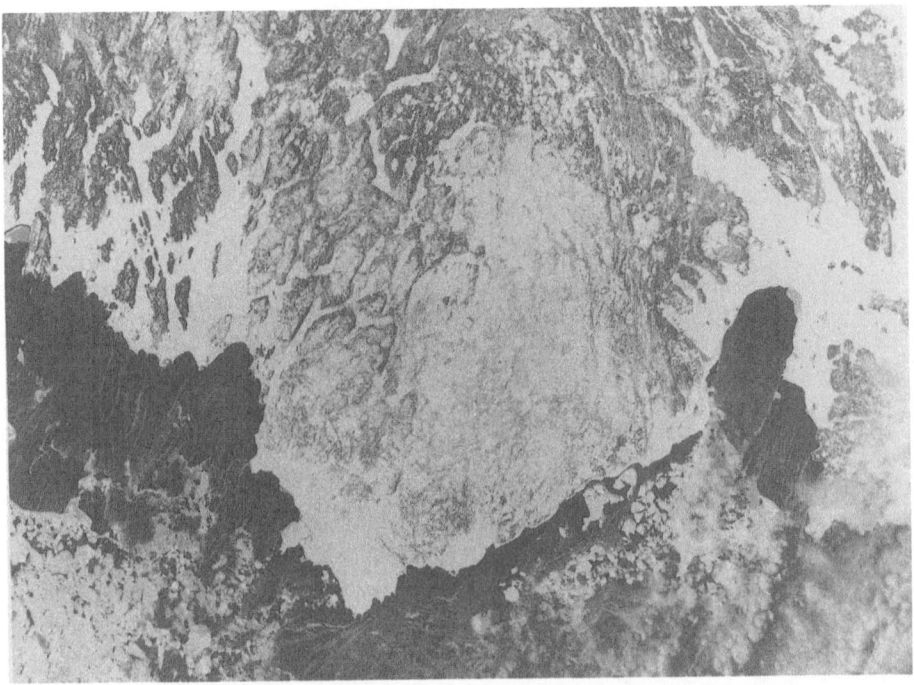

Fig. 24. Glaciated terrain of eastern Newfoundland, Canada. Space Shuttle photograph S36-151-040 is centered at about Lat. 49°N, Long 54°W. Shallow embayments are frozen and snow covered. Photograph courtesy of Lyndon B. Johnson Space Center, NASA, Houston

by earthquakes or weakening of fault traces by hydrothermal metamorphism (Tucholke 1992) can smooth the endogenic terrain. Transitions from endogenic to exogenic provinces are made possible on Earth by the presence of an atmosphere and hydrosphere that permit weathering by chemical and mechanical processes and facilitates erosion by winds, streams, glaciers, groundwater, waves, and currents. Changing relative levels of land and ocean lead to variations in the rates of production of clastic sediments and concentrations of dissolved chemicals brought from land to ocean (Worsley et al. 1986; Emery and Aubrey 1991, pp. 167–174). The ultimate sump for accumulation of these waste products is the ocean, in which clastic, chemical, and organic sediments bury irregular endogenic topography (Figs. 33–36). Although most of the source materials for exogenic provinces must be provided by endogenisis, both endogenic and exogenic processes operate simultaneously. Mountain building and erosion are concurrent, but the endogenic processes occur at faster rates and more continuously on the ocean floor than on land. The longer interval between episodes of endogenesis on continents along and near plate boundaries than on the ocean floor probably is the result of the damping effect caused by the greater thickness of crusts on continents.

Fig. 25. Glaciated terrain of Patagonia, Argentina. Space Shuttle photograph S48-151-074, centered at about Lat. 50°S, Long. 73°W, courtesy of Lyndon B. Johnson Space Center, NASA, Houston

The role of the biosphere in producing exogenic morphology must not be underestimated. The indirect role includes, most importantly, the ancient work of the biosphere in preventing a runaway greenhouse effect followed by loss of all the water (as for Venus). Without water, Earth would lack all the erosional and depositional morphologies produced by this medium. The presence of these features on Earth represents a marked difference from their absence on all other members of the Solar System except Mars, where morphological traces of the ancient presence of water remain. More direct production of morphology by the biosphere is the construction of coralgal reefs and the augmentation of detrital sediments by skeletal and organic remains in many waterlaid deposits of both freshwater and marine environments. All these morphological effects of the biosphere have added to the diversity of the Earth's surface and have augmented the variety of environments for habitation by different life forms – an illustration of Lovelock's (1979) concept of *Gaia*, a sort of Earth wisdom that is able to convert hostile environments into friendly ones through evolutionary changes in the biosphere.

A reversal of this adaptation may be taking place through the accelerating activities of humans, especially during the past century. These changes include

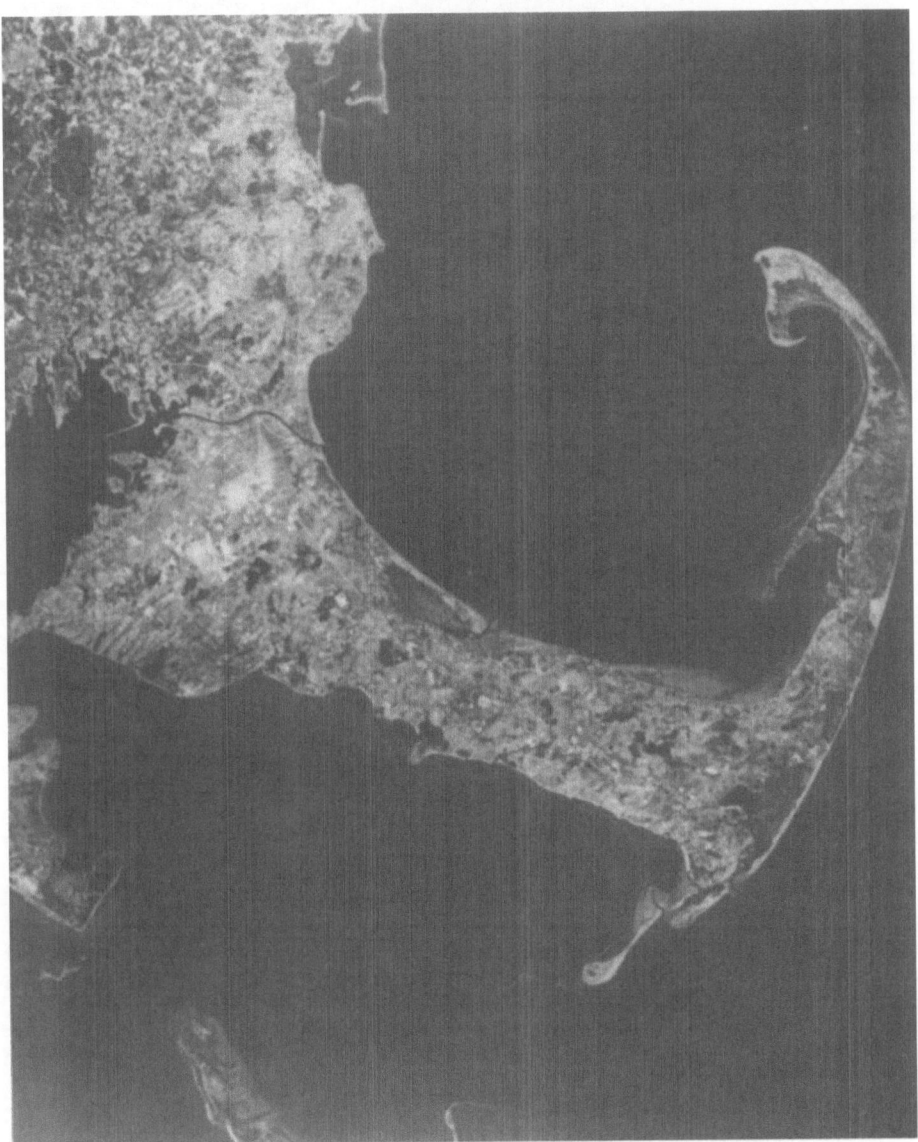

Fig. 26. Late Pleistocene (Wisconsinan) glaciated terrain of Cape Cod, MA, centered at Lat. 41°45′N, Long. 70°15′W and taken at an altitude of about 250 km by Capt. J. Creighton, USN (supported by Col. J. Caspar, USAF) aboard the space shuttle *Atlantis*. The channels entrained on the outwash plain are due to spring sapping. Photograph courtesy of R. E. Stevenson, Del Mar

Fig. 27. Nile River and Nile Delta in northern Egypt. West of the river is the Faiyum Depression, and east of it is the Gulf of Suez, Sinai Peninsula, and the Gulf of Aqaba (Elat) along the Dead Sea shear. The city west of the delta is Alexandria. Space Shuttle photograph S 36-79-090 facing northwestward from about Lat. 27°N, Long 33°E, courtesy of Lyndon B. Johnson Space Center, NASA, Houston

many new forms of morphology, but, more importantly, alterations of biological environments associated with 'improving' the welfare of humans by increasing the extraction of mineral resources, increasing the production of plants and animals for human food, clothing, and shelter, increasing the protection from acquisition of these resources by other humans, and the disposal of wastes that are harmful to the rest of the biosphere (including the dissipation of the ozone layer) and eventually to humans (Stern et al. 1992; and many others). These effects are happening rapidly because the human population on Earth is growing very fast with a doubling time of less than a generation. This fast rate of population growth not only leads to increasing demands on resources and space, but also causes increasing competition

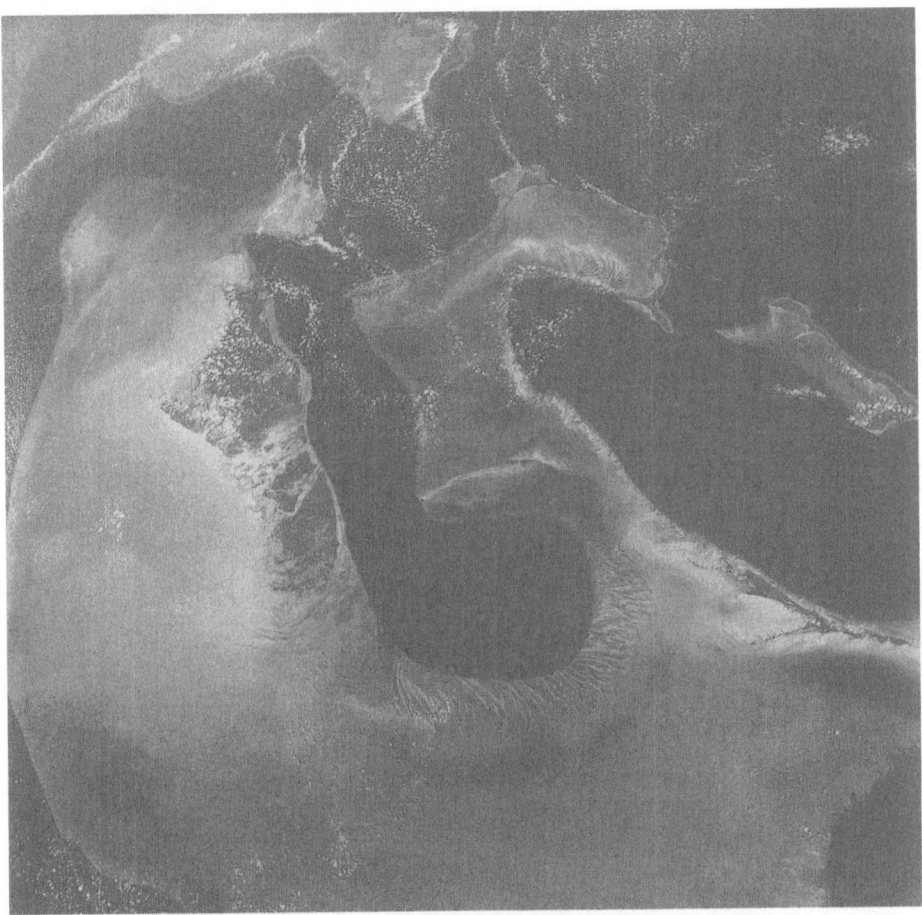

Fig. 28. Little and Great Bahama Banks, Mesozoic-Cenozoic carbonate constructional features atop a rifted continental/oceanic terrain formed during the opening of the present North Atlantic Ocean. According to Eberli and Ginsburg (1987), the northwest Great Bahama Bank was formed by coalescence of three smaller banks. The large island atop Great Bahama Bank is Andros Island, and the deep area next to it is The Tongue of the Ocean. The other deep area east of The Tongue is Exuma Sound. The bank tops are shallower than 15 m and their steep side slopes have reliefs of hundreds to thousands of meters. The linear features at the head of The Tongue of the Ocean are shaped by tidal currents. Space Shuttle photograph S29-90012 facing northward from about Lat. 22°N, Long. 77°W, courtesy of Lyndon B. Johnson Space Center, NASA, Houston

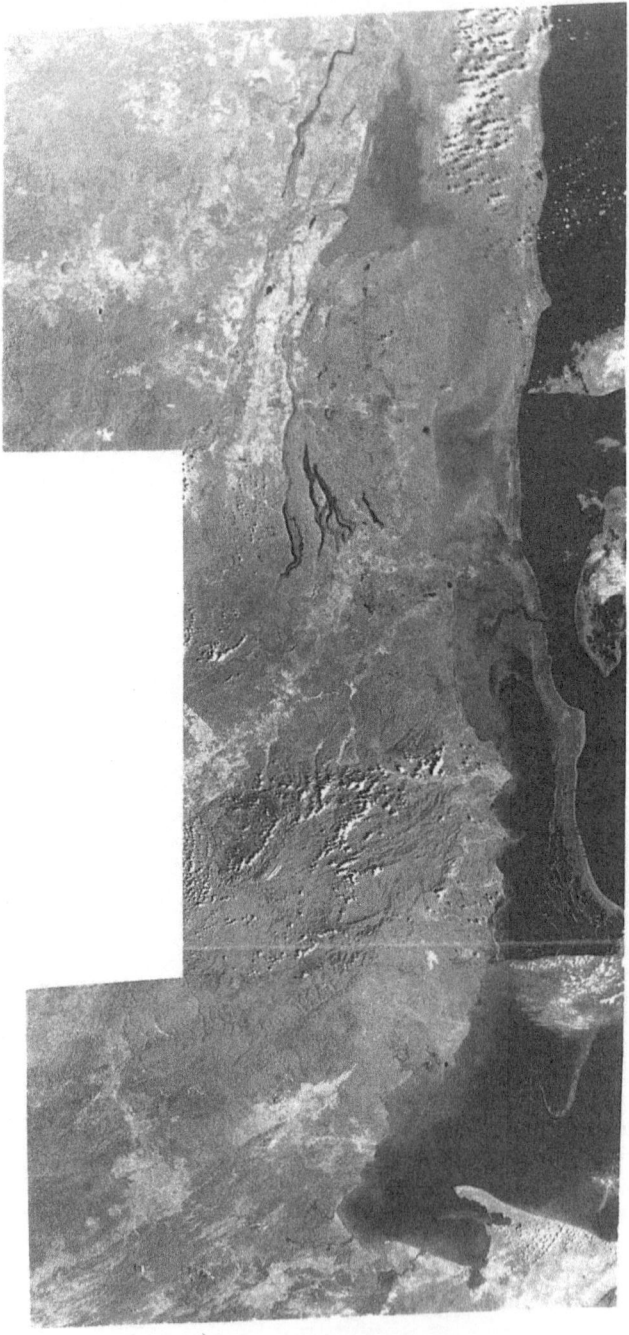

Fig. 29. Carbonate shelf off the Yucatan Peninsula at Belize showing coast, offshore reefs, and limestone coastal plain. LANDSAT photograph centered at about Lat. 18°N, Long. 88°W (Bloom 1986, p. 389)

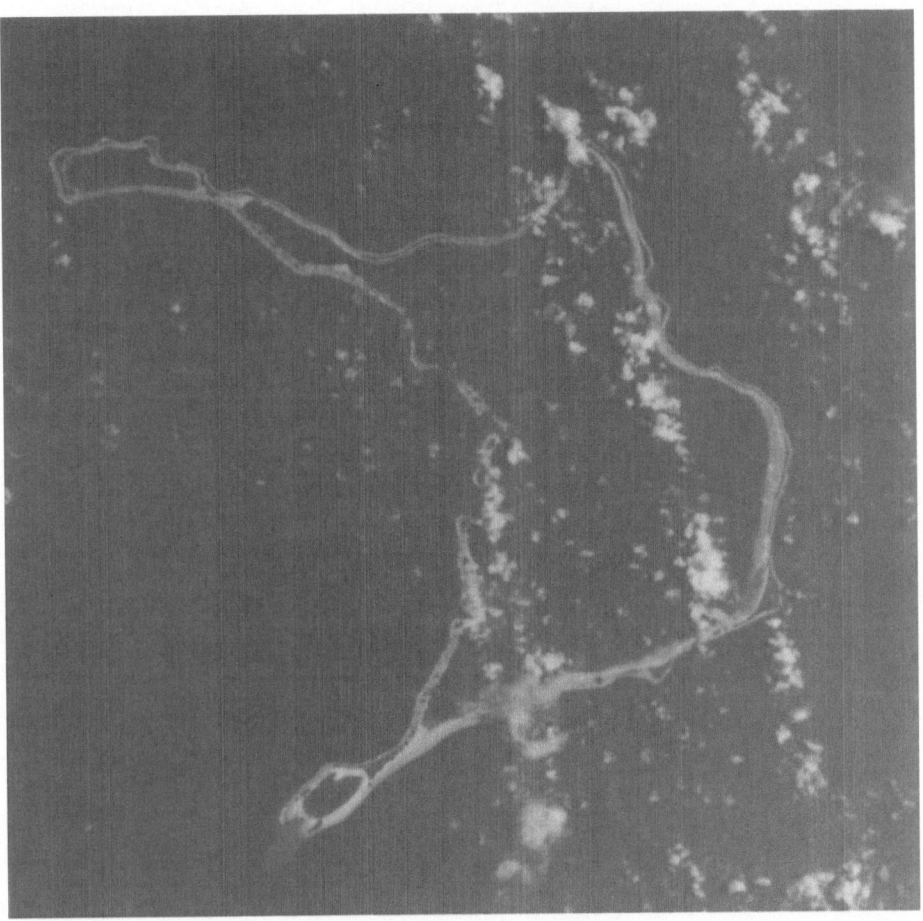

Fig. 30. Arno Atoll in the Marshall Islands, central Pacific Ocean at Lat. 7°10′N, Long. 171°50′E. This carbonate constructional feature was formed atop a subsiding volcano. As the volcanic peak gradually sank below sea level the carbonate buildup went through a fringing and barrier reef geometry, culminating in its present configuration as an atoll. Space Shuttle photograph S38-76-15 (top is north), courtesy of Lyndon B. Johnson Space Center, NASA, Houston

for them. An illustration is the construction of more than enough nuclear weapons during the past half-century to destroy all humans and much other life on Earth – this certainly is not the response to be expected of a true *Homo sapiens*.

4.4.3 Exotic

In addition to its endogenic and exogenic morphological provinces, the Earth has another kind of province that is far less important on Earth than on other rocky

Fig. 31. View of the dune field in the vicinity of Walvis Bay, Namibia, western Africa and centered about Lat. 23°S, Long. 15°E. Waves from near Antarctica drive sand along the coast, forming the northward-pointing spits and supplying sand for the longitudinal dunes. The dune field is limited inland by the northwesterly flowing Kuiseb River beyond which is the Precambrian shield of southern Africa. Space Shuttle photograph S35-603-42, courtesy of Lyndon B. Johnson Space Center, NASA, Houston

members of the Solar System. The characteristics of this province have been shaped by the impact of planetoids, asteroids, meteoroids, and perhaps comets. Grieve (1987, 1990) compiled a list of 120 impact sites from the presence of topographic craters, nickel-iron/stony meteorite fragments, geochemical anomalies, or stressed and brecciated country rock usually including shatter cones. One site contains 122 individual craters, but most have only one. At least five additional structures have been described since the compilations by Grieve. Hargraves et al. (1990) postulated that shatter cones in Proterozoic sandstones document a late Precambrian – early Paleozoic impact structure in southwestern Montana, the subsurface Ames Hole in

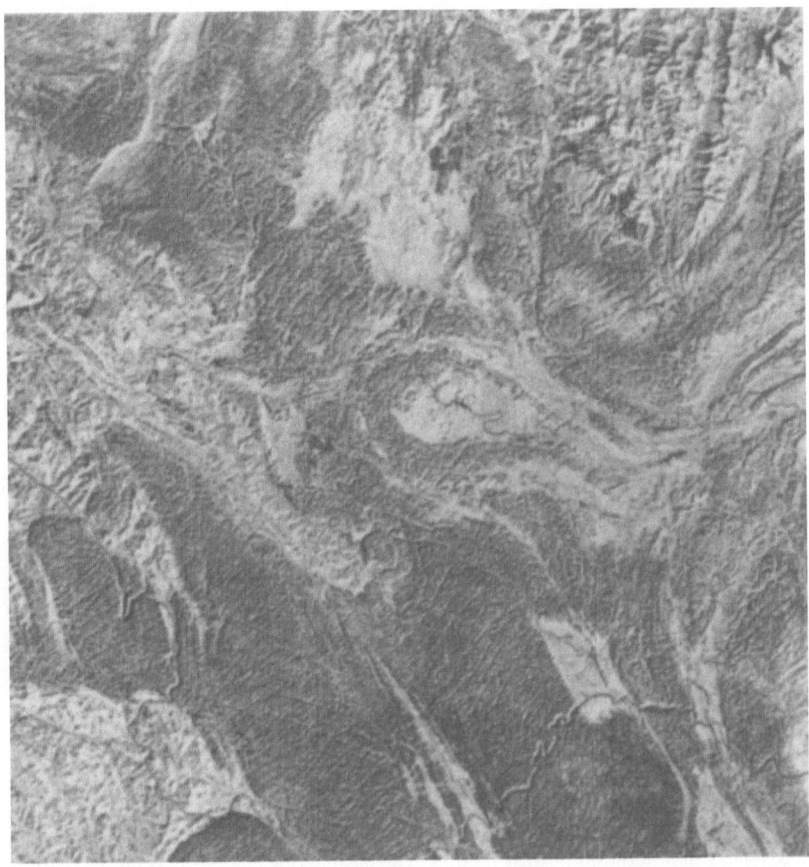

Fig. 32. A karst exogenic terrain developed on Devonian-Triassic (*dark-toned*) the Guangxi Province of southern China. LANDSAT photograph centered about Lat. 24°N, Long. 108°E. (Blair 1986, p. 415)

northern Oklahoma has been interpreted by Shirley (1992) as an impact structure blasted out of Cambro-Ordovician limestone during the Silurian: glassy microspheres from Upper Devonian sediments in South China are believed by Wang (1992) to result from an impact 365 Ma ago on the South China plate that caused a faunal extinction in eastern Gondwana, and the subsurface Avak structure in the Arctic coastal plain of Alaska has been interpreted by Kirschner et al. (1992) as having been caused by a meteor that struck a Late Cretaceous or Tertiary marine shelf or coastal plain. The glassy spherules near the Frasnian-Famennian boundary at Sanseilkles, Belgium, also may document the meteorite event that led to a worldwide benthic mass extinction in the Late Devonian (Claeys et al. 1992). The impactor that produced the spherules may have formed the Siljan Ring in Sweden or the Charlevoix structure in Quebec, Canada. Most impact sites are on continental shields or cratons of ancient rocks on those continents that are inhabited by

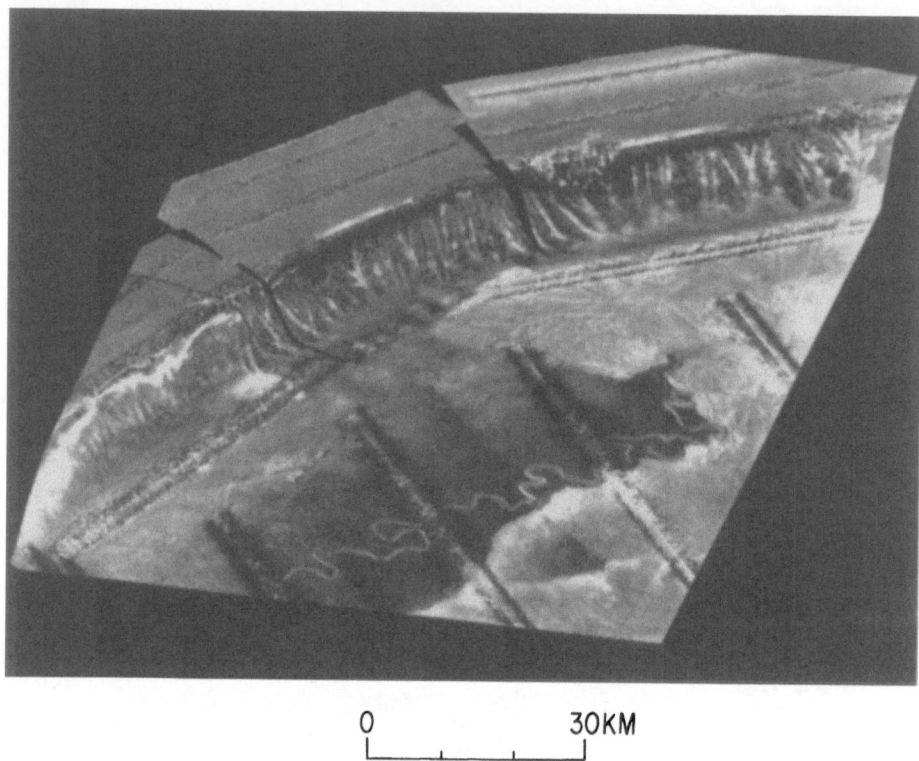

0   30KM

Fig. 33. Perspective view of the Florida Escarpment off western Florida. This scarp is the result of carbonate accretion and subsequent submarine erosion. Note the ravines and gullies cut into the scarp and the carbonate debris aprons at their mouths. The meandering channel beyond the scarp is on the terrigenous Mississippi Fan. Computer-generated image from three-dimensional GLORIA sonograph and bathymetry, looking northeastward from about Lat. 28°N, Long. 96°W, courtesy of D. C. Twichell of the Marine Branch, US Geological Survey, Woods Hole. (Twichell et al. 1990, fig. 3)

scientifically curious populations (Figs. 37–39). Probably most meteoroids impact on Earth at an angle of 45°; impacts at 15° from the horizontal or lower, and angles as high as 75° appear to be rare. Possibly oblong rimmed depressions having dimensions of 4 × 11 km in the Argentine Pampas may be products of such rare low-angle impacts. Schultz and Liaza (1992) believed that these depressions were caused by the low angle impact and ricochet of a chondritic body 150 to 300 m in diameter. Many of the impacts occurred in ancient times: 73 of the sites are in rocks

──▶

Fig. 35. A massive slump on the insular shelf of Puerto Rico. Schwab et al. (1991) believed that the slope failure was induced by tectonic overstepping of the slope. Imagery from three-dimensional GLORIA sonograph and bathymetry. The computer-generated image faces southward from about Lat. 19°N, Long 67°W. Photograph courtesy of W. C. Schwab, Marine Branch, US Geological Survey, Woods Hole

50 km

Fig. 34. Computer-generated image of Monterey Canyon off central California from a high resolution CD ROM compiled by NOAA and USGS. The up-canyon direction is northeast from about Lat. 36°33′N, Long. 122°15′W. Note meander on the deep-sea extension of canyon. Photograph courtesy of D. C. Twichell of the Marine Branch, US Geological Survey, Woods Hole

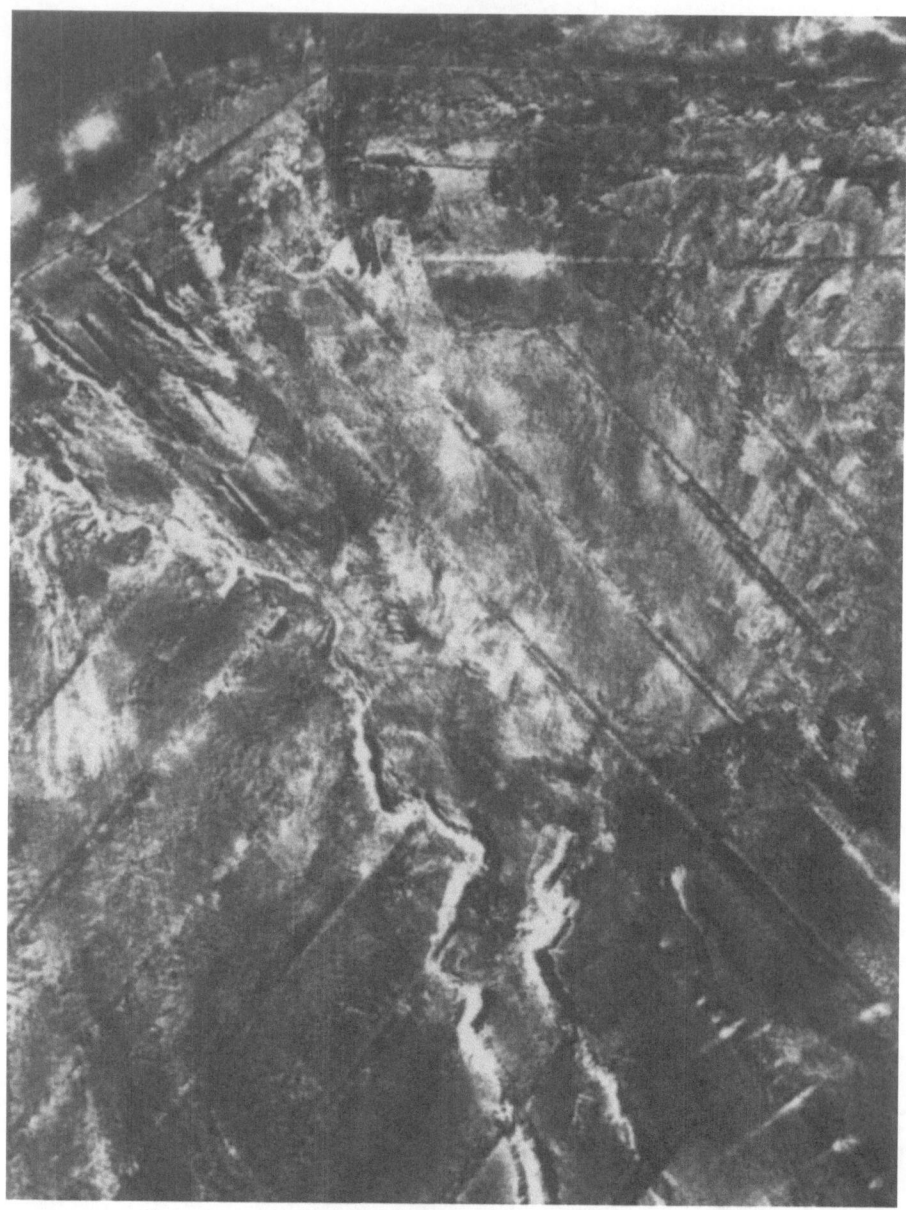

Fig. 36. Hudson Channel entrenched on the continental rise off the eastern United States. Note tributaries on the lower continental slope and upper rise. Photograph of mosaic images obtained by GLORIA side-scan sonar and centered about Lat. 38°N, Long. 71°W was provided by W. P. Dillon of the Marine Branch, US Geological Survey, Woods Hole. (EEZ-SCAN 87, Scientific Staff 1991, sheet 5, p. 24)

Fig. 37. Positions of known impact structures (*dots*) on the Earth – western hemisphere. *Diagonally lined areas* are continental shields. (After Grieve 1987, 1990; Jansa et al. 1989)

older than 600 Ma, 27 are in rocks of 600 to 400 Ma age, 10 are in rocks of 400 to 200 Ma age, and only 4 are in rocks younger than 200 Ma (Grieve 1990). The implication is that many more old craters have been removed by erosion or lie buried beneath later sediments. The total known area of impact sites on the Earth is about 0.07×10^6 km^2, only a trace compared with the areas of endogenic and exogenic morphological provinces.

Extraterrestrial impact sites are so rare on Earth that a long time elapsed before they were scientifically accepted even though many ancient accounts exist of the impact of meteorites. For example, the Bible (Acts 19: 35 written about A.D. 55) mentions 'that the city of Ephesus is a worshiper of the great goddess Diana and of the image that fell down from Jupiter.' Perhaps this image was the same meteorite that was brought to Rome from Emisa (Syria) when the child-priest Elagabalus became a short-lived emperor of Rome in the year A.D. 212. A stone in the northeast corner of the Kaaba at Mecca is an object of great veneration by Moslems; it has been reported to be possibly an ancient meteorite (Kahn 1938), but it may be an agate (Dietz and McHone 1974). Its celestial origin is implied by early

Fig. 38. Positions of known impact structures on the Earth – eastern hemisphere. *Symbols* and sources as for Fig. 37

Arab writers' belief that it was brought by the Angel Gabriel and installed by Abraham and Ishmael in the Kaaba when it was rebuilt after Noah's Flood (see also the Koran, Sura XIV). Other meteorites in Japan were reserved as a source of iron for swords of the emperor; some others have been preserved in Shinto and Buddhist shrines (Needham 1959, vol. III, p. 434; and others). Even though many small meteorites had been seen to fall and their fragments had long been recognized on Earth, Thomas Jefferson said that "It is easier to believe that Yankee professors would lie, than that stones would fall from heaven," discounting an account of a meteorite fall in Weston, Connecticut, during 1807 (Hartmann 1983, p. 160). Similar confusion occurred when a comet fragment destroyed 2000 km² of heavy forest when it exploded on 30 June 1908 high in the atmosphere above Tunguska, a thinly inhabited region in west-central Siberia (Kirova 1964; Turco et al. 1982). Its brilliant flash was seen for hundreds of kilometers, the sound of its explosion was heard for a 1000 km, and seismic waves from it traveled twice around the globe.

The best-known impact structure on the Earth is Barringer Crater (Meteor Crater of Arizona) because of extensive geological exploration, drilling, and

Fig. 39. Exotic morphology on Earth – astroblemes at Lac Aleau Claire, Canada. Interior rings, such as that in one of the depressions are common within impact craters of intermediate to large size on other members of the Solar System. Space Shuttle Hand-Held Earth Observation Photograph at Lat. 56°00′N, Long. 74°30′W. Challenger flight STS-61A, 30 Oct. to 6 Nov. 1985, altitude 290 km

application of theory of impact effects (Shoemaker 1960; Grieve 1987, p. 258; and Melosh 1989). The meteorite impacted the ground 25 000 years ago at about 15 km/s, entering and fusing the strata, exploding, and partly overturning outward the sides of the cavity. This explosion also vaporized most of the meteorite so that only small fragments have been found in the vicinity despite considerable effort by drilling to find a large mass of nickel-iron for mining. Structures made by more ancient meteorite impacts are more puzzling, but many have now been identified (Figs. 37 and 38), largely through the presence of shatter cones whose apexes point upward (Dietz 1947, 1959), thus indicating an intense force from above. For many years such sites had been attributed to volcanic activity (in the absence of volcanic rock they were termed cryptovolcanic), to gas explosions, or to solution collapse. Buried masses of nickel-iron are rare in impact structures, but at Sudbury (eastern Canada) the impact about 1850 Ma ago of a large meteoroid, or perhaps an asteroid, about 4 km in diameter (Grieve 1987, p. 260) apparently formed the ore in this major nickel-mining area (Dietz 1964, 1972). Grieve et al. (1991) stated that the composition of the igneous complex associated with the Sudbury structure corresponds with a mixture of Archean granite-greenstone terrain and possibly a small amount of Huronian cover rocks. Differentiation of the igneous complex is believed to be due to the great thickness of the impact melt and a slow cooling rate. The original size of the Sudbury structure is not known (Lowman 1992) and its present elliptical shape is the result of northwesterly ductile thrusting concentrated on the South Range and the interior of the igneous complex; apparently, the present complex represents only the central part of a much larger original structure.

Impact of a very large meteoroid, asteroid, or perhaps a comet, occurred at a site of unknown location, possibly in the Caribbean region (Hildebrand and Boynton 1990; Kerr 1990) and more specifically on the Yucatan Peninsula (Kerr 1991a; Pope et al. 1991; Smit et al. 1992). Its crater in Yucatan is now known as the Chicxulub crater (Hildebrand and Penfield 1990). This impact may have been the cause of the worldwide Cretaceous-Tertiary faunal extinctions (Alvarez et al. 1980; Alvarez and Asaro 1990), raising so much dust or ash that photosynthesis may have ceased for several years, reducing terrestrial and near-surface marine life, and imparting higher than normal concentrations of iridium into marine sediments. Smit et al. (1992) interpreted a 3-m-thick clastic unit at the K-T boundary along the Arroyo el Mimbral as a megawave or tsunami deposit produced by a planetesimal impact. A ripple bed at the top of the unit has an iridium anomaly of 921 ± 23 pg/g. Tektites having a platinum-group metal abundance anomaly and shocked minerals also have been reported from a marl bed along the K–T boundary in Haiti (Izett 1991). Glassy melt rock from the subsurface of Chicxulub yielded a ^{40}Ar/^{39}Ar age of 64.98 ± 0.08 Ma (Swisher et al. 1992). Comparable ages of 65.01 ± 0.08 and 65.07 ± 0.10 Ma also were obtained on tektite glass from Beloc, Haiti and Arroyo el Mimbral, Mexico, respectively. Officer et al. (1992), however, questioned this scenario where the impact responsible for the Cretaceous extinction occurred in the Caribbean. They proposed instead that the morphology of the Colombian Basin, the Massif de la Hotte (Haiti), the Cretaceous/Tertiary section in Haiti, the boulder bed in southwestern Cuba, the Isle of Pines (Cuba) fault pattern, and the Chicxulub structure in Yucatan can just as readily be explained by such terrestrial factors as changes in sea level, changes in climate and oceanic circulation, anoxic events, and volcanism, and they claim that there is no need to call upon an extraterrestrial event.

In Caravaca, southeastern Spain, spherules composed of finely crystallized, almost pure potassium feldspar in the structural form of sanidine occur at the Cretaceous-Tertiary boundary. Smit and Klaver (1981) believed that these spherules solidified from a melt derived from the impact body. According to them, the high potassium values indicate that the body was either a metal-sulphide-silicate planetesimal or a cometary body. Another climatic event that may have been the result of the impact is the terminal Eocene event about 34 Ma ago when winters became more severe, but summers were little affected. This event appears to correlate with the Popigai impact structure in Siberia at Lat. 71°30′N; Long. 111°00′E dated as 31 ± 6 Ma by fission (Grieve 1982) and with the formation of the North American tektite strewn field. O'Keefe (1980) proposed that tektites and microtektites which missed the Earth organized themselves into a Saturn-like ring structure. The observed cooling was due to the shadow of the ring on the winter hemisphere. O'Keefe estimated that the ring lasted between one and several million years. Alternately, extinctions at the Cretaceous-Tertiary boundary may have resulted from dust or ash from volcanism that produced the Deccan lavas in India at about this date (Baksi 1990; Courtillot 1990). The event is the source of much concern about the possibility of a long nuclear winter that might result from a future war.

Probably about three times as many craters have been formed on ocean floors as on the continents because of the larger area of the oceans, but impact sites in the ocean are very poorly known because of great difficulty in finding them, faster burial beneath marine sediments, and loss of ancient ocean floors by subduction and melting beneath continents. One such site was reported by Margolis et al. (1991) in the subantarctic South Pacific Ocean where cores of sediments deposited 2.3 Ma ago contain glassy spherules believed to have been formed by impact of an asteroid on the ocean floor. Another impact crater has been reported from the Scotian Shelf at a depth of 113 m (Jansa et al. 1989). Here the 45-m wide and 2.8-km deep circular structure is buried by 510 m of Tertiary and Quaternary sediments. A central uplift in the crater has a relief of 1.25 to possibly 2.7 km and its 11.5-km wide upper surface has a 3.5-km wide central pit. Cambrian-Ordovician rocks in the uplift display hairline fractures and planar features on quartz, indicative of a pressure of 8 to 12 GPa. The crater is partly filled by a seismically isotropic mass that Jansa et al. interpreted as a fallback breccia. Still another site, about 90 km off New Jersey must be near Deep Sea Drilling Project site 612 on the upper continental slope (Poag and Low 1987; Thein 1987; Glass 1990). At a depth of 175 m below the ocean floor is a layer of glassy materials, coesite, microtektites, and strained quartz and feldspar at the contact between middle and upper Eocene sediments. These authors attribute the materials to impact of a meteorite on the ocean floor about 40 Ma ago. The impact not only melted sediments and rocks at the site but the explosion created large waves that eroded and redeposited the debris over an area of at least 15 000 km^2 near the mouth of Chesapeake Bay as a conglomerate as thick as 60 m buried under about 400 m of later sediments (Poag et al. 1992).

Besides being impacted by large meteoroids, the Earth accumulates many small meteorites from outer space. Sizes range down to a fraction of a micrometer. Common in slowly deposited deep-ocean sediments are spheroids of nickel-iron (Murray and Renard 1891, pp. 327–336, plate XXIII), probably droplets from meteorites melted during passage through the atmosphere. The present total flux for all sizes is about 10^7 to 10^9 kg/yr (or about 500 tons/day) on Earth, and about 4 × 10^6 kg/yr on the Moon (Buddhue 1950, pp. 50–54; Short 1975, p. 31; Hartmann 1983, pp. 161–163). The impact rate was millions of times greater during the earliest history of the Earth and the Moon and, according to counts of craters on terrains having radiometrically measured ages on the Moon and their extrapolation to the Earth, has declined logarithmically to the present rates.

Fig. 40. Tektite strewn fields of the Earth, western hemisphere. (After O'Keefe 1976, fig. 3)

Impact of large meteoroids also is considered to be the origin of large fields of tektites, button- or dumbbell-shaped objects of glass produced by melting of country rocks and the glass then was blown thousands of kilometers away from the impact sites (King 1977). Four major strewn tektite fields are known on Earth (Figs. 40 and 41): the Australasian field encompasses Australia, Indonesia, Malaysia, Indochina, Thailand, southern China, the Philippines, and in deep-ocean sediments from the Indian Ocean, western Pacific Ocean, and the Philippine Sea; their ages range from 0.83 Ma to less than 20 000 years. The Ivory Coast tektites have been reported from the Gulf of Guinea and the eastern equatorial Atlantic Ocean, and they have a radiometric age of about 0.9 to 1.0 Ma. The Moldavites of Czechoslovakia have an age of 14.7 Ma. The North American field reported from the Caribbean Sea, Gulf of Mexico, the continental slope off New Jersey, Barbados, possibly from Cuba, Texas, Georgia, and Martha's Vineyard in Massachusetts has an age of about 34–35 Ma (O'Keefe 1976, pp. 17–32; Glass 1990). Other glasses that may represent tektites include the Darwin glasses from Tasmania that are associated with a small impact crater and which have ages similar to that of the Australasian tektites, the irghizites of southern Siberia found in the Zhamanshin

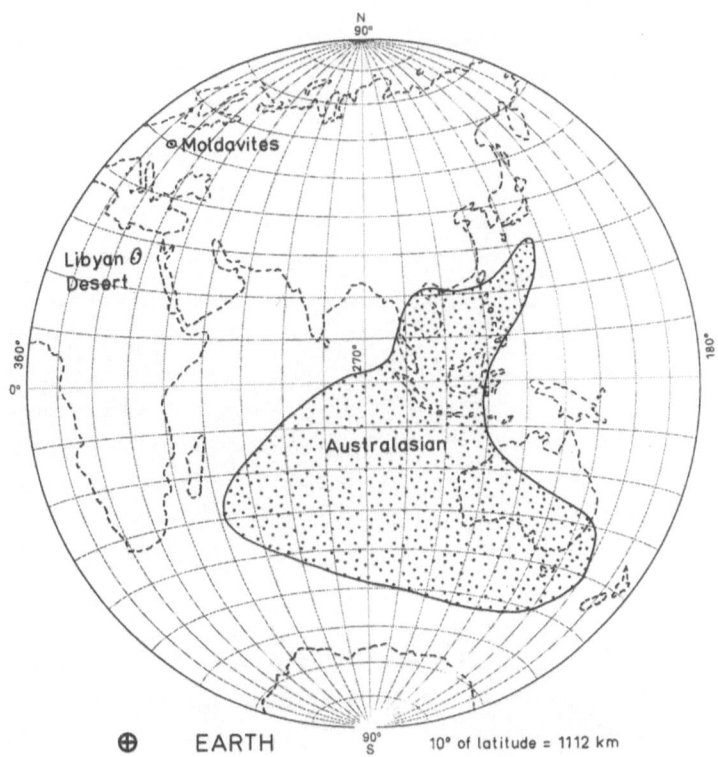

Fig. 41. Tektite strewn fields of the Earth, eastern hemisphere. (After O'Keefe 1976, fig. 3)

impact structure, the Aouelloul glass in Mauritania, and the 28-Ma-old glasses in the Libyan desert. The North American tektite strewn fields do not appear to be associated with any known crater, but the source of the Australasian field may be the Elgygytgyn Crater in Siberia (Dietz 1977).

5 Moon

5.1 Origin

The nearest extraterrestrial body to Earth is the Moon (only 356 410 to 406 697 km distant center-to-center, or about 30 Earth diameters). Earth's Moon, with its diameter of 3476 km (27% that of the Earth's diameter), is rather unique in that it has an unusually large ratio of satellite mass to planet mass, 0.12. Its mean density is much lower than that of Earth, 3.34 versus 5.52 g/cm^3. The higher density on Earth results from the large amount of iron and nickel concentrated in its core. Siderophile elements appear to be deficient in the Moon, volatile elements also are depleted, but the refractory elements are much richer than on Earth. This distribution indicates that the Moon was very hot during the past and its volatile elements were vaporized and escaped into space. Five models have been proposed for the origin of the Moon. They are: (1) fission from Earth, (2) collision ejection due to impact of a Mars-size planetesimal with Earth, (3) disintegration by tidal forces of a planetesimal that entered Earth's Roche Limit; the resulting debris formed an orbit around the Earth and served as the source for accumulation of the Moon, (4) coaccretion from planetesimals, and (5) capture (Brush 1986; Wood 1986).

In the fission theory it is postulated that tidal forces broke the Moon away from the Earth. G. H. Darwin, a son of Charles Darwin (1881, pp. 532–535), suggested that the Moon had been thrown out from the youthful Earth as a tidal bulge. Later, Osmond Fisher (see Brush 1986) proposed that the scar left by the separation of the Moon did not completely heal and is represented by the Pacific Ocean. The problem with the fission model is that there is no explanation of why the Earth had such an excess of angular momentum at the time the Moon was formed, and how the momentum was reduced after fission (Wood 1986). More recently, Wise (1966) modified this proposal, suggesting that the Moon separated from the Earth when iron drained inward to form the Earth's core. Such a migration supposedly increased the Earth's spin, and the increased centrifugal force allowed separation of the Moon from the Earth. That origin would explain the absence of iron in the Moon, as its total composition would be more comparable with that of the mantle. Initial analyses of isotopic and atomic abundances in Moon samples suggested that surface Moon rocks are not compatible with an Earth-mantle origin (Hartmann 1983, p. 150); however, inferred compositions of the sources of lunar lavas are consistent with a mantle origin (Binder 1974, 1975). This deduction also is supported by the findings of Brown (1978), who reported that the bulk composition of the Moon is similar to that of the Earth's mantle. In another approach, the precipitation hypothesis, the Moon supposedly was the result of Earth's fast accretion that caused its surface to heat to about 2000 K during which a hot silica-rich cloud spun off the Earth. As this cloud escaped into space, it lost its remaining condensates (Ringwood 1970). As pointed out by Hartmann (1983, p. 150), if such a theory were correct, Mercury, Mars, and Venus also should have similar large satellites. Wetherill (1985) ascribed the absence of a moon around Venus to a tidal evolution history that led either to its escape into heliocentric orbit or to its impact with Venus. Crashing on the planet could have been responsible for the retrograde motion of Venus and its inclination to the ecliptic. Any debris formed during the collision was reaccreted onto the planet. The absence of satellites around Mercury and Mars is the result of their sizes – they are too small to acquire a debris ring (Taylor 1982, pp. 427–428). Boss and Peale (1986) argued that the precipitation-fission model, like the fission model, does not provide a mechanism to eliminate the excess angular momentum and the difficulty of forming a protoearth close to dynamic instability in a region of massive planet-esimal impacts.

In an article published during 1946 in the *Proceedings of the American Philosophical Society*, and since forgotten, Daly proposed that the Moon was formed as the result of a tangential, slicing collision of an asteroid with an Earth whose surface still was liquid. This collision produced so much angular momentum as to create a separate revolving satellite, the Moon. During the collision liquid fragments from the surface of the Earth were exploded beyond the Roche Limit of the Earth. Many of these fragments became aggregated to the Moon with a smaller number falling back onto Earth. Baldwin and Wilhelms (1992) provided a review of this long-overlooked article. This concept of a glancing impact of a planet-size asteroid was proposed later and independently by Hartmann and Davis (1975), Wetherill (1976), Benz et al. (1989), and Melosh (1989, pp. 223–225), who suggested that our Moon was formed from Earth-mantle material, blown into space by impact of a planetesimal more than 4000 km in diameter after the Earth's core was formed, and from parts of the impacting body. This hot iron-poor expelled material lost its volatiles after expulsion into space and recondensed in a geocentric orbit around the Earth to form the Moon. Supposedly, this would account for the large angular momentum of the Earth-Moon system and for differences and similarities in their chemistry. Ringwood (1990), who believed that the single giant impact model cannot explain the siderophile chemistry of the Moon or the lack of the total melting of the Earth's mantle at a late stage of accretion, proposed an alternate collision model. Impact of a giant planetesimal at the time the Earth had accreted to about 60% of its present size provided the required angular momentum and led to melting and differentiation of the mantle, with subsequent accretion of small iron-rich planetesimals causing convective rehomogenization of the mantle in the absence of extensive melting. In this model protolunar material is believed to have been ejected from the Earth's mantle and placed into orbit at a very late stage of accretion by collisions of small high-velocity icy planetesimals injected into the inner Solar System from Jupiter in the outer system.

In the binary origin hypothesis, the Moon was formed from a cloud of debris that remained behind in orbit around the Earth with small contributions from distant parts of the Solar System and from debris ejected from the growing Earth (Taylor 1982, pp. 426–428). The most serious problem with this concept is that it does not explain the angular momentum of the Earth-Moon system. In the capture hypothesis the Moon was formed in a part of the Solar System more compatible with its chemistry, and subsequently was captured by the Earth. Wood (1986) stated that such a capture is improbable because the range of approach velocities and trajectories that would allow such a capture is extremely narrow. For the Moon to be captured in such a manner it must have been formed at a distance of 0.95 to 1.05 AU and have had an eccentricity of about 0.04. At this radial distance the Moon and Earth should have similar inventories of chemical elements and yet they do not. Although capture of the Moon by the Earth has a low probability, Singer (1986) believed such a capture to be feasible if seen within the restricted Sun-Earth-Moon system. Once the Moon was captured, an initial prograde orbit may have evolved into the present orbit of the Moon. Such a capture led to a decrease in rotational kinetic energy of the Earth by tidal friction that caused intense heating of the Earth; this heating in turn produced degassing of the planet and formation of its hydrosphere and atmosphere. Thus, degassing of the Earth is the result of either satellite impact (an impact that formed the Moon) or capture of the Moon. Whether the Moon is the result of collision or capture, its presence has had other considerable effects on the evolution of the Earth (Comins 1991). Without the Moon, tides on Earth would be much smaller and Earth's rotation rate would have decreased more slowly – factors that may have affected evolution of the biosphere. Another way that the Moon affected the Earth was by its very presence, protecting the Earth to some extent from the massive bombardment that the Solar System underwent early in its history. Wood and most other investigators believe, however, that the collisional model is the most plausible alternative at the present time. If the impact model is correct, the chemical composition of the Moon indicates that the impactor contributed at least 20% to the mass of the Moon (Jones and Hood 1990).

Since its formation and as angular momentum was transferred from the Earth's spin to lunar orbital motion, the distance between the Moon and Earth has slowly increased. Preservation of a 4.46-Ga-old crust on the Moon indicates that the satellite never came within the Roche Limit of the Earth ($\sim$ 2.9 Earth radii), because such a close encounter would have melted most of both bodies and reset their radiogenic clocks. In addition, the inclination of the lunar orbit has decreased, and its eccentricity has increased with time (Lambeck 1980). The current retreat of the Moon is about 5 cm/yr, and the rotation of the Earth is decreasing at a rate of about 5×10^{-22} rad/s. These changes in the Earth-Moon angular momentum system have produced changes in the length of the day and year (Condie 1989, pp. 36–37). Astronomical calculations and measurements of sequential layers secreted in bivalves and stromatolites indicate that 2 Ga ago the year was 800–900 days long, and 500–600 Ma ago it was 420–430 days long.

5.2 Composition

The Moon's patterns of cratered provinces versus maria long ago led to imagined portraits of a 'man in the moon,' a rabbit, or other objects. Speculations regarding the geology of the Moon go back at least as far as Anaxagoras (ca. 500–428 B.C.) who suggested that the satellite was made of stone and Democritus (ca. 460–370 B.C.) who proposed that its surface was mountainous (Taylor 1975, p. 1; Taylor 1982, pp. 4–5). Modern ideas are based primarily on recent scientific explorations of the Moon by American and Russian orbiting satellites, spacecraft that crashed (Ranger 7–9) or soft landed on the Moon (Surveyor, I, III, V, VI, and VII; Luna 13, 16, 20, and 24), unmanned lunar traverse vehicles (Lunokhod 1 and 2), and the six manned Apollo landings (11, 12, 14, 15, 16, and 17; Apollo 13 was aborted), all of which recovered a total of 381 kg of samples from the Moon between December 1966 and the end of August 1976 (Taylor 1975, pp. 4–9; 1982, pp. 3–5; Greeley 1987, pp. 74–75). Since 1987 no visits by manned probes have occurred (Moore 1990), although the recent flyby of Galileo in early December 1990 provided additional information (Johnson, 1991). Data from these missions indicate that accretion of the Moon probably was completed within 100 Ma of the isolation of the solar nebula and that crystal/liquid differentiation produced a felspathic crust, a mineralogically complex mantle, and a core (Taylor 1982, p. 435) A radiometric date of 4.44 ± 0.02 Ga from a ferro-anorthosite provides a lower limit for the age of the Moon and for crystallization of its crust (Carlson 1991). Other highland plutonic rocks display ages from 4.46 to 4.18 Ga and range in composition from magnesium-norite to gabbro-norite. These ages and compositional variations are believed to be consistent with petrologic models where the rocks were emplaced in distinct plutons by endogenous magmatism (Walker 1983; Carlson 1991). Other workers have proposed that the uplands represent upper flotation layers of a magma cumulate series formed during solidification of a 'magma' ocean (Walker 1983). A third model accounts for the uplands by accumulation of cosmic sediment from the last stage of accretion, a model no longer considered feasible.

By 4.4 Ga ago the interior of the Moon was crystallized and a felspathic crust was fully developed, with ferro-anorthosites at the top grading downward to norites at the base of the crust. Partial anatexis (melting of preexisting rocks) of the anorthositic norites and norites of the lower crust led to formation of plutons that supposedly intruded into the cooler upper crust (Binder 1980). During the next 500 Ma the surface of the Moon underwent a heavy bombardment by planetesimals and smaller bodies, producing the present chaotic crust, crustal chemical complexities, and morphology. By 3.8 Ga ago, the bulk of this heavy bombardment, which affected the entire Solar System, was finished, although episodic impacts continue to the present. Early intense cratering in the first 10 to 100 million years of this bombardment may have inhibited formation of a crust, and the lunar highlands may not have begun as a coherent crust but represented megaregoliths of pulverized rocks and dust tens of kilometers thick with the loose rubble underlain by loosely welded breccia and melt rocks of impact origin (Hartmann 1980). As the crust solidified, coherent igneous rocks accumulated beneath the megaregolith. Cratering on this lower layer led to further thickening of the megaregolith and to exhumation of various igneous clasts. Partial melting of the Moon's interior from 4 to 3 Ga ago produced basalts that flooded the craters, with the intense volcanism lasting more than 800 Ma, from 3.9 to 3.1 Ga ago (Greeley 1985, p. 83). However, studies by Schultz and Spudis (1983), Head et al. (1991c), Greeley et al. (1991), Hawke et al. (1991), and Pieters et al. (1991) indicate that mare volcanism was important prior to cataclastic basin formation. Some volcanism also may have continued to as recently as 1 Ga ago (Schultz and Spudis 1983). Thus lavas that filled the basins have been extruded during long periods of time. The oldest mare basalt sample recovered to-date, at the Apollo 11 site, yielded an age of 3.85 ± 0.03 Ga. Other samples from this site are 3.55 Ga old, and the basalts from Apollo 17 site have ages of 3.79 to 3.69 Ga with one of 3.84 ± 0.02 Ga. These dates provide a younger limit for the time of formation of multiringed basins and highland craters (Taylor 1982, pp. 266–267). The thickness of the basalt flows ranges from 2 to 4 km in Mare Crisium to 1.5–2.0 km near the edge of Crisium. A Lunar Sounder experiment yielded a minimum thickness of 1.4 km at Mare Serenitatis (Taylor 1982, pp. 265–266). Tycho and other large craters show evidence of impact-generated basaltic melts at the surface (Taylor 1982, pp. 190–191; Taylor 1991). These impact melts may blanket the base of the ringed basins, and they indicate a basaltic lunar crust.

Relative to the Earth the Moon is almost seismically inert with moonquakes being less than 3 on the Richter scale; fewer than than 3000 events occur annually. Some hypocenters are at a depth of about 100 km and others at 1000 km are at the transition between the lithosphere and asthenosphere. Thus, the Moon's lithosphere is an order of magnitude thicker than the Earth's and is quite stable. The shallow events may be associated with thermal and tidal stresses along the deep fractures around the peripheries of the multiring basins (Taylor 1982, p. 355). Although the present magnetic field of the Moon is very small, studies during the Apollo missions show that the surface is extensively magnetized, indicating that the Moon had a magnetic field in the past (Russell 1991). This ancient field may be related to impact processes, small pockets of molten iron emplaced during differentiation and which acted as small dynamos, proximity to the Earth during the past, a solar field that was more than 100 times more intense during the past than the present 10-gamma level, or accretion of the Moon from a sphere of gas and dust, or from a fluid core. Of this wide variety of hypotheses only the last merits serious consideration. However, whether a core that is only 488 km in radius (1 to 4% of the lunar volume) can produce a dynamo is debatable (Taylor 1982, p. 370). Some observed anomalies appear to correlate with an ejecta blanket from the multiring basins and others with swirl structures that may be related to impacts. According to Taylor, surface-field variability may be caused by impact-induced enhancement of a preexisting weak ambient field. Heat-flow data from the Moon are quite sparse; values at the Apollo 15 Hadley Rille and Apollo Taurus-Littrow sites are 21 and 16 mW/m^2, about one-fourth those on the Earth.

The Moon's crust generally appears to be gravitationally compensated and displays no effects from its pre-4.2 Ga bombardment. Its crust is thick enough and its interior cool enough to support prominent positive topographic elements; for example, the Apennine Mountains that rise 7 km above the surface of Mare Imbrium are uncompensated. As summarized by Taylor (1982, pp. 348–350): 1) negative Bouguer anomalies occur on young ray craters (caused by mass excavation and low-density breccia falling back into the crater), and on young craters filled with lava; 2) the older highlands are in isostatic equilibrium and the pre-Imbrian craters on the highland have no anomalies, suggesting that there was no mantle rebound and 3) positive anomalies are associated with the mountain ranges around the edges of the multiring basins. The mountains are uncompensated and supported by rigid lithosphere, the circular maria on the near side, and the Marius Hills that are believed to be volcanic domes resting on basaltic maria. The excess masses (mascons) represented by large positive anomalies in the maria appear to be the result not of the filling of a hole by 16 km of basalt, but by a 2-km fill and a 12-km rise of the lunar Mohorovičić discontinuity (Bowin et al. 1975).

Processes now active on the surface of the Moon are mainly cosmic, ranging from large to micrometeoritic impacts (that produce craters and erosion of particles by impact, sputtering (to throw out fluid or solid particles by slight or irregular explosions and production of submicron-sized pits) and cosmic ray and solar-flare particles that have effects on an atomic scale with the surficial material being saturated with solar-wind gases (Taylor 1982, p. 117). The erosion rate is about 1 Å/yr by sputtering and ten times as much by micrometeoritic impact (Taylor 1982, p. 158). The lunar atmospheric gases detected by the mass spectrometer deployed during Apollo 17 probably were caused by radioactive decay and solar wind (Taylor 1982, p. 107). The ALSEP (Apollo Lunar Surface Experiments Package) Suprathermal Ion experiment at the Apollo 14 site also detected a mass signature characteristic of the vapor group (Hills and Freeman 1991). Horanyi (1991) investigated the possibility of electrostatic erosion and transportation of lunar dust. Evidence for such transport is indicated by large boulders embedded in the soil with protruding surfaces that are virtually sediment free, the horizon glow observed aboard the Surveyor 5, 6, and 7 spacecrafts, the high-altitude streams noted by the Apollo 17 crew, and the Lunar Ejecta and Meteorite experiments deployed from Apollo 17. Although these exogenic processes appear to be of scientific interest, they had no significant role in forming the present morphology of the Moon.

Exploration at the Moon's surface and geophysical measurements (mainly seismic) indicate that the thickness of the regolith averages 4 to 5 m over the maria and 10 m over the highlands. Regolith thicknesses of as much as 37 m and 20 m have been reported from Taurus-Littrow Valley, the Cayley Plain, and on the Descartes Mountains (Taylor 1982, p. 119). The accumulation rate of this loose debris averages about 1.5 m/Ga, with the rate being higher between 4 and 3.5 Ga ago. The regolith consists of a wide variety of glasses produced by melting during impact, glass-cemented aggregates of glassy rock, and mineral fragments formed during micrometeorite impacts. Some of this debris has ages greater than 4.6 Ga, older than the Solar System itself (Taylor 1982, pp. 123–126); such anomalous ages must be the result of numerous meteoritic impacts that led to small losses of volatile rubidium relative to strontium during impact melting of the crust. Beneath the regolith is about 1 km of shattered rock. The mare basalts are 1.5 to 4 km thick, and lunar crustal thicknesses range from an average of 64 km on the near side (recent studies indicate that the crust on the near side may only be 50 km thick; Y. Nakamura, in Taylor 1982, p. 356) to 86 km on the far side; the lunar average thickness is 74 km (Wilhelms 1987, pp. 12–14). A discontinuity at 20-km depth in the crust may indicate closure of cracks produced by large impacts. Beneath the crust is a partly or totally differentiated mantle. Taylor (1982, p. 360) favored a model where the entire Moon has been differentiated, with the mantle extending to a depth of 1250 km and composed of cumulates (olivine, orthopyroxene, and clinopyroxene). Beneath the mantle is a small core having a 488-km radius.

Other compositional models that have been proposed include ones having a molten upper mantle underlain by a primitive undifferentiated lower mantle and no core, and a model based on seismic data consisting of a solid rigid upper mantle extending to 1100 km and which is divided by a discontinuity at between 400 and 800 km depth. Most moonquakes originate from the base of this upper mantle; Taylor (1975, p. 281) believed that 'lithosphere' is a better term for this layer because it avoids the chemical, mineralogical, and tectonic connotations implied by the name 'mantle.' An attenuated zone (lower mantle) extends to a depth of 1400 km ('asthenosphere'). In such a model the core is only 338 km in radius. Seismic data from the impact of a 1-ton meteorite that struck the far side of the Moon in July 1972 suggest that the attenuated zone may be a region of partial melting (Taylor 1982, pp. 291–292, 360). Observations that the induced magnetic field does not decay appreciably when the Moon crosses the magnetic tail of the Earth suggests that the Moon has a highly conducting core (Russell 1991). This core below the 'lunar asthenosphere' comprises 0.5 to 2.0% of the lunar mass and is believed to be sulfur-poor, with some investigators suggesting that it also is nickel-poor while others suggest that it is nickel-rich. If the core is nickel-rich, the bulk nickel concentrations of the Moon would be more than twice those in the Earth's mantle (Hirschmann 1991).

5.3 Hypsometry

Several attempts to determine the hypsometry of the near side of the Moon were made before satellite altimetry became available, mainly by estimating the heights of mountains from the lengths of their shadows. One of the most thorough studies was by Joksch (1957) on the basis of 2000 measurements and a review of other attempts made as early as 1899. Joksch's results and those by Hedérvári (1960) suggest the presence of two hypsometric levels, one for the visible maria and one for the mountains. Hedérváric considered these levels akin to the effects of continental versus oceanic regions of the Earth. Later curves based on subsequent but similar measurements showed less evidence of two levels (Neiman 1964; Dietz and Holden 1965 – from data of Baldwin 1963), or they indicated only a single level (Gabrilov

1972). When detailed measurements were obtained for the entire Moon's surface from the orbits of Apollo 15, 16, and 17, a new level of understanding became possible. Contours of the general topography of the Moon were obtained from harmonic analysis (12th degree) by Bills and Ferrari (1977) of detailed measurements by laser altimetry from the orbits supplemented by orbital and Earth-bound photogrammetry and limb profile measurements for the entire Moon. The principal axes of the Moon were found to be: $X_1 = 1738.43$ km; $X_2 = 1737.50$ km; and $X_3 = 1736.66$ km (the latter indicating minor polar flattening). The published 1-km interval contours relative to a mean sphere of 1737.53-km radius were transferred to an equal-area projection (Figs. 42 and 43), which shows that the surface of the near side is predominantly lower than that reference level and that the surface of the far side is mainly higher than it. The far side is the more irregular and appears to contain the highest and lowest elevations on the Moon. Differences in relief may reflect differences in crustal thicknesses, with the crust on the far side being several

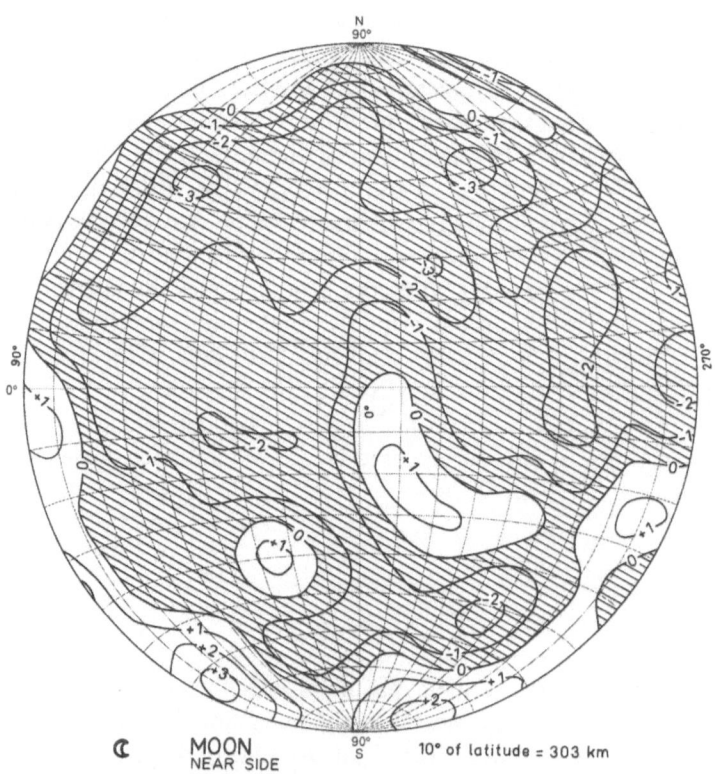

Fig. 42. General topographic contours of the Moon – near side (facing Earth, Long. 270°W is on the *right*, Long. 0° is in the middle, Long. 90°W is on the *left*; 1-km contour intervals). Based on harmonic analysis to degree-12 on maps by Bills and Ferrari (1977) and replotted on a Lambert azimuthal equal-area projection. *Diagonal lines*, areas lower than that of the mean reference sphere having a radius of 1737.53 km

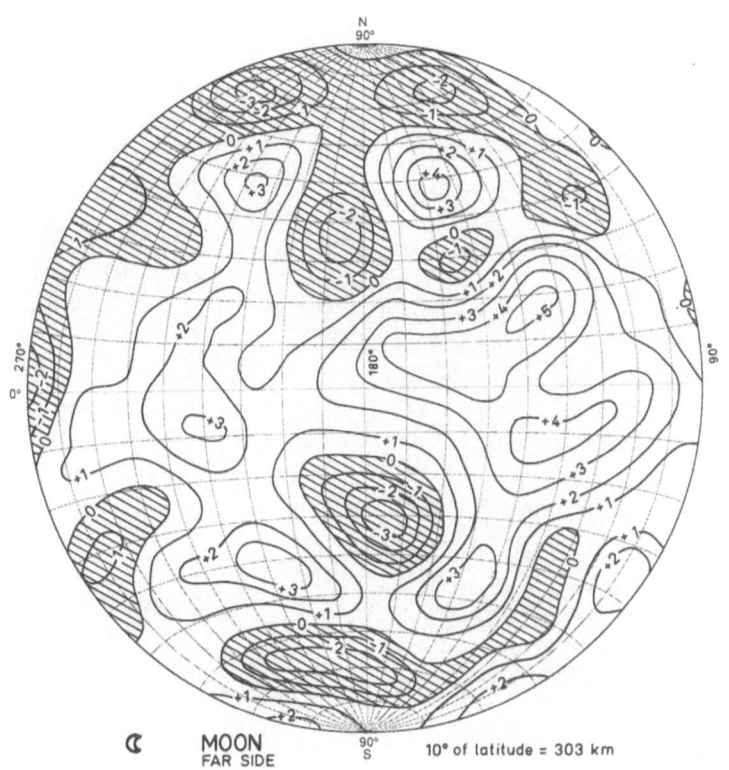

Fig. 43. General topographic contours of the Moon–far side, Long. 90°W is on the right, Long. 180°W is in the *middle*, Long. 270°W is on the *left*. *Symbols* and source as for Fig. 42. Note the generally higher elevations than on the near side (Fig. 42)

tens of kilometers thicker. Measurements of areas between the various contours provided the necessary information for construction of a hypsometric curve for the entire surface of the Moon (Table 3; Fig. 5). This new curve shows only a single topographic frequency level, in keeping with the simple one-crust composition and structure of the Moon's lithosphere. However, the spacing of average elevations within 30° circles located on a graticule at 30° of latitude and longitude (degree-12 harmonics) may be too far apart to show a break in topography caused by two crusts of different density in contrast with closer spacing of average elevations to be contoured (compare the three hypsometric curves for the Earth, Fig. 5).

5.4 Morphology

5.4.1 Exotic

The surface of the Moon is made of three distinct terrains, the rugged highlands or terrae, basins and basin-related terrains, and the flat maria (Figs. 44 and 45).

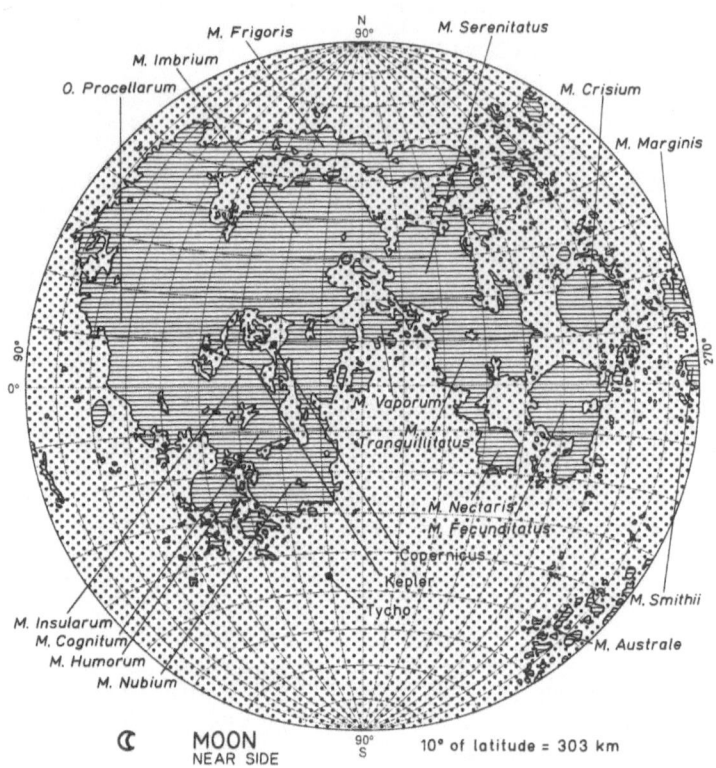

Fig. 44. Morphological provinces of the Moon – near side. With little or no exception, the morphology is exotic in origin and is the result of planetoid, asteroid, or meteoroid impacts. Many thousand overlapping craters within the *dotted area* were produced by explosion on impact. *Horizontally lined* areas represent maria, relatively flat regions that are blanketed by lava flows and to a lesser extent by country rock melted by impact of large missiles. The maria contain many secondary craters derived from later impact craters. *Italic type* names of maria; normal type names of three large Copernican craters having rays

Impact structures can be divided arbitrarily into craters having diameters less than 200 km and basins having diameters larger than 200 km. Structures larger than 100 km show a gradual transition from a central peak surrounded by an irregular ring of peaks, through a single concentric ring of peaks, to multiple concentric rings for the largest structures (Strom 1984). Structures having a central peak and an irregular ring of peaks are known as central-peak basins, and those having a concentric ring of peaks are known as peak-ring basins (Strom 1984, and references therein; Fig. 46). The central peaks are believed to be the result of push mechanisms due to the collapse of crater walls to form terraces, or to the presence of subcrater fault blocks. These allochthonous blocks converge on the center of the crater and push up the peak. Alternately, in a pull mechanism the central peak is formed by rebound of the subcrater substratum that, in turn, pulls the walls, causing them to

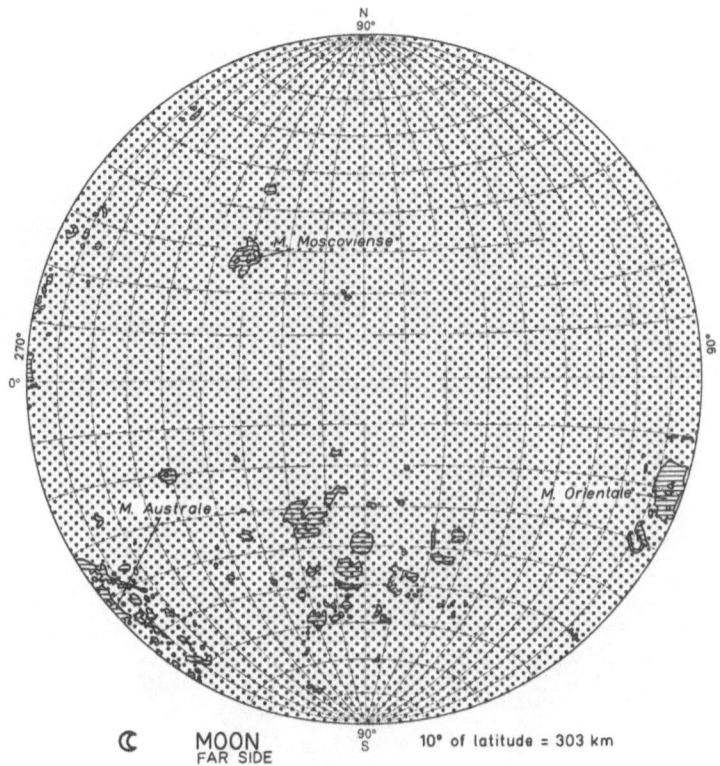

Fig. 45. Morphological provinces of the Moon – far side (not seen from Earth). *Symbols* and source as for Fig. 44

collapse. Wilhelm (1984) inferred the following sequence of events that led to formation of the rings: impact and formation of a simple crater, lesser deepening and lateral expansion and ejection of massive debris at low angles, rebound of the interior, and ejection of knobby debris at higher angles and lower velocities. Several oscillations occur at successively lower amplitudes and extend to smaller radii; the hinge line of each oscillation is recorded by an interior ring. Exterior to the structures and sloping away from them is a debris apron (Fig. 47). The lower parts of this apron are entrained with grooves, pits, and subcircular secondary craters that become more numerous beyond one-radius distance from the rim of the central structure, where they dominate the intercrater terrain (Wilhelms 1984).

The Highlands is a cratered terrain characterized by rugged relief, high albedo, and it consists of impact-ejecta deposits and remnants of crater rims (Fig. 48). This cratered terrain with its highly brecciated rock masses, derived from an anorthositic crust, resulted from a heavy meteoritic bombardment that churned up the lunar crust as the crust crystallized (Greeley 1985, p. 73). The high-albedo relatively smooth plains (i.e. Cayley Plains) in the Highlands represent impact-generated

Fig. 46. Line of impact craters on the Moon. *Top to bottom* Ptolomaeus, Alphonsus, and Arsachel. Between Alphonsus and Arsachel is the smaller structure Alpetragius. The dark spots on Alphonsus (arrows) may represent pyroclastic vents (Greeley 1987, fig. 4.25a). Photograph from Lunar Orbiter IV-108-H2. (Bowker and Hughes 1971, p. 301)

Fig. 47. Ejecta (Hevelius Formation) of the Moon's Orientale Basin. According to Greeley (1987, p. 81), the formation has overriden craters and buried the crater on the lower left by 'deceleration dunes' when the ejecta terminated against the far wall of the crater. Photograph from Lunar Orbiter IV-170-H2, courtesy of Lunar Orbiter IV Principal Investigator L. J. Kosofsky and NASA

ejecta deposits emplaced during the major basin-forming events or during formation of the secondary craters. The lack of relief in these plains is ascribed to the liquid behavior of the ejecta when subjected to seismic shaking fluidization (Greeley 1985, p. 81). Basins having diameters larger than 220 km typically display

Fig. 48. Highland and highland plains (Cayley Plains) in the area of Apollo 16 landing site on the Moon. Photograph AS16-M-0439, courtesy of Apollo Experiment Principal Investigator F. J. Doyle and NASA

concentric rings and were formed during a long period prior to 3.8 Ga ago during which older structures were destroyed (Carr 1983).

Only in the seventeenth century could aspects of the surface begin to be detailed with telescopes, and the first telephotographs were made in 1840 (Fisher 1943, p. 108). No areas of large folds, like those of mountain ranges on Earth, are present, as noted by Chamberlain (1945), but straight escarpments near the edges of maria have been attributed to faults (Dietz 1946) probably produced by movements of crusted lava fields. Most unusual was the observed presence of tens of thousands of large (to about 270 km) and small craters that led to long controversy about their origin. Hooke (1665, pp. 240–245) had made similar model craters by dropping lead musket balls into soft clay, but he could not imagine the source of bodies to impact the Moon and thus proposed the origin of the Moon's craters by analogy with gas bubbles rising to the surface of a pot of boiling chalk and water. Similarity to craters on Earth first favored volcanism (Dana 1846; Shaler 1903; Barrell 1927; Jeffreys 1929; Spurr 1944), and Herschel (1787) even thought that he had seen a glowing lava flow on the Moon during 20 April 1787. However, volcanic origin is opposed by the absence of large broadly conical mountains beneath the craters, the absence of extinct issuing lava flows, and the large diameters and great depths (to 4 km) of some craters. Many craters larger than 50 km contain central mountains, as do volcanic calderas (large craters formed by collapse or explosion) on the Earth, but experiments by Gilbert (1893) and Wegener (1921) showed that impacts of stones, paste, and water drops on sand or mud surfaces also produce central cones. The results of these experiments were extended by Melosh (1989, pp. 126–162) using the theory of impact.

Favoring an origin by impact explosions were the circular shape, low volume of rock in the rims and floors of the craters above their surroundings, and presence of long (to more than 2000 km for Tycho) arcs of light-reflecting material radiating from large recent craters and thinly covering the surrounding topography. These arcs are considered to be thin deposits of debris blown from the craters by explosive volcanism (such as at Tambora in 1815 and at Krakatoa in 1883). During the 1930s the question of origin of the craters still was open, but information from crypto-volcanic structures on the Earth gained by Boon and Albritton (1936), Daly (1946), Dietz (1946, 1947), and others favoring an impact origin, gathered many adherents for a similar origin of most craters on the Moon, especially after Dietz's (1946) cogent comparisons and discussion. After the Moon landings of the Apollo Project between 1969 and 1972 and examinations of rocks brought back to Earth by astronauts and unmanned landers, the impact origin for the larger craters became well accepted. Some small craters, closely spaced along long wavy lines, still suggested a volcanic origin, but now they are considered to be collapse features of lava tubes in maria composed of melt from large impacts, a concept that is supported by the presence of secondary craters produced by large low-angle ejecta from primary impact sites. According to Fryer (1971, table I), evidence exists for crater formation on the Moon by the following processes: (1) single explosions (meteoritic impact, high velocity secondary impact, volcanic explosion); (2) multiple explosions (meteoritic multiple primary impacts, multiple high-velocity secondary impacts, volcanic processes); (3) low-velocity excavations (low-velocity secondary impacts, tertiary impacts); (4) collapses (quiescent caldera subsidence, rubble drainage into cavities and fissures, collapse of lava surface into tubes; and (5) emissions of material (cinder cones, shield volcanoes, ring dikes).

In 1945 Emery became interested in some of the larger and later craters of the Moon through a study of the excellent photographs made at Mount Wilson, California. The Moon, as a companion of the Earth, is forced by its tri-axial shape and Earth's gravitational pull to always present the same side to the Earth (period of axial rotation equal to period of revolution around the Earth). If the impact craters were formed after the Moon came under the influence of Earth's gravity, there should have been more impacts on the far side of the Moon than on the near side, because the Earth's gravity should have intercepted or deflected many of the planetoids or meteoroids that passed near the Earth en route toward the Moon. In the 1940s there were no observations of the topography on the far side of the Moon, but study of the rays, or debris arcs, that extend from the largest and freshest craters (especially Tycho, Copernicus, and Kepler) yielded some pertinent information. Plots of these rays on a gnomonic projection and examination of their appearance on a photograph of the moon printed on a glass sphere (Wright 1935) showed that many of the longest rays are nearly straight lines (great circles) radiating from large craters like the spokes of a wheel (similar to patterns of debris cast out by large man-made explosions on Earth). Later studies (Wilhelms 1987, p. 29) showed that the shorter nonradial rays consist of elements that are radial to the primary crater but originate at topographic irregularities. The radial patterns indicate that the rays had formed when the Moon rotated slowly. If they had formed at a possible time when the Moon's rotation might have been as fast as that of the Earth (perhaps before becoming a companion of the Earth), the rays would have curved about 15° along a 2000-km path from Tycho as the Moon rotated, while the impact ejecta followed paths above the turning surface of the Moon. A large curvature because of this Coriolis effect is not evident; thus the Moon could have been a companion of the Earth when the craters were formed, and many more planetoids and meteoroids would have struck the Earth than the Moon. As there are now relatively few impact craters on the Earth's surface, they must have been destroyed by erosion or become buried under later sediments, and therefore the ones on the Moon are very ancient (Precambrian). At the time, Emery was not satisfied by this indirect evidence for the age of the Moon's craters and went on to other research without publishing his results. Subsequent dating of rock samples brought from the Moon indeed show that the craters having long ray patterns were formed about 1 Ga ago.

5.4.2 Endogenic and Exogenic

Among terrains of endogenic origin are the maria, volcanic domes, dark halo craters that may be pyroclastic features, small shield volcanoes – possibly spatter cones, sinuous linear rills, and grabens. The maria (dark and relatively flat plains) cover about 6.3×10^6 km^2, about 17% of the surface of the Moon (Head 1976) as lava flows (Figs. 49 and 50). They resemble terrestrial basalts but lack hydrous minerals and are slightly higher in iron, magnesium, and titanium. They are believed to have been derived from within the mantle at depths of 150 to as much as 450 km and were erupted 3.9 to 3.1 Ga ago during an 800-Ga period when the Moon underwent an extensive phase of volcanism (Greeley 1985, p. 83). According to Schultz and Spudis (1983), volcanism may have had a longer duration, extending from prior to 4 Ga to about 1 Ga. That these effusive units were not the result of impact is indicated by the following: (1) volcanism generated by impact would be only temporary; (2) volcanism contemporaneous with Tycho occurred at least 100 km from its rim, but an impact on Tycho should not generate volcanism so far from the impact area; (3) impact melts would tend to seal subsurface fractures, preventing a large backdrain noted on Tycho and Aristarchus; (4) the anorthositic crust of the highlands cannot produce the mare basalts by partial or complete melting, or by fractionation after the crust was melted; and (5) stratigraphic evidence demonstrates that eruption of the lavas generally occurred much later than formation of the impact structures (Strom and Fielder 1971; Taylor 1982, pp. 322–323). These volcanic rocks were able to flow freely onto the floors of the basins produced by the impacts, and through gaps in basin rims onto surrounding

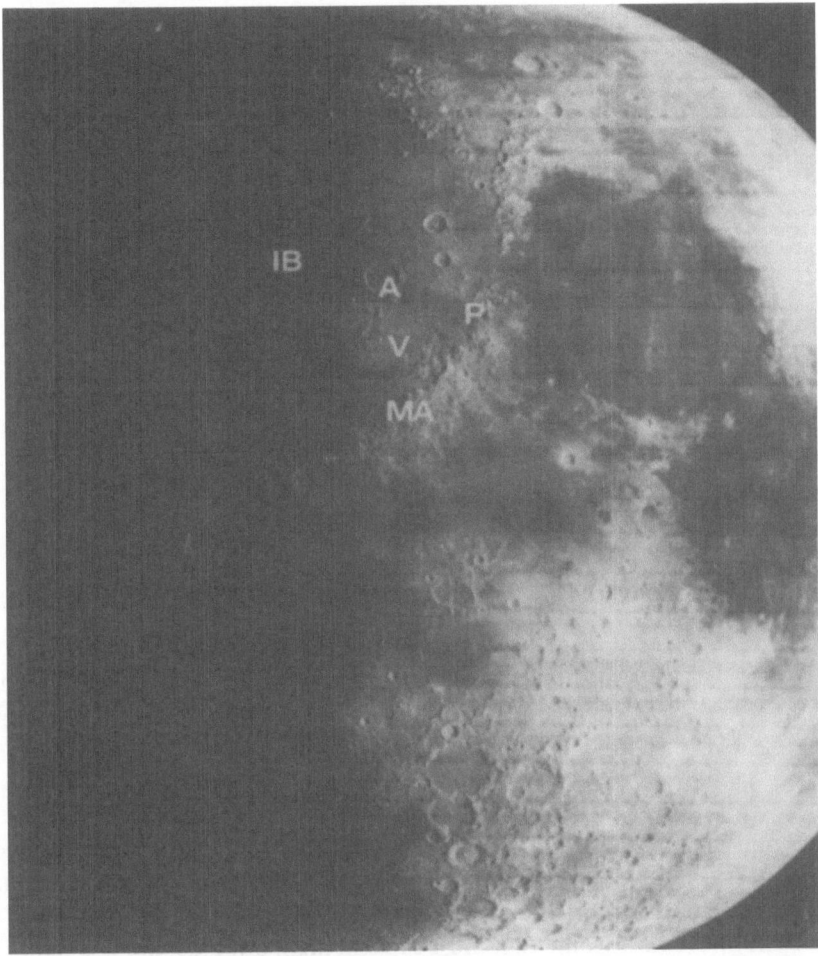

Fig. 49. The Imbrium Basin (**IB**) region of the Moon showing Montes Apennines (**MA**), a volcanic basin-fill unit (**V**), Archimedes Crater (**A**), and Apollo 15 landing site (**P**). Photograph courtesy of Lunar Orbiter Principal Investigator L. J. Kosofsky and NASA

low topography including the surfaces of previous maria and their impressed craters. The diameters of the 46 basins that were flooded with basalt exceed 300 km and range up to 3200 km (Wilhelms 1987, pp. 64–65). According to Greeley (1985, p. 83), these flows formed vast lava lakes or seas that required hundreds, or possibly thousands of years to cool.

At least three eruptive phases have been recognized. The oldest lava unit is rich in titanium, and erupted at high rates to form massive flow units in the centers of basins. The second eruptive phase was characterized by lavas having a lower titanium content, was emplaced at a lower rate than the first unit and has lava

Fig. 50. View of Mare Imbrium of the Moon looking toward Copernicus Crater and Montes Carpatus, a major ring of the Imbrium Basin. Features on the maria were produced by secondary impacts from Copernicus. Photograph AS17-M2444, courtesy of Apollo Experiment Principal Investigator F. J. Doyle and NASA

channels and lava tubes. High-titanium lavas of the third phase were erupted at lower volumes. In the Maria Imbrium such lavas extruded from a vent in the outer basin rim and consisted of three flows that reached distances of 1200, 600, and 400 km. They have flow fronts that are 1 to 90 m high, but most are less than 15 m high (Schaber 1973; Taylor 1982, pp. 270–272). The area covered by the flows exceeds 2×10^5 km^2, an area equal to the Columbia River basalts. The youngest series of flows was extruded via a fissure about 20 km long near the crater Euler located at the intersection of two concentric ring systems. To-date, 1296 vents have been mapped on the Moon.

Some mare lavas display benches along their margins, perhaps representing high stands of lava. Also associated with these flows are numerous sinuous rilles displaying abrupt discontinuities and a crater-form depression at their heads. They are believed to be former lava channels and/or collapsed lava tubes (Fig. 51; Murray 1971; Greeley 1985, p. 87). According to Murray, rilles are distinctly concentrated around the edges of Maria Imbrium and Maria Orientale and possibly around the peripheral regions of circular maria. They tend to be particularly abundant in the areas of the Marius Hills and the Aristarchus Plateau. Hadley Rille in the Imbrium Maria, investigated by Apollo 15, is more than 130 km long, and it originated in a depression about 10 km long and 5 km wide at the base

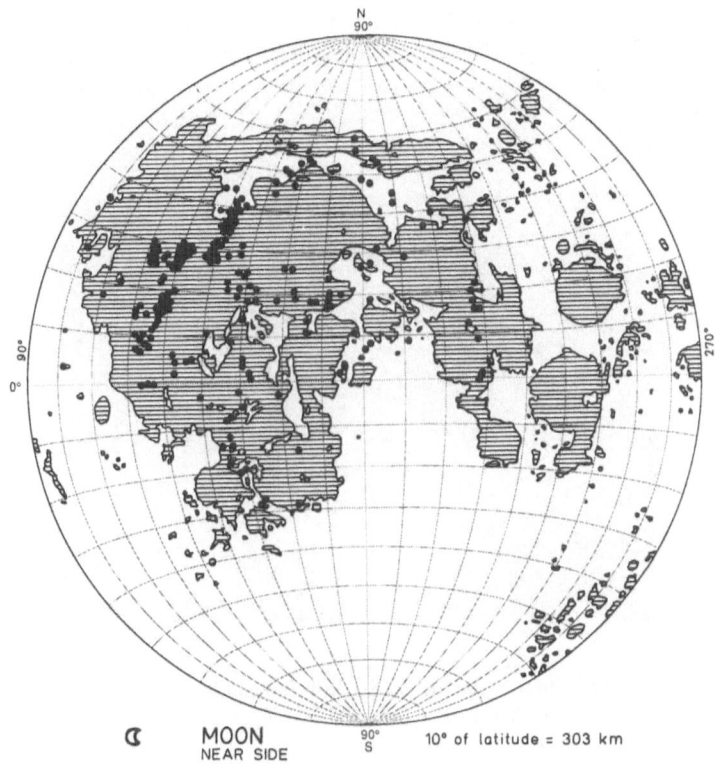

Fig. 51. Distribution of sinuous rilles on the Moon – near side. *Dots* represent the crater sources or, where no source can be discerned, the center of the track of the rilles. Most are on the maria (*horizontally lined areas* as in Fig. 44). (Murray 1971, fig. 45)

of the Apennine Mountains (Fig. 52). Strom and Fielder (1971) stated that there is evidence of drainback of lava and subsequent subsidence of the floors of Tycho and Aristarchus, indicating that both lows are underlain by chambers or an extensive network of fissures that serve as reservoirs of a large volume of magma. Another prominent topographic feature on the maria surfaces is wrinkled or mare ridges, and broad arches capped by sharply crenulated, sinuous morphologic features (Greeley 1985, p. 87). On the centers of some of the flows also are channels bordered by 20 to 50-m-high levees that are larger than their terrestrial counterparts (Taylor 1982, p. 269). Some ridges appear to postdate formation of the maria, because they intersect impact craters – evidence that the lava was solid enough at the time the ridge was formed to preserve the impact structures. Others display flow features, indicating that lava was being extruded at the time the ridge was formed, and thus the ridges were contemporaneous with emplacement of the lava. Some of these positive elements may have been produced by faulting, others may reflect the underlying topography, and still others may be due to lava squeeze ups (Greeley 1985, pp. 87 and 91).

Fig. 52. Apollo 15 landing site showing the Hadley Rille (**H**) and Monte Apennines (**A**) on the Moon. Photograph courtesy of Apollo Experiment Principal Investigator F. J. Doyle and NASA

Other endogenic volcanic features on the Moon include pyroclastic cones, lava flows, domes, calderas, collapse craters, and fissure eruptions from the Marius Hills and from Maria Imbrium (Figs. 53 and 54; Fielder and Fielder 1971; Guest 1971a; Murray 1971). Evidence of volcanic activity also has been noted on the Aristarchus Plateau. Although features produced by extrusive processes are uncommon on the Moon (Table 4), they do exist (Table 5). Their rarity is considered to result from the small size of the Moon and the greater thickness of its lithosphere as compared with lithospheres of larger bodies. More than 200 volcanic domes have been mapped on the Moon. Some features of volcanic origin are low, relatively flat-topped, circular, and commonly with a summit crater; Head and Gifford (1980) ascribed these features to low rates of effusion. Other types of dome appear to be nonvolcanic in origin, representing structural highs that were not buried by younger mare lava deposits.

Tectonic nonvolcanic endogenic morphologic features reported from the Moon include numerous linear valleys that criss-cross the lunar surface (Figs. 55 and 56); they have the general form of grabens and appear to be relatively deep-seated (Golombek 1979). These structures were formed after effusion of the early mare lavas and appear to be aligned along reactivated older structures. These older structures reflect a tensional field generated early in the history of the Moon, and they have persisted throughout geological time. This structural pattern or lunar grid consists of linear faults, ridges, valleys, rilles, joints, crater chains, crater central

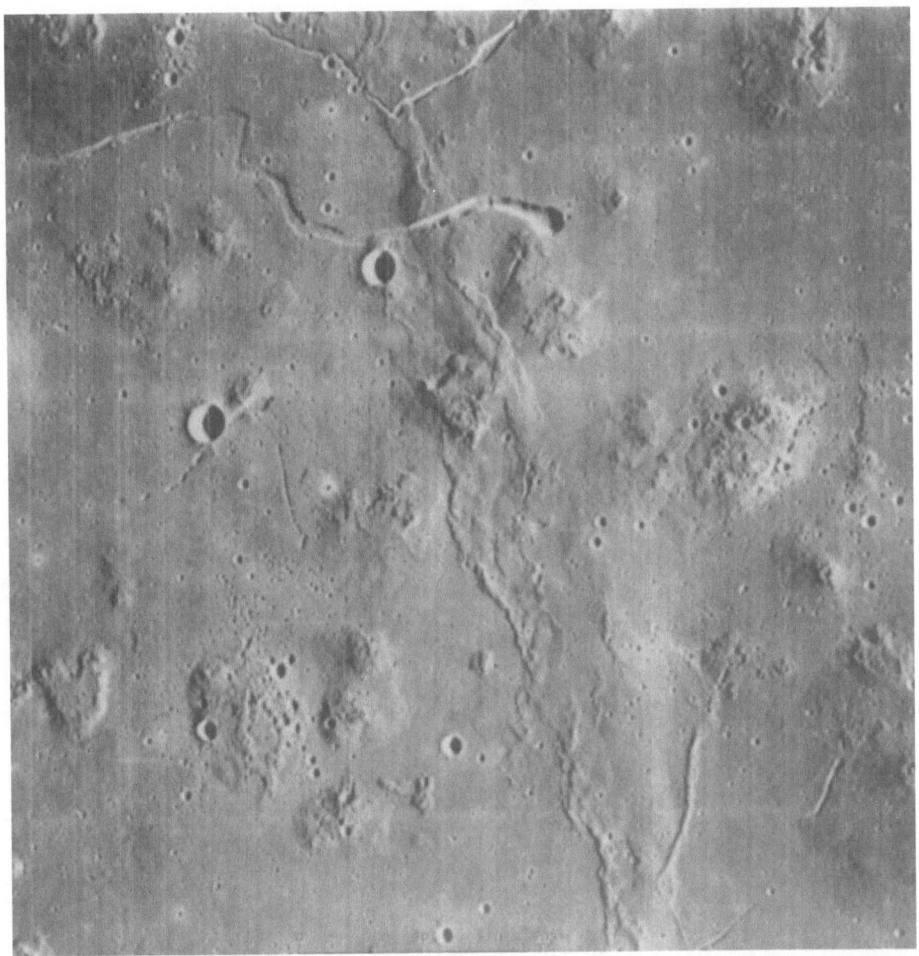

Fig. 53. Marius Hills, a volcanic complex of domes, cones, and rilles in Oceanus Procellarum of the Moon. (Wilhelms 1987, fig. 5.7)

peaks, and polygonal crater rims that intersect and cross each other and display southwest-northeast (system A), southeast-northwest (B), weak north-south (C), and faint east-west (D) trends, and a joint system radial to the Imbrium Basin termed R_1 (Mutch 1972, pp. 247–254; Hale 1980). Magmatic intrusions and emplacement of dikes appear to have fractured the floors of many craters, but some of these structures could have been produced by viscous relaxation of crater topography (Hall et al. 1981).

Lack of a hydrosphere and atmosphere caused exogenic processes on the Moon to be restricted to gravitational ones (Guest 1971b). Possible landslides, rockfalls and associated boulder tracks, and surface creep that produced a treebark-like texture have been reported by Greeley (1985, pp. 100 and 101).

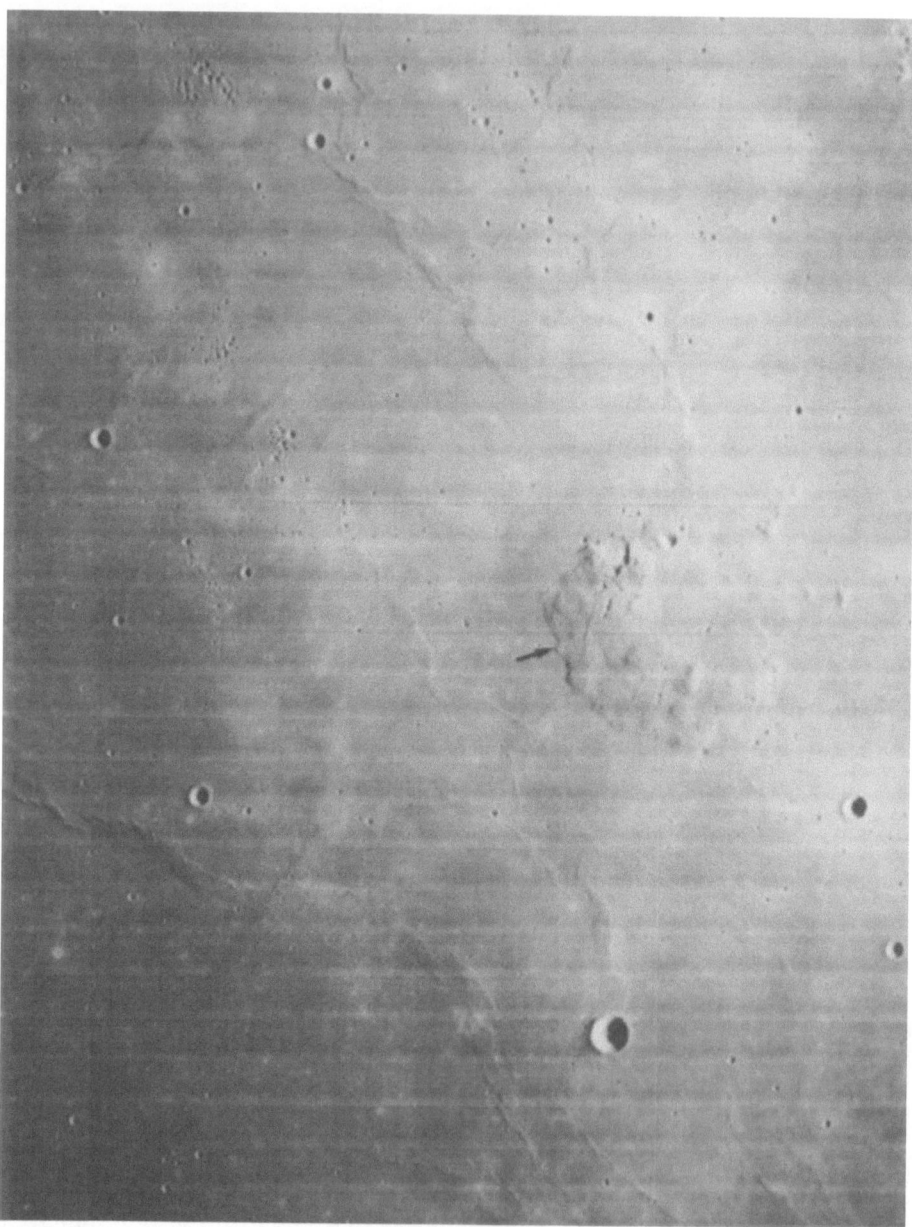

Fig. 54. Rumker Hills (*arrow*), a volcanic platform northwest of Oceanus Procellarum of the Moon. Note tectonic or volcanic ridges on the maria. Photograph from Lunar Orbiter IV-170-H2, courtesy of Lunar Orbiter IV Principal Investigator L. J. Kosofsky and NASA

Fig. 55. Vallis Alpes, a feature of the Moon that Greeley (1987, fig. 4.26) believed to be tectonic in origin and associated with the formation of the Imbrium Basin. The graben apparently was filled with lava by the rille along the axis of the low. Note the two rilles left of the graben. Photograph from Lunar Orbiter IV-170-H2, courtesy of the Lunar Orbiter IV Principal Investigator L. J. Kosofsky and NASA

5.5 Chronology

The geological history of the Earth has been established mainly from sedimentary rocks, through their superposition, distribution, content of fossil plants and animals, and the composition and shape of their mineral grains. The absence on the

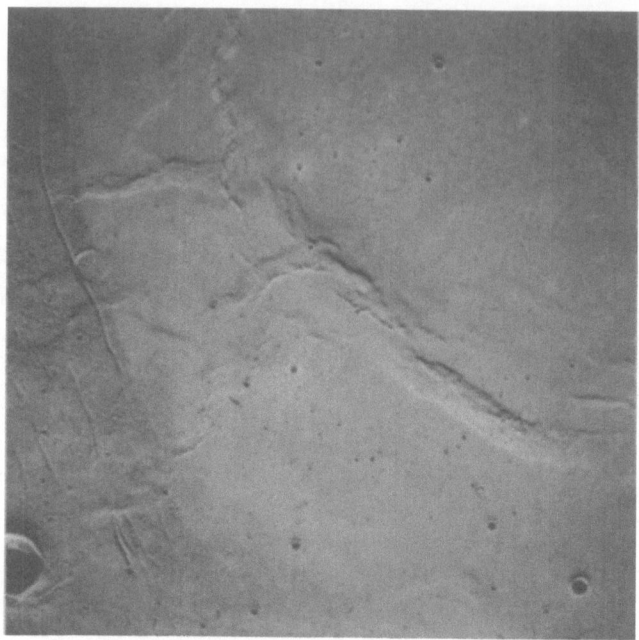

Fig. 56. Ridges in Maria Serenitatis. Some ridges cross from the younger (lighter) less cratered mare into the older darker maria. These ridges are believed to have been formed by tectonic and volcanic processes on the Moon. Note linear features at the *left* in the dark maria; they may be grabens. Photograph AS17-0451, courtesy of Apollo Experiment Principal Investigator F. J. Doyle and NASA

Moon of an atmosphere and hydrosphere precludes that approach to stratigraphy, but the formation of many impact craters by early bombardment, the production of broad maria by fluid basaltic flows, the invasion of older craters by these maria, the production of craters atop the maria by later meteorite impacts, the size distributions of the craters, the wide patterns of ejecta from the basins and large craters, and the physiographic stages of erosion by subsequent impacts of smaller meteorites serve as clues to the sequential history of the Moon. This approach was utilized by several workers and especially by Shoemaker (1962), whose long study of Moon topography led to classification of Moon stratigraphy into systems and series akin to those of the Earth (Shoemaker and Hackman 1962; Shoemaker 1964). His classification has been reinforced by many other investigators and especially by radiometric age determinations, as summarized by Wilhelms (1987). About 300 radiometric age determinations were made on rock samples brought to Earth from the Moon. Nearly two-thirds were based on $^{40}Ar/^{39}Ar$, nearly one-fourth were on Rb/Sr, and the rest were on Sm/Nd, $^{83}Kr/^{81}Kr$, and U/Pb/Th isotopes.

Some results of studies of the morphology, rocks, and datings are summarized in Table 6 for an interpretation of the Moon's history. The earliest impacts cannot be dated because the Moon's surface had no crust able to support and retain

Table 6. Moon stratigraphy

Ga B.P.	System or series	Total area (10^6 km^2)	Number of craters > 30 km diameter (10^6 km^2)
	Copernican	Craters	44
1.1	Eratosthenes	12.1	88
3.26			
	Upper Imbrian	35.3	174
3.72			
	Lower Imbrian	14.2	195
3.85			
	Nectarian	19.5	1300
3.92			
4.2	Crust	38.0	?
4.55			Origin

craters until about 4.2 Ga ago. About 3.92 Ga ago the Nectarian event began with formation of the Nectaris, Humorum, and other basins, the emplacement of the ejecta in the Janssen Formation and eruptions of mare lavas rich in aluminum, potassium, rare earths, and phosphorus. This volcanism began 4.1 Ga ago and continued into early to mid-Imbrian time (Greeley 1985, p. 102). It covered many earlier craters and much of the original crustal debris from previous impacts. Both the maria and remaining terrae began to be impacted by subsequent meteoroids until a series of much larger impacts 3.85 Ga ago produced a new horizon, the Lower Imbrian. The general sequence was repeated by the Upper Imbrian 3.72 Ga ago, and by the Eratosthenes maria 3.26 Ga ago. The Orientale Basin was formed at this time, and two-thirds of the mare basalts were extruded during Imbrian time (Greeley 1985, p. 102). One-third of the remaining visible maria were emplaced during Eratosthenian time. Finally, beginning 1.1 Ga ago, several large impacts produced large craters that still are sharp and mostly accompanied by long broad, thin, rays of debris, the Copernican system. These were the rays that Emery studied from photographs in 1945; they remain because too little time has elapsed since their formation for them to have been erased by subsequent small impacts as have rays from older craters of earlier systems. Superposition of mare lava flows on Copernican-age craters indicate that lunar volcanism extended to about 1 Ga ago. This chronology, however, has been questioned recently. According to Dalrymple and Ryder (1992), high resolution ^{40}Ar/^{39}Ar age spectra on single fragments from lunar impact melt rocks indicate that most of the ejecta is coeval or predates the Imbrium impact. These authors believe that their data indicate that the Imbrium impact is no older than 3.870 Ga and probably no older than 3.836 Ga. They further postulated that their data show no evidence for lunar bombardment prior to 3.9 Ga. According to Hartmann (quoted in Kerr 1992), on the other hand, the absence of material older than 3.9 Ga is not due to the absence of impacts at that time, but rather that the impacts were so frequent then that they obliterated earlier events.

The presence of dated maria allow the grouping of impact craters into a sequence in which craters are observed to be progressively more eroded (rounded and subdued) with age. The areas of terrae also contain craters of various ages denoted by different extents of subsequent impact erosion. Maria that once extended into the areas of present terrae have become so churned by impacts that they are recognizable only as rock fragments in the debris. Note from Tables 4 and 6 the variation in total areas of maria that belong to the different systems, the different time intervals between the systems, and the decreasing numbers of large craters adjusted to the entire area of the Moon for each interval. The time intervals between the maria (Table 5) suggest the advent of several separate series of impacting events rather than a long period of sweeping up the debris from only a single event that may have produced the Moon itself. In fact, Ryder (1990) advocated impacts at about 3.9 Ga during which the Earth may have been struck by a large planetoid that produced an Earth-centered orbiting disk whose solid components were rapidly collected to form the Moon.

Surrounding many large craters and basins to a distance of a few hundred kilometers are many small craters that are considered as secondary, formed by material ejected during the impact that created the main craters or basins. Also present are lineaments, gashes, and irregular or intermittent grooves that radiate outward from some basins and craters. These are considered to be scars formed by fragments blown at low angles from the impact sites. Other finer ejecta traveled much farther, some in ray patterns that reached many hundreds of kilometers from their sources. During early development of a crust on the Moon tectonic folds and faults probably formed, but all traces of them have been destroyed by contemporary and subsequent bombardment of the Moon's surface by extralunar missiles (planetoids, asteroids, meteoroids, and meteorites). The incessant pounding of the Moon's surface by these missiles has produced large quantities of very fine dust that would have dissolved quickly on Earth but which accumulates on the water-free Moon as thin layers and filling of interstices. Except for far-reaching explosion debris from impacts, the only sediments on the Moon appear to be those of landslides or mass movements down steep slopes of craters. Thus, about 84% of the morphology was slightly modified by exotic agents and the remaining 16% was modified by extrusion of lava flows: terrae and maria of the present topography little modified by exogenic processes. Areas of maria indicated in Table 4 omit the former presence of maria later converted into terrae by dense production of new craters.

Mapping of the exotic and endogenic morphological provinces on the Moon (Figs. 44 and 45) was based on generalized maps by Wilhelms (1987), although much more detail is given on maps for the near side (Wilhelms and McCauley 1971), the east side (Wilhelms and El-Baz 1977), the west side (Scott et al. 1977), central far side (Stuart-Alexander 1978), and the north side (Lucchitta 1978). The mapping shows that maria occupy about 30% of the total area of the near side and only 2% of the area of the far side of the Moon. Thus, the maria constitute 16% of the total 38×10^6 km^2 of the Moon, and the terrae – 84 percent, or more than five times the area of maria. The maria of the near side also are larger than those of the far side. Most are in the northern hemisphere of the near side and the southern hemisphere of the far side, an antipodal distribution pattern. Largest among these basins according to Wilhelms (1987, pp. 64–65) are: Procellarum centered at Lat. 26°N, Long. 15°W, 3200 km (pre-Nectarian); South Pole-Aitken, Lat. 56°S, Long. 180°W, 2500 km (pre-Nectarian); Imbrium, Lat. 33°N, Long. 18°W, 1160 km (Lower Imbrium); Crisium, Lat. 17.5°N, Long. 301.5°W, 1060 km (Nectarian); Fecunditatis, Lat. 4°S, Long. 308°W, 990 km (pre-Nectarian); and Orientale, Lat. 20°S, Long. 95°W, 930 km (Lower Imbrium). The about 2000 km diameter South Pole-Aitken Basin was imaged by the probe Galileo using new equipment during a flyby on 8 and 9 December 1990 while the probe was en route to Jupiter (Belton 1991; Head et al. 1991c). The main basin is about 2000 km southwest of Orientale basin and probably is pre-Nectarian in age. Compositional information obtained by the Galileo spacecraft during its flyby of the Earth-Moon system on the inside of the far side South Pole-Aitken Basin indicates that this impact depression penetrated deep into the lunar crust or lunar mantle or that the basin contains ancient mare basalts that later became mantled by highland ejecta (Belton et al. 1992). Similar measurements also indicated that the ejecta of the Orientale Basin (900 km diameter) and outside the Cordillera Mountains was excavated only from the crust and that it rests on pre-Orientale terrain of highland materials and wide terrains of ancient maria. Note that the positions of the centers of the basins differ from the centers of their remaining maria, especially for the much eroded older basins.

The present orientation of the Moon with respect to the Earth is quite stable, but about 59% of the Moon's surface can be seen from the Earth over time only because of changing viewpoints provided by the geometry of the Earth-Moon movements (librations; Fisher 1943, pp. 40–42). The present positions of the maria are asymmetrical (northern hemisphere on the near side, and southern hemisphere on the far side). An original, equatorially symmetrical distribution would be expected because probably most of the missiles that formed the craters, which later were flooded with lava, belonged to the Solar System and had orbits in or near the plane of the ecliptic (path of the Earth during its revolution around the Sun). If the original equator can be identified according to the most symmetrical distribution of maria (Figs. 44 and 45), its maximum departure from the present equator would be at about Lat. 30°N, Long. 0° and Lat. 30°S, Long. 180°W. Conceivably, the present asymmetry means that the axis of rotation shifted sometime after the end of the heavy bombardment had formed most of the impact craters.

MOON | LUNAR ORBITER   29 USA

6 Mercury

6.1 Observations

The innermost planet of the solar system, Mercury, was known to the Sumerians about 2500 BC, and to the Egyptians it was known as Sebegu (Strouhal, 1992, p. 240). Its speed in orbit (revolution of 88 days, rotation of 59 days, diameter of 4879 km; Table 1) and difficulty of observation (maximum angle of only 27°45′ from the Sun) may have been the reason for attributes to Hermes (Greek) and Mercury (Roman) with winged sandals and hat and identification as a fleet messenger for Zeus (Greek) and a patron of business (Roman). In the Middle Ages astrologers assigned to those persons under the influence of the planet a lively or mercurial temperament, and alchemists named the metal mercury (quicksilver) after it. Details of Mercury's topography could not be examined from the Earth by telescope or radar because of its small size, great distance, and angular nearness to the Sun. In fact, the most detailed maps of Mercury made prior to about 1970 (Murray et al., 1972) bear no resemblance to the truth. Only in 1965 was even its period of rotation found to be exactly two-thirds of its period of revolution about the Sun. The only spacecraft to visit it has been Mariner 10 – in three passes during orbits around the Sun (29 March 1974, 21 September 1974, and 24 March 1975) after which contact was lost. During these three encounters, only the same half of Mercury was sunlit with the rest in darkness (Moore 1990), so that we know the general morphology of less than half of the planet but have insufficient information to draw contours of elevation (Fig. 57). Names assigned by NASA to most of the topographic features, especially to craters and their adjacent ejecta aprons and the plains, were intended to commemorate authors and composers even though few, if any, of them are known to have had any interest in the topography of Mercury or of any other planet or moon.

The observed mean density of Mercury is 5.43 g/cm³ (Table 1; uncompressed – about 5.3 g/cm³; the Earth has a mean density of 5.52 g/cm³, which corresponds with an uncompressed density of about 4.45 g/cm³). This density indicates that Mercury has an iron content of about 60 to 70%, much greater than that of any other rocky planet or satellite (Hartmann 1983, p. 282; Cameron et al. 1988; Schubert et al. 1988). As pointed out by Lewis (1988), no plausible sampling of the products of the solar nebula could account for this density and metallic iron content. Such a density may be the result of: (1) vaporization of the silicate portion of the planet and its removal by the solar wind during the early history of the Sun;

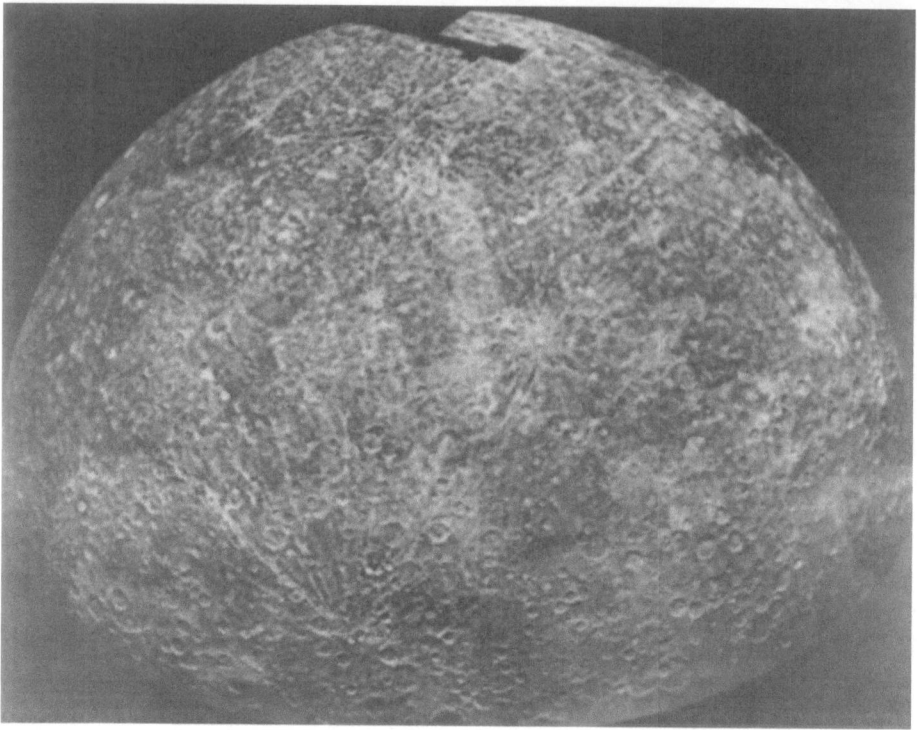

Fig. 57. The southern hemisphere of Mercury centered at Long. 110°W. Note the numerous impact craters, many with ray patterns. Mosaic 6A (Davies et al. 1978, p. 24)

(2) condensation under conditions of chemical equilibrium so that for Mercury very close to the Sun only iron and refractory silicates condensed; (3) formation from building blocks closely related to enstatite blocks with mechanical iron/silicate fractionation enhancing the proportion of iron; or (4) a giant impact that stripped the crust and upper mantle from a previously differentiated planet (Chapman 1988; Lewis 1988; Wasson 1988; Wetherill 1988). According to Chapman, the Wetherill impact model or an intermediate model of Goettel (1988) may be the more viable concept for impact. If the impact model is correct, the FeO content in Mercury would reflect the oxidation state at the time of accretion, and ejection of a felspar crust would have depleted calcium and aluminum along with the volatile alkali metals (Lewis 1988). The planet's magnetic field is weak compared with that of the Earth (about 500 gammas), but it is strong enough to form a magnetosheath and magnetosphere (Strom 1984). It is dipolar, inclined about 11° to the axis of rotation and has the same polarity as the Earth's present field (Strom 1984; Connerney and Ness 1988). The field is generated probably by convection within the thin liquid shell of the outer core that must contain a small amount of sulfur.

  Mercury's thermal history indicates a hot start (Schubert et al. 1988) with early core differentiation and gradual cooling via mantle convection. As the solid inner core grew, it transferred its sulfur to the liquid outer core. Under this global thermal regime as the planet cooled it contracted by 6 to 10 km in radius – a contraction that exceeded the observed limit of 2 km at the lobate scarps (Melosh and McKinnon 1988). Schubert et al. estimated that the lithosphere on Mercury was 50 km thick at the end of heavy planetesimal bombardment and is about 150 km thick now. Reflectance spectral measurements from Earth and Mariner 10 are comparable with those for the Moon and are attributed to Fe- and Ti-bearing agglutinates in the regolith (Vilas 1988). As a result of gravitational focusing of meteoritic material by the Sun and of Mercury's higher surface temperature, melt production on the surface of Mercury is about three times that at the hottest point of the Moon (Cintala 1981). Leinert et al. (1981) stated that the flux of extra-mercurian particles is at least eight times greater than on the Moon; consequently, regolith production must have been much faster on Mercury. Greater production of regolith, however, does not necessarily increase the agglutinate fraction (Vilas 1988). A weak absorption feature attributed to Fe^{+2} in orthopyroxene within the surface material was noted in two recent high-quality spectra but was absent in a third – a difference that is believed not to be caused by differences in terrain morphology. Its absence may indicate that some of the surface is highly reduced (Vilas 1988).

In a recent article, Cintala (1992) stated that a model of impact heating and calculations of impact fluxes and surface temperatures suggest that the rates of impact melting and vaporization on Mercury were much greater than on the Moon, and that 14 times more melt and 20 times more vapor were produced on Mercury than on the Moon. He further stated that because the stronger gravity field of Mercury offsets its greater kinetic energy, craters formed on the Moon and Mercury by identical impactors should be similar in size, thus regolith mixing in both is about the same. However, as impact melt production on Mercury exceeds that on the Moon by a factor of two, soils on Mercury would tend to mature faster than those on the Moon. The much larger agglutinate population on Mercury is believed by Cintala to be responsible for the absence of optical contrasts caused by variations in Fe^{+2} content on the surface of Mercury. The abundance of agglutinates also indictates that most of the pyroxenes have been fused and included in the glasses, attenuating the Fe^{+2} crystal-field absorption near 0.9 mm. Recent spectral measurements of Mercury support this speculation. Particle density in the gaseous atmosphere on Mercury is so low that the planet's surface at the base of the exosphere is the site of more frequent collision of atoms than within the atmosphere itself. Hydrogen, helium, and oxygen were detected by the ultraviolet spectrometer on Mariner 10, and sodium and potassium have been discovered by Earth-based instruments (Hunten et al. 1988).

6.2 Hypsometry

Radar altimetry data obtained by the Arecibo Observatory indicate that when the mercurian equatorial topography is plotted on an absolute altitude scale with a datum defined by a 2439.0-km radius reference sphere, the histogram, like those of the Moon and Venus, is unimodal (Fig. 5; Clark et al. 1988; Harmon and Campbell 1988). The mean of altitudes is + 0.7 km and a histogram of altitudes displays a slight peak at + 0.3 km. The typical relief between highlands and lowlands on Mercury is about 3 km, with an extreme range of altitudes of about 7 km (− 2.4 to + 4.6 km) from the lowest crater floors to the highest plateaus. On the Moon the maximum relief is 10 km, or about 7 km of relief exclusive of craters (Harmon and Campbell 1988). The hypsometric curve for the entire planet, measured by radar from Earth, is fairly smooth with only a single flattening (Table 3, Fig. 5). The curve is similar to that of the Moon except for more irregularity in the high portion.

The equatorial region of Mercury displays two topographic highs; one at Long. 10°W has an abrupt scarp on its west side associated with a fault system. Another topographic high occurs south of Caloris Basin between Long. 160°W and 240°W with topographic lows at Long. 180°W and 210°W that correspond with plains regions. A third high occurs near Long. 310°W in the unimaged hemisphere. According to Harmon and Campbell, the long axis of the planet's dynamical figure is aligned with its perihelion subpolar points at Long. 0° and 180°W. Although the two largest bulges are aligned along Long. 10°W–190°W axis, the planet's topography also is aligned with its dynamical figure. Mercury rotates three times (58.65-day rotational period) during every two orbits (orbital period 88 days) around the Sun. Its 3 to 2 resonance corresponds with variations of about 100 m in the dynamic figure. This dynamic figure of Mercury may be caused by the long-wavelength component of uncompensated topography at the bulges or to the lunar-like mascons on the smooth plains in and around Caloris Basin (Melosh and Dzurisin 1978a; Harmon and Campbell 1988). According to these authors, the altitude profiles on the plains are consistent with their subsidence under a load which, if only partly uncompensated, could influence Mercury's dynamic figure.

6.3 Morphology

6.3.1 Exotic

The morphology of Mercury, unlike that of the Moon, is dominated primarily by features of endogenic origin and secondarily by ones of exotic origin (Table 4; Fig. 58). What few exogenic features are present are induced by impact ejecta and mass movements. Conceivably, some erosional/depositional morphology was initiated by water and ice, but thorough bombardment by meteorites and other bodies has destroyed the evidence. The mere possibility of the former existence of such features is suggested by the presence of ice beneath the planet surface at both poles, as indicated by maps made by radar, which can penetrate the ground surface, unlike conventional telescope imaging (Slade et al. 1991). Polar surface temperatures are as low as − 173°C (100 K), in contrast with equatorial ones that can

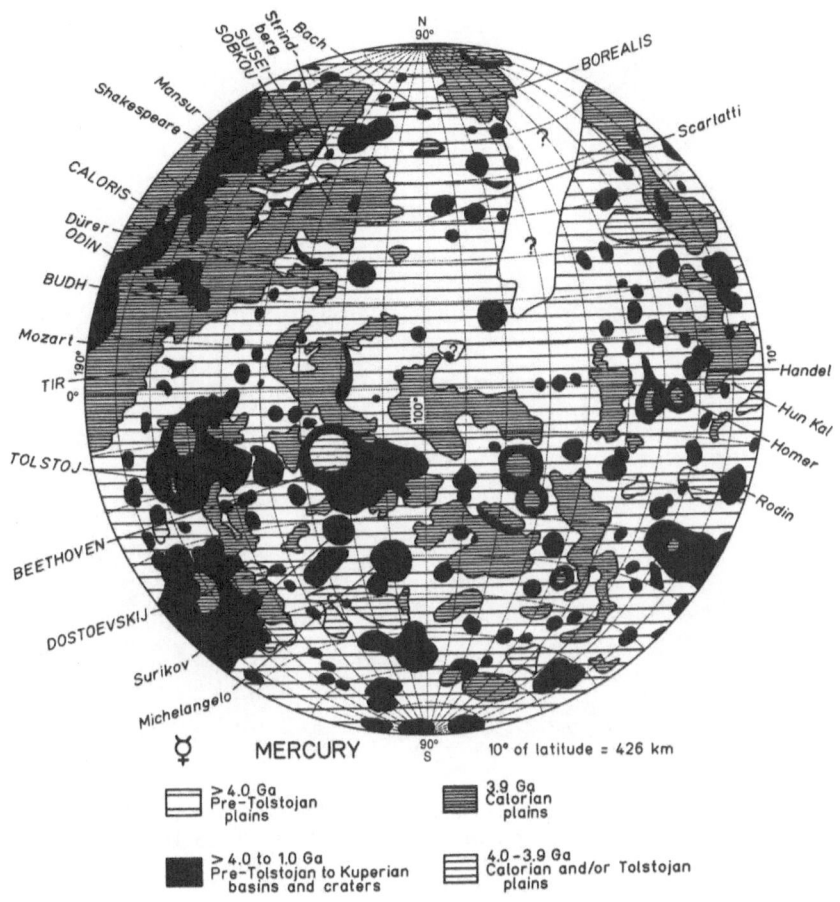

Fig. 58. Morphological provinces of Mercury-all that has been imaged to 1991. Long. 10°W is on the *right*, Long. 100°W is in the *middle*, Long. 190°W is on the *left*. Reckoning of longitude is from a 1.4-km diameter crater named Hun Kal (20 in Maya language) at Lat. 0°00′, Long. 20°00′W (Murray et al. 1974, figs. 1, 4). Names are those of craters and associated plains referred to in the text; the names in *capital letters* refer to the largest plains. (US Geological Survey 1979; Moore 1990, p. 50)

reach 427°C (700 K; Table 2). Possibly, the ice has remained protected from sublimation for millennia because of its subsurface position.

Although heavily cratered, the highlands of Mercury are not as heavily cratered as the lunar ones (Fig. 59; Spudis and Guest 1988). This is due to the wide distribution of intercrater plains that are present but not as extensive on the Moon. Overlapping large craters and basins like those on lunar highlands also are absent on Mercury. These observations led Spudis and Guest to conclude that Mercury underwent some type of resurfacing that was more intense than on the Moon. Mercurian craters display some morphological features comparable with those on the Moon: the smaller structures are bowl shaped, with increasing size they exhibit

Fig. 59. Heavily cratered terrain and intercrater plains of Mercury. Note the ridge (compressional feature) that meanders across the cratered terrain. Photograph FDS 27325 (Davies et al. 1978, p. 38)

rims, central peaks, and terraces on the inner wall, the ejecta blankets display a hummocky radial facies and swarms of secondary impact craters, and the fresher craters have dark or bright haloes and well-developed ray systems. The craters differ from those on the Moon in that ejecta and secondary craters are less extensive for a given rim diameter than are comparable structures on the Moon – a difference due to the 2.5 times higher gravity of Mercury than of the Moon (Table 1; Gault et al. 1975). From a study of widths of terraces on the inner walls of craters coupled with a gravity-driven slump model, Leith and McKinnon (1991) also concluded that differences in crater morphology between Mercury and the Moon must be caused by differences in surface gravity. As a result of much tighter clustering of secondary craters, degradation also is much greater on Mercury than on the Moon. The general similarity of craters, however, indicates that those on both bodies may be presumed to have the same origin and general age span. Neither Moon nor Mercury has an atmosphere, so erosion is limited to impacts of missiles and to mass movements, and thus exogenic morphology has a negligible percentage area on both bodies (Table 4). Strom and Neukum (1988) stated that the objects responsible for the late heavy bombardment on Mercury and the other rocky planets were accretional remnants, whereas those on the gaseous outer solar planets were produced by objects in planetocentric orbits.

In a study of 447 mercurian craters, Pike (1988) found that the morphology of the craters is more complex with their increasing size, and that depth, rim height and width, peak and floor diameter, rim wall complexity, frequency and spacing of concentric rings, presence or absence of a bowl-shaped interior, central peak, scalloped rim crest, slump deposits, and rim wall terraces are strongly size dependent. The onset of crater formation results from inertial forces (central-peak recoil) followed by gravity (wall failure). Substrate contrasts appear to have a minor influence on crater morphology and affect only the smaller structures. The impact itself is the primary control for ring spacing and perhaps ring emplacement. Schultz (1988) believed also that crater morphology denotes signatures of both the size and velocity of the impact bodies and that contrasting crustal properties have only a secondary function. Slopes of complex craters on Mercury are much steeper than on the Moon and Earth, an observation that remains unexplained. Pike (1988) recognized seven classes of craters that resemble lunar classes, but the size ranges are more like those on Mars. These classes in order of increasing diameter and topographic complexity are: simple, modified-simple, immature-complex, mature complex, protobasins, two-ring, and multi-ring basins. According to Strom and Neukum (1988), although the mercurian highlands (like those on Mars and the Moon) are heavily cratered and show similar crater distribution, they have few craters with diameters less than 50 km. This paucity is ascribed to obliteration of craters by growth of intercrater plains. These plains range in age from 4.2 to 4.0 Ga and were emplaced near the end of the late heavy bombardment, much earlier than on the Moon, meaning that the bombardment ended sooner on Mercury. Solomon (1978), Strom (1984), and Strom and Neukum (1988) attributed this difference from the Moon to the formation of Mercury's larger core, which produced extensive melting, global expansion, and crustal tension during the period of heavy bombardment. Subsequent cooling of the lithosphere led to global contraction that shut off the magma sources.

Impact structures on Mercury, like those on the Moon, have been dated on the basis of their degree of degradation and of such components as rays, secondary craters, ejecta facies, rim sharpness, inner terraces, and central peaks. Five classes of craters have been recognized with C_5 being the youngest and C_1 the oldest (Murray et al. 1975; Spudis and Guest 1988). Although such criteria have proven useful for mapping Mercury, some difficulties have been encountered in application – difficulties that can be resolved only when Mercury is imaged at different lighting angles (Spudis and Guest 1988). The structures on Mercury can be classified into two groups: those in regions of high crater density that underwent early resurfacing by intercrater plains, and those craters that postdate the intercrater plains. Among the younger basins are several large impact structures. Dostoevskij is the most degraded and oldest, probably one of the oldest on the planet; Tolstoj is a multi-ring depression whose interior is covered by a smooth plain and its lineated ejecta extend as much as 500 km from the basin but are absent from the northern and western sectors; Beethoven Basin, with possibly only one ring and ejecta that

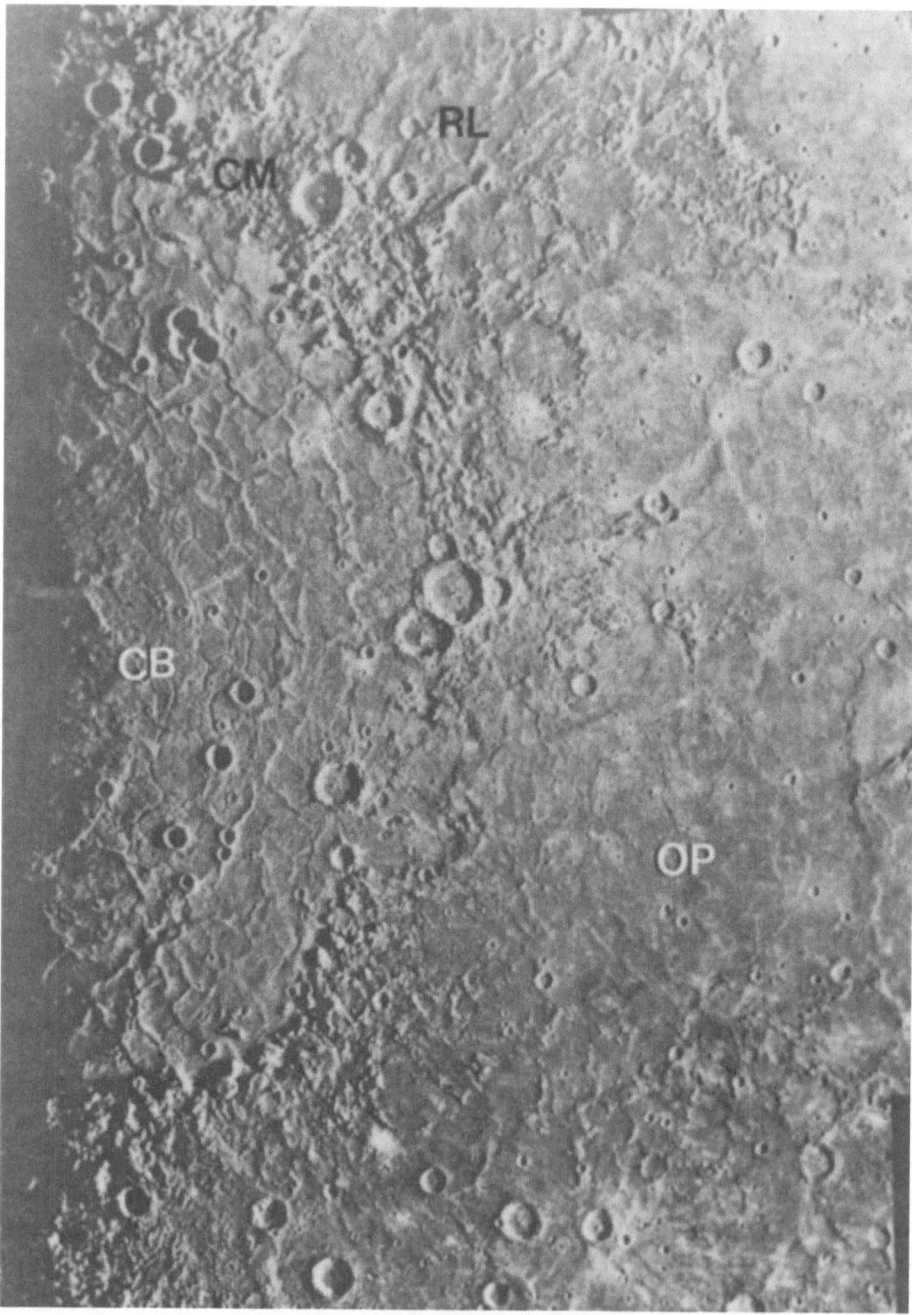

Fig. 60. Caloris Basin (**CB**) at Lat. 35°N, Long 190°W on Mercury, showing ridges and grooved smooth plains of the basin, Caloris Montes (**CM**), radial lineated facies (**RL**) of the Van Eydt Formation beyond the montes, and the intercrater Odin Planitia (**OP**) southeast of Caloris Basin. Mosaic 3-F31 (Davies et al. 1978, p. 56)

extend as much as 500 km from the crater, may postdate Tolstoj and may even be younger than Caloris Basin, and Caloris Basin only half of which was imaged by Mariner 10 (Fig. 60). The floor of Caloris is covered by extensive plains that are disturbed by sinuous ridges and fractures (Schaber and McCauley 1980; Guest and Greeley 1983). The materials of these plains may be volcanic in origin and extruded by the release of magma as a consequence of the impact or they may represent impact melts (Spudis and Guest 1988). Caloris Basin (Fig. 60) is surrounded by a ring of mountains having a diameter of about 1300 km and with a relief of 1 or 2 km with the main ring around the basin containing a structural block 100 to 159 km wide, 30 to 50 km long, and a relief of 1 or 2 km (Trask and Guest 1975). Associated with it are peripheral secondary craters, hummocky plains, and lineaments that extend several hundred kilometers beyond the outermost of the concentric rings of Caloris Basin. The presence of chains of small craters, grooves, and flat-floored open-ended interrupted valleys like those on the Moon supports their origin through erosion by low-trajectory ejecta from craters. Secondary

Fig. 61. Irregular (weird) terrain at Lat. 27 S, Long. 24 W on Mercury antipodal to Caloris Basin. This deformation is believed to have been produced by surface and compressive waves caused by an asteroid impact on the Caloris site that focused in the region antipodal to the basin. Note that the fill in the 150-km diameter Petrach Basin is undeformed, indicating that filling of this basin must have occurred after formation of the weird terrain. Photograph FDS 27370 (Davies et al. 1978, p. 103)

cratering and associated farther-reaching ejecta on Mercury represent more a mixing with local material than a deposition of a new discrete layer.

The intermontane plains that separate the blocks of the Caloris Basin rings are quite rugged and consist probably of a mixture of fallback material and impact melt. Two ejecta units have been identified beyond the Caloris Mountains: the Caloris lineated terrain extending 1200 km from the foot of the mountains and which consists of low hilly ridges and grooves subradial to the Caloris Basin and representing ejecta from the secondary craters of the Caloris Basin, and hummocky plains 800 km from the mountains consisting of hills 0.3 to 1 km across and from tens to a few hundred meters high.

An unusual exotic feature present on Mercury is a swirl-like structure near crater Handel (Fig. 58) that may have been produced by impact of a cometary coma and scouring of the surface by hot gases and high-energy plasma (Schultz and Srnka 1980). The hills and depressions that disrupt earlier terrains and form a slightly elongate area antipodal to Caloris Basin may have resulted from focused seismic waves initiated by the impact that formed Caloris Basin (Fig. 61; Schultz and Gault 1975). Similar terrains also have been identified antipodal to the Imbrium and Orientale basins on the Moon.

6.3.2 Endogenic

Endogenic provinces on Mercury consist of volcanic plains and features of tectonic origin. Topographically, two types of plains can be recognized on the planet. One type is the intercrater plains on the highlands, a level to gently-rolling terrain between and around the cratered terrain on which are superimposed many craters having diameters less than 10 km (Spudis and Guest 1988; Strom et al. 1990b). These secondary impact structures are elongate, shallow, and may be open at one side; crater clusters and chains also are common. Some craters on the heavily cratered terrain are embayed or covered by intercrater plains (Guest and O'Donnell 1977; Strom 1977), thus some of the intercrater plains are older than the craters and others are younger. Leake (1982) reported that the plains were deposited during a period represented by C-5 (C-5 of the US Geological Survey is the youngest and C-1 is the oldest, the opposite of the nomenclature of Murray et al. 1975; Spudis and Guest 1988 wherein C-1 is the youngest and C-5 is the oldest) through C-3 craters, that the areas covered by the plains decrease with decreasing age, and that secondary craters on the intercrater plains were derived from C-1 to C-3 craters and C-4 basins of the heavily cratered terrain.

Some investigators believed that the plains were formed by ejecta material (Malin and Dzurisin 1977) and pointed out that the lineated terrain that modified cratered terrain and intercrater plains (consisting of areas cut by straight and parallel valleys as much as 10 km wide with scalloped margins) resemble the lunar Imbrium morphology that was formed by secondary impacts. Plots of size frequency of craters are more closely akin to those of impact than for volcanoes, according to Oberbeck et al. (1977), and the crater rims that are eroded by near-horizontal paths of ejecta from primary impact sites are nearest the impact basins. These craters also contain the flattest and thickest smooth plains. The thickest supposed ejecta (estimated to be more than 2 km) have surfaces that contain far fewer irregular, clustered, and chained craters (formed by secondary impacts) than do areas having ejecta thinner than 2 km, implying deep burial of the many eroded surfaces on Mercury.

Other investigators pointed out that there is no source of ejecta to form the widespread intercrater plains and ascribed the plains to volcanic origin (Spudis and Guest 1988, and references therein). These lava flows supposedly were extruded along fractures formed by expansion of Mercury when its core was formed. This global volcanic event supposedly destroyed most of the early impact craters (Murray et al. 1975). Because fracture pathways cannot be seen on the planet's surface, they must have been covered during emplacement of the plains. Subsequent impact cratering may have enhanced obliteration of the fractures (Strom 1984). Soon after formation of Caloris Basin and the younger plains global compression associated with cooling closed the magma conduits, thus inhibiting further surface volcanism.

A second type of plain is relatively flat, sparsely cratered, and covers wide regions of Mercury (Strom et al. 1975; Trask and Guest 1975). In contrast to the maria on the Moon, these plains have albedo characteristics similar to those of intercrater plains. Many are named for composers, writers, and artists (Fig. 58), but other names are more descriptive: Caloris for hot in Latin, Sobkou for desert in Japanese, and Suisei for the planet Mercury in Japanese (Figs. 62 and 63). They are flat to gently undulating, display numerous lunar-mare-type wrinkle ridges, and tend to fill depressions that range in size from regional troughs to crater floors (Spudis and Guest 1988). These plains cover about 27×10^6 km^2, or almost 40% of the area imaged by Mariner 10. They are not confined to the distal margins of Caloris Basin, and their wide extent led Spudis and Guest to suggest that the plains are volcanic in origin. The absence of volcanic features associated with the plains may result from the limited Mariner 10 coverage, and the lack of contrast between the plains and the cratered terrain may reflect a lower FeO and TiO$_2$ content of the mercurian plains relative to lunar maria.

Although Mercury is not now tectonically active, it was during the past. Melosh and McKinnon (1988) ascribed this early tectonism to small changes in the shape of the lithosphere by tidal despinning and shrinkage during cooling. They believed that the planet's small size coupled with a deficiency of radioactive elements (which gave Mercury a low surface heat flow and a thick lithosphere) explains its subdued tectonic history. Mercury's tectonic history goes back to the time when the lithosphere became solid enough to retain structural features that have been preserved to the present. The most ancient of these features is Mercury's global tectonic grid expressed by lineaments or preferred orientation of much later features. This grid has a preferred orientation of northeast-southwest in the equatorial regions and it changes toward an east-west direction near the poles. These fracture directions were produced by despinning by solar tides early in the

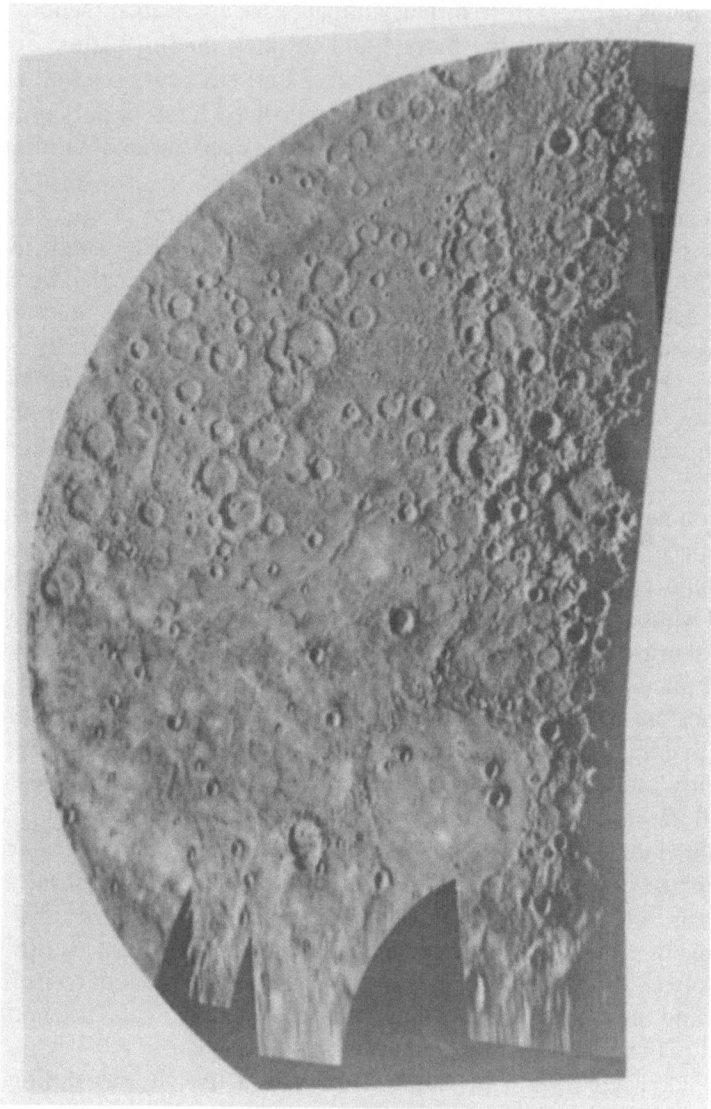

Fig. 62. Borealis Planitia in the north polar region of Mercury. *Left* The crater flooded by a smooth plain is Goethe Basin with a diameter of 340 km. Mosaic 8E. (Davies et al. 1978, p. 27)

history of the planet from 20 to 59 days and by subsequent relaxation of the equatorial bulge (Melosh and Dzurisin 1978a; Melosh and McKinnon 1988; Thomas et al. 1988).

Disrupting the surface of Mercury are numerous lobate to arcuate to irregular scarps (Fig. 64). The lobate linear features range in length from 20 to 500 km, have reliefs of a few hundred meters to 1–2 km, and have rounded crests with the terrain

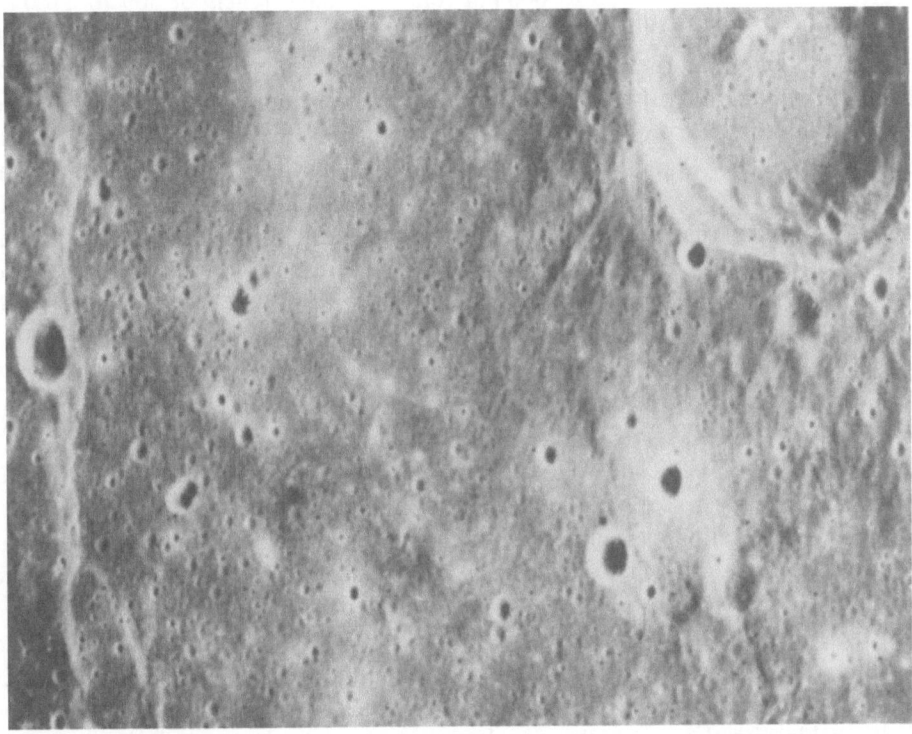

Fig. 63. Budh Planitia, a smooth plain east of Caloris Basin on Mercury. Note lobate scarp on the left. (Davies et al. 1978, p. 93)

Fig. 64. A segment of Discovery Rupe on Mercury, a 500-km long and about 2-km high scarp, cutting through two previous impact craters. This linear feature is believed to be a thrust or reverse fault indicative of a mercurian compressional phase. (Davies et al. 1978, p. 107)

behind the scarps varying from flat to sloping away from the scarp. The arcuate
scarps are smooth arcs 100 to 600 km long and have reliefs of 500 to 1100 m.
Irregular scarps are confined to the interior of craters and are less than 100 km long
and 400 m high (Dzurisin 1978; Harmon et al. 1986). These various features have
been interpreted as thrust faults (Strom et al. 1975; Solomon 1977) formed during
the interval from late pre-Tolstojan until the end of middle to late Calorian (Spudis
and Guest 1988). Some investigators believed that they were caused by tidal
slowing of Mercury's rotation (Melosh 1977; Melosh and Dzurisin 1978a), whereas
others proposed that they were produced by cooling of the mantle and partial
solidification of the core, leading to a 1- to 2-km decrease in Mercury's radius. Also
present on the intercrater plains are several linear ridges 50 to 350 km long, 100 to
1000 m high, asymmetrical in cross section, rounded and convex upward (Dzurisin
1978). They are aligned with the mercurian tectonic global grid and could be
dike-like volcanic extrusions or compressional features. Other tectonic features
on the planet, such as ring structures in the centers of basins, are impact related.
Thomas et al. (1988), however, ascribed the features in the Caloris Basin to
interaction between core cooling and basin formation. They also related an
elongate bulge with outer arc extensional features near Tolstoj Basin to local
tectonics, and an area of subsidence near the Phidias Crater to a volcano-tectonic
origin.

Except for the graben structures on the floor of Caloris Basin and in the
position antipodal to the basin (both impact induced), the surface of the planet
imaged by Mariner 10 is free of tensional structures. No tensional structures
comparable with those near the peripheries of Moon basins or with the system of
radial grabens on Mars have been documented on Mercury. Thus the tectonic
regime of Mercury appears to have been dominated by a compressional rather than
a tensional regime (Strom 1984).

In some respects the geohistory of Mercury reflects that of the Moon. Like the
Moon, Mercury may have undergone early global crustal melting of at least the
outer few hundred kilometers leading to concentration of low-density plagioclase
feldspar in the upper crust. A crust composed largely of anorthositic rocks appears
to be consistent with the limited amount of full-disk spectra (McCord and Clark
1979). These data are in direct contradiction to the supposition of other workers
who stated that Mercury's high density is the result of removal of the crust and
upper mantle by some extramercurian process. The earliest surface of Mercury was
characterized by multi-ring basins, a morphology that resembles the lunar highlands (pre-Tolstojan, pre-4.0 Ga). This terrain essentially was destroyed by emplacement of intercrater plains of volcanic origin – a volcanic event that occurred
prior to emplacement of maria on the Moon. That volcanic surface was brecciated
by a subsequent cratering event (Tolstojan period, 4.0–3.9 Ga). The most prominent and widespread stratigraphic datum on Mercury is the Caloris period (3.9 Ga),
a datum that clearly divided the stratigraphic succession into two parts. This
cratering event was followed by another massive effusion of lavas to form
the mercurian smooth plains. Subsequent declining rates of impact during the

Mansurian (3.5–3.0 Ga) and Kuiperian (1.0 Ga), which continue to the present, have had minimal effect on the surface of Mercury.

6.4 Chronology

Spudis and Guest (1988) and others (Murray et al. 1975; King and Scott 1990; Strom et al. 1975, 1990a) described a chronostratigraphic classification of the morphologic features on Mercury (Table 7), a system similar to that developed on the Moon (Table 6). Examples of the Pre-Tolstojan system, the oldest one and formed before creation of the Tolstoj cratered basin that consists of remnants of ancient multi-ring basins, are randomly distributed and largely were erased by later intercrater plains. They are counterparts of the lunar pre-Nectorian system and may be older than 4 Ga. The plains that resurfaced the planet after the pre-Tolstojan system may indicate a widespread volcanic episode that destroyed all craters smaller than 300 to 500 km in diameter. It has an area of about $17 \times 10^6 \text{ km}^2$ (Fig. 58). The pre-Tolstojan Dostoevskij, Shakespeare, Homer, and Surikov craters (Fig. 58) were emplaced after this volcanic event. The Tolstojan system is represented by the Tolstoj Basin, its ejecta (the Goya Formation) extending as far as 500 km from the crater, numerous other large craters and basins including Scarlatti, Durer, double-ring structures, smooth plains emplaced during the late Tolstojan and which are scarred by more craters than the later plains, and the Beethoven crater (Fig. 58) formed at the end of the Tolstojan Period. This system is believed to be 4.0–3.9 Ga old and equivalent to the Nectorian system in the Moon.

The Calorian system, which is a prominent geological marker, consists of the Caloris Basin, and its ejecta Caloris Montes (mountain material), Nervo (intermontane plains), Odin (hummocky plains), and Van Eyck (lineated plains) formations. The center of Caloris Basin is covered by a plain that may be volcanic in

Table 7. Chronostratigraphy of Mercury. (Spudis and Guest 1988)

System	Units	Approximate age of base of system, Ga	Lunar counterparts
Kuiperian	Crater materials	1.0	Copernican
Mansurian	Crater materials	3.0–3.5	Eratosthenian
Calorian	Caloris Group, cratered plains, small-basin materials	3.9	Imbrian
Tolstojan	Goya Formation; crater, small-basin, plains materials	3.9–4.0	Nectarian
pre-Tolstojan	Intercrater plains, multi-ring basin, crater materials	pre-4.0	pre-Nectorian

origin and which was disrupted by postbasin tectonism. Numerous two-ring craters, including Strindberg, Rodin, Michelangelo, and Bach (Fig. 58), also were formed at this time. Many of these craters are partly or completely covered by plains materials. Included with the Calorian system are extensive smooth plains of volcanic origin having an area of 9×10^6 km^2 plus part of the 1×10^6 km^2 of the undifferentiated area composed of Calorian or Tolstojan plains (Fig. 58). The Calorian system may be 3.9 Ga old and it is equivalent to the lunar Imbrian system. Global focusing of seismic waves associated with the Caloris craters may be responsible for the hilly and fractured terrain near the antipodal position of the Caloris Basin (Schultz and Gault 1975; Hughes et al. 1977). The post-smooth-plain systems consist of the Mansurian and the Kuiperian systems. The 3.5- to 3.0-Ga old Mansurian system (named after the crater Mansur and equivalent to the lunar Eratosthenian system) contains relatively fresh craters and some minor plains material confined to the crater floors. These materials probably are impact melt and fallback ejecta. The 1.0-Ga-old Kuiperian system is coeval with the lunar Copernican system and, typified by the crater Kuiper, it includes all the fresh-rayed craters and their associated ejecta.

7 Venus

7.1 Early History and Structure

7.1.1 Planetary Setting

Venus, the second planet from the Sun, is the brightest object in the Solar System after the Sun and Moon. Because of its prominence in the evening sky after sunset and its brightness at dawn Venus also is known as the Evening Star and the Morning Star. Venus (Latin) also was known as Aphrodite (Greek), the Goddess of Love. The planet was known as Astarte to the Phoenicians, as Ishtar to the Babylonians, and as Tai-pe (the Beautiful One) to the Chinese. Observations of the planet go back as far as 1900 B.C. when they were recorded by the Babylonians on the cuniform Venus Tablets (Saunders and Carr 1984). With the aid of a telescope Galileo determined in A.D. 1610 that Venus had lunar-like phases, lending support to the Copernican Sun-centered concept of the Solar System (Greeley 1985, p. 132). Some early observers using a telescope reported that Venus was featureless, but others later noted that the planet displayed bright patches. These patches were used to determine the length of a venusian day. In 1897 Lowell even published a map of canals that he believed he saw on Venus (Saunders and Carr 1984). Another feature noted by Father Johannes Riccoli during the seventeenth century was the Ashen Light. This faint phosphorescence of the night hemisphere when Venus appears as a thin crescent is believed to be due to extensive twilight or to electrical phenomena. Early observers were particularly interested in timing the transit of Venus across the Sun; the most famous attempt probably is the one made by Captain James Cook of HMS Endeavour, and he was given the responsibility for making these measurements by the Royal Society. The observations were made on 3 June 1769 from Tahiti; Bligh, future captain of HMS Bounty, was a member of Cook's crew. Speculations regarding a possible atmosphere on Venus go as far back as 1796 (Schroter in Saunders and Carr 1984). During the midnineteenth century astronomers noted that Venus displayed a dark halo when silhouetted against the Sun, an observation that led Lomonosov to infer that the planet had an atmosphere. Visual and photographic spectroscopic data obtained during the third decade of the twentieth century led to the conclusion that the principal component of the venusian atmosphere is carbon dioxide. Infrared measurements by Kuiper (1962) showed nearly 40 carbon dioxide absorption bands, and 300 times the concentration of the gas in the venusian atmosphere than in the Earth's atmosphere; thus, Venus is in a stage of experiencing an intense greenhouse effect.

Recent investigations of Venus are based on radar observations from Earth and spacecraft missions to the planet. Radar images have yielded information on surface roughness at Freyja and Akna montes north of Lakshmi Planum and Maxwell, Theia, and Rhea montes (Figs. 65–68). Other features imaged by radar include large, smooth, roughly circular features having diameters between 200 and 1300 km, bright regions with diameters less than 300 km, and banded terrains of possible tectonic origin (Head and Campbell 1982; Campbell et al. 1989). Spacecraft sources of Venus data include: Mariner 2 flyby in 1962; Venera 4 with a hard-lander probe on the night side and Mariner 5 flyby in 1967; Venera 5 and Venera 6 with a hard-lander probe on the night side in 1969; Venera 7 with a hard-lander probe on the night side in 1970; Venera 8 with a burn-up 'soft'-lander probe on the day side in 1972; Mariner 10 flyby in 1974; Venera 9 orbiter and 'soft'-lander probe on the day side in 1975; Venera 10 orbiter with 'soft'-lander probe on the day side in 1975; Pioneer Venus 1 and Venus 2 with four hard-lander probes on day and night

Fig. 65. Topography of Venus from Longs. 0° to 180°E. The planetary datum level is 6051.0 km. Contours are at 1-km intervals, and *diagonally lined areas* are below the planetary datum. Because Venus has a retrograde motion (the Sun rises in the west and sets in the east; opposite to the direction on Earth), longitude increases to the east according to the International Astronomical Union (IAU) convention; zero longitude at the *left* on the map is compatible with current IAU coordinates. (After Masursky et al. 1980, fig. 2; Anonymous 1989)

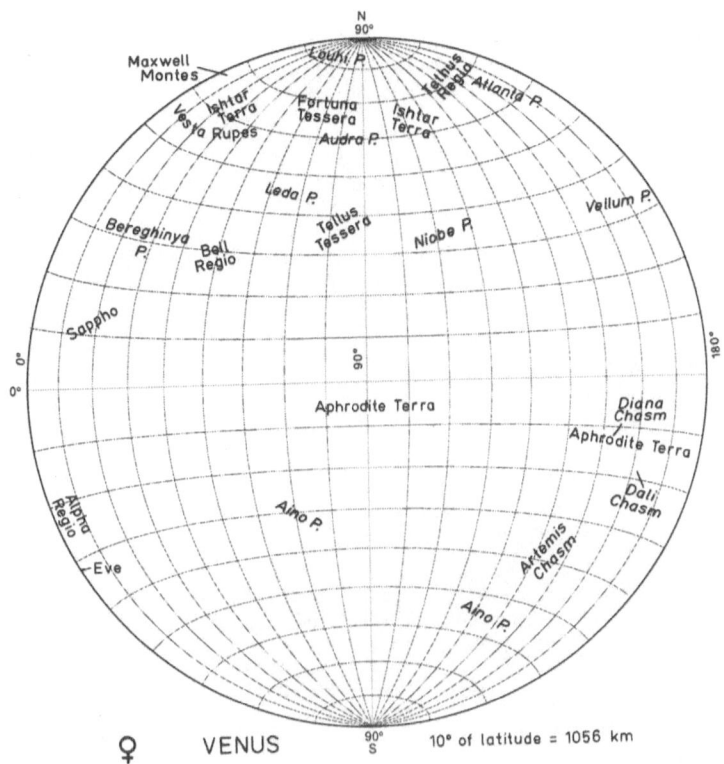

Fig. 66. Names of larger morphological features of Venus. See also Fig. 65

sides in 1978; Venera 11 with 'soft'-lander probe on the day side in 1978; Venera 12 with 'soft'-lander probe on the day side in 1978; Venera 13 and Venera 14 with 'soft'-lander probes on the day side in 1982; and Venera 15 and Venera 16 orbiters with radar imaging in 1983. On 4 May 1989 the Magellan spacecraft was launched from a space shuttle, entering orbit around Venus on 10 August 1990. Its mission was to produce global high-resolution image, altimetry, and gravity data for Venus–a mission that still continues (Saunders and Pettengill 1991). All topographic features on Venus have been named after female mythological characters or famous modern women scientists or artists.

Venus has an orbital period of 224 days, 16 h, 48 min, an eccentricity of 0.006, a retrograde rotation of 242.6 ± 0.9 days, a rotation axis inclined at $177°$ to the plane of the ecliptic, a diameter of 12 104 km, and a density of 5.27 g/cm^3 (Table 1). It possesses the highest albedo of any planet in the Solar System, absorbing much less solar energy than the Earth (150 versus 240 W/m^2; Prinn and Fegley 1987). Like Earth and Mars, Venus has an atmosphere with clouds that circulate in response to solar energy, a motion modulated by surface friction and planetary spin. This atmosphere consists mainly of carbon dioxide and nitrogen, with a 100-fold excess of nonradiogenic argon and a deficit of radiogenic ^{40}Ar relative to the Earth's

$\math女$ VENUS $\begin{smallmatrix}90°\\S\end{smallmatrix}$ 10° of latitude = 1056 km

Fig. 67. Topography of Venus from Long. 180° to 360°E. *Symbols* as for Fig. 65

atmosphere (Hoffman et al. 1980). The radiogenic argon probably came from degassing of the upper 6 km of an enriched crust. Kaula (1990a) stated that the low concentration of radiogenic argon (one-third as much as on Earth) implies that there has been more difficulty for volatiles to rise to the surface of Venus than to the surface of Earth since early in the history of the two planets; ^{40}K has a half-life of 1.47 Ga, thus half of the argon would have been generated quite early in the development of the planets. Turcotte and Schubert (1988) speculated that the low concentrations of ^{40}Ar can be attributed to the absence of plate tectonics, the lack of which led to less efficient degassing of Venus than of Earth. Other detected constituents include helium, neon, and possibly hydrogen sulfide; oxygen may be less than 30 ppm, and water vapor may be as high as 1000 ppm. The ratio of sulfur dioxide and water evaporated from cloud droplets is equivalent to that expected from H_2SO_4 at a concentration of 85%, and the rate of loss of helium from Venus' atmosphere is nearly equal to the flux of helium through Earth's atmosphere. The high concentration of rare gases has been interpreted as indicating that Venus has been degassed less than the Earth or that these gases are related to the nature of planetary accretion. Earth was hit by a very large body, possibly Mars-size, whereas Venus was impacted by only smaller planetesimals (Kaula 1990a, b). The

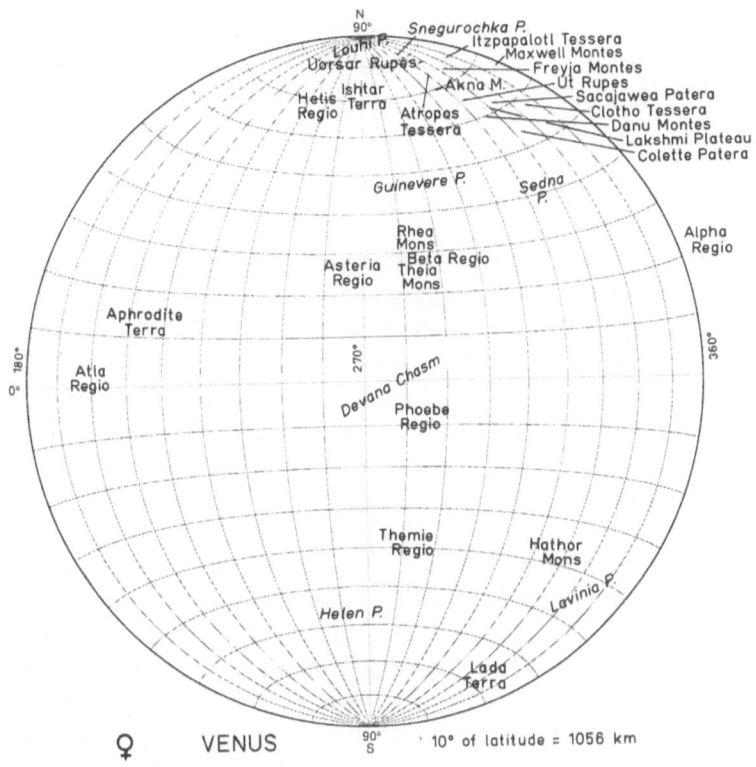

Fig. 68. Names of larger morphological features of Venus. See also Fig. 67

shock wave created by the impact of this large body on Earth raised the temperature by a few thousand kelvins, with solar wind driving off the volatiles.

Surface temperatures as high as 730 K and a maximum pressure of 90 bar have been recorded on the planet. Such high surface temperatures suggest that the crust of Venus may be less resistant to creep than the crust of the Earth (McGill et al. 1983). The high venusian surface temperatures can be ascribed to the greenhouse effect of its dense atmosphere, the prime sources of infrared opacity being CO_2, H_2O, cloud particles, and SO_2, with CO and HCl having very minor roles (Pollack et al. 1980). The dominant motion of the lower atmosphere is within a few degrees of west, but in the radially heated cloud layer between altitudes of 50 and 55 km there is an equatorial flow of 1 to 7 m/s. Ambient wind velocities near the surface range from 100 m/s at 65 km to 1 m/s or less (Counselman et al. 1980). Wind velocities between 0.4 and 1.3 m/s were measured at Venera sites 9 and 10 east of Beta Regio (McGill et al. 1983), and aerodynamic noise recorded by Veneras 13 and 14 northwest of Alpha Regio could have been generated by wind having a velocity of 0.3 to 0.6 m/s (Moroz 1983). McGill et al. stated that the in situ measurements at Venera sites 9 and 10 indicate that winds are capable of transporting sediment particles ranging from a few micrometers to several millimeters in diameter. Comparison of a few sequential images at Venera sites 13 and 14 appear to indicate some eolian sediment transport (Moroz 1983); however, bulk densities measured by Russian landers show that these eolian deposits are limited in extent.

7.1.2 Hydrosphere

Although Venus does not have a hydrosphere at present, it may have had one during the past. This hydrosphere could have been acquired by internal degassing or by cometary input (Grinspoon and Lewis 1988). The high deuterium/hydrogen ratio of Venus' atmosphere led Donahue et al. (1982) to state that water degassed from Venus was only 0.3% of that from Earth, although possibly more may have been extruded. Kasting and Pollack (1983) placed the volume of any former hydrosphere at less than 1% of the Earth's, with dehydration of the crust and upper mantle being completed by about 1 Ga. Early surface temperatures that were at least 353 to 373 K (80 to 100 °C), allowed the water vapor to become intermixed with the entire atmosphere rather than with just the lower atmosphere as on Earth, and solar radiation in the ultraviolet end of the spectrum removed the hydrogen from the vapor in the upper atmosphere (Kasting et al. 1984). It is estimated that a volume equivalent to Earth's oceans could have been removed from Venus in only 600 Ma. Watson et al. (1984) suggested a 'runaway greenhouse' as the process that removed the hydrosphere, with the hydrogen escaping into space and the residual oxygen reacting with surface lavas. In time, recycling of the crust returned this oxygen to the mantle. Other investigators questioned the validity of a primeval-ocean concept. They pointed out that if such an ocean was hydrodynamically lost why did not such a process affect the inert gases and why was cometary flux so much greater during the past? Hence, there is the alternate scenario that the water still remains in the interior, because Venus did not undergo the impact of a large planetesimal as Earth did (Kaula 1990a, b). Because of the absence of a hydro-sphere, the surface of Venus could have been paved thickly with crustal material, and if this crust could have maintained its identity it may have prevented mantle convection from reaching the surface – therefore no plate tectonics (Kaula 1990a, b). As a result of a Venus mantle being enriched in water, crustal-forming melts and volatiles would be inhibited from rising to the surface, leading to a dry stiff upper mantle and a thin stable crust (Williams and Pan 1990). It is the absence of a hydrosphere (the main difference between Earth and Venus) that tends to make the venusian upper mantle more viscous so that tectonism of the planet would be driven by the bulk mantle rather than by an intermediate boundary layer (Kaula 1990b). If crustal accretion on Venus has occurred at rates comparable with those on Earth (about 19 km^3/yr; Table 5) and there was no recycling by plate tectonics, the crust on Venus must be thicker than 100 km. High pressures at the base of such a crust, however, would change the basalt (or gabbro) to eclogite, a rock type denser than the ferromagnesian rocks in the mantle. This denser layer would sink into the mantle; by this process of 'delamination' the crust on Venus could be recycled (Kaula 1990a; Turcotte 1991). Such a process would tend to limit crustal thickness to less than 100 km.

7.1.3 Core, Mantle, and Lithosphere

Measurements of radioactivity made by Venera 8, which landed on a rolling plain east of Phoebe Regio that has steep-sided domes and a possible collapsed caldera (Figs. 65–68), appear to indicate granitic rock, whereas those at the landing site of Veneras 9 and 10, east of Beta Regio, are tholeiitic and alkaline basalts (Surkov 1983; Nikolaveya 1990). More recently, Basilevsky et al. (1992) proposed that the rocks at Venera 8 site may be comparable to lamprophyes. Terrestrial analogs of the venusian rocks at this site display an enrichment of their magma in incompatible elements, indicating contamination by crustal material, thus suggesting that the rocks on the Venera 8 site also may be contaminated by crustal material. Rocks at Venera sites 13 and 14 northwest of Alpha Regio, are characteristic of leucitic basalt (Venera 13) and tholeiitic basalt (Venera 14; Moroz 1983). The panorama imaged at Venera site 9 was the side of a topographic high with the surface consisting of slab-like clasts of 0.5 to 0.7 m diameter, whereas on the Venera 10 landing area the surface was much smoother and consisted of soil and sheet-like exposures of rock (Florenskiy et al. 1983). The panorama imaged by Veneras 13 and 14 is of a rocky terrain with a slabby or platy texture, soil, and some granular material (Moroz 1983).

After being corrected for the effects of compression Venus' bulk density of 5.27 g/cm^3 is about 2% less than if it had the same composition as Earth. Removing the effect of thermal expansion caused by its higher near-surface temperature (surface temperature of about 700°K and pressure of 100 bars) decreases this uncompressed density difference by only 1.7%. Thus density differences between Earth and Venus may be explained by the condensation theory, whereby Venus incorporated less sulfur than Earth and has a smaller ratio of sulfur to silicate. This theory also suggests that the core in Venus has less iron sulfide and formed later than the core in Earth (Hsui and Toksöz 1977), because under the equilibrium condensation model FeO would not condense at the position of Venus (Lewis and Kreimendahl 1980). Goettel et al. (1981) stated that the difference in density between Earth and Venus is caused by a different temperature structure and a deeper phase change of basalt to eclogite. The low content of FeO in Venus also could account for the lower uncorrected density of Venus, because the phase transitions involving olivine to spinel and perovskite would occur at greater depths than on Earth. Because iron-poor olivines have higher melting temperatures and higher viscosities at any given temperature than iron-rich olivines, interior temperatures on Venus may be higher, and temperature profiles may stabilize at higher temperatures in the iron-poor Venus mantle than in the iron-rich Earth mantle (Saunders and Carr 1984). This equilibrium condensation sequence and the lesser concentration of initial water than on Earth also led to a rise in the melting point of the mantle and to the formation of a lithosphere that may be twice as thick as that on Earth (Warner and Morrison 1978). This postulated greater lithospheric thickness is reflected in Venus' surface morphology (Hartmann 1983, p. 279).

Solomon and Head (1991) proposed a much thinner crust. They believed that the crust under the venusian plains is typically only 10 to 30 km thick, but in the tessera terrain and mountain belts it may be several tens of kilometers thicker. Evidence for such a thin crust beneath the plains is based on measured depths of craters, which indicate little viscous relaxation of relief since crater formation, a fact that requires a layer having the strength of mantle at depths of 10 to 20 km, and the observation that a number of tectonic features on the planet display characteristic width or spacing scales of either 10 to 20 km or 100 to 300 km. Solomon and Head (1991) speculated that these two scales are controlled by the thickness of a strong upper crustal layer and the depth to and the thickness of a strong upper mantle layer. They stated that there is quantitative agreement between observed topographic dimensions and crustal thickness in plains underlain by crusts thinner than about 30 km. The somewhat thicker crust in the tessera terrain and mountain belts may be due to sinking into the mantle of a tertiary crust produced by basal melting or recycling of lower crust through conversion to a dense eclogite. An average crustal thickness of 10 to 20 km in the plains, the global relief, and simple isostatic considerations suggest that the crustal volume in Venus is 10^{10} km^3, a volume comparable with that on Earth (Solomon and Head 1991). From this calculation it can be concluded that if no, or very little, crust has been recycled, the average rate of production of crust on Venus is much slower than on Earth, or if production rates have been comparable some type of recycling must have occurred in Venus during the past. The rates of addition by volcanic flux and associated magmatic intrusion (on Earth the ratio by volume of intrusives to extrusives is as high as 5 to 10; Table 5) and by melt production, estimated from thermal and dynamic models, have yet to be determined.

Although Venus may have a liquid core of moderate size, it does not appear to have a magnetic field – an absence that may result either from the planet's low rotational velocity or because its core is stably stratified (Merrill and McElhinny 1983, p. 350). Venus' gravity amplitudes are similar to those on Earth, but they correlate better with the topography, at least in the 100-km size range of topographic features (Sjogren et al. 1980; Bowin 1983). Whereas most of the major geoid anomalies on Earth are caused by deep sources, major mass anomalies on Venus are mainly within the upper 200 km of the planet. According to Bowin (1985), the greater range of anomalies and the close correlation of gravity anomalies with topography on Venus can be explained by the thicker crust on Venus (70 to 80 km) than on Earth (5 to > 70 km). Williams and Gaddis (1991) stated that gravity models agree with a belief that most of the topographic support of the highland region is provided by a thick crust that isostatically compensates the low surface. Regions of high gravity-to-topography response (+ 100 mgal) over Beta Regio and eastern Aphrodite Terra may indicate a combination of crustal loading by recent volcanism and uplift by thermal expansion from heating at depth (Bowin 1983, 1985).

7.1.4 Tectonic Models

Three tectonic models have been proposed to explain the morphology of Venus: one where large horizontal displacements occur and where deformation is within narrow belts (plate tectonics); a second with large horizontal displacements and where deformation is accommodated over a wide area (plate tectonics and mantle plumes, i.e. Mid-Atlantic Ridge/Iceland); and a third having limited surface horizontal displacement where deformation is associated mainly with mantle plumes (Grimm and Solomon 1989).

When evaluating the validity of the plate-tectonic concept, one must remember that conditions on Venus are so different from those on Earth that a one-on-one comparison may not be possible (Brass and Harrison 1982; Solomon and Head 1982a). As pointed out by Phillips and Malin (1983), we tend to approach the study of the morphology of other planets with a terrestrial bias, assuming that plate boundaries are universally the same, which is a concept of dubious value; similar specific processes may not necessarily produce the same topographic forms. For example, as a result of the high temperature on Venus, spreading belts may have less relief than on Earth. Kiefer (1990) speculated that the high temperature would limit the maximum relief of a spreading ridge on Venus to about 1.5 km and the maximum geoid irregularity would be 8 m, both much less than those reported from Venus to date. Some workers have equated the topographic style of the venusian equatorial highlands to plate tectonics, pointing out that Aphrodite Terra displays topographic features found on terrestrial divergent margins of Earth. Such features include 2000- to 4000-km long and 100- to 200-km wide structural discontinuities trending at high angles to the topographic trend in western Aphrodite Terra (Figs. 65–68) that resemble oceanic fracture zones and strike-slip faults on Earth. Terrains between these linear features display bilateral symmetry indicative of having been created at rise crests and then moving laterally to their present positions (Crumpler et al. 1987; Head and Crumpler 1987). High-resolution images obtained by the Arecibo radar system in Puerto Rico also show parallel 10- to 15-km-wide bands of radar back scattering that parallel the ranges of Ishtar Terra, which resemble faults or folds on Earth (Campbell et al. 1983). Sotin et al. (1989) speculated that the 1000-km wide topographic grain of western Aphrodite Terra that is symmetrical about an axis and is less than 1000 km wide may be the result of spreading at half rates of 0.3–0.5 cm/yr during the past 200 Ma over a mantle whose temperature is 1870 K. Head et al. (1990) stated that the main topographic elements in western Ishtar Terra (Danu, Akna, Preyja, and Maxwell montes) suggest convergence, underthrusting, and possible subduction of lowland plains along the margins of a preexisting tessera plateau. Comparison of images recorded by GLORIA on terrain of active plate boundaries on Earth, using a sound source having a wavelength of 23 cm, with SAR images of Venus recorded by radar having a wavelength of 12.6 cm, led McKenzie et al. (1992b, 1992c) to propose that the topography in the Artemis Corona region of Venus is comparable with that on spreading ridges on Earth. They further stated that features comparable to trenches (belts of subduction) on Earth are widespread on Venus (Dali and Diana chasmata). Sandwell and Schubert (1992) also proposed that subduction may be occurring on the margins of large venusian coronae (Eithinoha, Heng-O, Artemis, and Latona) with compensating back-arc extension taking place in the expanding coronae interiors. Solomon (quoted by Kerr 1992) suspected that only the large coronae have a fabric comparable to plate tectonics on Earth; if so, this would imply that heat loss on Venus is not be a plate-tectonic mechanism.

There are numerous objections to such a plate-tectonic scenario for Venus. For example, if crustal spreading is occurring on Venus, why do not the hot-spot structures form chains as they do on Earth (Stofan and Saunders 1990)? Kaula and Phillips (1981) also stated that the morphology of Venus is not comparable with divergent accreted terrain, and Kiefer (1990) believed that the evidence for trans-form faults on the radar images made from Earth is not conclusive. Herrick et al. (1989) and Herrick and Grimm (1991) stated that the depth of compensation across Aphrodite Terra west of Long. 135°E is 70 km but east of Long. 135°E it is 230 km. Superimposed on this trend are four well-defined peaks and one region where compensation depths are poorly constrained. This trend in depth of compensation is difficult to reconcile with a terrestrial type of plate divergence. Phillips and Malin (1983) also believed that the major ridges on Venus may not be spreading ridges, because they decay much more rapidly with distance from their axial highs and do not form a global system; the higher buoyancy of the venusian lithosphere also would make subduction less effective on Venus. They proposed that if spreading is occurring it is confined to the median plains, and if the spreading were rather fast, these spreading ridges would not be detectable in the Pioneer altimetry data. Phillips and Malin also noted that the distinct topography associated with zones of convergence (trenches and adjacent ridges or mountain ranges) are missing from Venus. Although some features (Ut Rupes at Lat. 55°N, Long. 315°E, a ridge parallel to the southwestern margin of Ishtar Terra) do resemble a convergent province, they do not have an associated trench. On the other hand, Artemis Chasma (Lat. 49°S, Long. 135°E), which resembles the Java and Aleutian trenches on Earth, does not have an associated ridge. Anderson (1980, 1981) pointed out that the high temperature at the surface and in the upper mantle of Venus must tend to deepen the eclogite stability, leading to formation of a thick (possibly 100 to 170 km) basaltic crust. Such a thick crust would tend to prevent deep subduction as on Earth, and crustal recycling would not be possible. Anderson postulated that the tectonic style on Venus is similar to that on Earth during the Archean, when temperatures and pressure gradients were higher than now. Yet, Condie (1989, p. 342) suggested that during the Archean the Earth's oceanic lithosphere was recycled into the mantle at sinks and that the plate-tectonic regime was similar to modern lava lakes.

If Venus' surface morphology were due to plate tectonics, why is its surface topography not bimodal like on Earth? Could it be obscured by extrusion of massive volcanics? Kaula and Muradian (1982) questioned the concept that a morphology created by plate tectonics on Venus is not recognizable because of being concealed by a large volume of volcanics. They pointed out that if volcanic rocks were extruded in a large enough volume to hide the plate tectonic regime, there would be no need for plate tectonics, as volcanism would be more than sufficient for transfer of heat to the surface. Is it possible that bimodality has been destroyed by erosion of the highs and deposition on the lows? Venus lacks a hydrosphere, but depressions formed by plate divergence conceivably could have become completely filled by products of erosion from topographic highs by eolian processes (assuming that surface wind velocities are greater than those reported to date), thus removing two topographic levels such as those on Earth. On Earth, the sea level limits the depth of erosion that continents can undergo and limits the height of filling of ocean basins. Extensive sedimentation in oceanic basins on Earth can occur only during relative falls in sea level when erosion in coastal regions is rapid. Phillips and Malin (1983) claimed that if there were no ocean on Earth, rapid erosion of the continents and other highs could lead to a single-level topography having only a few meters of relief, but such a scenario is unrealistic because two levels of topography must always have an ocean on well-watered planets. In the absence of water, the eolian processes and the limited amount of weathering that are occurring on Venus are incapable of forming a single tier topography from a bimodal one. Such processes are so slow that plate tectonics that have a periodicity ranging from 210 to 350 Ma (at least on Earth) would create and destroy ocean basins on Venus before they could become filled. Possibly plate tectonics are no longer active on Venus, having ceased as a result of elimination of a hydrosphere and/or creation of an increased surface temperature caused by a severe greenhouse effect (Phillips and Malin 1983). If plate tectonics ceased as early as several hundred million years after accretion, about 4.5 Ga ago, there would have been enough time to fill the lows even by such a slow process as eolian sedimentation. Finally, if Venus has been dry for all or most of its history, sialic crust (which is responsible for the Earth's topographic bimodality) may be limited or nonexistent because abundant water may be necessary to form large volumes of sial by differentiation along subduction belts (Anderson 1980; 1981; McGill et al. 1983).

The second tectonic model also is characterized by large horizontal tectonic transport, but, in contrast with the plate divergent model where deformation is restricted to a narrow zone, in this model the lithosphere is easily deformed and the strain is broadly accommodated. Grimm and Solomon (1989) called this model the distributed deformation model. In it, divergence and mantle plume activity occur at the same locality to form a plateau or volcanic rise such as Iceland on the Mid-Atlantic Ridge. Later, these volcanic rises may have split, and the two halves became transported laterally to form the tessera terrain on the venusian plains (Head 1990a; Head and Crumpler 1990; Roberts and Head 1990a). An example of such a mantle-plume plateau is Aphrodite Terra. Grimm and Solomon indicated that the fit of Aphrodite Terra to a thermal boundary is poor and that its variations in topography probably are caused by dynamic uplift or variations in crustal thickness. Such topography may manifest mantle plume activity. Western Ishtar Terra also may be an example of compression and crustal thickening above a cylindrical mantle downwelling (Bindschadler et al. 1990). Grimm and Phillips (1990) proposed that Lakshmi Planum is the result of a mantle plume rising beneath a prior tessera terrain. The depth of compensation of this topographic high can be accounted for by a combination of deformation and subsurface melting of a preexisting tessera and subsurface melting along a convergent plate boundary and a mantle plume (Roberts and Head 1990a, 1990b).

The third model, the hot-spot or mantle plume one, was proposed by Phillips et al. (1981), who believed that lateral transport on Venus is limited. Under this model tectonics on Venus are dominated by plumes rising from the mantle and impinging on the lithosphere to produce hot spots. A study of convective stress coupled with an elastic lithosphere as it applies to tectonism on Venus, led Phillips (1990) to speculate that dynamic uplift of Venus' surface must be accompanied by extensive brittle failure of the top of the lithosphere and viscous flow of the lower part. Coupling of mantle flow to the base of the lithosphere could produce only limited lateral transport of the lithosphere. As on Earth, mantle plumes on Venus should be short lived geologically. They should be characterized by broad topographic highs and geoid rises, extensional tectonics, pressure-release melting, surface volcanism, and collapse and relaxation phenomena as the plume dies (Solomon and Head 1991). Phillips et al. estimated that crustal production under this model probably is only about 0.7 km^3/yr. In contrast, crustal generation along the Mid-Ocean Ridge on Earth is about 19 km^3/yr (Table 5), with interplate and island-arc volcanism contributing an additional 2 km^3/yr. Thickening of the crust by volcanic resurfacing should lead to delamination (basalt/eclogite phase) and recycling of the lower crust (Kaula 1990a, b; Turcotte 1991). This recycling, which returns heat-producing (from radioactivity) crust to the mantle, would lead in turn to mantle convection and ascending plumes that produce pressure melting and volcanism. Such resurfacing suggests that the age of the surface of Venus ranges from 800 to 0 Ma, with the maximum age reflecting the time needed to resurface Venus once (Schaber 1991). The tectonic model proposed by Nikishin (1990) is somewhat similar to this mantle plume model. He believed that Venus went through two stages of deformation wherein plate tectonics have no role, but where mantle plumes are numerous. During stage I (4-1 Ga ago) the soft ductile crust was deformed by rising mantle plumes, with crustal extension occurring above the rising plumes, leading to formation of broad plains and accompanied by crustal thickening and uplift where the plumes converged. When the lithosphere cooled and thickened it underwent a second phase of deformation (stage II) that formed 'weakened uplift belts' and mantle hot spots, belts of compressive ridges, and extrusion of basalts to blanket the plains. It is during this second stage that hot-spot structures, domes, coronae, supercoronae, and rifts developed.

The contention that cylindrical plumes are the predominant form of convection on Venus (Schubert et al. 1990) is supported by possible motions in the mantle based on numerical calculations of three-dimensional convection in constant spherical shells heated both internally and from below. The lower surface of each shell is assumed to be isothermal and stress free and its upper surface is assumed to be rigid and isothermal, the ratios of the inner and outer radii being compatible with the possible core size of Venus, and with Rayleigh numbers about 100 times the critical numbers for initiation of convection. Similar conditions exist on Mars. On Earth, the top boundary of the shell is free of shear stress, enhancing extensive convection systems to drive plates, but a stiffer lithosphere on Venus and Mars promotes a style of cylindrical convection. Mantles of planets having rigid upper boundaries should have higher temperatures than those with a stress-free upper boundary. Schubert et al. (1990) believed that this is why Venus not only has a higher surface temperature, but also may have a hotter mantle than Earth. Spacing of plumes is believed to be a function of the relative amounts of basal and internal heating. With appropriate parameters for the mantle and with 20% of the heat being provided from below, it is estimated that about 20 plumes are active at any one time (Schubert et al. 1990). Mantle plumes tend to be concentrated in regions where the crust is thinner or is fractured, permitting magma and heat transport across the lithosphere. Such a mechanism may explain the grouping of coronae on Venus.

These pre-Magellan tectonic speculations about Venus reflect two internal planetary conditions (Solomon and Head 1991). In one global scenario (the Earth-scale mantle scenario), where the average heat flux is about 70 mW/m^2, most of the heat is transported through the lithosphere by conduction (see also Solomon and Head 1990). This is in contrast to postulations by Morgan and Phillips (1983) and Phillips and Malin (1983) that this heat transfer originally may have been via plate tectonics. Later, with the loss of a hydrosphere and termination of plate tectonics, volcanism and plume activity took over the heat transfer. According to Solomon and Head (1991), the resulting high mantle temperatures would have caused a significant volume of melt to be produced atop mantle plumes and spreading belts, if they existed. Because of high thermal gradients the lower part of the crust should be weak, particularly at topographic highs where the crust may be thicker than 10 km – thus their suggestion that thin-skinned tectonics should be present in the higher regions. This scenario, however, is contradicted by the surface density of impact structures and the short- to intermediate-wavelength positive features that indicate a thick crust. In the second scenario (the cool mantle model) the mantle temperature and heat flux is less than on Earth, the fraction of heat flux from crustal radioactivity is larger, and magmatic activity is limited. Comparatively low thermal gradients in the lithosphere would make it easier for the lithosphere to support high surface relief. A low temperature and high viscosity in the mantle also would explain the strong correlation between gravity and topography and the large depths of compensation. According to Solomon and Head, the problem with this model is the need to find an explanation for a cool mantle. They proposed that a low temperature of the venusian mantle may be a result of a much smaller dimension of the largest planetesimal involved in the accretion process than of that involved in accretion of the Earth. If the colliding bodies on Venus were significantly smaller than those on Earth, Venus could have had a lower internal temperature. Such a mechanism was proposed by Kaula (1990a, b) to explain the greater abundance of nonradiogenic gases on Venus as well as the absence of a satellite comparable to the Moon.

7.2 Hypsometry

Radar-altimetry data acquired by Pioneer Venus between Lat. 74°N and 63°S and by Veneras 15 and 16 north of Lat. 30°N (Masursky et al. 1980; Anonymous 1989) were compiled into a topographic map with a 1-km contour interval with the planetary datum level at 6051.84-km radii (Figs. 65–68). Elevations on the map are accurate within about 200 m and the spatial resolution is about 30 km, according to Masursky et al. Most of the surface displayed in Figs. 65 and 67 lies at altitudes near 6051.0 km (Fig. 5) and has a maximum relief of 13 km, nearly two-thirds of that on Earth. The highest point at Maxwell Montes is at an elevation of 6062.1 km or 11.1 km above the planetary datum. Average slopes steeper than 30° have been reported from the southwestern side of Maxwell Montes, the southern side of Danu Montes, and the chasmata east of Thetis Regio (Ford and Pettengill 1992).

The lowest point is at Diana Chasma with an elevation of 6049 km, or 2 km below the planetary datum. Venusian topography is unimodal with a mean of 6051.84, a median of 6051.4, and a mode of 6052.1 km; more than 80% of the venusian surface lies within 1 km of the mean radius. As a result of the extremely low average relief of the surface of Venus, the peak in a topographic histogram of elevation versus area is very narrow. In fact, the surface of Venus is so smooth that nearly 90% of its surface lies within an elevation range of only 3 km. In contrast with Earth, where today the continental masses and oceanic basins trend north-south (except for Asia and Australia that trend east-west, and Antarctica that is centered on the South Pole), the topographic features on Venus trend in a general east-west direction. A comparable trend did occur on Earth, however, during the Late Proterozoic 800 to 600 Ma ago when an east-west trending supercontinent dominated the Earth (Morel and Irving 1978). The global morphological fabric on Venus (broad terrain belts and aligned regional slope features within the plains, elongate planitia, highland margins, chasmas, and linear features within the highlands) trend at less than 45° to the equator. According to Sharpton and Head (1986), this fabric, which is quite different from the complex terrestrial morphology of plate boundaries on Earth and the more subdued lunar and mercurian grids, indicates that the major heat transfer on Venus is due neither to simple conduction, as on the Moon, nor to plate recycling, as on Earth. The tectonic style of Venus must be controlled by its own unique pattern of heat loss – mantle plumes.

Topographic highs on Venus also differ from those of Earth in their small areal extent. Comparison of a histogram of oceanic basin depths on Earth with one for Venus also shows distinct differences. The histogram of oceanic regions on Earth is asymmetrical in cross section. The gradual slope on the left of the major peak indicates increasing depth as the oceanic basement cools with time, and the abrupt drop to the right of the peak means that deep lows (such as trenches) contribute little to the topography of the ocean floor (Phillips and Malin 1983). In contrast, the histogram of Venus is quite symmetrical in cross section. If plate tectonics are, or have been, important on Venus, the peak in elevation indicates either a period of excessive plate formation that created a secondary peak in the histogram (if the peak is removed, both histograms are comparable), or there has been a preferential consumption of older oceanic basement on Venus. Phillips and Malin (1983) noted that the total spreading elevation on the histograms of Earth and Venus is about the same and, if the histogram for Venus were caused by crustal plate spreading, such a distribution would demand an implausibly thicker lithosphere on Venus to compensate for the absence of a hydrosphere overburden and the higher surface temperature on Venus. The absence of bimodality may be due to: (1) predominance of one crustal type; (2) if an erosional base level is present, it is near the topographic mode; (3) erosion has been effective in reducing an old bimodal topography; (4) endogenic and exogenic processes are in equilibrium; or (5) the lithosphere is uniformly thin, tending to reduce the planetary relief (Phillips and Malin 1983; Head 1990c).

7.3 Morphology

7.3.1 Exotic

7.3.1.1 *Impact Craters*

Venera 15–16 radar images north of Lat. 30°N revealed 146 probable impact structures having diameters larger than 8 km (Basilevsky et al. 1987). Detailed mapping by Magellan of about 89% of the surface of Venus displays 842 craters of probable impact origin ranging from 1.5 to 280 km in diameter (Arvidson et al. 1992; Plaut 1992; Schaber et al. 1992; Fig. 69). Schaber et al. classified the impact structures into six types that largely are related to size. They are: (1) multiple ring craters; (2) double ring craters; (3) craters with central peaks; (4) structures with structureless floors; (5) irregular craters; and (6) multiple types. Although some impact structures are disrupted by tectonic features or are embayed by plains materials, most are undisturbed (Fig. 70). Head et al. (1992) suggested that the distribution of impact craters on Venus is controlled by regional resurfacing through volcanism and tectonism, whereas Arvidson et al. proposed that impact craters began to accumulate after resurfacing decreased to a rate too small to erase the craters in most areas. This decrease in resurfacing supposedly began about 500 Ma ago, but Venus continues to be volcanically active. This activity is documented by the dynamically supported long-wave topography indicative of mantle up-welling and partial melting, coronae formed by upwelling that display various stages of formation from novae (uplifted fractured areas) to collapsed degraded features, and volcanic flows that exhibit various stages of degradation (Arvidson et al. 1992; Izenberg et al. 1992). Crater density suggests that the volcanic zone in the Atla to Beta Regio areas and the Aphrodite Terra zone are younger than the surrounding regions. Schaber (1991) stated that the surface of Venus is less cratered than the postmare surface of the Moon and the youngest volcanic plains of Mars.

Schaber et al. (1992) reported that the spatial distribution of craters on Venus is uniform, that craters having diameters larger than about 35 km display a size/density distribution comparable to the young crater populations on other rocky

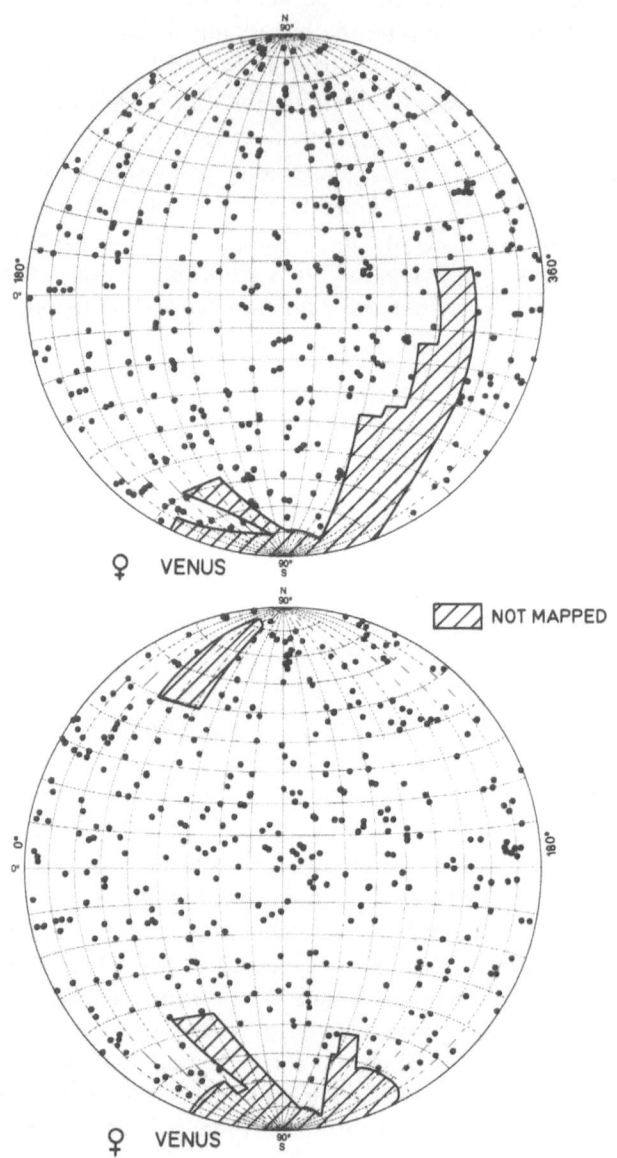

Fig. 69. Distribution of impact structures (842) on Venus after 89% of the venusian surface was imaged through Magellan orbit 2578. (After Schaber et al. 1992, fig. 1a)

planets but at a lower density, which they believed indicate an average age of 0.5 Ga. Because of atmospheric filtering, craters smaller than 35 km in diameter decline rapidly. The catastrophic failure of the smaller asteroids must lead to collision of the residual wake with the venusian surface, atmospheric cratering, creating craterless radar-bright scour zones. According to Schultz (1992), the

Fig. 70. Impact crater in a volcanic plain of Venus extending from Lats. 22° to 37°N, and from Long. 20° to 340°E. See Fig. 85 for location of structure. Note that the ejecta are partly covered by plains materials. At the left of the structure is a lava channel that terminates on a fan. Photograph courtesy of the Magellan Experiment Principal Investigator G. H. Pettengill, the Magellan Project, and NASA

maximum diameters of such structures are in the order of 1 to 3 km. Ivanov et al. (1992), who believed that the high atmospheric density and high surface temperature on Venus influenced the morphology of craters, proposed that as an asteroid fell considerable energy was transferred to the thick venusian atmosphere. This dynamic gas flow resembled a strong explosion and a significant wake of low density formed behind the asteroid, influencing not only the shape of the impact crater but also the distribution of ejecta facies. In contrast with other planets, crater modification ranges from pristine (62%), embayed by lavas (4%), with the rest being modified by tectonism; none of the structures is markedly modified. Schaber et al. also reported that the larger structures, as on other planets, display a progression in morphologies and that smaller impacts generally are irregular or multiple rather than simple bowl shaped. Other features displayed by venusian craters include diffuse radar-bright and dark features around some craters, and radar-bright or dark parabolic arcs opening westward, and that outflows (Fig. 70) originate in the ejecta. Craters imaged to date also are largely unmodified by volcanism in their exteriors; the only exception may be a crater south of Mylitta

Fluctus that has been breached and flooded (Head et al. 1991a). The smooth plains within the interiors of some craters may represent impact melt, volcanic flooding, or impact-triggered volcanism.

Structures larger than 15-km diameter, like those on other planets, are typically circular, and exhibit some or all the features typical of complex craters. The structures display a general progression with increasing size from no central peak, to one central peak, to multipeak or peak ring. The complex crater Daniolova exhibits a multiple central peak structure (Phillips et al. 1991), and Cleopatra Patera on Maxwell Montes (an impact structure with a peak ring superimposed on faults of the montes) is tectonically underformed, suggesting that the impact was very recent or that Maxwell Montes has not been active for a long time. According to Phillips et al. (1991), a steep-walled sinuous channel a few kilometers wide cuts across the hummocky rim of Cleopatra Patera and debouches onto the adjacent plain. Thus, the channel may have been a source for the strata that comprise the plain in Fortunas Tessera east of Maxwell Montes. Possibly the channel served as a conduit for shock melt generated at Cleopatra Patera (Basilevsky and Ivanov 1990). Craters smaller than 15 km are irregular with noncircular rims and multiple hummocky floors, and they tend to overlap or have rim contact. This morphology is believed to be the result of Venus' dense atmosphere, which broke up and dispersed incoming meteoroids, thus causing near-simultaneous impact of the fragments and production of multiple and irregular craters (Phillips et al. 1991). That the dense atmosphere of Venus has had an important role in impact processes also is suggested by the bright haloes surrounding the craters to a width of tens to hundreds of kilometers (Plaut 1992). Wispy surface streaks near the craters may have been caused by atmospheric turbulence associated with the impacts that apparently mobilized fine-grained sediments near the craters, and radar-dark circular 'shadows' are believed to have been caused by air-bursting impactors (Zahnle 1992). Experiments by Schultz (1992) demonstrated that ambient atmospheric pressure should affect gravity-scaling relations on Venus, be minor on Earth, and be insignificant on Mars. In the absence of pressure effects, aerodynamic drag could be significant in reducing crater size.

According to Garvin (1990), there is no substantial impact regolith on Venus (unlike on the Moon, Mercury, and Mars), because more than 99% of impact material was deposited only near the crater sources. Two types of ejecta have been mapped to date. Neither show any evidence of weathering or volcanic resurfacing, and they are brighter in the rays than on the surrounding plains. One type, a hummocky ejecta facies, consists of ridges and furrows that generally tend to be concentric with the rim crater and are best developed near the rim. This ejecta, which may consist of blocks larger than many radar wavelengths, extends from 0.5 to 0.8 crater radii beyond the crater rims. Ivanov et al. (1992) believed that the shock wave associated with falling asteroids had a strong influence on the distribution and types of ejecta and suggested that the hummocky facies may correspond with the boundary of flow of dense gas beyond the low-density wake created by the falling asteroid. They further postulated that the collapse of the gas-rock debris column in the high and low density gas regions may create a turbidity-like flow outside the region of hummocky ejecta. The distal edge of this outer ejecta tends to be lobate to slightly pointed – a petal-like appearance (Fig. 70). For impact craters larger than 15 km the outer ejecta facies extends as far as 2.5 crater radii from the impact structure. Its outer range appears to exceed the limit predicted by ballistic emplacement, suggesting that some other process is involved in emplacement; its distribution apparently is influenced by atmospheric interactions. Even the inner hummocky ejecta probably has undergone outward flow as it became entrained into the atmosphere. Beyond the hummocky facies is an outer ejecta facies, whose distribution pattern appears to be asymmetrical, indicating oblique impact with the missing sectors located uprange For impact craters larger than 15 km the outer ejecta facies extends as far as 2.5 crater radii from the impact structure. Its outer range appears to exceed the limit predicted by ballistic emplacement, suggesting that some other process is involved in emplacement; its distribution apparently is influenced by atmospheric interactions. Even the inner hummocky ejecta probably has undergone outward flow as it became entrained into the atmosphere.

Secondary craters are best developed beyond the outer ejecta facies associated with craters formed by oblique impact. Some of these craters also possess flow-like ejecta suggestive of low-viscosity material (Fig. 70); some of it was fluid enough to flow around obstacles such as volcanic shields. These flows, which are estimated to be in the order of meters thick, could represent impact melt or the more distal runoff of the finer ejecta caused by enhanced thermal turbulance produced by impact-generated melt and vapor (Phillips et al. 1991). As a result of large gravitational acceleration and impact velocity such melts should be much larger on Venus than on the Moon. According to Ivanov et al. (1992), melt generation on Venus is about 20 to 40% higher than on Earth for the same crater size, and that the molten flows stay fluid for an order of magnitude longer time on Venus than on Earth. Simow and Wood (1992) suggest that possibly two mechanisms produced these outflows – an erosive channel-forming process and a depositional one. According to them, the erosive fluid may have been an impact melt and the depositional one a fluidized solid debris, vaporized material, and/or melt. Schultz (1992) proposed that the outflows are formed from ejecta associated with the larger asteroids that survive entry into the venusian atmosphere. This debris is entrained into the winds formed in response to the impact and it is transported downwind as ground-hugging flows with the flow being maintained by atmospheric turbulence. This leads to the creation of radar-dark lobes at the distal ends of the flows.

About half of the craters are partly or completely surrounded by areas of low radar backscatter. These 'dark margins' may be the result of removal of wavelength structures or of cover by a blanket of fine material produced by ablation of metereoids during atmospheric transit or impact. In contrast, Ivanov et al. (1992) ascribed the dark margins to 'back venting' by shock wave pressures and depressuring in ground pores. Also associated with the craters are large parabolic features. Near Carson Crater (Lat. 24.2°S; Long 344.1°E) they consist of alternating darker and slightly lighter bands curving around the eastern side of craters and extending to the northwest and southwest for several hundred kilometers. These bands may represent fines thrown high into the atmosphere during impact and transported westward during their descent. Campbell et al. (1992) ascribed the parabolas to the injection of material into the upper atmosphere during impact, with the particles being transported westward by zonal winds. They estimated that the largest particles in the parabolas have diameters of 1 to 2 cm and that this debris took about 2 h to settle through the atmosphere, allowing material to be transported several hundred kilometers downwind (westward) before being deposited on the venusian surface.

Cumulative size distribution of impact craters on Venus is nearly identical with that of the North American and European cratons, perhaps indicating that the two surfaces have the same average crater production age or retention age (Schaber 1991). The relative youthfulness of both surfaces is documented from impact distribution – an age that is less than the post-mare population on the Moon after 3.5–3.0 Ga ago. Impact craters on Venus do not appear to be uniformly distributed. For example, the Sappho region (an area of approximately 5×10^6 km^2) is devoid of craters, indicating that the region must have undergone resurfacing. Volcanism probably had a major role in resurfacing Sappho, but elsewhere tectonic processes may have been more important.

Three resurfacing models have been proposed to explain the surface distribution of craters: vertical equilibrium or 'leaky planet', regional resurfacing 'collage', and catastrophic models (Head et al. 1992). In the 'leaky planet' model volcanism is uniformly distributed and proceeds relatively uniformly with time. Phillips et al. (1991, 1992) speculated that cratering on Venus is random in space and time, with the structures being preserved in regions that are quiescent and destroyed in regions where rates of surfacing by tectonism and volcanism are high. This may explain why few impact structures are in the process of being destroyed and why craters are absent in other areas. In this model the production of impact craters can be used only to determine the average age of the planet's resurfacing, and it has no meaning on a regional basis – a model that is binary with craters being pristine at one end and eradicated at the other end. In this scenario the surface of Venus ranges in age from 800 to 0 Ma, with the youngest surface in the Sappho region and the oldest one reflecting the time required to resurface the planet once (Schaber 1991). According to Plaut (1992), the areal density of venusian craters is suggestive of a surface that is on the order of hundreds of millions of years old. In the regional resurfacing 'collage' model, as described by Head et al. (1992), volcanic features are produced in different parts of Venus at different times. Evidence for such serial volcanism is the large number of features, shield volcanoes, coronae, arachnoids, novae, large volcanoes, and cratering by mantle plumes or hot spots. In the catastrophic resurfacing model a volcanic pulse destroyed the entire crater population and produced a pristine surface on which another crater population is imprinted. Schaber et al. (1992) proposed that the venusian impact-crater morphologies and distribution result from a global resurfacing event or events, the latest of which ended 0.5 Ga ago. Afterward this volcanism declined but did not cease. Since this event, about 10% of Venus has been resurfaced and about 4% of the craters have been destroyed. They also pointed out that the dense venusian atmosphere has affected the crater production with intermediate size impactors being fragmented, leading to the production of overlapping or multiple craters. A narrow size range of impactors has produced shock-induced 'splotches' and no craters, and the smallest impactors have had no visible effect on the venusian surface. Large impactors were not affected by the dense atmosphere.

7.3.2 Endogenic

7.3.2.1 Plains

The largest endogenic provinces on Venus are the plains, which constitute more than 90% of the surface of the planet and are associated with downwellings in the mantle convection pattern (Herrick and Phillips 1992; Fig. 71). These low-relief areas are constructed of volcanic deposits (Solomon et al. 1991; Guest et al. 1992). Guest et al. stated that the volcanic plains material is mainly basalt emplaced by massive eruptions from deep-seated sources and from multiple small eruptions from shallow reservoirs. Some of the eruptions that formed domes, however, may consist of silicic lavas, and a few plains areas seemed to have had a history of pyroclastic deposition. Masursky et al. (1980) subdivided the plains into two units: the upland rolling plains that range in elevation from 0 to +2 km and which comprise 65% of the surface, and the lowland areas or planitias that are lower than the datum level and constitute 27% of the surface. The gentle rolling upland plains have a radar brightness intermediate between the dark lowlands and the bright highland provinces (Ford and Senske 1990). Most of the plains imaged by Magellan display considerable evidence of deformation (Fig. 72). These and similar deformations on Mars and Earth are caused by a small amount of crustal shortening with the observed geometries being comparable with those predicted by the Coulomb–Anderson model (Watters 1991). This model predicts fractures parallel to the planes of maximum and minimum stresses. An extensively deformed plains segment with an area of 10 000 km^2 centered near Lat. 32°N is marked by hundreds of northwest-trending lineations, some of which represent paired normal faults bounding flat-floored grabens spaced 0.5 to 3 km apart. Differences in their reflectivity suggest that more than one generation of extensional structures may be present.

Plains development appears to be the result of an interplay between deformation and volcanism. This interplay is indicated by disruption of northeast-trending lineations by volcanic pits and domes. Some domes in turn appear to be cut by lineations, a geometry indicative of pre- and post-volcanic faulting. The intensity of deformation coupled with a structural pattern extending hundreds of kilometers and a volcanic resurfacing indicate that the broad rise or saddle that separates Sednas Planitia to the north from Guinevere Planitia to the south (Figs. 67 and 68) is the result of mantle convection that thermally uplifted and stretched the plain (Solomon et al. 1991). Another striking structural feature observed in the plains is a

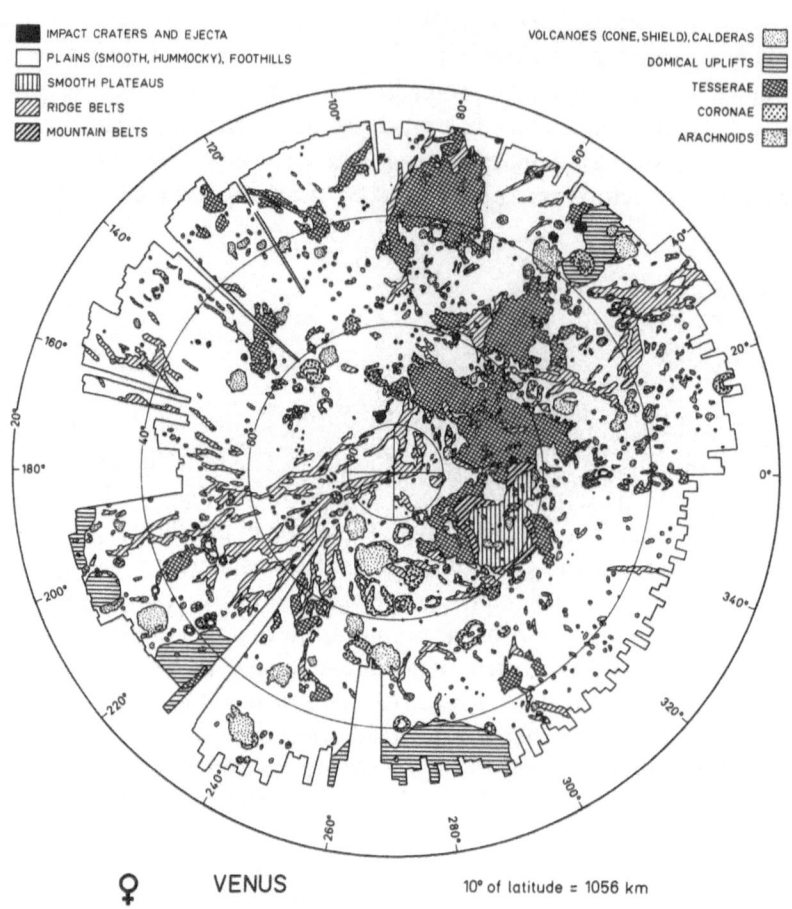

♀   VENUS   10° of latitude ≈ 1056 km

Fig. 71. Geomorphic provinces of the northern hemisphere of Venus. Compiled from data from Veneras 15 and 16 by Sukhanov et al. (1989) and Senske (1990, fig. 5), omitting many very small features. Except for the few impact craters and their local ejecta, the morphology is endogenic; exogenic forms are thin and local, formed only by wind and mass movements

regular and nearly orthogonal pattern of cross-cutting lineations (Fig. 73). The brighter of these lineations trends northwestward, is more or less continuous, and is spaced 2 to 3 km apart. The fainter set trends northeastward and is spaced about 1.2 km apart. Solomon et al. (1991) interpreted these trends as caused by deformation of a near-surface layer atop a nearly molten ductile layer, with the nearly right angle trend of the lineations resulting from a change in the relative magnitudes of two principal horizontal extensional stresses. A similar conclusion was suggested by Watters (1991), who proposed that the observed geometries are caused by a small amount of crustal shortening. Other areas of the plains display a complex pattern of radar bright lineations commonly associated with volcanoes and their deposits and with length scales similar to the spacing of the faint lineations in the

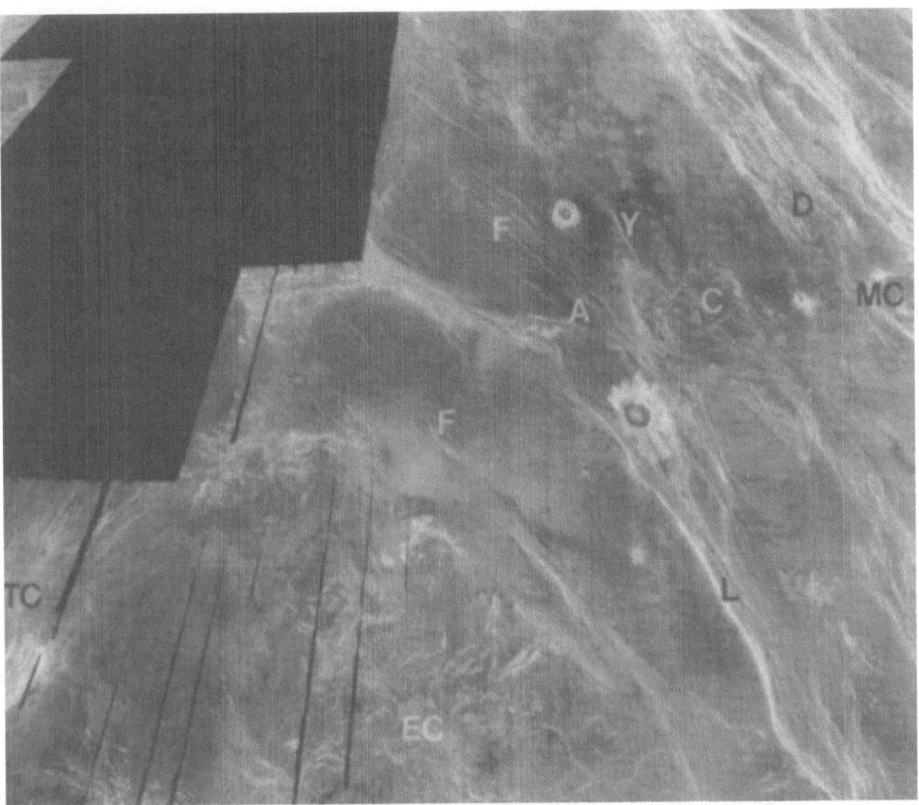

Fig. 72. Deformed plains in the northern hemisphere of Venus. Note folds (**F**) and linear ridges Dennitsa (**D**), Lukelong (**L**), and Yumyn-udyr (**Y**), dorsae of folded and/or faulted origin. Superimposed on the terrain are four coronae (**C**) unnamed structure; **EC** Earhart Corona; **TC** Tushol Corona; **MC** Maslenitsa Corona), and one arachnoid (**A**). Note that the distribution of ejecta on the large impact in the center of the photograph is partly controlled by the structure of Lukelong Dorsa. Photograph C175N164 extending from Lat. 67° to 82°N and from Long. 97° to 231°E. Courtesy of the Magellan Experiment Principal Investigator G. H. Pettengill, the Magellan Project, and NASA

cross-cutting structural pattern. Such an irregular pattern may have been produced by deformation of a thin volcanic layer shortly after extrusion, when its properties or the forces acting upon it were not uniform in either space or time (Solomon et al. 1991).

The plains appear to be the result of deposition and deformation (McGill 1991). The oldest structural elements are closely spaced grabens and horsts that are 1 km or less in width, and even narrower anastomosing fractures that transect the tesserae. Locally, they are cut by an arcuate set of grabens having maximum widths of 15 km. Truncating all these structures are plains materials. Other preplain structures are lobate ridges in geographical association with younger and much larger ridge belts. Plains materials onlap these older ridges that, in turn, are

Fig. 73. Orthogonal lineations centered at Lat. 30°N, Long. 222°E. (After Solomon et al. 1991, fig. 3). Photograph courtesy of S. C. Solomon, copyright 1992 by the AAAS

complexly deformed into younger ridge complexes. Associated with the younger ridges are fracture complexes. Deformations of the plain and ridge formations appear to have been episodic. Shear zones in the ridges truncated by uplifted plains represent early phases of deformation, and long, narrow, closely-spaced fractures (1 km apart) cutting across the ridges represent a more recent tectonic episode.

Lobate lava flows and associated channels diverted by structural highs are still younger. Local fractures and ridges related to coronae formation superposed on the lobate flows are among the youngest structures on Venus (McGill 1991).

Another type of deformation in the plains is a polygonal pattern of bright lineations (Fig. 76). Johnson and Sandwell (1992) ascribed these features either to a cooling lava, or to increased heat flux at the base of the lithosphere leading to thermal stresses that produced tension in the upper lithosphere and compression in the lower one. This second explanation appears to be more compatible with the kilometer size scale of the venusian polygons. Also present in the plains are linear regularly spaced ridges and lineations. The ridges may have been produced by horizontal compression and shortening (Barsukov et al. 1986; Frank and Head 1990; Head 1990b; Zuber 1990). Lineations having spacing between 1 and 2.5 km are believed to be tension fractures in the upper brittle layers of the volcanic plains (Banerdt and Sammis 1992). This layer and the stronger one beneath it were subjected to stress greater or equal to the tensile stress of the upper layer, causing the upper one to fracture. A horizontal detachment along the top surface of the lower layer allowed frictionally resisted displacement of the upper layer and stress relaxation near the fracture. With continued tension another fracture formed at some distance from the previous one. Banerdt and Sammis believed that such a tectonic/stratigraphic setting can account for the regular spacing of the lineations. The style of deformation ranges from broad arches having a maximum width of 9 km to narrow segments with no arches. The several-hundred-m-high deformation belts consist of curvilinear anastomosing subparallel ridges and grooves parallel to the margins of the belts. These ridges, which can be followed for hundreds of kilometers, are spaced 10 to 20 km apart.

In Lavinia Planitia (Fig. 67) at Lat. 50°S, Long. 340°E the belts are radar bright, intensively deformed, and separated by darker and smoother lower plains. According to Solomon et al. (1991), this ridge morphology is similar in plan form and dimensions to features in lunar mare that have been interpreted as compressional in origin. The belt in Lavinia Planitia formed because of northeast-southwest compression. One of the belts, Hippolyta Linea, contains evidence of volcanism, with the volcanic material ponded in topographic lows. Faulting in the region of the belts appears to be extensional in origin, being marked by numerous shallow grabens. Concentration of grabens along the crest of the structural high indicates that some of this extension is due to flexure along the crest of broad anticlines.

Other types of faulting and fracturing associated with the folding include tensional fractures and deep-seated thrust faults. One of the deformation belts 800 km south of Hippolyta Linea (in Lavinia Planitia) is cross-cut by northwest-trending 500-m-wide grooves that are continuous for more than 100 km, and another set of less prominent grooves displays distinct changes in trend within the belt. The first set postdates the formation of the belt, and most of the second set predates it. Changes in trend of the older lineations are believed to result from shear that accompanied the crustal shortening and formation of the belt. Different morphologies of the belts may reflect differences in formation mechanisms. The belts were caused by buckling or fault-bend folding of component surface layers and groove belts by extension and fracturing of the lithosphere as it stretched across a broad uplift. Small scale morphological differences may indicate a superposed stress field, with maximum horizontal compressive stress oriented west-northwest to east-southeast and the least stress field being north-northeast to west-southwest (Solomon et al. 1991). Recently, Squyres et al. (1992a) proposed that the orthogonal features (tens to hundreds of kilometers wide, 1000 km long and hundreds of meters high) transecting Lavinia Planitia are the result of horizontal stresses produced by convection circulation in the mantle coupled to the crust. According to Squyres et al., the small-scale wavelengths of deformation noted within individual ridge belts suggest that the lower limit on the venusian thermal gradient in the Lavinia Planitia region is about 30 K/km, a value somewhat higher than previously suggested.

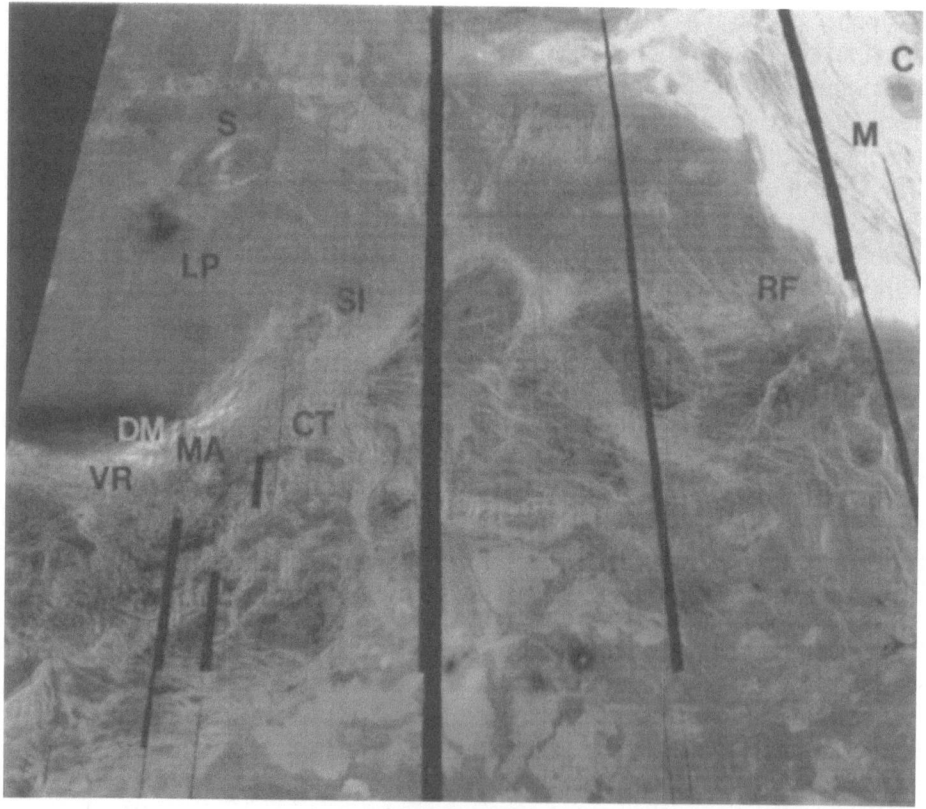

Fig. 74. Western end of Ishtar Terra (from Lat. 52° to 67°N, Long. 9° to 323°E; Figs. 65–68) of Venus. From *left to right* are Lakshmi Planum (**LP**), Sacajawea Patera (**S**), Danu Montes (**DM**), Manani Crater (**MA**), Vesta Rupes (**VR**), Clothe Tessera (**CT**), Siddons Crater (**SI**), Auska Dorsum (**A**), Rangrid Fossae (**FR**), Maxwell Montes (**M**), and Cleopatra Crater (**C**). Photograph C160N347 courtesy of the Magellan Experiment Principal Investigator G. H. Pettengill, the Magellan Project, and NASA

7.3.2.2 Tesserae

Scattered throughout the plains are isolated inliers of radar-bright terrain known as tesserae (Figs. 74–76). They are associated with negative lithospheric density anomalies that Herrick and Phillips (1992) believed are areas of thickened crust. These highs are characterized by two or more sets of intersecting linear features that produce patterns ranging between orthogonal, diagonal, chevron, and chaotic (Head 1990a). The ridges are several meters high and 5 to 20 km apart. These features are believed to have been formed by gravity sliding, interaction between mantle plumes and crustal spreading axes, late-stage deformation of uplifted regions of thick crust, or a combination of compression and subsequent gravitational relaxation (Head 1990a; Solomon et al. 1991). They occur in small patches within large regions that total approximately 30% of the region north of Lat. 30°N

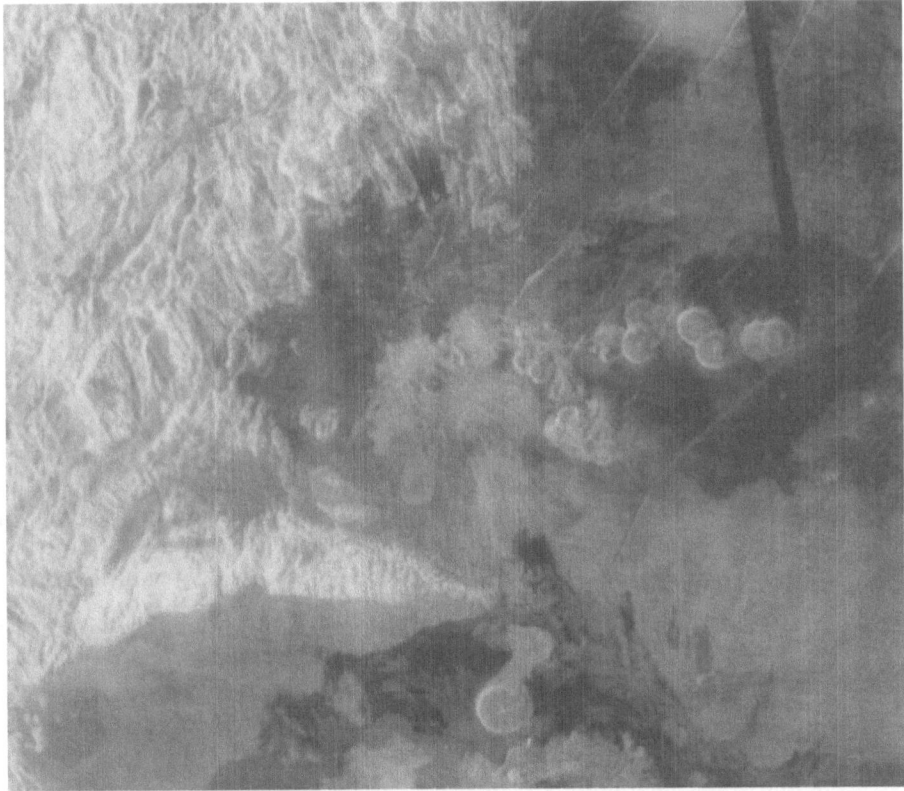

Fig. 75. Eastern edge of Alpha Regio (Lat. 27° to 32°S, Long. 6° to 13°E) of Venus. Note siliceous domes east of region amongst a volcanic plain (Aino Planitia). Oblique photograph F305009, published with permission of the Magellan Experiment Principal Investigator G. H. Pettengill, the Magellan Project, and NASA

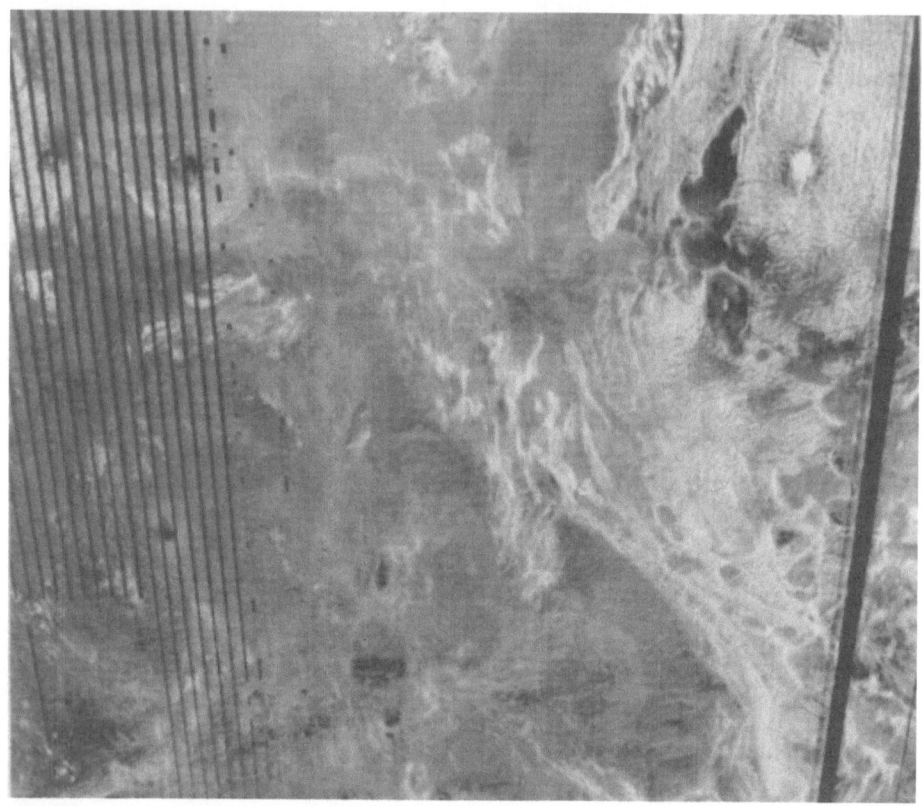

Fig. 76. Eastern edge of Tellus Regio and Niobe Planitia (Lat. 37° to 52°N, Long. 81° to 110°E; Figs. 65 and 66) west (*right*) of the high. Note siliceous domes, complex fracture pattern on the plain, and near-circular structure with radiating fractures. Photograph C145N096 courtesy of the Magellan Experiment Principal Investigator G. H. Pettengill, the Magellan Project, and NASA

that was imaged by Veneras 15 and 16 (Fig. 71). They also have been observed at lower northern and southern latitudes by the Arecibo Observatory, and Pioneer Venus data suggest that they are widespread throughout the planet. They rise 1 to 2 km above the plains and have a surface microrelief of 1 m or less. They display a relatively shallow (< 100 km) depth of compensation, which Solomon et al. (1991) interpreted to mean that most of the features are compensated by enhanced crustal thickness rather than by mantle dynamics. The largest structure mapped to date by Magellan is Alpha Regio, a polygonal feature about 1300 km wide near Lat. 24°S, Long. 356°E (Figs. 67 and 68). It has a maximum relief of 5 km, but generally it rises 1 to 2 km above the surrounding plains. The western edge of the high consists of broad ridges and scarps 10 to 20 km apart entrained by grooves and ridges 200 to 800 m apart and embayed by volcanic plains along their western edges; patches of plains also occur at various levels farther east. Thus plains volcanism postdates the latest phase of formation of the tesserae.

7.3.2.3 Highlands

Highland areas having elevations higher than 2 km constitute 8% of the surface of Venus (Figs. 65–68; Masursky et al., 1980). They occur in three areas: Ishtar Terra centered at Lat. 65°N, Long. 350°E; Aphrodite Terra near the equator between Longs. 60° and 205°E, and Beta Regio between Lat. 15° and 40°N and Long. 275° and 296°E. Included with Ishtar Terra are Lakshmi Planum (a western plateau), Maxwell Montes (a central high), and a complex eastern terrain (Fig. 74). Lakshmi Planum is bounded along its southern margin by the Vesta Rupes scarp with slopes in excess of 20° and by Akna Montes and Freyja Montes near its western and northern margins. Many of the steep slopes of this region contain closely spaced troughs and lineations perpendicular to the downslope direction. Smrekar and Solomon (1992) interpreted these features as caused by gravitational spreading. Similar extensional features also occur within actively converging margins. In this type of terrain the extensional faulting is cut by older folds and thrust faults in some regions (Danu, Akna, Freyja, and southern Maxwell montes), and elsewhere some of the structures have been deformed by later shortening and horizontal shearing.

These observations led Smrekar and Solomon to propose that gravitational spreading in belts of convergence occurred at the same time that mountain building was active. Finite modeling of gravitational spreading predicts that spreading on Ishtar Terra should occur first on the Lakshmi Plateau, followed by extension on the marginal slopes, and that sometimes thrusting may spread to the adjacent plains. These same models also predict that for reasonable estimates of crustal thickness and thermal gradient, gravitational spreading of high topography by flow in the lower crust and subsequent failure and relaxation of relief should occur in a geologically short time. For a thermal gradient of 15 K/km, crustal thicknesses of 30, 20, and 10 km, topographic relief of 3 km, and a slope of 3° the topography relaxes 25% in 10^5, 10^8, and 10^{10} years. The high relief of Ishtar Terra, thus, must reflect the presence of an anomalous crust or the topographic high is being maintained or was maintained until times more recent than the 500 Ma average crater retention age on Venus. Smrekar and Solomon (1992) postulated that a large depth of compensation and abundant volcanism suggests that the mountain belts and steep plateau of Ishtar Terra are less than 10 Ma old and that tectonics may be ongoing. Extending for more than 1200 km along the southern and southeastern sides of Lakshmi Planum inboard of Vesta Rupes are the Danu Montes that rise 3 km above the adjacent plains especially along its northeastern arm. Lakshmi Planum has a relief of about 4 to 5 km, similar to that of the Tibetan Plateau on Earth, is a site of volcanism and has two major volcanic edifices – Colette and Sacajawea paterae (Solomon et al. 1991).

Formation of this plateau has been ascribed to hot-spot origin and to underthrusting of crust beneath a tessera region with subsequent melting and volcanism (Head et al. 1991a). Maxwell Montes east of Lakshmi Planum trends east-west and rises 8 km above the plateau and 11 km above datum (Fig. 74). The surface of this high is rough both at the scales of the Earth-based Arecibo wavelength and the Pioneer wavelength. East of Maxwell Montes between Lat. 50°

and 75°N and Long. 40° and 75°E is a complex of ridges, troughs, and closed depressions having reliefs of 2 to 3 km. These topographic features are discontinuous, relatively small, and randomly distributed. One of the lows drops to an elevation of 1.8 km below planetary datum – one of the lowest region of Venus (Masursky et al. 1980). Lenardie et al. (1991) proposed a two-phase model to explain the structure of Ishtar Terrain. During the first phase a proto-Lakshmi Planum was created by crustal convergence over a mantle downwelling. During the second phase a transient plume was triggered by the thickened crustal block to form a planum leading to a slab-driven tectonic pulse and low-angle subduction and orogenesis of the edges of the block. As discussed by Kaula et al. (1992), Ishtar Terra could be the creation of mantle upflow or downflow with upflow being favored by the extensive volcanic plain of Lakshmi Plateau and its high geoid, and a downflow origin by the intense deformation of mountain belts and the absence of rifts. Possibly both are occurring or have recently occurred with Lakshmi being the locus of upflow and Maxwell Montes a region of downflow. According to Kaula et al., Ishtar Terra is the product of a history marked by drastic changes in tectonic styles during several 100 Ma. Features created during these changes were formed in response to a competent layer less than 10 km thick and flows of 100 km or broader, reflecting heterogeneity in the mantle.

Aphrodite Terra in the equatorial region of Venus (Figs. 77 and 78) has an area of 2.9×10^7 km^2, comparable in size to Africa. It consists of three mountainous regions separated by lower terrains: the western mountains with an elevation of 5.5 km and a relief of 4 km, the more complex central mountains with an elevation of nearly 4 km and which rise 3 km above their surroundings, and the slightly higher eastern mountains with an elevation of 5.7 km. Connecting the eastern and central parts of Aphrodite Terra is a complex of ridges and troughs 1.0 to 2.5 km above datum level. Diana and Dali chasmata are in this topographic complex. Southwest of the ridge-and-trough region is a broad platform having an elevation of 1.5 to 2.5 km. Artemis Chasma with a diameter of 2.6 km consists of two asymmetrical ridges separated by a trough that resembles a rift within the plateau. Ovda Regio in western Aphrodite Terra is a plateau-shaped high having a relief of 4 km and an area of 3000 by 2000 km; it has a rugged surface with local relief varying more than 1 km over distances of a few tens of kilometers (Bindschadler et al. 1992a). Its margins tend to be relatively steep. A 7.5-km long pile of debris noted off a steep scarp during a second pass by Magellan, but not observed during a previous pass, suggested that Aphrodite Terra may be seismically active (Cole 1991; Kerr 1991c). Originally, it was proposed that a venusquake may have displaced approximately 1 km^3 of the plateau's edge into the adjacent trough. Later, it was suggested that this supposed visual evidence of active tectonics may not be real but is merely the result of distortion of the data (Cole 1991).

Seven regions at low to middle latitudes are Atla, Eistla, Bell, Thetis, Phoebe, Beta, and Alpha regios. Both Atla and Eistla regios are at major tectonic junctions (Senske et al. 1992). Atla is at Lat. 4°N, Long. 200°E centered on convergent zones of deformation, contains volcanoes, and has a gravity anomaly with a depth of compensation of 200 ± 23 km. Ozza Mons, a 7.5-km high volcano in the central

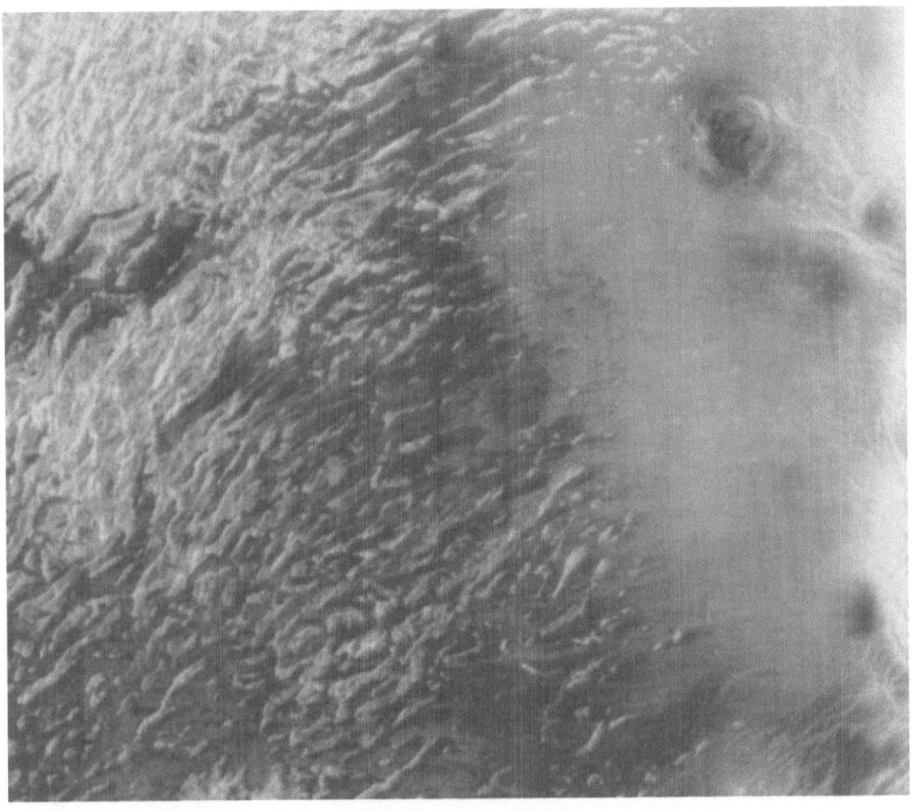

Fig. 77. Ridged terrain of Aphrodite Terra (Lat. 2°S to 2°N, Long. 67° to 73°E; Figs. 65 and 66) of Venus. Note patera on the right. Photograph F00N070 courtesy of the Magellan Experiment Principal Investigator G. H. Pettengill, the Magellan Project, and NASA

part of Atla is located at the point where Canis, Dali, Parga, and two other rifts intersect. Capping Ozza is a 100 by 50-km radar-dark plateau having a relief of 1.5 km and whose surface is scarred by numerous pits and collapse structures. North of Ozza is a field of volcanic domes having diameters of 10 km and whose lava flows are superposed and cut by faults. Around the periphery of Atla Regio are radar-dark plains on which are located several isolated areas of tesserae. Adjacent to Atla is Maat Mons along the northern flank of Dali Chasma. Unlike Ozza, this volcanic high is not disrupted by faulting. Eistla Regio between Lats. 10° and 25°N and Long. 0° and 55°E consists of a series of discontinuous rises trending about eastsoutheast-westnorthwest (Grimm and Phillips 1992; Solomon et al. 1992). The western and central portions of Eistla that have been imaged to date contain broad rises, rift zones, large volcanoes, high positive gravity anomalies, and a 100 to 200-km depth of compensation. Western Eistla region at Lat. 20°N and Long. 0° is a 2300 by 2000-km high having a relief of 1.4 km. Along its crest are Sif and Gula

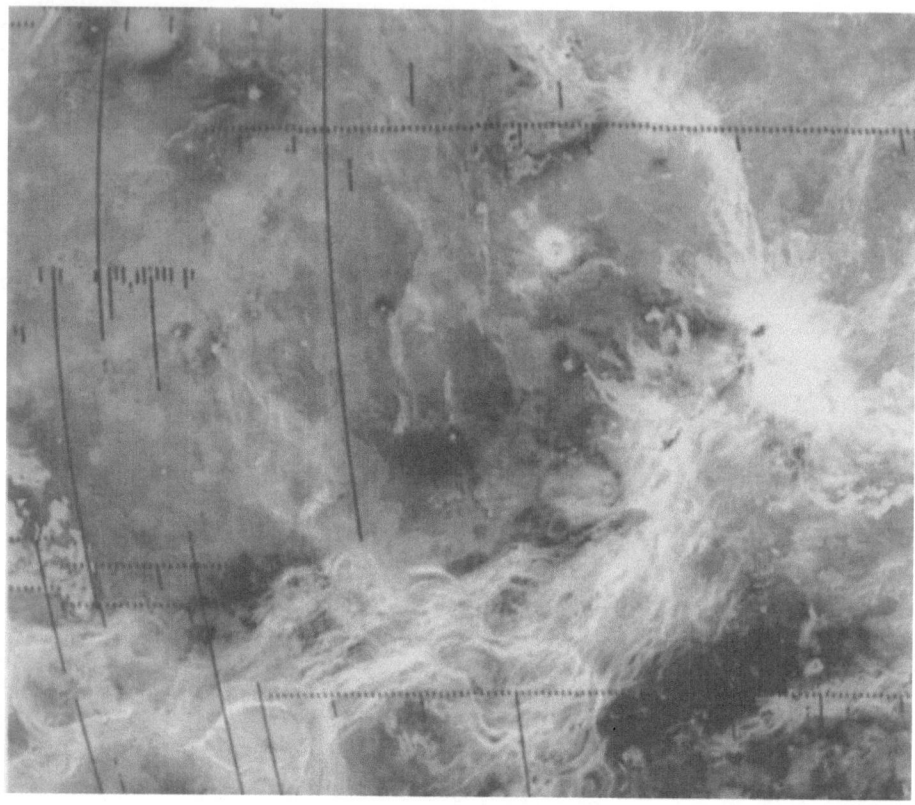

Fig. 78. Aphrodite Terra (Lats. 22°S to 22°N, Longs. 154° to 211°E; Figs. 65–68) on Venus. The dark area at the *lower right* is Dali Chasm. Photograph C200N183 courtesy of the Magellan Experiment Principal Investigator G. H. Pettengill, the Magellan Project, and NASA

montes. Southeast of Gula Mons is the northwest-trending Goor Linea, a collection of scarps and fracture zones (Senske et al. 1992). Bell Regio northwest of Aphrodite Terra at Lat. 32°N, Long 50°E is a domical high having a relief of 1.5 to 2.0 km and containing a corona-like feature (Nefertiti) and Tepev Mons (a 500-km diameter shield volcano). Thetis Regio at Lat. 10°S, Long. 130°E is a plateau-shaped high having a relief of 4 km. Its surface is quite rugged, but its side slopes are less steep than those of Ovda Regio. Like Ishtar Terra and Alpha Regio, but unlike Ovda Regio, its highest elevations occur along its margins and its interior is 2 km below its surrounding peaks (Bindschadler et al. 1992a). Phoebe Regio at Lat. 10°S, Long. 280°E is an elongate 1700 by 100-km plateau with an arc-shaped ridged terrain extending westward from Phoebe.

Beta Regio (Figs. 68 and 74) of Aphrodite Terra between Lat. 15° and 40°N and Long. 275° and 290°E is a complex region of radar brightness, roughness, and relief. This topographic high is dominated by Theia and Thea mons with elevations of more than 4.5 km above the datum. South of Beta Regio is a region of en echelon ridges that have an average elevation of 2 km and axial troughs. A canyon complex in the region consisting of an axial trough in the disrupted region of ridges and troughs, the trough offset to the east of Theia Mons, and the south-trending radar-bright 'tail' of Theia Mons that has been named Devana Chasma by Masursky et al. (1980) and is about 4000 km long. South of Beta Regio are north-south trending highs (Phoebe Regio and Themis Regio) and west of Beta is a group of north-south aligned inliers and low-relief discontinuous linear features that are radar bright. One of these features extends about 9000 km. A discontinuous 3000-km long north-south-trending ridge-trough complex occurs along the eastern margin of Beta Regio. Masursky et al. (1980) reported that another radar-bright linear feature far south of Beta Regio begins at Themis Regio, trends west-northwest, and terminates on the southwest-trending linear zone east of Aphrodite Terra.

Alpha Regio is a 1300 by 1500-km high centered at Lat. 25°S, Long. 3°E with a relief of about 2 km. It has a rugged interior and relatively steep side slopes (Bindschadler et al. 1992b). Adjacent and southwest of the high is Eve Corona and surrounding most of Alpha Regio are relatively radar-dark plains. Structures within Alpha consist of fine-scale ridges, broad linear ridges, broad arcuate ridges, linear disruption belts, and grabens. Bindschadler et al. stated that the occurrence of compressional features along and parallel to the margins of the high and the absence of any high-elevation ring of extensional features is more compatible with a cold-spot to mantle downwelling origin, than to a hot-spot upwelling model – an origin that also is applicable to other upland areas such as Ovda and Thetis regions with their complex ridged terrain.

The most recent tectonic episode along the Danu Montes on the Lakshmi Planum is reflected in the high relief of its northeastern arm. Thrust faults and wrinkle ridges in the adjacent plains indicate that they too were involved in the deformation. Images recorded by Magellan reveal evidence of widespread extension in both Danu Montes and Vespa Rupes. The 1- to 5-km spacing of the extensional features indicates that crustal attenuation occurred primarily in the upper lithosphere rather than in the whole lithosphere. Two collapse features suggest that this extension was accompanied by magmatism. Some collapse features grade downslope into sinuous depressions, which Solomon et al. (1991) attribute to withdrawal and channeling of magma toward the plains. Because these features cut across the thrust faults, they must have developed during the final phases of mountain building. Some extensional features, like those in the Himalayas, may result from gravitational spreading of the upper crust over the lower one – a crustal creep caused by thickening and uplift of the crust. The high temperature on Venus (730 K) should enhance such gravitational spreading on the topographic highs (Smrekar 1991). Other structural features may have been formed by collapse.

The broad ridges and troughs in Freyja Montes display various structural styles including linear fracture-like discontinuities, closely spaced fractures, and linear graben-like lows several kilometers wide. Structural relationships between these features suggest transcurrent motion in a right-lateral sense. A serrated pattern parallel to the topographic grain appears to represent fault-bound depressions that resemble pull-apart basins associated with transcurrent faults on Earth (Solomon et al. 1991). If so, they were formed by right-lateral shear motion, indicating that the most recent compressive tectonic event in central Freyja Montes was oblique to the topographic trend. Possibly the rhombs were formed by gravitational collapse of the mountain range, with their geometry being controlled by preexisting structures. South of the Freyja Montes are perched plains of possible volcanic origin. The asymmetrical scarps (steep slope facing southward toward the Lakshmi Planum) on the plain parallel to the front of Freyja Montes are the traces of low-angle thrust faults formed during tilting of the volcanic plains contemporaneously with crustal thickening and mountain formation (Solomon et al. 1991).

The Itzpapalotl Tessera north of the Freyja Montes displays considerable evidence of volcanism including sinuous channels hundreds of kilometers long that originate in the higher parts of the tessera and debouch onto the adjacent plains. The plains outboard of the foredeep at the base of Ursar Rupes are volcanic in origin (Head et al. 1991a). They show little evidence of deformation and they onlap the faulted mountains. Thus, some of the volcanism postdates formation of the mountains. Solomon et al. (1991) interpreted the recent volcanism on the Itzpapalotl Tessera at an elevation comparable with Lakshmi Planum but separated from the plateau by the intervening mountain range, as indicating that some of the magmatism in Ishtar Terra is the result of remelting of thickened crust rather than pressure-release melting in a mantle plume beneath the plateau. These authors also reported that there is no significant evidence of volcanic deposits or structures on Freyja Montes. The north-south trending eastern Freyja Montes are marked by extensional structures, indicating that the region no longer is under compression but has been subjected to recent gravitational collapse, whereas the east-west trending central segment of Freyja Montes presents evidence of folding and shear as though crustal shortening in this segment continued until geologically recent time.

The rim and flanks of the eastern edge of Lakshmi Planum south of Freyja Montes are characterized by grabens parallel to the topographic grain and volcanic plains onlapping them at the base of the eastern margin of the plateau. These plains, which were involved in the deformation of Freyja Montes, had their sources in central Lakshmi Planum. A 100×150-km dome-like structure, which is disrupted by intersecting graben structures spaced 2 to 5 km apart and flanked and surrounded by ridges that may be folds or thrust faults, may represent a collapsed high. Ridges that disrupt the grabens may be the result of thrusting along the periphery of the gravitationally collapsed high (Solomon et al. 1991).

The topographic rises in the equatorial region (Aphrodite Terra and Asteria, Beta, Eistla, and Phoebe regios) are products of rifting and volcanism and they contain rift zones 100 to 300 km wide and thousands of kilometers long and large volcanic constructional features. If these highs have isostatic compensation depths that are typical of volcanic rises (100 to 200 km), then the topography in the region is partly supported by mantle dynamic processes (Solomon et al. 1991). The volcanic rises were constructed by mantle upwelling (Kiefer and Hagler 1991). The only volcanic rise mapped at the time that Solomon et al. (1991) and Head et al. (1991a) published their articles and Senske and Stofan (1991) published their abstract was western Eistla Regio. This high is capped by two volcanic structures (Sif and Gula mons) that are linked by northwest-trending rifts cutting across the rise and having a relief of 2 to 4 km above the planetary datum. The broad distribution of volcanism, including the large number of domes within the adjacent plains, suggests the presence of a wide region of melting in the subsurface region.

Bindschadler et al. (1992a) believed that the highland regions of Venus can be divided into two groups: volcanic rises and plateau-shaped highlands. Volcanic rises such as Western Eistla, Beta, Atla, and Bell Regios and possibly Central Eistla, Eastern Eistla, and Imdr regiones are characterized by shield volcanoes, extensive lava flows, extensional tectonics, and large positive gravity and geoid anomalies centered on the highland. These positive features probably are related to large mantle plumes. Plateau-shaped highlands such as the Ovda and Thetis regios and probably Tellus Regio, Alpha Regio, Phoebe Regio, Western Ishtar Terra, and possibly Fortuna and Laima tesserae were created by mantle downwelling. Atalanta and Lavinia planitae (bowl-shaped depressions containing compressional features and lacking such hot-spot features as coronae and large shield volcanoes) also may result from mantle downwelling. If mantle downwelling can deform and thicken the crust, these lowlands could evolve into plateau-shaped highlands.

7.3.2.4 Volcanoes and Intrusives

Prior to the Magellan expedition approximately 20 000 small domes of probable volcanic origin were imaged by Veneras 15 and 16 over the northern 25% of the surface of Venus (Fig. 79, Aubele and Slyuta 1990). Garvin and Williams (1990), using Arecibo radar images suggested that these features are venusian analogs of Icelandic lava shields. The more than 90% coverage of Venus by Magellan revealed at least 550 shield volcanoes with individuals having diameters of less than 20 km, 274 intermediate volcanoes having diameters of 20 to 100 km, 156 large volcanoes with diameters larger than 100 km (including anemone structures characterized by petal-like flows or radiating radar-bright digitate patterns; steep-sided domes; ticks having circular interiors, rims, and radial ridges or spurs extending away from the rim), and large volcanoes with diameters larger than 100 km. There are 86 caldera-like structures independent of those associated with volcanoes and having diameters of 60 to 80 km, 362 coronae, 259 arachnoids, 50 novae, 53 lava flood-flow fields, and 50 lava channels (Figs. 80–82, Head et al. 1992; Stofan et al. 1992). It appears that at least 80% of the venusian surface is paved with volcanic units; the remainder largely consists of tesserae which Head et al. believe may be deformed lava flows. The vast majority of these units may be basaltic in composition, but morphological variations in the igneous features denote a wider range of compositions. Associated with these extrusive and other intrusive structures are bright linear fractures resembling dike swarms on Earth. Such fractures were formed

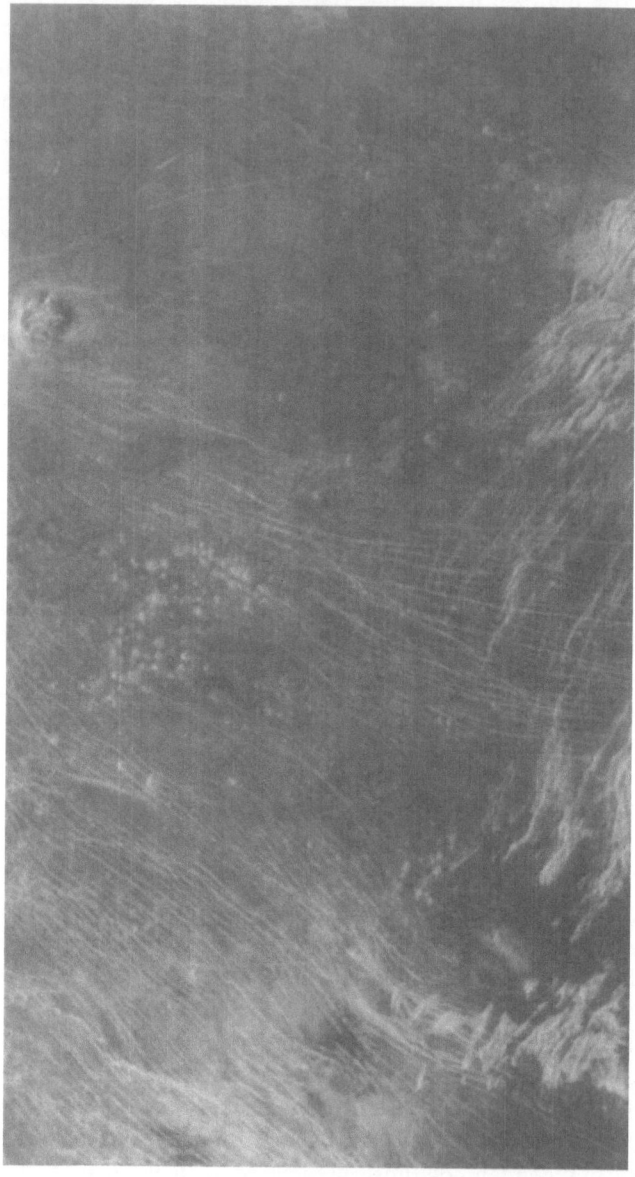

Fig. 79. Lineated plains and volcanic shields in the region of Sedna and Guinevere planitae (Figs. 67 and 68) on Venus. Photograph P38304 was provided by S. C. Solomon of Massachusetts Institute of Technology; courtesy of the Magellan Experiment Principal Investigator G. H. Pettengill, the Magellan Project, and NASA

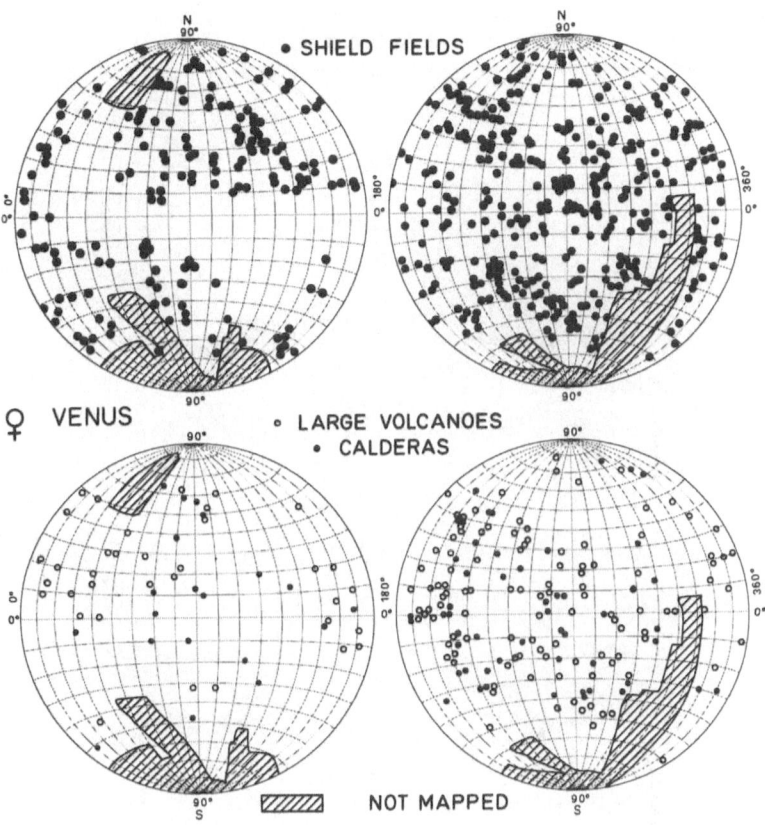

Fig. 80. Distribution of shield volcanoes, large volcanoes, and calderas on Venus. (After Head et al. 1992, figs. 2f, 4e, and 5d)

probably because the dikes failed to reach the venusian surface or the magma in the features was partially drained (McKenzie et al. 1992c). The large volcanic structures are associated with upwellings in the mantle convection patttern, but the coronae are correlated with neither upwelling nor downwelling patterns. Herrick and Phillips (1992) believe that the coronae are passive features associated with rifting caused by mantle convection.

Volcanic activity is concentrated near the equator in an area 10 000 km across, having more than 1500 volcanic centers (Kerr 1991d; Crumpler et al. 1992). These volcanic centers have diameters larger than 25 km and consist of more than 500 extrusive volcanoes (Crumpler et al. 1992), 553 fields of small shield volcanoes (Fig. 79; Aubele et al. 1992), and fewer than 500 structures representing surface expression of magma intrusion. These distribution centers with significant concentrations in the Beta-Atla-Themis region and possibly in the Alpha-Tellus-Tethus region are analogous to hot-spot distributions on Earth. Kerr (1991d) reported that the volcanic mass may originally have been emplaced far from the equator, but that

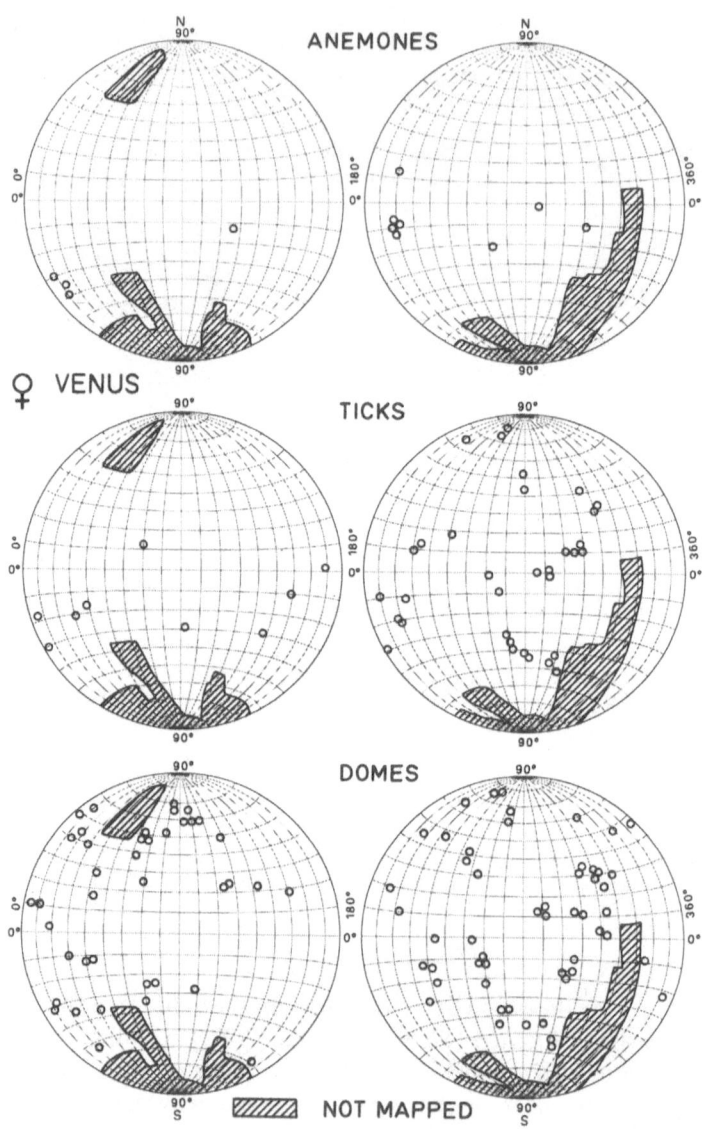

Fig. 81. Distribution of intermediate volcano types on Venus. *Symbol* for steep-sided domes represents clusters rather than individual structures. (After Head et al. 1992, figs. 3d, 3g, and 4k)

the added mass caused Venus to dip until the cluster arrived near the equator, a more stable position for such a large mass. Magellan data revealed four basic constructions: shield-shaped, flat-topped or table-like with steep flank slopes, dome shaped (steep, nonuniform slopes and broad top), and cone shaped. The most common type (the shield) is less than 100 to 200 m high, with a diameter of 2 to 8 km with some of the structures reaching dimensions of 28 km; they comprise

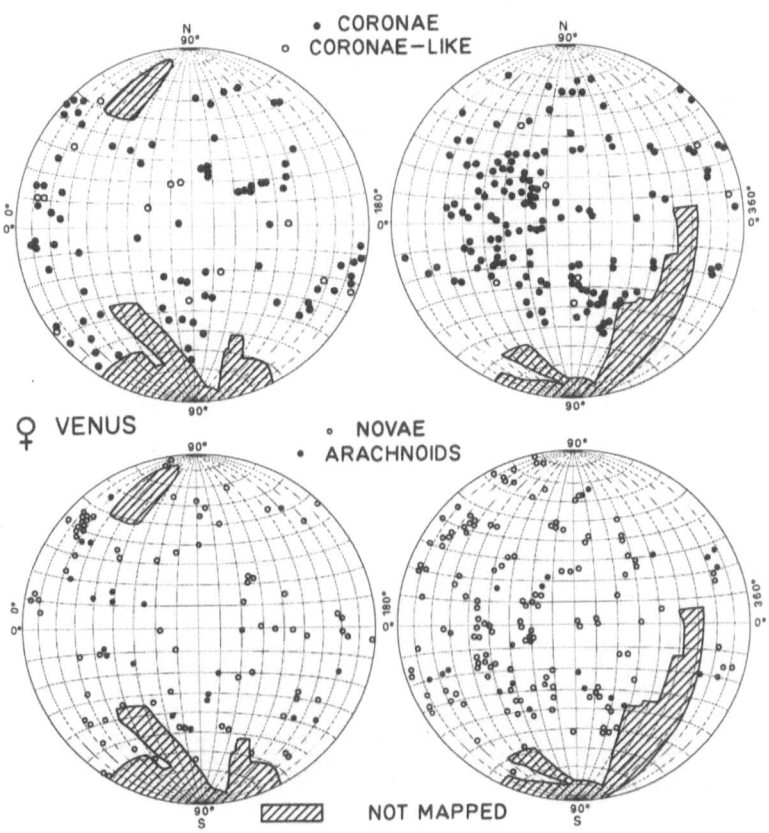

Fig. 82. Distribution of coronae, novae, and arachnoids on Venus. (After Stofan et al. 1992, fig. 1; Head et al. 1992, figs. 7c, and 8c)

more than 90% of the population, and most of the remainder are flat topped (Head et al. 1991a, b; Aubele et al. 1992). At least 90% of the shield volcanoes have single summit craters or pits with an average diameter of 700 m, but some contain double pits or summit domes (Fig. 83). Absence of a flexural moat caused by volcanic loading of the lithosphere around the periphery of the larger structures suggests gradual moat filling and coverage by radial flows (McGovern and Solomon 1992). Occurrences of partially filled depressions around Tepev Mons and possibly Maat Mons suggest that this infilling process is continuing. Analytical plate models indicate an elastic plate thickness of 10 to 20 km for the lithosphere at Tepev Mons. Smaller structures occur both as isolated structures and clusters. For example, in Guinevere Planitia is a cluster of 55 edifices 1.3 to 6.5 km in diameter superimposed on a system of small grabens. These shields are the most likely source of local intershield dark plains. Associated with some of the pits west of Alpha Regio are 1-km-wide channels that extend 10 km downslope.

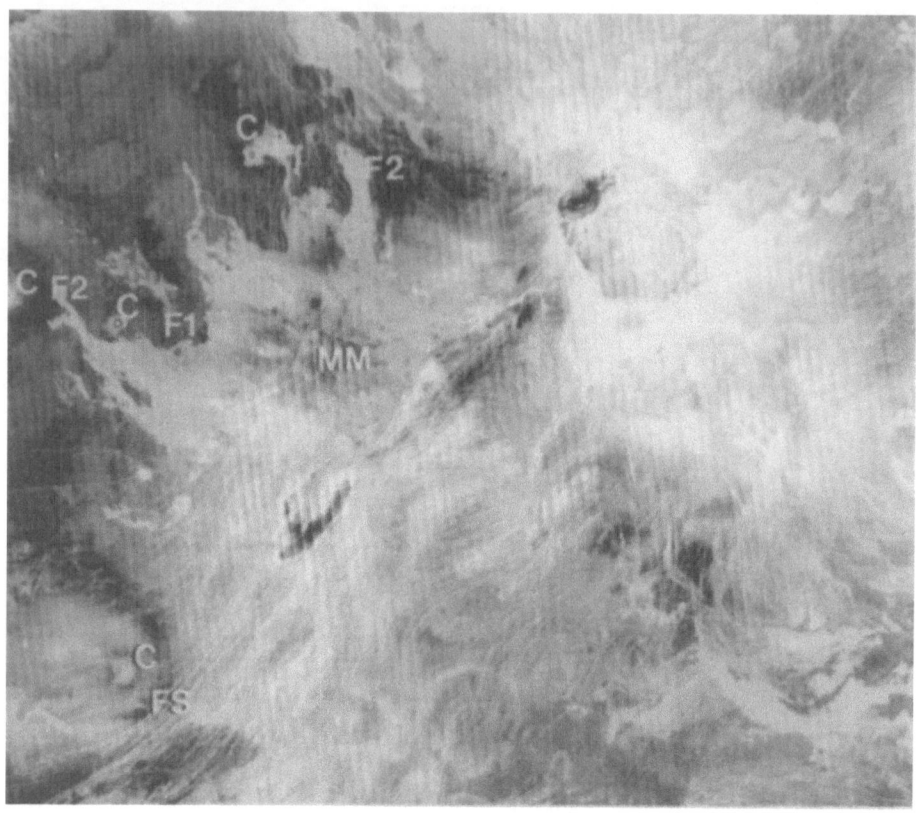

Fig. 83. Maat Mons (**MM**), a volcanic high at Lat. 1°S, Long. 194°E. This peak is the second highest topographic feature on Venus, rising 8 km. It is located along an extensive fracture system (**FS**). Note well preserved impact craters (**C**) and volcanic-flow lobes. The younger lobes (**F2**) are superimposed on older flows (**F3**). Photograph C100N197 extending from Lats. 7°S to 7°N and from Longs. 288° to 206°E (Figs. 67 and 68), courtesy of the Magellan Experiment Principal Investigator G. H. Pettengill, the Magellan Project, and NASA

The large edifices that have diameters of as much as several hundred kilo meters display massive outpourings of lavas tens of kilometers wide and hundreds of kilometers long. That the flows had extremely high effusion rates or very low viscosities is indicated by the long distances that they extend and the way that they flowed around topographic barriers. A 40- to 50-km diameter caldera on the summit of Sif Mons appears to be filled to its rim with lava, some of which perhaps overflowed the rim. This large structure was formed probably by large scale evacuation of the magma linked to intrusions along northwest-trending rifts or extrusion of large flows on the flanks (Head et al. 1991a). The smaller 3- to 10-km diameter calderas were formed probably by local collapse associated with the smaller volume near-summit flows. The Sif Mons edifice may have been constructed by voluminous melts and numerous small eruptions associated with a thermal plume that uplifted and fractured western Eistla Regio. The progression of roughness indicates that the large effusive deposits on Sif and Gula montes consist of superposed flows or multiple vents (Campbell and Campbell 1990). Sif Mons, with a relief of 1.7 km, is at the southern end of a northwest-trending deformed zone. It has a 40-km diameter caldera that contains numerous pits and depressions, reflecting multiple episodes of collapse, and the radar-bright lava flows associated with it extend 300 to 600 km (Senske and Stofan 1991). The 2.3-km-high Gula Mons, the southernmost of a group of three volcanic centers, contains radar-dark flows, has a northeast-trending summit rift zone, and is at the northern end of a northwest-trending deformed zone of faults, grabens, and scarps (Guor Linea). This zone cross-cuts and is onlapped by the flows, indicating that rifting and volcanism were contemporaneous (Senske and Stofan 1991). Also on the flanks of these volcanic highs are abundant smaller volcanic domes and mottled dark lava flows. From the density of impact craters Senske and Stofan estimated that the age of this volcanism is greater than the 1.0 to 0.5 Ga calculated for volcanism in other parts of Venus.

The lava field of Mylitta Fluctus in Lavinia Planitia is dominated by radar-bright lobate flows atop darker plains deposits and older lava flows, and it extends about 800 km north-south and 380 km east-west (Roberts et al. 1991). The source of these flows is a 40×20 km caldera located within a deformation belt of extensional origin at the northern edge of Lada Terra (Roberts et al. 1992). The younger flows are ponded behind a ridge belt within Lavinia Planitia. These flows are about 400 to 1000 km long, 30 to 100 km wide, and 10 to 50 m thick with individual flows having a volume of 100 km^2. On the proximal end of the flows is a complex braided and branching channel system; at Sachs Patera south of Lakshmi Planum is an elliptical depression 40 km in diameter and 130 m deep with a 30-km-diameter arcuate region to the north. This depression appears to have undergone two stages of collapse, with a 7-km structure along the southwestern rim of the caldera possibly representing a third stage. The structure formed probably by sagging, with much of the magma leaving the area via dikes and fissure eruptions at the south. Sacajawea Patera (Figs. 68 and 74) in Lakshmi Planum is a caldera having a diameter of 200–300 km and a depth of 1 to 2 km; it is the largest such structure mapped to date. Its northern rim averages 600 m higher than its southern rim (Head et al. 1991a). An annulus of volcanic plains 120×215 km within the caldera is surrounded by brighter deposits. Along the caldera's margin is a wide belt of concentric faults many of which are 4 to 100 km long and have the form of 1- to 2-km-wide grabens. In turn, these fractures are cut farther west and east by numerous lineaments. Zatla, a 6-km impact crater northwest of Sacajawea, post-dates the grabens. Extending southeastward from Sacajawea Patera is a system of linear features with a 12-km shield volcano lying along one of the structures. These structures were interpreted by Head et al. (1991a) as a flanking rift zone along which magma injection and extrusion occurred. From images recorded by Magellan, Head et al. suggested the following history for the region: (1) magma rising from depth reached neutral buoyancy in the upper crust of eastern Lakshmi

Planum and formed a magma chamber; (2) lateral eruption partly emptied the reservoir, leading to sagging of the roof of the chamber to produce an annulus of graben structures suggestive of a brittle crust several kilometers thick; (3) concurrent with the sagging was dike injection and occasional floods of lava on the outer edge of the annulus; and (4) during the last stages of collapse the caldera was flooded with lava several kilometers thick, possibly forming a lake. Sacajawea and Sachs paterae, with their annuli of grabens and lack of significant edifice deposits, are different from more typical calderas in such basaltic shields as Sif Mons.

Unless the volatile content exceeds 4% the high surface pressure should inhibit pyroclastic activity on Venus (Head and Wilson 1986). However, at least one such pyroclastic site has been tentatively identified. This one is in Guinevere Planitia (Figs. 67 and 68) nearly 0.5 km below the datum, where the pressure is between 9.0 and 9.5 MPa (Head et al. 1991a). If this identification is correct, the extensive blanket of pyroclastics (20 km or more in diameter) is suggestive of a Plinian eruption. Radar-bright areas south of these craters are believed by Head et al. to be the result of erosion of a fine-grained blanket by eddies induced in prevailing winds by craters and volcanic shields in the region. Local darkening in the bright areas suggests that the pyroclastics are thicker at the edges of the eroded areas than elsewhere. On Earth, only 10 to 20% of magmas are extruded to form flows and edifices. Head et al. speculated that this ratio of intrusion to extrusion may be less on Venus because: (1) depth to a specific magma compaction is shallower on Venus; (2) high atmospheric pressures on Venus would tend to decrease differences in melt/rock densities; and (3) high-density crust is more abundant on Venus than on Earth. That magmas reach shallower depths on Venus may be reflected in the abundance of fractures and grabens, with the fractures serving as avenues for dike injection. Another type of injection occurs in one of the three flat-topped domes in Sedna Planitia (Fig. 68; Lat. 36°N, Long. 330°E). The top of this dome is fractured in a star-like pattern that Head et al. believed was formed by updoming of plains units in a manner similar to that of laccoliths and cryodomes on Earth.

Lava flows on Venus display a wide variety of forms. Those near the source may have central channels with levees, but the flows farther downslope are broader, have braided channels, levees, and breakouts, and they display lateral textural changes suggestive of both tube-fed and sheet flows. Sinuous channels that terminate on low areas and form large plains deposits also have been noted. Their morphology implies low-viscosity lava or lava effusion rates that allow turbulence and erosion to incise channels into preexisting plains deposits. Such erosion may be enhanced by the high temperatures on Venus. Baker et al. (1992) classified the venusian channels into simple, complex, and compound. Simple channels are long, single main types; complex forms include anastomosing or braided types; compound channels consist of both simple and complex types. Simple channels include sinuous rilles (that resemble their lunar counterparts) and canali (long sinuous channels of high width-to-depth ratio and constant width). Most canali are on volcanic plains. Baker et al. suggested that the channels were eroded by ultramafic silicate melts, sulfur, and carbonate lavas. Channelling processes also indicate the possibility that some may have an aqueous origin, a possibility that cannot be excluded with absolute certainty. Massive extrusion has created an extensive drainage system in Lada Terra (Lat. 50°S, Long. 21°E; Fig. 68) including islands shaped by fluid dynamics, streamlined structures, braiding, and both mechanical and thermal erosion. The morphology suggests that fluid ferrobasalts and possibly even komatiites (high-temperature Mg-rich basalts) may be common on Venus. However, the resurfacing rate by this massive and impressive volcanism may be quite low. Head et al. (1991b) estimated that the flow is less than 2 km³/yr; thus, the mode of crustal recycling has yet to be resolved.

Steep-sided domes, known as 'pancakes' or farrum, have been noted in Magellan images (Fig. 75; Lat. 30°S, Long. 11° to 13°E, east of Alpha Regio). To date, after a survey of more than 95% of the venusian surface, 145 such structures have been mapped (Pavri et al. 1992). They have diameters that range from less than 10 km to almost 100 km, a mean diameter of 23.8 km, and a mean relief of 700 m. They are in sharp contact with surrounding plains and display well-defined fracture patterns, radial cracks on the outer steep slopes, an annulus of radial and concentric ridges and cracks on the outer edge of the flat top, and an inner network of radial and concentric cracks. Cracks on the surface of the domes probably were caused by the tensions created by spreading viscous currents, and radial cracks in the plains beyond the domes by stress induced by this viscous flow in the plains material (McKenzie et al. 1992a). Morphologically, these features resemble dacite-rhyolite domes on Earth. Gamma-spectrometric measurements of Venera 8 at Lat. 10°S, Long. 335°E) on or near one of these structures indicate that the rocks at the landing site are geochemically close to quartz monzonite/quartz syenite, appearing to verify this interpretation (Head et al. 1991a). The siliceous structures of Venus, however, have volumes of 50 to 250 km³, whereas most such structures on Earth are typically < 1 km³. Pavri et al. (1992) proposed two models for emplacement of the domes. In one model, the compositionally evolved magma model, basaltic magma is differentiated to produce siliceous magmas that then erupted through fissures; a related possibility is that the domes are ignimbrite eruptions. In the second model, the basaltic bubble enhancement model, the domes result from extrusion of basaltic foams following volatile enhancement of the upper part of the magma chamber.

Another site of siliceous volcanism on Venus may be in eastern Aino Planitia (Fig. 66; Moore and Schenk 1992; Moore et al. 1992). Magellan images from this region show a flow complex of three overlapping phases of flow festoons or surface ridges. The flows extend as far as 260 km and their margins are rather steep, suggesting viscous flows. Flow viscosities estimated from the spacing of parallel ridges are on the order of 10^{11} P. On Earth, such viscosities and morphologies are characteristic of flows of siliceous composition: andesite, dacite, and even rhyolite. The radar-bright surfaces that extend 360 to 400 km from the volcano sources may represent very thin deposits of low viscosity lavas or are the result of an explosive eruption. Moore et al. (1992) proposed that these bright surfaces and the lavas extruded by the 0.5–1.1-km-high domical structures were derived from a magma differentiated within the crust or mantle, a differentiation that produced a gas-rich low viscosity phase, high viscosity lavas, and a residual primary magma. A similar ridged flow complex also may be present in the mountainous terrain of eastern Ovda Regio where relict topography influenced the morphology of the flows.

Another type of viscous flow on Venus is noted in the form of ridges having surface areas of about 48 000 km² and thicknesses of 56 to several hundred meters in Ovda Regio. These two ridged flows are dominated by ridge and trough segments and lobate and digitate flow lobes. The surface morphology of the flows resembles andesitic to dacitic flows on Earth. They probably are less silicic than the

Fig. 84. Quetzalpetlatl Corona with an 800-km diameter is near the bottom left of the photograph. Three similar structures are associated with this feature and are along the edges of the photograph that extends from Lat. 67° to 52°S and Long. 37° to 351°E (Figs. 65 and 66) on Venus. Photograph C160S014, courtesy of the Magellan Experiment Principal Investigator G. H. Pettengill, the Magellan Project, and NASA. Similar-type structures also are present on Earth, where they are much smaller with diameters ranging from 15 to 50 m and are the result of vertical pressures caused by ice blocks moving up and down by tidal motion (Dionne 1992). They display well-defined circles with an outer ring having a relief of a few centimeters and a width of 25–60 cm. Some of the structures also display an inner ring and a chaotic terrain of alternating small ridges and depressions. They form every winter in the mud tidal flats of the St. Lawrence estuary and are capable of being fossilized and preserved in the geological column

pancake domes, but more silicic than the ridge flows near Artemis. A much thinner ridged flow at Lat. 51°N, Long. 147°E is considered by Moore and Schenk (1992) to be more mafic than the flow in Ovda Regio. Flows having a more mafic composition also were mapped near a small ring caldera at Lat. 53°N, Long 98°E, on the flanks of Renpet Mons, and at Lat. 10°N, Long. 273°E, Lat. 16°N, Long. 68°E, and Lat. 64°N, Long. 120°E.

Unusual igneous structures recognized only on Venus are coronae (Figs. 82, 84, and 85), circular to elliptical structures (Head et al. 1991a; Solomon et al. 1991). To date, 362 of these structures have been mapped on the planet (Stofan 1992; Stofan et al. 1992). They are 60 to more than 2000 km in diameter, completely or partially surrounded by annuli of concentric ridges, commonly with a raised

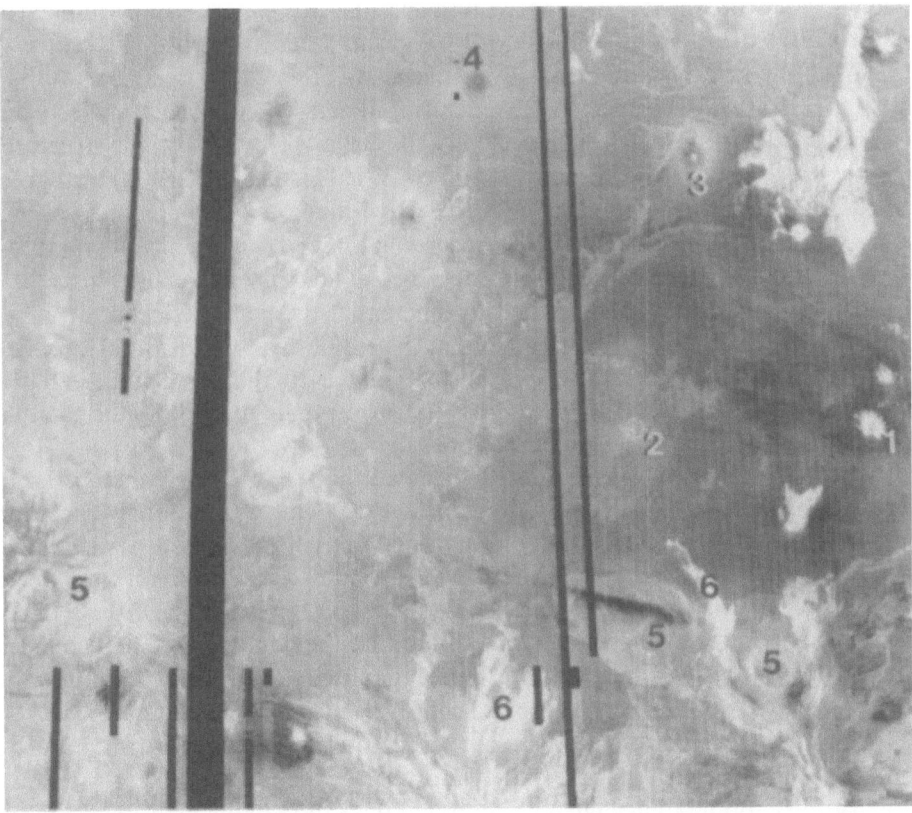

Fig. 85. Corona structures (5), lava lobes (6), and impact craters (1–4) shown on imagery of Venus (Lat. 22° to 37°N, Long. 20° to 340°E; Figs. 65–68). Impact crater **1** is shown in more detail on Fig. 70. If structure **2** is an impact crater the extent of ejecta and its lobe-like terminations indicate a very fluid state. Note that the lava lobes on the *bottom right* appear to originate from coronae. Photograph courtesy of the Magellan Experiment Principal Investigator G. H. Pettengill, the Magellan Project, and NASA

interior topography, a peripheral trough or moat, and numerous associated volcanic and tectonic features. The interiors of many are covered by lava flows and are accompanied by calderas, shields, domes, and lava channels – features that are believed to have been produced by mantle upwelling. According to Squyres et al. (1992), coronae are among the dominant tectonic forms on Venus. One of the largest coronae is Quetzalpetlatl (Fig. 84) with an 800 km diameter centered at about Lat. 68°, Long. 355°E in northern Lada Terra. The approximately 1060 km diameter Heng-O Chasma (Stofan et al. 1992, table 1; in the western Tinatin Planitia), Aditi, and Penardum are other coronae analogous to those recognized from Veneras 15 and 16 and Magellan images (Plaut et al. 1990; Sandwell and Schubert 1991). A quasi-circular corona-like structure having a diameter of 130 km also has been observed at Lat. 59°S, Long. 350°E, about 65 km west of the source of Mylitta Fluctus. Topographic, structural, and stratigraphic observations suggest that both structures formed at the same time. They display contrasting styles of magma activity along the rift zone, with the corona-like features being dominated by structural deformation and minor volcanism, whereas Mylitta Fluctus was dominated by volcanism to create a circular area with a sharp rim. According to Squyres et al. (1992), volcanism occurred during all phases of corona formation, but it may have diminished as relaxation followed. The diversity of volcanic structures and the presence of steep-sided domes and cones led Head et al. (1991a) to suggest that eruption of some viscous magmas may have changed with time; thus magma composition also may have changed. According to Sandwell and Schubert (1991), Heng-O, Aditi, and Penardum display evidence of lithospheric flexure along their perimeters – this flexure is attributed to differential thermal subsidence at either side of the coronal rim. They estimated that the elastic layer is at least 30 km thick with a 40-km-thick plate providing a close match to the observations. If the temperature at the base of the elastic layers corresponds to 1070 K, the thermal gradient outside Heng-O is 8 to 11°C/km.

Stofan et al. (1991) and Stofan (1992) stated that the major characteristics of coronae can best be explained by hot-spot activity with gravitational relaxation being responsible for modification of the primary structure. Squyres et al. (1992b) believed that the coronae were formed by the ascent of a mantle diapir that forced the lithosphere upward to form a dome. As the diapir impinged on the base of the lithosphere it flattened and spread laterally so that the dome relaxed and changed to a more flat-topped shape. As the flattened diapir cooled, the raised lithospheric segment relaxed still further to form a central sag, a raised rim, and a moat.

Janes et al. (1992) also proposed that a corona results from a rising mantle diapir impinging against the base of the lithosphere, which spreads laterally to form a plateau-like topographic high. Gravitational relaxation of the plateau as a result of the cooling of the thinned near-surface diapir and the hotspot-thinned lithosphere produced morphological features characteristic of a corona including a raised rim, an annular moat with its concentric extensional features, and a central region at an elevation below the rim but higher than the surrounding terrain. In scenarios discussed by Sandwell and Schubert (1992) the ridge, trench, and outer rim topography of the corona are caused by tectonism or thermal subsidence. In the tectonic model the head of the plume thins and weakens as it approaches the lithosphere and spreads laterally. As melt and hot mantle material accumulate on the surface of the lithosphere above the diapir, the lithosphere collapses under its load. At the same time the older and denser lithosphere surrounding the circular zone of subsidence sinks into the mantle, forming a circular subduction zone that increases in radius with time. As the interior spreads laterally to fill the growing zone of subduction, a trench and outer rise is created. Examples of such origin are Artemis and Latona. In the thermal model the trench and outer rises are the result of a plume that spreads radially as it reaches the base of the lithosphere, and thermal conduction and/or advection thins the lithosphere to create a circular area with a sharp rim. As the hot interior cools and subsides relative to the cooler exterior, a ridge, trench, and rise topography is formed. Sandwell and Schubert (1992) stated that the topography of Heng-O matches this thermal subsidence model.

Structures, which Janes and Squyres (1991) named novae (Fig. 82; 50 mapped to date), have only radial fracturing and a dome that rises as much as 1 km above the surrounding plains. Janes and Squyres believed that the novae are young coronae or coronae that were arrested during their development. Sukhanov et al. (1989) named somewhat similar circular complexes with gently sloping concentric rims and radial lineaments arachnoids (Fig. 71). They consist of concentric or circular patterns of fractures or ridges that extend outward for several radii (Head et al., 1992). The radial ridges, some of which appear to be transitional to graben-like features, may indicate crustal shortening. Of the 259 arachnoid structures mapped by Head et al., most have diameters between 50 and 175 km, peaking between 100 and 125 km. They tend to be associated frequently with coronae, suggesting that both have a similar volcanic/tectonic origin, but at different scales. Head et al. believed that arachnoids were formed by intrusion, that the radial features represent dike intrusions, and that the compressional ridges are caused by subsidence-induced tectonism. The limited evidence of volcanism suggests that extrusive activity is minor in comparison with intrusion in the formation of arachnoids.

7.3.3 Exogenic

A study of 3400 wind streaks by Greeley et al. (1992) indicated that long and narrow forms are the most abundant, that they are most common between Lats. 17° to 30°S and Lats. 5° to 53°N, and that they are associated with certain impact craters and deformed terrains that provide debris that can be entrained by the low-velocity winds of Venus. Other eolian features described by Greeley et al. from Venus are the Aglaonice transverse dune field centered at Lat. 25°S, Long. 340°E, a 1290-km^2 field of transverse dunes associated with an ejecta flow channel from the Aglaonice Crater, the Meshkenet field at Lat. 67°N, Long. 90°E in a valley between Ishtar Terra (Fig. 66) and Meshkenet Tessera, and possible yardangs at Lat. 9°N, Long. 60.5°E about 300 km southeast of Meade Crater. Possible dunes may be present in the region of the 'crater farm', a set of four craters with Carson being the easternmost. These dunes are roughly perpendicular to the dome-related streaks and they may be the transverse type emplaced by westerly winds. Eolian processes evidently occur widely over Venus.

Arvidson et al. (1991) analyzed Magellan images to determine the rate and history of exogenic processes on Venus. The Pioneer Venus data showed that the average Fresnel reflectivity was about 0.14, indicating that soil cover on most of Venus is less than tens of centimeters thick. However, similar measurements over the tesserae reveal the presence of a thicker soil. These reflectivity measurements have been verified by Magellan data. Low emissivities were recorded by Magellan on Maxwell Montes, Danu Montes, Gula Mons, and Sif Mons (Fig. 68). On Maxwell Montes the low emissivity correlates better with large-scale slope than with elevation. Arvidson et al. (1991) and Klose and Wood (1991) interpreted these results as indicating that the material on the surface is reacting with the dense atmosphere to form products having low dielectric constants. Klose et al. (1992) proposed that the low emissivity (the complement of power reflectivity) was produced by secondary weathered surface material, pyrite – an electrical semi-conductor mineral. Arvidson et al. (1992) reported that surface degradation is produced by in situ weathering, that this weathering is elevation dependent, that surface modification is orders of magnitude slower than on Earth, but that it may be comparable to the estimated rate of weathering and erosion of bedrock on Mars. The weathered surface material could be transported down slopes, exposing fresh rocks that in turn could undergo weathering. Such downslope displacements by rock slumps, rock and/or block slides, rock avalanches, and debris flows have been mapped in areas of high relief and steep slopes on Venus (Malin 1992). The gravitational structures appear to be larger than the subaerial ones on Earth, but smaller than those on Mars. Malin suggested that about one major slide per year may occur on Venus if the planet is as seismically and volcanically active as the Earth.

As previously mentioned, parabolic low-emissivity streaks also have been noted within 30° latitude of the equator near seven or eight craters. These streaks are open toward the west with their axes parallel to latitude, and they have east-west dimensions of 500 to 1000 km (Arvidson et al. 1991). Each parabola is associated with an impact crater located near the focus of each parabolic streak. Their low reflectivity is believed to be due to smoothing of the terrain by mass wasting and deposition of eolian dust, silt, and sand. Streaks several kilometers long and associated with the 38-km-diameter impact Carson Crater are believed to be the result of perturbation of wind flow by neighboring domes, with erosion taking place downwind and deposition occurring along the edge of the erosion zone (Arvidson et al. 1991). Some streaks northeast of Carson Crater have a backscatter cross section that is 9 dB higher than that of their surroundings, suggesting that surficial ejecta has been eroded and underlying material exposed. Some streaks also have backscatter perimeters probably resulting from deposition of material eroded from the parabolic streaks. The dimensions of the parabolic streaks suggest that they are the result of high retrograde westward winds (velocities of 100 m/s have been measured at a height of 60 to 70 km) that disperse material westward. Campbell et al. (1992) proposed that this material was injected into the upper atmosphere during meteoric impacts and it took about 2 h to settle. Material ejected eastward would tend to be transported toward the crater, and material ejected north and south would be deflected westward – creating the parabolic images. Arvidson et al. (1991) noted that each streak covers about 100 times the area of its crater source. As no crater is capable of producing enough fine-grained ejecta to be carried 1000 km westward, they suggested that the streaks are due to long-term weathering and eolian transport by surface winds having a strong westerly component. Local topographic irregularities have exerted a strong influence on the distribution, so some of the streaks trend north to north-northwest parallel to the topographic trend and others are oriented west-northwest and northeast at right angles to the topographic grain. Dome-related streaks associated with pyroclastic debris also occur between Lat. 25° and 30°N and appear to be caused by southwest to south-southwest winds. These observations together with those from the Carson impact crater indicate that wind flow on Venus is both zonal and meridional, a flow that is modified by local topography. As a result of the high surface temperatures on Venus, these eolian sediments may tend to adhere to the surface to form a thin layer of accreted material, an accretion that is due to 'cold welding' processes (Marshall et al. 1991), perhaps similar to the formation of ignimbrites of ash falls on Earth.

Fractured plains also appear to have been modified by exogenic processes. If the average retention age of the plains is the same as the age of average craters, the more than 400 Ma of exposure to exogenic processes have been long enough to subdue or remove meters of low relief and heights associated with blocks (Arvidson et al. 1991). During a short time the plains probably were modified by eolian processes, but during a longer period homogenization must have been due to chemical weathering. Such weathering can be seen in the tesserae. The rough tectonic relief apparently enhanced mass wasting and formation of thick soil. Evidence of such mass wasting has been observed in some of the Magellan images, and the process is reflected in the irregular contacts between tessera blocks and surrounding plains. Mass wasting also may be responsible for the degraded nature of the crater on a tessera at Lat. 64°S, Long. 120°E. The plains also were modified by volcanic flow erosion, leading to the formation of channels, streamlined erosional remnants, ponding, overtopping of topographic highs, and the construction of anastomosing patterns. According to Arvidson et al., no unequivocal evidence of fluvial, lacustrine, or marine landforms has been seen on any of the Magellan images. There is no evidence to indicate that Venus had a different climate during the past 800 Ma; consequently, for the past 800 Ma Venus has been affected by the same exogetic conditions that exist today. The rate and intensity of these processes is so low that some endogenic and exotic terrains may be as old as 800 Ma.

7.3.4 Summary

As demonstrated above, the surface morphology of Venus is essentially a product of internal or endogenic forces. Features due to exotic processes constitute probably less than 2% of the morphology of the planet, and external or exogenic processes have not modified endogenic/exotic terrains to any significant degree. Whatever morphologic changes have occurred are the result of endogenic processes. Volcanic resurfacing on Venus, for example, is so effective that very little terrain older than 1 Ga apparently has survived. Only the highlands may contain records of older deformation episodes. As pointed out by Solomon et al. (1991), in the absence of exogenic processes the only way that topographic highs on Venus can be destroyed is by ductile flow of the lower crust that partly supports the topography once compressional forces have ceased. Numerical models suggest that these positive elements probably have a life span of only 100 Ma.

Endogenic processes on Venus can be attributed to two mechanisms. Features which are a few hundred kilometers long are likely to be products of mantle convection and its dynamic processes, and smaller features are likely to be results of internal deformation of the crust or of tectonic deformation of a thin upper crustal layer decoupled from the rest of the crust (Solomon et al. 1991). Volcanic features such as coronae are the result of mantle plumes, whereas broad topographic highs are products of broad extension and greater production of magma. The steep scarps associated with Lakshmi Planum are clear evidence that these forces have been active in recent geological time. To date no clear evidence of a plate tectonic regime, where deformation is restricted to zones a few to tens of kilometers wide along plate boundaries has been recognized on Venus. No analogs comparable with trenches, fracture zones, or spreading ridges found on Earth have been imaged on Venus. Why is there no plate activity? Is it because the lithosphere is so thick that it cannot be broken into flexible plates, or is it because the venusian lithosphere is so thin (the higher surface implies such a lithosphere), hot, and light that it cannot be subducted? Solomon et al. (1992) ascribe the unique tectonic fabric of Venus to three factors: (1) the absence of exogenic processes so that different episodes of deformation are preserved; (2) a high surface temperature that facilitates ductile behavior in the mid- to lower-crust, giving rise to a variety of small-scale endogenic features, and (3) because Venus lacks a low viscosity layer in its mantle, convection in the mantle is so strongly coupled to the lower lithosphere that crustal stress fields propagate for long distances.

8 Mars

8.1 Planetary Setting

The planet Mars was well recognized in very ancient times because of its brightness and red color. About 1000 B.C. it was named Nergal by the Babylonians after their god of death and pestilence, and Horus the Red by the Egyptians. Later, the Greeks named it Ares for their god of battle (who supported the Trojans), and the Romans renamed it Mars for their god of war (thus the term, martial). The symbol for Mars is a shield and spear. Mars' two moons, discovered in 1877, are known appropriately as Phobos (Fear) and Deimos (Panic) from the two horses of Mars' warchariot. Perhaps because of their similar brightness, Mars and Venus are associated in mythology with Harmony, born of the union of strife and love; similarly, Ares and Aphrodite became the parents of Eros (Cupid). Mars has a mean equatorial diameter of 6794 km (0.51 times that of the Earth), a mass of 6.418×10^{26} g, a mean density of 3.93 g/cm^3 (Earth has a density of 5.52 g/cm^3), an elliptical orbit of 780 Earth days, a daily rotation rate of 24 h and 37 min, and perihelion and aphelion distances from the Sun of 1.381 and 1.666 AU, respectively (Table 1; Carr 1984). Mars' present obliquity of 25.1° may have changed during the past because of waxing and waning of polar ice caps (Rubincam 1992). These glacial advances and retreats tend to change Mars' dynamical flattening, which is out of phase with the Sun. This in turn produces an annual solar torque on Mars that varies the angle between the equatorial and orbital planes. Rubincam estimated that climatic changes in martian history were capable of secularly increasing the planet's obliquity by 1° or 2° since the Solar System was formed. Such a change, in turn, would enhance the martian seasons. Additional change in the obliquity of Mars may have been caused by creation of the Tharsis bulge, an increase of $+7°$ (Ward 1979). Bills (1990), however, indicated that until we know the dynamical history of the planets we cannot know the exact effect that the bulge had on the obliquity of Mars.

The first telescopic observations of Mars by Galileo Galilei in 1610 were followed in 1659 by a drawing of Mars by Christiaan Huygens; the drawing showed a triangular dark patch,the Syrtis Major, and more details were added to it by W. R. Davis in 1864–1865. More detailed maps produced by Giovanni Schiaparelli between 1877 and 1888 showed other dark patches that he connected by straight lines, which many later viewers were encouraged to interpret as irrigation canals leading from seasonally white polar areas – ice caps that had been discovered by

Giovanni Cassini in 1666. Other maps showing versions of these linear markings
were made by several astronomers, especially Percival Lowell (who established the
Lowell Observatory in Arizona) who, between 1894 and 1909, claimed 500 such
canals between oases (Fig. 86) that were discribed and named by his associate,
W. H. Pickering, also of Harvard Observatory. Their detailed maps (Mutch et al.

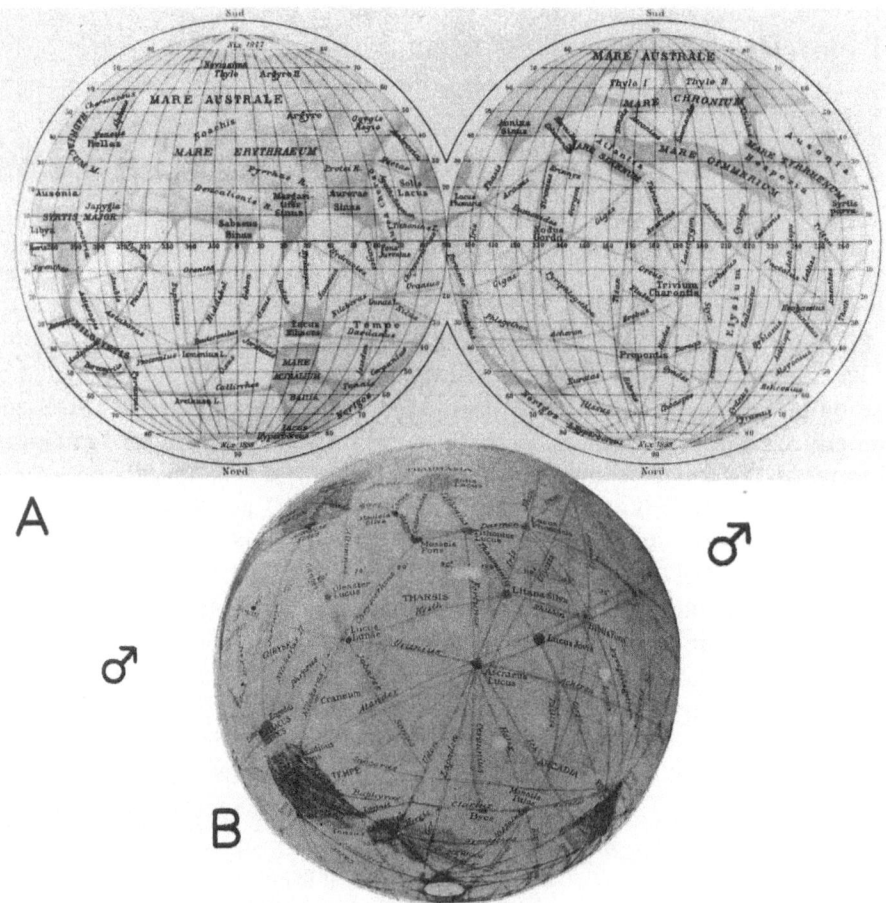

Fig. 86. Early maps of Mars made by viewing from telescopes and thus oriented with south at the top,
unlike the other maps in this book which were made by other methods **A** Schiaparelli's map of 1877.
Note that Syrtis Major, a prominent dark feature, is at the far left, Tempe is near the *middle*, and
Elysium is at the *right*. For comparison see Fig. 92. These and other large features are connected by a
network of straight lines termed *canali* grooves by Schiaparelli (Mutch et al. 1976, fig. 1.10). **B** Lowell's
1908 map of Mars, showing Tempe at the *left*, Tharsis above the midpoint, and Arcadia at the *right*. The
map by Lowell is similar to but different in detail from the one by Schiaparelli. Lowell considered the
canali to be man-made waterways or canals leading from ice caps (*white*) in the polar regions to once-
inhabited oases thereby leading to his cultural and political inferences about the martians

1976, pp. 20–25; Hartmann 1983, pp. 27–28; Wilford 1990, pp. 18–35) led to widespread belief in advanced life on Mars and to the writing of many science-fiction stories as well, including especially 11 novels by Edgar Rice Burroughs, who also was famous for his series of Tarzan books, although he visited neither Mars nor Africa. An adaptation of H. G. Wells' (1898) book *War of the Worlds* was radio-broadcast on 30 October 1938 by Orson Welles and unintentionally spread considerable panic in the United States (Wilford 1990, pp. 36–57). All this imaginary topography and population vanished when successful flybys, orbiters, and landers began sending information, beginning on 14 July 1965 (Moore 1990, p. 125). These probes were the American Mariners 4, 6, 7, and 9 and the Soviet Mars 4, 5, and 6 plus Phobos 2 that photographed one of Mars' moons. The only landers that transmitted information were the American Vikings 1 and 2.

The gravitational field and the general topography permit the calculation of crustal thickness for a tenth-degree model with an assumed crustal density of 2.9 g/cm^3, a mean crustal thickness of 40 km, and a crust-mantle density contrast of 0.6 g/cm^3. A point of ground truth was provided by seismic data from Viking 2 at Lat. 47°53′N, Long. 225°52′W. The greatest computed crustal thickness of 77 km is in the Tharsis volcanic region, and the thinnest crustal thickness of 8 km is at Hellas Planitia (Bills and Ferrari 1978). On the basis of the early bombardment of the martian crust, terrestrial and lunar impact sites, explosion and ejection data, and hydraulic theory, MacKinnon and Tanaka (1989) concluded that the crust of a fractured layer averaging 10 km thick is capped by an ejecta layer 1 to 2 km thick. They believed that this two-layer model is compatible with erosional and mechanical discontinuities at depths of 1 to 3 km, noted in several broad areas of Mars, and the initiation of fluvial and debris flows and of mass wasting that have produced certain landforms on the martian surface. According to Johnson and Toksöz (1977), Mars' mantle is enriched in FeO relative to that of the Earth, corresponding with an olivine composition of 75% forsterite and 25% fayalite and is essentially homogenous, with the olivine-to-spinel phase transition occurring at depths between 1200 and 1500 m. Acoustic velocities in the upper mantle typically range from 7.64 to 7.80 km/s with a density inversion caused by dominance of thermal expansion over compression, possibly producing an asthenosphere at a depth of about 250 km (Johnson and Toksöz 1977). Local partial melting could occur in the asthenosphere if a small amount of water is present in the upper mantle system.

The martian thermal history, as summarized by Hartmann (1983, pp. 285-286), is that Mars may have started with a magma ocean that produced several ancient lava-flooded impact structures, that the mantle reached the melting point of iron before silicates melted and that downward drainage of iron formed a core before an asthenosphere of molten silicates was created, and that the asthenosphere appeared 3 to 0.5 Ga ago at a depth of about 400 km. The magma ocean, which was the source of the lava plains and huge volcanic cones, has largely disappeared, leaving a lithosphere at least several hundred kilometers thick. Johnson and Toksz stated that Mars has a core with a radius of 1500–2000 km (about 40% of the radius of the planet; Hartmann 1983, p. 284) and a composition on the Fe-rich side of the Fe–FeS zero pressure eutectic composition. Its lower density, 0.71 times that of

Earth, suggests that the core has a lower percentage of total iron than that of Earth and that it has a higher proportion of volatiles (Arvidson et al. 1980). The presence of a molten core is supported by Mars' very weak magnetic field with a dipole of 3 $\times 10^{-4}$ times that of the Earth's (Smith et al. 1965; Dolginov et al. 1973). According to Intriligator and Smith (1979), this planetary field corresponds with a surface field of ~ 20 gammas and a dipole moment of $\sim 8 \times 10^{21}$ gauss/cm^3. The low magnetic field may be caused either by stratification within the core or by the presence of less conductive material than on Earth (Johnston and Toksöz 1977; Merrill and McElhinny 1983, pp. 344–346).

In contrast with the Earth, the low-degree gravity field and topography on Mars display a high degree of correlation (Lorell et al. 1972). Gravity data indicate that the regional topography of the planet can be classified in two groups, an older one that is isostatically compensated at relatively shallow depths and a younger group composed mainly of the Tharsis plateau and adjacent low areas of Chryse and Amazonis planitias that are only partly compensated, with the depth of partial compensation also shallow (Phillips and Saunders 1975). Analysis of the gravity data indicates that the depth of complete compensation of the old terrain is no more than 100 km and the depth of partial compensation of the younger terrain is no more than 150 km. According to Phillips and Saunders, it is reasonable to assume that the depth of compensation for both terrains probably is the same, no more than 100 km. Mars' topography apparently is consistent with a model of viscous relaxation in a regime of approximate thermal steady state. Density anomalies in the crust and lithosphere contribute more to the long-wave gravity field on Mars than do such anomalies on Earth (Lambeck 1979). Gravity data indicate that the martian crust and lithosphere generally are thicker than on Earth and consequently they are able to support larger uncompensated loads than on Earth (Phillips and Lambeck 1980; Smith et al. 1990). A 12th degree and order spherical harmonic model for the martian gravitational potential, based on two-way Doppler data collected by ten Deep Space Network stations during Mariner 9 and Vikings 1 and 2 missions, show complete compensation on a global scale, supporting the contention that most of the geological activity on Mars terminated 1 to 2 Ga ago (Christensen and Balmino 1979).

According to Davis and Golombek (1990), photoclinometric profiles across 125 erosional features and 141 grabens in the equatorial region display three discontinuities within the shallow crust. The levels at depths of 0.3–0.6, 1.0, and 2–3 km in the pits, troughs, and valley walls within Noctis Labyrinthus and Valles Marineris, and the escarpments within the fretted terrane of Sacra Fossae and Kasei Valles may represent: (1) the contact between the layered rock making up the ridged plains and the underlying regolith; (2) the base of the ground-ice layer; and (3) the deepest discontinuity corresponding with the base of the martian regolith and the contact between the overlying ejecta and fractured basement rocks. The 1-km deep discontinuity (which may represent the contact between ice-laden and dry regolith, ice-laden and water-laden regolith, or pristine and cemented regolith) limits the depth of sapping canyons (Davis and Golombek 1990). High-resolution Viking orbiter images indicate that a heavily-cratered terrain forms the basement of Chryse Planitia; above this surface are deposits of eolian, fluvial, and volcanic origin. The floor of the plain has been modified by at least two sets of channels, lunar-like mare ridges, mesas and plateaus, and fields of knobs (Greeley et al. 1977).

Viking seismic investigations indicate that Mars is an inactive planet (Tittman 1979). No definite seismic signal was identified during the 2100 hours that a three-axis short-period seismometer operated on the surface of Mars in the Utopia Planitia region (Anderson et al. 1977). A possible local seismic event was detected on Sol 80, but no wind data were obtained at that time, so wind disturbance cannot be ruled out as the cause of the event. In contrast, Golombek et al. (1992) believe that the level of structural/volcanic features on the martian surface suggest that Mars is active at present. They believed that stresses caused by cooling of the martian lithosphere should result in seismic activity, and that this seismicity is greater than on the Moon but less than on the Earth. As they indicate, if Mars is active, these seismic events can be used to map the planet's interior. As postulated by Solomon (1978), a single thick rigid immobile lithospheric plate encloses the entire planet of Mars now and likely has done so since the creation of the oldest preserved surface geological unit. Downwelling of such a plate would be short lived and incapable of disrupting the lithosphere; thus, a mid-ocean ridge system comparable with that on Earth does not exist on Mars. According to Schubert et al. (1990), convection models of the mantle of Mars indicate that cylindrical plumes are the prominent form of upwelling when sufficient heat enters the mantle from the core. The model predicts several to about ten major mantle plumes in the planet, yet there are only two major volcanic centers, Tharsis and Elysium. Schubert et al. (1990) ascribed this discrepancy to: (1) the model is not realistic enough to predict the number of hot spots; (2) the properties of the lithosphere on Mars are such that it may restrict mantle plume activity to one or two surface manifestations, and (3) plume activity is restricted beneath Tharsis because the lithosphere in that region is thinner or extensively fractured – facilitating magma and heat transport across the lithosphere.

The atmospheric structure measured by the Viking landers (Lander 1 in Chryse Planitia at Lat. 22°N to 27°N, Long. 47°W to 97°W; Lander 2 in Utopia Planitia at Lat. 47°N to 67°N, Long. 225°W to 74°W) from an altitude of 120 km showed that the mean temperature decreased with a lapse rate of 1.6 K/km. Surface pressures and temperatures at Viking sites 1 and 2 were 7.62 and 7.81 mbar and 238 and 226 K, respectively (Seiff and Kirk 1977). The martian atmosphere consists of more than 95% carbon dioxide, 2.7% nitrogen, and 1.6% argon. The remainder, in decreasing order of abundance, consists of oxygen, carbon monoxide, neon, krypton, xenon, and ozone; the water-vapor content is variable, possibly reaching several percent during summer and in the order of only a part-per-millon during winter (Owen et al. 1977; Snyder 1979). Isotopic ratios for carbon and oxygen in the martian atmosphere are similar to terrestrial ones, but anomalous ratios were found for nitrogen, argon, and xenon – anomalies that have been interpreted as the result of depletion of light isotopes by a higher rate of escape from the exosphere. Although ozone is a minor species in the martian atmosphere, it influences the atmospheric structure and consequently the deposition of carbon dioxide and stability of the permanent ice cap of dry ice in the southern hemisphere (Kuhn et al. 1979). The present atmosphere apparently represents only a small fraction of the total volume degassed from Mars. Owen et al. (1977) estimated that at least ten times more nitrogen and 20 times more carbon dioxide originally were degassed on Mars than the martian atmosphere contains now. Leovy (1983) stated that the total amount of carbon dioxide in the atmosphere probably represents about one-tenth the amount degassed and suggested that the total amount of gas that has been vented on Mars is less than the amount vented on Earth by a factor of many hundreds. Models based on the ratio of nitrogen isotopes indicate that degassing on Mars began early in its history and that the total volume of nitrogen gas vented was perhaps 100 times the amount now present.

Although the amount of water in the present martian atmosphere is small in comparison with that on Earth, the relative humidity is high – as attested by the water in recurrent ice clouds near topographic features seen by Mariner 9, the occasional morning fogs imaged by Viking, and the north-polar hood composed of water-ice clouds noted by Mariner 9 (Davies 1979). The best known of the recurrent clouds are those associated with Tharsis, Elysium, Nix Olympica, and Hellas (Snyder 1979). Numerous lee-wave clouds produced by air flow over topographic barriers also have been described (Pickersgill and Hunt 1979). Except where the lapse rate is great and the water is confined to the lowest 1 or 2 km, or during very dusty conditions when a large temperature inversion exists from 10 to 20 km (above which the vapor is present), or when the water over carbon dioxide frost in the spring polar regions is above the temperature inversion at ~ 2 km, the vapor is well mixed to an elevation of at least 10 km (Davies 1979).

Dust appears to be the major contributor to atmospheric opacity on Mars; its presence gives the sky on Mars its yellowish-brown color. Major dust storms resulting in the formation of yellow clouds were observed from Earth in 1922, 1924, 1943, 1956, and 1958 (Snyder 1979). The 1971 storm commenced 53 days before the Pioneer 9 Mars encounter and hid most of the planet; other storms occurred during 1973 (one storm), 1977 (two storms), and 1979 (one storm). The dust storms apparently begin at one or several locations, dust rises for several days (most strongly at noon) and then suddenly the dust intensifies and covers most of the planet (Snyder 1979, and references therein). The principal sources for the major storms are the general vicinity of Solis Planum at Lat. 25°S, Long. 90°W, Hellespontus at Lat. 30°S, Long. 320°W, and Isidis Planitia at Lat. 15°N, Long. 265°W. The dust not only changes the opacity of the atmosphere, but also changes the atmospheric temperature structure (as noted by the 15-μm channel of the infrared radiometers on the Viking orbiter). Wherever the dust went, the temperature of the upper atmosphere increased and diurnal pressure variations occurred at the two Viking lander sites. During one of the storms average wind speeds of 17.7 m/s were recorded near sunset at the Lander site 1, with gusts reaching velocities above 25 m/s (Snyder 1979). The dust extends to a height of 30 km and consists of particles that are equidimensional but nonspherical, have rough surfaces, and have cross sections with weighted mean radii of about 9.4 μm. The principal opaque mineral is magnetite (Pollack et al. 1977). Dust and associated water particles tend to be removed from the atmosphere, in part by acting as nucleii for condensation of CO_2-ice particles within the winter polar regions. The resultant CO_2-H_2O-dust aggregates are much larger and have faster settling velocities than the mineral particles alone and, consequently, they are removed faster from the atmosphere. The storms appear to be caused by an increase in insolation near perihelion, leading to sublimation of the polar cap and formation of sharp horizontal gradients between the polar cap and adjacent clear surfaces.

Wind patterns on Mars consist of a midlatitude belt of westerlies, high-altitude jet streams, traveling storms, and an equatorial summer regime that, except for gradual changes associated with the change in seasons, is free of day-to-day variations (Leovy 1983). In the equatorial region the wind flow is controlled by the daily change in solar heating and wind interaction with local topography. The daily solar heating also may create planet-wide oscillations, and because Mars is a desert planet, the oscillations should be stronger than on Earth. Tidal winds near the surface of Mars should be five to ten times faster than those of Earth, where average velocities are only 1 m/s. These velocities increase with altitude; on Earth tidal winds reach velocities of 50 m/s at elevations of 80 and 100 km. That tidal winds are stronger on Mars is indicated by large variations in pressure noted at the two Viking lander sites; large temperature excursions documented by the Viking landers at higher altitudes, and the 120-km height of the turbopause that is nearly 20 km higher than on Earth (Leovy 1983). These tidal winds may be the chief cause of the dust storms on Mars. As discussed by Leovy, the strength of the tidal winds is sensitive to solar heating of the atmosphere, a heating that depends not only on the amount of solar energy received but also on the opacity of the atmosphere. If the atmosphere is dusty, the dust particles can absorb sunlight and, in turn, can heat the surrounding atmosphere. Thus, if the dust content is high enough, the tidal winds must intensify and lead to suspension of additional surface dust. On Earth, winds as slow as 6 or 7 m/s can transport sand grains, but the low density of the atmosphere on Mars requires wind speeds of at least 30 to 60 m/s to transport this material. Leovy (1983) believed that the dust storms begin in localized areas where a combination of tidal winds and local topographic winds place some of the dust in suspension. Since there is no rain on Mars, once the dust is suspended it can remain in suspension for weeks or months. As this dust spreads, it enhances heating of the atmosphere which, in turn, increases the velocity of the tidal winds and that tends to increase the amount and extent of the dust clouds.

The amount of water degassed from Mars has been deduced using several methods. Estimates based on the abundance of ^{36}Ar indicate that a layer of water about 10 m deep for the entire surface of the planet was produced during degassing (Owen et al. 1977). Preferential degassing of water could have produced a 40-m layer, with some of the water being bound in rocks and not exchangeable with the atmospheric carbon dioxide. There still is no adequate model for a climatic change that would permit Mars' surface to be warm enough to allow this amount of water to be released. Pioneer data indicate that use of the ^{36}Ar abundance in the atmosphere to estimate the amount of water degassed from Mars (if both Earth and Mars have similar $^{36}Ar/H_2O$ ratios) may be unwarranted (Squyres 1984; Kasting and Toon 1989). Other methods that have been used to estimate the amount of degassed water include the measured abundance of ^{36}Ar and the ^{36}Ar/H ratio of ordinary chondrites (on the assumption that the chondrites formed at a position similar to Mars in the solar nebula), yielding a figure of 600 g/cm^2 (a 6-m layer), the K/H$_2$O ratio (on the assumption that the ratio is the same on Earth and Mars), yielding a degassed value of 10^4 g/cm^2 (a 100-m layer) of water, and assuming an N/H$_2$O ratio like that on Earth, yielding a value of 1.3×10^4 g/cm^2 of water (Squyres 1984, and references therein). Squyres estimated that the water degassed early in the history of Mars may have been sufficient to produce a layer 10 to 100 m deep. In contrast, Liu (1988) proposed that water equal to a layer 3 km deep was degassed from the outer 230 km of the planet during consolidation of the crust. That water had a major role in the geological history of Mars is demonstrated by the presence of surficial outflow channels, valley systems, and a variety of morphologic features which indicate that a significant amount of water may be present beneath the surface. Possible sinks for this water may be ground-ice, groundwater, the present atmosphere, loss to space by dissociation and escape, chemical incorporation in silicates, adsorption on grains in the regolith, and trapping in the polar deposits. Although fluvial features are extensive on the surface of Mars, there is no evidence of a hydrologic cycle comparable to that on Earth. The fluvial features were the result of brief isolated flows that ceased as soon as the local supply of water was exhausted (Squyres 1984).

The most prominent features of Mars are its polar caps whose advances and retreats have been documented from Earth for more than two centuries. As a result of orbital eccentricity the fall and winter seasons in the north are 75 days shorter and warmer than those in the south. Thus, the south cap is much larger during its winter, disappears more rapidly in the spring, and is smaller in the summer than the north cap. The angular dimensions of the north cap may range from a maximum width between 53° and 80° to a minimum of about 6°, and the south cap may range from a maximum width of 100° to a minimum of 5° or less (Snyder 1979). The polar regions contain layered deposits free of impacts and cut by a spiral pattern of deep arcuate troughs at the base, perennial ice that at the north reaches almost to the southern edge of the layered deposits, and a seasonal ice cap that extends from the pole down to Latitudes 50° to 45°N. The seasonal cap is a thin layer of solid carbon dioxide. Over the years there has been considerable disagreement as to the nature of the perennial ice, with some advocating dry ice (Leighton and Murray 1966); others believing that although the residual frost caps of Mars probably are water–ice, a permanent reservoir of solid CO$_2$ is present within the north residual cap comprising two to five times that within the present atmosphere of Mars (Murray and Malin 1973). Still others belilieve that the permanent cap consists of water–ice. Late summer temperatures of 205 K, measured by the Viking orbiter using an infrared thermal mapper at the north pole clearly are higher than the 148 K saturation temperature of CO$_2$ at the mean martian surface pressure of 6.1 mbar – a temperature that negates the possibility of dry ice on the surface. Thus, the north cap must consist of water–ice (Kieffer et al. 1976). Mariner 9 infrared interferometer, spectrometer spectra, and television images used in conjunction with multispectral thermal emission models were used to determine the temperatures of the dark bare ground and bright frost regions; they indicate that residual CO_2 was present in the south polar cap during the summer of 1971–1972 (Paige et al. 1990). These observations do not rule out the possibility that water–ice is a major constituent of the south cap. As pointed out by Jakosky and Farmer (1982), the basic question is whether the residual water–ice does or does not retain its dry-ice frost cover during the summer. Earlier Earth-based observations indicate that the cover is not retained every summer, and whether or not it is retained depends on the presence or absence of global dust storms.

Has Mars had its present climate throughout its history? Ward (1973) proposed that, as on Earth, large variations of the orbital obliquity, produced by gravitational perturbations of other planets, and the precession of its spin axis, produced from the solar torque exerted on the planet's equatorial bulge, must have affected its climate during the past. According to Ward, the obliquity oscillates at a scale of 1.2×10^5 years, with the amplitude of oscillation varying at a time scale of 1.2×10^6 years. The present obliquity is 25.1°, with past maximum variations ranging from 14.9° to 35.5°. Kargel and Strom (1992), who have attributed to climate change some anomalous features south of Lat. 33°S and in the northern plains, believed that the planet once was covered by vast ice sheets – a glacial episode that crater density indicates occurred late in martian history. Thus the possibility exists that Mars may have had a warm, moist, and dense atmosphere much later than previously proposed.

The nature of the surface geology at Viking lander sites 1 and 2 has provided the only information on composition of the lithosphere on Mars. A wide variety of rock types at site 1 on a volcanic terrain in Chryse Planitia suggests in situ weathering of extrusive and near-surface basaltic rocks along a linear vent (Binder et al. 1977). Viking lander 2 site on Utopia Planitia consists of fine-grained sediment interspersed with boulders (Mutch et al. 1977). The finer grained sediment may have been transported from the north-polar region, and the boulders could be either residue of ejecta from crater Mie or remnants of lavas that previously covered the site. Filled polygonal cracks at the site may have been formed by ice wedging. However, temperatures at Utopia Planitia are always below the freezing point of water, leading Arvidson et al. (1983) to propose that the cracks were formed by desiccation of clays. The large size of the polygons, a few to about 10 km across and two orders of magnitude larger than those of Earth, led Pechmann (1980) to propose that the cracks were formed by tensional tectonics caused by large-scale warping of the surface. Recently, McGill and Hills (1992) proposed that the polygons were produced by dessication of sediment or cooling of lava flows coupled with differential compaction over buried topography. Polygonal size may be related to spacing of highs in the buried topography. Data from an X-ray fluorescence experiment indicate that fine materials at both Viking lander sites consist mainly of silica and iron containing 1% water by weight (Clark et al. 1977; Arvidson et al. 1983). This similarity in composition at two sites 6460 km apart suggests that the sampled material is part of a thin fine-grained eolian blanket that covers much of the planet's surface (Squyres 1984). The next most abundant elements were found to be magnesium, calcium, sulfur, aluminum, chlorine, and titanium. The sulfur content was 100 times higher than the average content in the Earth's crust, and potassium was at least five times lower. Sulfur and chlorine, probably in the form of water-soluble salts, may have been deposited at the surface by waters diffusing from below (Snyder 1979, and references therein). The fine-grained sediment, which can be characterized as an iron-rich clay, probably was derived from decomposition of basic rock such as basalt. The high iron and low potassium and aluminum contents suggest that the degree of differentiation of basalts on Mars is less than on Earth (Toulmin et al. 1977; Arvidson et al. 1978).

Squyres (1984) stated that chemical data from the fine-grained sediment fit well with a mixture of iron-rich smectite clays or palagonite, and that they may be products of the interaction of basalt with ground-ice; the clays are less likely to result from impact-induced hydrothermal alteration (Newson 1980). High spatial and spectral resolution near-infrared telescope charge-couple device images obtained from the Pic-di-Midi Observatory in France indicate the presence of mafic minerals suggestive of a basaltic crust and that the bright spots on the planets are more altered than are the dark strips (Pinet and Chevrel 1990). The magnetic mineral adhering to the magnets on the landers probably was maghemitite whose yellowish brown or reddish color may contribute to the color of Mars (Arvidson et al. 1983). Coey et al. (1990) indicated that soil analysis from the Viking mission shows that the martian regolith was derived from basalt, and that the mineral responsible for magnetism on the soils of Mars is titanomagnetite. Spectral observations from the Mauna Kea Observatory indicate that hematite accounts for the spectral characteristics of the soil and airborne dust (Bell et al. 1990). Resolved spectra in the 2.2–2.4 μ range show regional variations that reflect local or regional variations in geology and not the homogenous globally distributed eolian dust, and that minerals in the scapolite family fit the spectral structure (Clark et al. 1990).

Meteoritic influx, which may contribute as much as 2 to 29% of the soil mass (Flynn and McKay 1990), tends to perturb their chemical compositions, so that the composition of martian soil is a reflection of a mixture of crust/meteoritic debris rather than of only the planet's crust. Analyses of fine-grained sediment for organic compounds by a mass chromatograph-mass spectrometer indicate that life similar to terrestrial biota does not exist, at least at the two Viking lander sites (Biemann et al. 1977). Carbon fixation from the atmosphere (Horowitz et al. 1977), and the evolution of oxygen when soil samples were humidified (Oyama and Berdahl 1977) probably was chemical, not biological, in origin. Results from the Labeled Release life detection experiment (where an attempt was made to detect heterotrophic metabolism in the radioactive gas after the injection of a nutrient containing seven ^{14}C-labeled organic substrates to the soil samples) also proved to be inconclusive. However, Leven and Straat (1977) stated that the results obtained during the experiment are consistent with a biological response and narrow the number of possible nonorganic chemical reactions. These results tend to leave the question of life on Mars as yet to be resolved. This biosphere, even if it is or was present, did not have a role in exogenic processes on Mars as it has on Earth.

8.2 Satellites

Phobos, the inner satellite of Mars, has major diameters of 27, 21, and 19 km, a volume of 5340 km^3, a density of 1.9 ± 0.5 g/cm^3, a surface temperature of ~ 225 K, a sidereal period of revolution of 7 h, 39 min, and 25 s, and an equatorial surface gravity of 0.3–0.5 m/s^2 (Table 1; Snyder 1979; Greeley 1985, p. 16, table 2.1). Deimos, the outer satellite, has diameters of 15, 12 and 11 km, a volume of 1225 km^3, a density of 1.5 ± 0.5 g/cm^3, a surface temperature of ~ 225 K, and an equatorial surface gravity of 0.3 cm/s^2. The satellites were imaged during the Mariner 9 mission in 1971–1972 and during the Viking mission in 1976 when about 50 high-resolution images were obtained by the Viking orbiter cameras (Duxbury and Veverka 1977; Veverka and Duxbury 1977). According to Snyder (1979), both satellites are irregular, are very dark with albedos of 6 or 7%, and have crater densities near limiting equilibrium density. Their color suggests that they consist of the same material as carbonaceous chondrites found about 3 AU from the Sun – a composition indicating that they are captured objects. Hunten (1979) proposed that their capture was the result of drag in an extended proto-atmosphere of solar composition, and that this atmosphere was either rotating rapidly or was a slowly rotating condensation within the solar nebula but with a density one to two orders of magnitude greater than densities usually associated with slowly rotating atmospheres. Hunten estimated that regularization of the orbits of the satellites probably took a few years, and the removal of the proto-atmosphere after the nebular pressure was eliminated also took a few years. If Phobos and Deimos were captured in this manner, they must have arrived during the final phases of planetary accretion (Snyder 1979). Stevenson et al. (1986) questioned the capture hypothesis based on the supposed low mean densities and dark surfaces of the satellites, indicating that the ability to identify materials by remote sensing is limited. Thus Phobos and Deimos may not resemble carbonaceous chondrites and may have a composition comparable with that of the inner main Asteroid Belt or with Mars itself, except that they are undifferentiated. Thus the proposal was made that the moons originated from collisional debris that accumulated around Mars as the planet grew.

Surface features on Phobos can be classified into three categories: (1) elongate rill-like features associated with crater Stickney, the largest crater on Phobos (Fig. 87); (2) chains and clusters of irregular elongate depressions; and (3) parallel lineations (Veverka and Duxbury 1977). Except for category 2, similar features also were imaged on Deimos (Fig. 87). That category 2 features were not recognized on the Deimos images may be a reflection of lower image resolution there rather than of their absence. The largest of the elongate features on Phobos is 10 km, is quite linear, and has the appearance of a trough; it may represent a fracture produced by an impact on Phobos. Another feature extending more or less continuously from the rim of Stickney appears to consist of coalescing depressions and terminates in an arcuate trough at the far limb of Stickney. No such features were noted in the

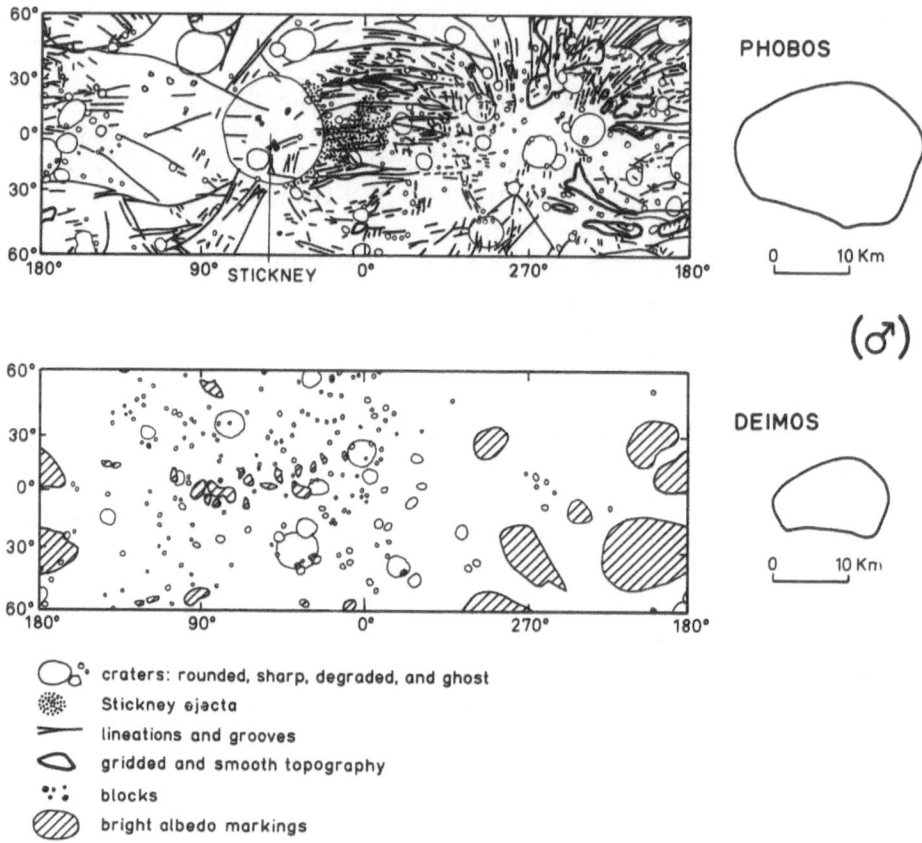

Fig. 87. Silhouettes and morphology of the martian satellites, Phobos and Deimos. (After Thomas et al. 1986, fig. 4 and maps of Phobos and Deimos)

Roche impact site, a 5-km crater near the north pole. As no data were collected in the area of Hall (a 6-km structure near the south pole), it is not possible to determine whether such lineations are associated with this structure. The crater chains of category 2 consist of irregular craters ~ 50–200 m in diameter that locally cluster in a herringbone pattern that is typical of secondary ejecta. Some of them are superimposed on Stickney, indicating that they are younger than this large crater, and on Phobos the chains are parallel to the orbital (equatorial) plane of the satellite. Veverka and Duxbury (1977) proposed that these secondary craters were produced by ejecta that was thrown out of Phobos, traveled in an orbit around Mars, and later reimpacted on the satellite. An alternate explanation for the secondaries is that they were produced by ejecta that originated from Deimos. The striations of category 3 are 150 to 200 m wide, are rather shallow, are separated from adjacent structures by distances comparable with their widths, and can be followed for distances of more than 5 km. At least two families of striations inclined about 10° to each other can be distinguished. They cross the large craters near the north pole and are overlain by smaller craters. Thus, as pointed out by Veverka and Duxbury, they are neither the oldest nor the youngest features on the satellites. They appear to lie along small circles perpendicular to the Mars-Phobos direction (parallel with the satellites' orbital velocity vector) and appear to be concentrated at high northern latitudes near Long. 180°W. Veverka and Duxbury proposed three possible origins for the striations: (1) layering within Phobos; (2) rows of impact craters; or (3) fractures caused by extensional stresses related to tidal stresses (such stresses would distort a body at the distance of Phobos from Mars), impact, or local degassing. More recently, Thomas et al. (1986) suggested that the grooves are fractures caused by the same impact which produced Crater Stickney with tidal stresses having a secondary role.

Crater counts on Deimos, when compared with the saturation curve of the lunar highlands (the curve defining the crater density where crater production and crater obliteration balance each other; Hartman 1973), indicate that the surface of Deimos is saturated with craters. Veverka and Duxbury estimated that a minimum time of 1.5 Ga was needed for the satellite to reach its crater-saturation point. Crater counts on Phobos indicate that for craters of $\geqslant 0.3$ km diameter this satellite also is close to saturation. Similar saturation was noted inside the large 4.9 km crater Roche, which Veverka and Duxbury (1977) interpreted as indicating that such erosional mechanisms as slumping of the rim walls are unimportant, an observation that is understandable because the equatorial gravity on Phobos is so weak (Greeley 1985, p. 16, table 2.1).

8.3 Hypsometry

The best photographic coverage of Mars up to 1992 was the series of 7329 pictures obtained by the orbiter Mariner 9 between 13 November 1971 and 27 October 1972. These, plus associated occultations and spectral measurements from Mariner 9 and Earth-based radar data, provide the best available information about the topography of Mars (Christensen 1975; US Geological Survey 1976; Bills and Ferrari 1978). A map of the general topography, based upon elevations that were coordinated in a degree-16 harmonic analysis by Bills and Ferrari, has contours at 2-km intervals with some 1-km ones all referenced to the level of the 6.1-mbar areoid. These contours were transferred to our equal-area projection base map (Figs. 88 and 89) from which one can see that 74% of the area of Mars lies above the level of the 6.1-mbar reference. According to Bills and Ferrari, the 6.1-mbar level corresponds with a gravity model having a radius of 3382.95 km, in contrast with the mean radius for Mars' topography of 3389.92 km ($X_1 = 3394.5$ km, $X_2 = 3399.2$ km, $X_3 = 3376.1$ km; the latter indicates polar flattening). Measurements of the areas between the contours of Figs. 88 and 89 were used to construct a hypsometric curve for Mars (Table 3; Fig. 5, wide line). The single flattish part of the curve may indicate the absence of an equivalent of Earth's two crusts (continental and oceanic), but it could as well be the result of a too wide spacing of average elevations (within 14.2° circles centered on a graticule of 22.5° of latitude and longitude across the surface of Mars) for a degree-16 harmonic analysis to be able to identify a slope between surfaces of different density crusts. The latter point is illustrated by the fact that the highest contour on Figures 88 and 89 is only 10 km, whereas four peaks of the Tharsis Montes and Olympus Mons region (centering at about Lat. 5°N, Long. 120°W) reach elevations of about 27 km, corrected to about 23 km, according to more detailed contours of the US Geological Survey (1976) using stereophotometric methods.

The images from Mariner 9 include many pairs that covered identical areas of Mars but were made from different directions during orbits of the satellite. These pairs made possible an estimation of the topography by stereophotogrammetric methods, as described by Wu et al. (1973). In this way, topographic details on a

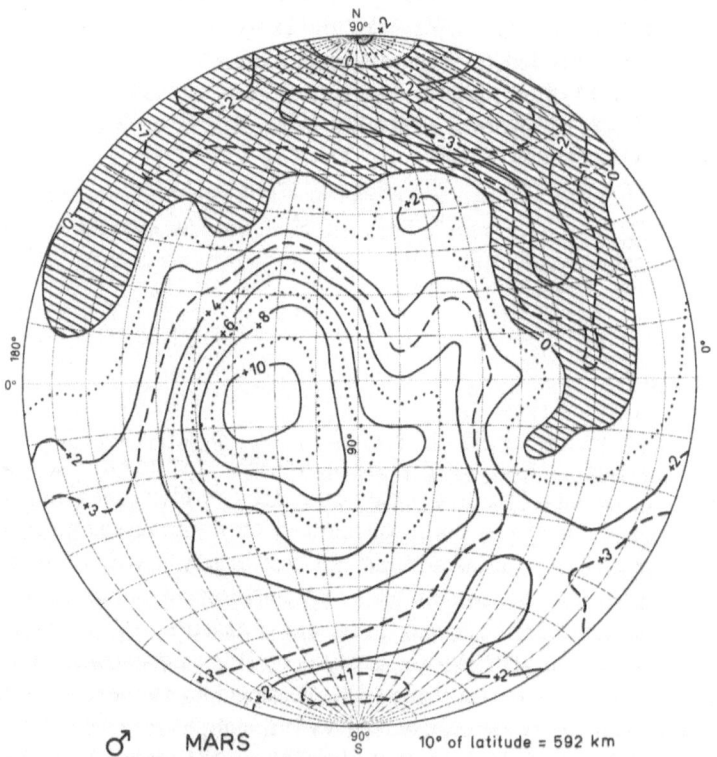

Fig. 88. General topographic contours of Mars; Long. 0° is on the *right*, Long. 90°W is in the *middle*, and Long. 180°W is on the *left*. Longitude is reckoned from a small crater named Airy at Lat. 5°11'S, Long. 0°00' (Mutch et al. 1976, p. 57; US Geological Survey 1976). Contour interval is 2 km (*continuous lines*) with selected 1-km contours (*dashed lines*). Based on harmonic analysis to degree-16 on maps by Bills and Ferrari (1978), and replotted on a Lambert azimuthal equal-area projection with some 1-km contours (*dotted lines*) interpolated by us to permit production of a hypsometric curve for Mars (Fig. 5). *Diagonally lined pattern* denotes areas lower than that of the 6.1-mbar reference areoid

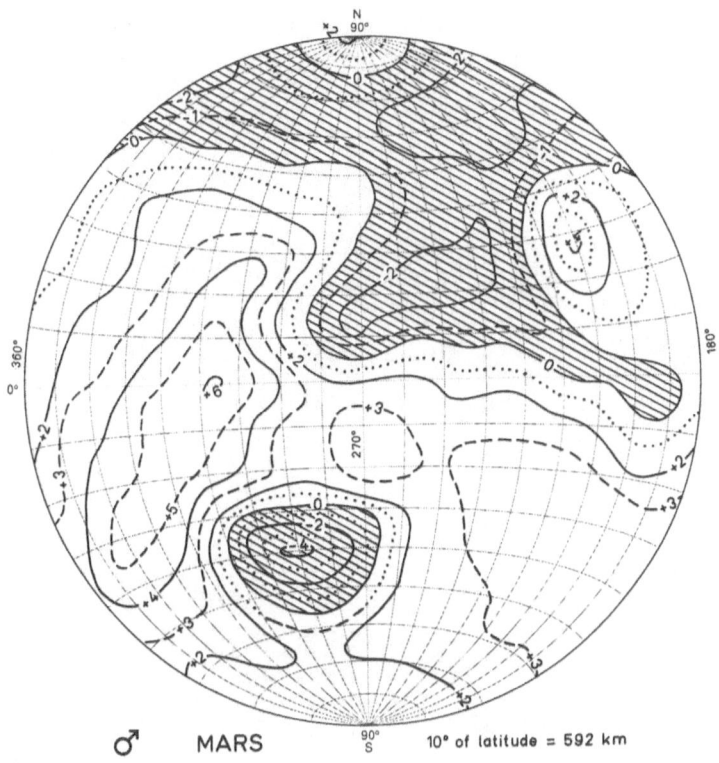

Fig. 89. General topographic contours of Mars, between Long. 180° and 360°W. Long. 180°W is on the *right*, Long. 270°W is in the *middle*, and Long. 360°W (0°) is on the *left*. *Symbols* and source as for Fig. 88

composite air-brushed map could be fitted to the known measurements of elevation to produce a far more detailed topographic map (US Geological Survey 1976) than that which resulted from harmonic analysis of elevations relative to a mean sphere or above the 6.1-mbar level that was chosen as a base for mapping. The resulting map of martian topography (Figs. 90 and 91) illustrates many different landforms such as volcanoes, volcanic flows, broad low craters produced by meteoroid, asteroid, planetoid, or comet impacts, canyons having tributaries, cross sections, and general shapes somewhat similar to stream-cut canyons on Earth, straight cliffs presumed to be fault scarps, and plains deposited by wind, by ejecta from impact craters, and possibly by streams. The areas between the contours of Figs. 90 and 91 were measured in the same way as were those of Figs. 88 and 89 to serve as the basis for an improved hypsometric curve of martian topography. Curves from both sets of contours are presented in Fig. 5 (data of Table 3), from which one can see that the more detailed contours from US Geological Survey (1976) reveal two slight flattenings somewhat akin to those on the Earth. On Mars their mean elevations lie at about + 3.5 km and about − 0.5 km, about 4.0 km apart, similar to the 4.5-km separations of the two levels on the Earth (Fig. 3), but

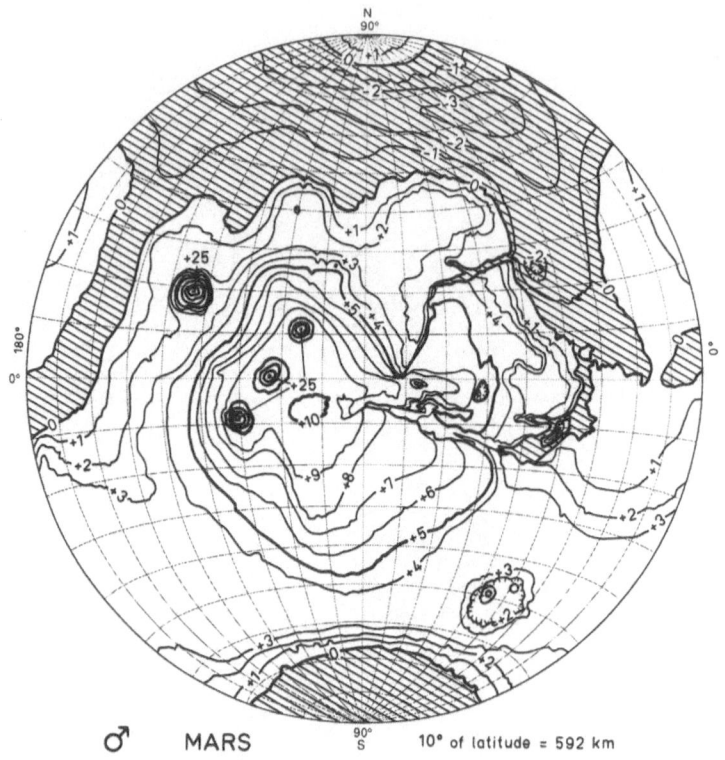

Fig. 90. General topographic contours of Mars, between Long. 0° and 180°W. Contour intervals are 1 km up to 10 km elevation and 5 km between 10 and 25 km elevation. Based on stereophotometric methods applied to images obtained on board Mariner 9, as interpreted by US Geological Survey (1976). Projection and *symbols* as for Fig. 88

the two flattenings on Mars are much less pronounced than on the Earth. Very similar results for Mars were obtained by Coradini et al. (1980) using the US Geological Survey topography, and a close approach by Lipskii et al. (1975) is for the part of the area of Mars between Lats. 20°N and 50°S. Later, a much more detailed topographic map of Mars was published by the US Geological Survey (1989), but its detail was more than we deemed necessary for comparison with the general topographies of other members of the Solar System.

8.4 Morphology

8.4.1 General

Mars' surface consists of two distinct morphologic provinces of nearly hemispheric extent. The southern hemisphere is heavily cratered, contains the older geological

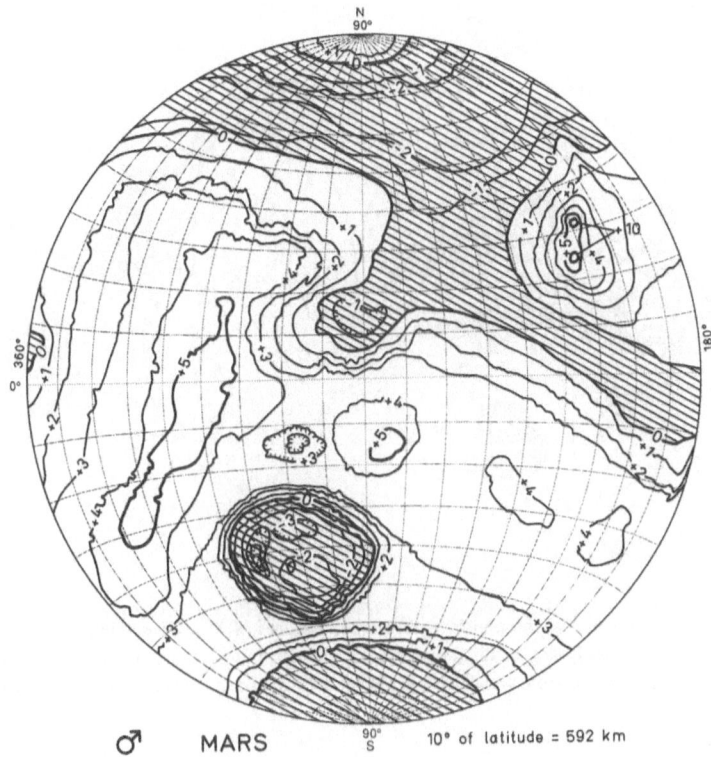

Fig. 91. General topographic contours of Mars, between Long. 180° and 360°W. Source and methods as for Fig. 90, and *symbols* as for Fig. 89

units, and resembles the lunar and mercurian highlands (Wilhelms and Squyres 1984). The northern hemisphere, which lies 3 km lower than the southern one, is only lightly cratered, and it contains the younger geological units including the Tharsis and Elysium volcanic regions. As described by Wilhelms and Squyres, these two provinces are in isostatic equilibrium but are separated by a scarp or sloping transitional zone as wide as 700 km where the highland morphology is partly replaced by lowland deposits. That both provinces are in isostatic equilibrium across the transitional zone indicates that a different type or thickness of crust underlies the two hemispheres (Wise et al. 1979).

The gross morphology of Mars in Figs. 88–91 only dimly indicates secondary features such as plains, mountains, valleys, and impact craters. Detailed mapping of these features by the US Geological Survey (1976) presents a panorama well worth serious study but in too large a format to be included within this book. Therefore, a simplified version, adapted from Mutch et al. (1976), is presented in Fig. 92. This figure shows the distribution and names of the major features to aid during the discussion of their geological origins. The total relief on Mars exceeds 30 km, with a maximum elevation of 27 km at the summit of Olympus Mons and a minimum

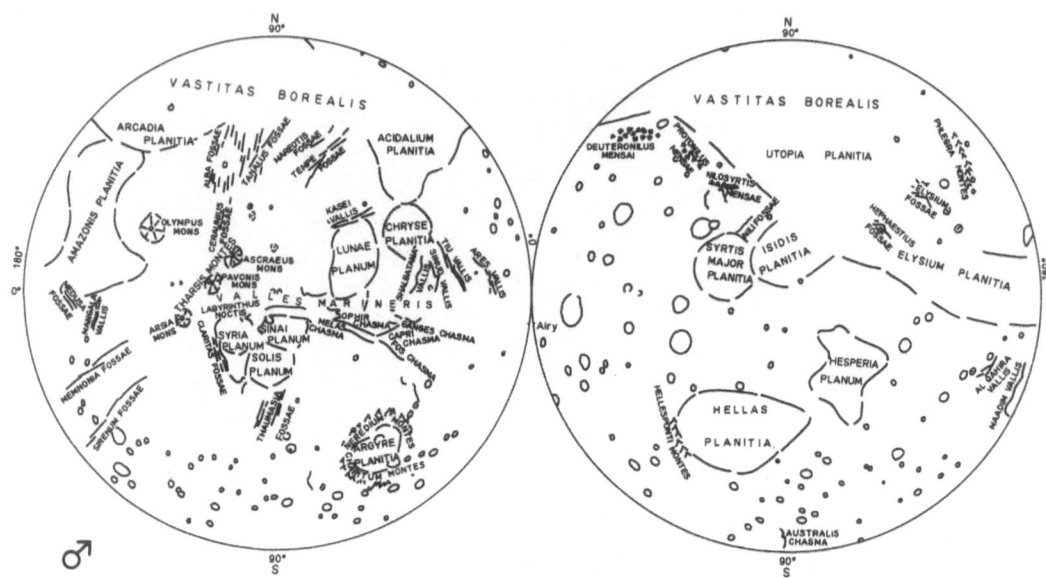

Fig. 92. Map showing some of the major topographic features of Mars. Included are plains (vastitas, planitias, planums), valleys (fossae, vallis), mountains (montes, mensae, chasmatas), and impact craters (not named here). The projection and limiting meridians are the same as for Figs. 90 and 91. (After Mutch et al. 1976, fig. 3.1)

elevation of − 4 km on the floor of Hellas Basin. The major terrains on Mars display an asymmetrical distribution, with the sparsely cratered plains of the northern hemisphere being separated from the cratered highlands of the southern hemisphere by a young plain that trends 30° to the equator in the western hemisphere. Superimposed on these plains are the elevated regions of Tharsis and Elysium and their volcanoes (Greeley 1985, p. 153). The Tharsis Plateau or bulge straddling the equator at Long. 110°W extends 4000 km from north to south, 3000 km from east to west, and rises about 6 km above the surrounding plains (Mutch et al. 1976, p. 61). The major topographic features of the plateau are three large volcanoes (the Tharsis Montes). Elysium Planitia is a 1500-km wide broad dome with 4 to 5 km relief located at Lat. 25°N, Long. 210°W. Like the Tharsis Plateau, this high contains major volcanoes and lava plains. Another major positive element of Mars is the Thaumasia Plateau located at Lat. 40°S, Long. 90°W. It is 2000 km wide, 4 km high, and is characterized by immense fracturing rather than by volcanism. Whereas the Tharsis Plateau is not in isostatic equilibrium, the Thaumasia Plateau is (Phillips and Saunders 1975).

Among the planetary topographic lows are Hellas, Argyre, and Isidis planitias having rim structures suggestive of impact origin, and the lows in Chryse and Amazonis planitias near Tharsis (Mutch et al. 1976; Fig. 92). Chryse Trough extends more than 5000 km from the north rim of Argyre Basin to the plains of Maria Acidalium region. A discussion of the morphology of Mars is complicated by evident large and long-term global changes of morphological agents on that planet. The effects of these agents start with the densely cratered terrain left at the end of the early heavy bombardment (Fig. 93A). This morphology is similar to that of the present surfaces of the Moon and of Mercury (Figs. 44, 45, and 58), where the absence of much volcanism and of atmospheres and hydrospheres preserved the cratered surfaces, unlike on Earth. The next stage on Mars is tectonic: it involves the doming and resulting radial fracturing of the Tharsis region (Fig. 93B), which was followed by extrusion of igneous rock to form volcanoes and broad adjacent volcanic plains (Fig. 93C). Renewed uplift and both radial and concentric fracturing (Fig. 93D) was followed by emplacement of additional volcanic plains (Fig. 93E). Lastly, erosion by wind and probably by water, ice, mass movements, and the poleward movement of the debris by wind produced thick layered plains at high latitudes (Fig. 93F). A subsequent change in wind direction removed much of these layered deposits, transporting reworked debris equatorward to between Lats. 30° and 40°N and S.

8.4.2 Chronology

Lacking rock samples for radiometric dating or dates based upon fossils, the timing of successive epochs of impact cratering, tectonism, volcanism, erosion, and deposition cannot be established with certainty. However, the size distribution and areal frequency of impact craters provide guides to the relative ages of the different kinds of surfaces (such dates may have been paralleled elsewhere in the Solar System); this involves the effects of heavy bombardment early in the history of planets and moons and the subsequent diminution in intensity of bombardment (Hartmann 1977). This search led to investigations of the degree of saturation of areas by impact craters, the ratio of diameters and numbers of craters, and the degree of degradation by subsequent smaller impacts. Prior to collecting and radiometric dating of rock samples from the Moon, these methods were employed there; close confirmation by later radiometric ages of lunar rocks led to the expanded use of terrain analysis for subsequent explorations of other bodies.

The surficial rock units in Mars are assigned to three time-stratigraphic systems (Scott and Carr 1978). The oldest is the Noachian system named after the Noachis quadrangle. It includes hilly and cratered material, basin and rim material, and mountain material. All these units represent highly brecciated and faulted crustal rocks formed during an early period of heavy bombardment of Mars (Table 8). Knobby material may represent degraded and reworked material of late Noachian age, but its morphology was developed during the Hesperian. The Hesperian system consists of ridge plains material at the base; it is partly streaked with light and dark plumes representing secondary modifications of the basal unit. The ridged unit and its possible equivalent streaked plains material exhibit features comparable to the lunar maria, lobate scarps, rilles, and wrinkle ridges that are interpreted as indicating basalt flows. Included within the system also are rolled plains and volcanic material of intermediate age. Also present in the Hesperian

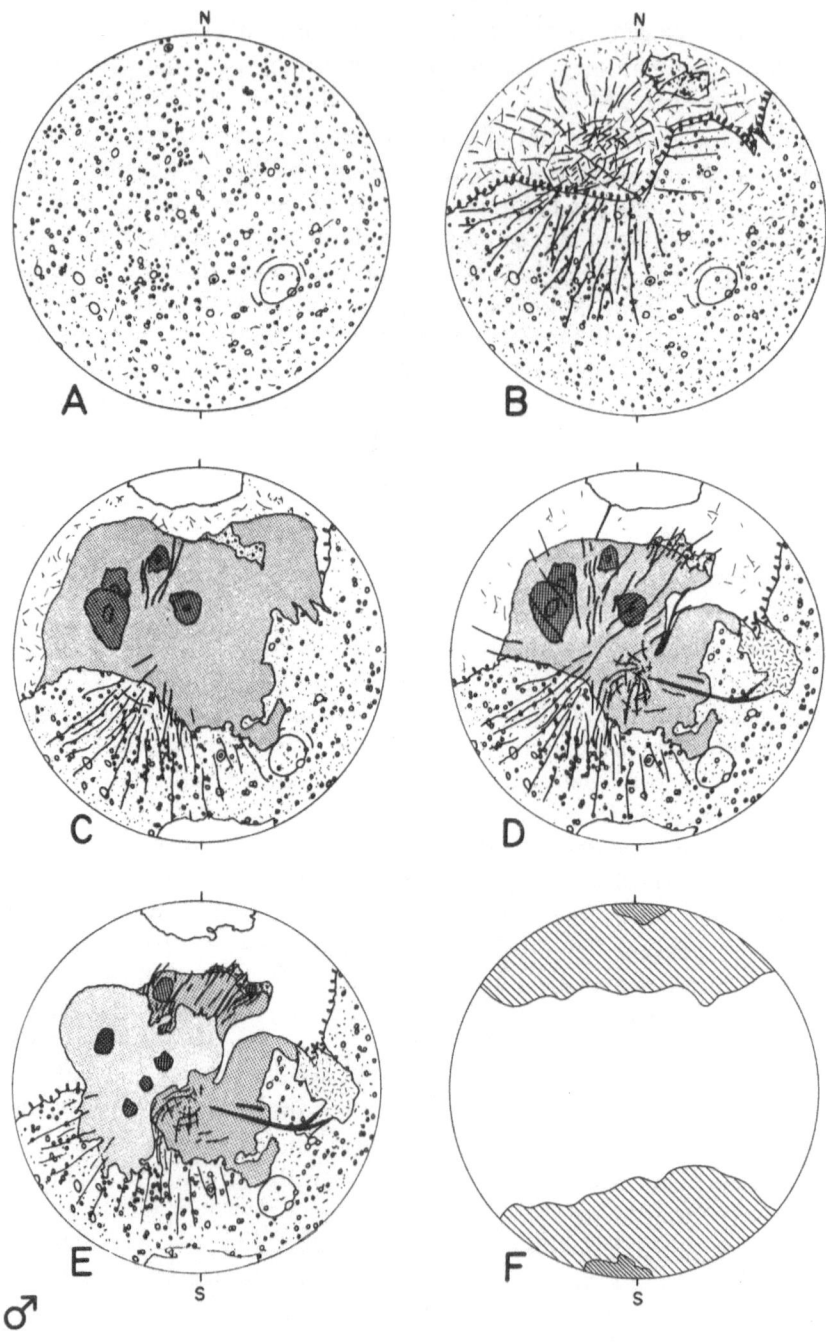

Fig. 93. Stages of morphological evolution of the martian hemisphere between Long. 0° and 180°W (Mutch et al. 1976, fig. 9.1; Figs. 117 and 118 of this book). **A** Last stage of early dense crater formation by impact of planetoids and meteorites. **B** Early uplift of the Tharsis region with radial fracturing of

Table 8. Relative ages of martian features. (Barlow 1990; Gulick and Baker 1990)

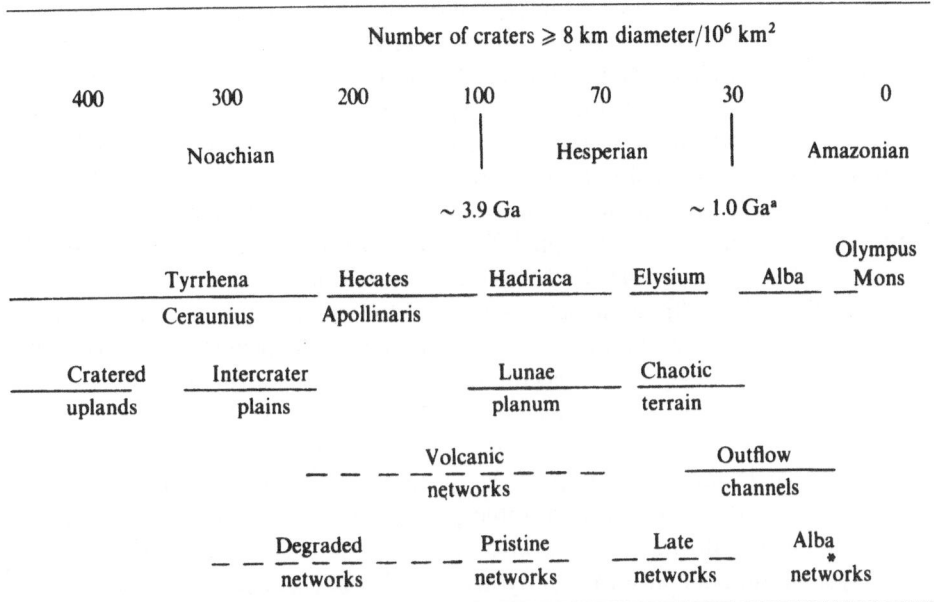

Number of craters ≥ 8 km diameter/10^6 km²						
400	300	200	100	70	30	0
	Noachian			Hesperian		Amazonian
			~ 3.9 Ga		~ 1.0 Ga[a]	
	Tyrrhena	Hecates	Hadriaca	Elysium	Alba	Olympus Mons
	Ceraunius	Apollinaris				
Cratered uplands	Intercrater plains		Lunae planum	Chaotic terrain		
		Volcanic networks			Outflow channels	
	Degraded networks		Pristine networks	Late networks	Alba networks	

[a] Tentative ages assigned to formations by the authors from information discussed in text.

system are deflation plains. These are a sequence of eroded surfaces beneath plains and plateau material of Amazonian and Hesperian age at which lower Hesperian and Noachian units are exposed. Channel deposits may have a wide range of ages: from Noachian to Amazonian. The youngest unit (the Amazonian system) includes cratered plains at the base, smooth plains, volcanic plains of the Tharsis region, parts of the canyon floors, bedded polar sheets and overlying ice caps. Fresh volcanoes (small cratered domes) of the Tharsis region also are included in this system. Rocks that have higher albedos than the older systems are mostly volcanic, alluvium, and eolian deposits; they are variable in thickness, occur in low areas of regional extent in minor lows in the uplands, and mantle large areas of the Tharsis region. The crater population of this unit is less than that of the lunar maria.

8.4.3 Exotic

Like the surfaces of the Moon and Mercury, the surface of Mars is scarred by impact structures that display both simple and complex crater morphologies. Their morphology depends upon surface gravity, structure of the lithosphere, and for the

---

crust. C Emplacement of volcanoes in the Tharsis region with widespread deposition of volcanic plains. D Renewed uplift and radial fracturing in the Tharsis region. E Formation of additional volcanoes and emplacement of later volcanic plains. F Deposition of eolian plains in polar regions and their subsequent eolian erosion and deposition at adjacent middle latitudes

largest impacts on planetary curvature (Schultz and Frey 1990). As a result of eolian erosion, rayed craters are rare; ejecta blankets and secondary craters also are less abundant on Mars than on the Moon. On Mars the transition from simple to complex to multiringed basins is in the 5- to 10-km diameter range, whereas on the Moon the transition is at 20 km (Carr 1984, and references therein). Most craters smaller than 15 km in diameter are bowl-shaped, have raised rims, and have smooth walls and floors (Mutch et al. 1976, p. 116). Craters larger than 15 km tend to be flat floored, vary in shape from circular to polygonal, and many contain central peaks with pronounced pits. These pits are believed to be the result of ground-ice that was volatilized by impact-generated heat and vented to form the pits (Wood et al. 1978). In structures larger than 100 km the central peaks are replaced by central rings, and in the largest of impacts the peaks are replaced by a group of concentric rings. In a recent survey of multiring structures Schultz and Frey (1990) recognized three subclasses: those less than $\sim$ 1850 km in diameter have an Orientale-type (Moon) concentric structure, those larger than Argyre have rugged concentric annuli, and impacts larger than Chryse exhibit multiple concentric rings and very shallow topographic profiles. The morpho/tectonic fabric of Hellas and Isidis multiring impact craters indicates a four-stage sequential development: (1) formation of concentric 'canyons' outside the basin boundary scarp at the time or near the time that the basin was formed; (2) formation of radial troughs; (3) a hiatus followed by concentric graben formation; and (4) emplacement of volcanic plains between the massif ring and the boundary scarp at one side of the basin. This geometry results from dynamic effects of transient cavity collapse or elastic flexure under basin ejecta or transient loads, and it is widened by early modification processes that include lithospheric flexure during isostatic basin uplift, and elastic flexure under central basin loads (Wichman and Schultz 1989).

Mouginis–Mark (1979) identified six crater types on the basis of the morphology of the ejecta facies associated with fluidized craters. Type 1 crater ejecta is lobate and illustrates a surface-flow emplacement mechanism; type 2 has an inner facies with a convex distal edge and an outer facies that is lobate; and type 3 displays multiple ejecta facies related to the lobate facies of types 1 and 2. Types 1 to 3 possess distal ridges on their outermost ejecta facies that previously were described as rampart craters (Carr et al. 1977). Type 4 has a radial pattern of grooves and ridges superimposed on sheets or platelets of ejecta and the distal ridge is missing; type 5 craters display a complex morphology of massive flow units, transverse ridging, pitted material, and smooth material superimposed on lobate facies and secondary craters; and type 6 (known as a pancake crater, or a pedestal crater; Arvidson et al. 1976; Guest et al. 1977) is a simple bowl-shaped crater less than 5 km in diameter with a single facies extending outward as much as six crater radii and having convex outer edges. Type 1 tends to grade into types 4 and 5 with increasing size, indicating that crater size controls crater morphology for any single target material. Except for pancake craters, fluidized impacts do not display any latitudinal bias, but the distribution of the different classes, the incidence of central peaks and terraces, and the preservation of secondary peaks do appear to be influenced by the nature of the impact terrain (Mouginis–Mark 1979). Pancake craters tend to occur on fractured terrain, old lavas, and channel deposits. Radially textured craters (type 4) are found on lavas and rolling plains; these craters appear to have been formed on plains that have been eroded. Apparently, the surrounding plains were degraded by eolian erosion, but the craters that were protected by block ejecta survived as pedestals (Greeley 1985, p. 179).

Craters formed on recent lavas are accompanied by secondary craters, central peaks, and terraces, whereas craters formed on ancient terrains tend to be much simpler, perhaps because of subsequent erosion. As on the Moon, some craters on ancient cratered uplands of Mars provided avenues for localized igneous activity (Schultz and Glicken 1979). This igneous activity probably led to local melting of ground-ice. If such thawed material were confined and then released catastrophically, it may have formed chaotic terrains and outflow channels (Schultz and Glicken 1979). That the craters provided local zones of weakness is suggested by modification of the craters' morphology comparable to volcanically modified lunar craters. Modification is localized within the craters' interior; it is selective and not simply the result of relief; it affects a wide range of crater sizes and ages and is unrelated to regional processes. The largest impact craters may have provided pathways for igneous extrusions comparable with the emplacement of lunar maria. Such pathways may be preserved along the borders of the fretted and chaotic terrains with what may be a 2000-km-diameter impact structure in Syria Planum that has undergone extensive structural and igneous modification. Schultz and Frey (1990) also proposed that the spatial association of the multiring impact at Tharsis and Elysium basins and Mars' hemispheric terrain dichotomy may indicate that the two levels are the result of interaction of the impacts with mantle processes that helped localize volcanic activity within these two regions. Other theories proposed to explain the differences in age and elevation include an internal origin and a single mega-impact. Barlow (1990) stated that multiple impacts for the origin of the hemispheric dichotomy are not supported by crater-size frequency distribution data. Analysis of the cratering record in the knobby terrain (the oldest unit preserved in the northern hemisphere) indicates that the hemispheric dichotomy occurred during the heavy bombardment intermediate epoch of the middle to late Noachian on the stratigraphic scale (Table 8).

The earliest topography of impact craters on Mars became modified during subsequent eons by burial beneath volcanoes and volcanic flows (especially in the Tharsis/Olympus Montes region), by erosion through impact of later smaller meteorites (a sort of sand-blasting effect), erosion by wind, ice, and water (especially in the Valles Mariner region), burial beneath wind-laid deposits (especially at high latitudes), and continued formation of new impact craters (everywhere on Mars). Some of these ancient impact structures are represented only by grooved features spaced 1 to 10 km apart. Such features, which occur in clustered sets of a dozen or more, have been interpreted by Craddock et al. (1990) as impact material that originated from a crater near the center of Daedalia Planum, an early Noachian basin with four interior rings of 1100, 1500, 2200, and 3200 km, and an exterior ring of 6400 km diameter. The impact has both concentric and radial basin-related faults. Early Hesperian and early Amazonian volcanic terrain nearby may have been extruded via deep-seated interior faults in the low. Furrowed and pitted wall craters tend to be more abundant in a latitudinal belt just south of the equator, a distribution that Mutch et al. (1976, p. 147) equated to water erosion in the equatorial region. This also is the region where there is a concentration of channels and furrows. However, Soderblom et al. (1974) proposed that fluvial features are pervasive throughout Mars and that they have been preferentially exposed in the equatorial region by subsequent erosion of an overlying sediment cover by eolian action. The areal distribution of breached wall craters indicates a geographically randomly operated process, and the global plot of central peak craters is suggestive of sapping and stripping processes, or of an impact mechanism associated with rebound (Mutch et al. 1976, p. 147).

The earliest impact craters tend to be larger than later ones, with the largest and one of the earliest basins (Hellas) being about 1600×2000 km and having a rim that is 50 to 400 km wide. This, the largest impact basin in the entire Solar System, is marked by a concentric pattern of weakness that extends about 1600 km from the center of the crater (Greeley 1985, p. 153). According to Greeley, the floor of Hellas is filled with volcanic plains and mantled with eolian deposits. The floor of the 900-km wide Argyre Basin in the southern hemisphere also is mantled with eolian deposits, and the 1100-km wide Isidis Basin in the northern hemisphere is filled with younger lavas that poured out eastward from the basin. All three of these topographic lows are surrounded by massifs less than 230 km wide and by inward-facing scarps. Radial ejecta deposits are absent from the periphery of these uplifted crustal blocks (Greeley 1985, p. 158). Obviously, the largest and highest-rimmed craters were less easily destroyed by erosion and/or burial than were smaller ones, so that a chart of the present distribution of craters larger than 15 km in diameter is a composite (Figs. 94 and 95).

Analyses of crater data obtained by Mariners 4, 6, and 7 led investigators to postulate that the large craters on Mars have existed for most of martian history, that they have been subjected to some process of removal that destroyed or modified most craters smaller than 15–20 km in diameter, and that this process of obliteration has diminished with time, allowing a population of younger craters to be superimposed on the older eroded structures (Mutch et al. 1976, p. 126). Data from the subsequent Mariner 9 mission led to two observations. The crater densities on the martian satellites Phobos and Deimos (which have no atmosphere) are greater than that of Mars, indicating that they have retained a more complete record than that of Mars. A crater frequency plot from the martian cratered terrain displays three linear segments, one of which reflects a period of terminal accretion; erosion possibly associated with a temporary dense atmosphere during or after this period reduced the number of craters below saturation values; a second segment identified with craters of 5- to 30-km diameter suggests a long period of constant degradation, probably by eolian processes; and a third segment indicates a recent period of reduced erosion rates during which the impact structures retained their configuration (Hartmann 1973).

Early investigations of asteroid distributions led Hartmann (1973) to believe that the influx rate on Mars is ten times greater than on the Moon. Because of its

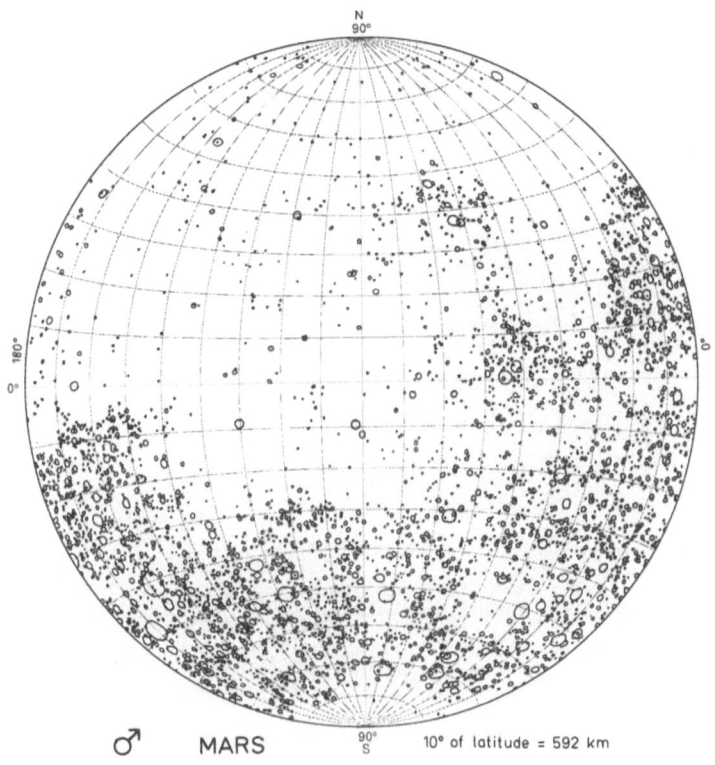

Fig. 94. Distribution of exposed martian impact craters larger than 15 km in diameter in the hemisphere between Long. 0° and 180°W. (Mutch et al. 1976, fig. 4.46)

atmosphere, impact velocities on Mars are five-sevenths of those on the Moon. This would lead to the production of smaller structures on Mars, and if cratering rates on Mars differ from those on the Moon by a factor of six, the martian chronology may be as follows: (1) the heavily cratered regions have an exposure age of 3.5 Ga; (2) intermediate deposition and erosion continued until 600 Ma ago; (3) the volcanic plains in the northern hemisphere are 300 Ma old; (4) the volcanic flanks of Arsia Mons are 200 Ma old; (5) those of Olympus Mons and Ascraeus Mons are 100 Ma old; and (6) Pavonis Mons is 80 Ma old. In the Hartmann model high cratering and high obliteration rates occurred concurrently. Barlow (1990) proposed that the period of high obliteration rates terminated during the declining stages of heavy bombardment at the time when the ridged plains were formed. Chapman (1974), Jones (1974), and Soderholm et al. (1974) agreed that the period of increased erosion was followed by a period of diminished rate of obliteration. In the model proposed by Chapman and Jones, however, obliteration occurred some time after cratering rates declined. Jones used crater diameter-frequency distributions not only for the total population but also to distinguish between four morphologic classes (fresh, slightly degraded, moderately degraded, and highly degraded) to

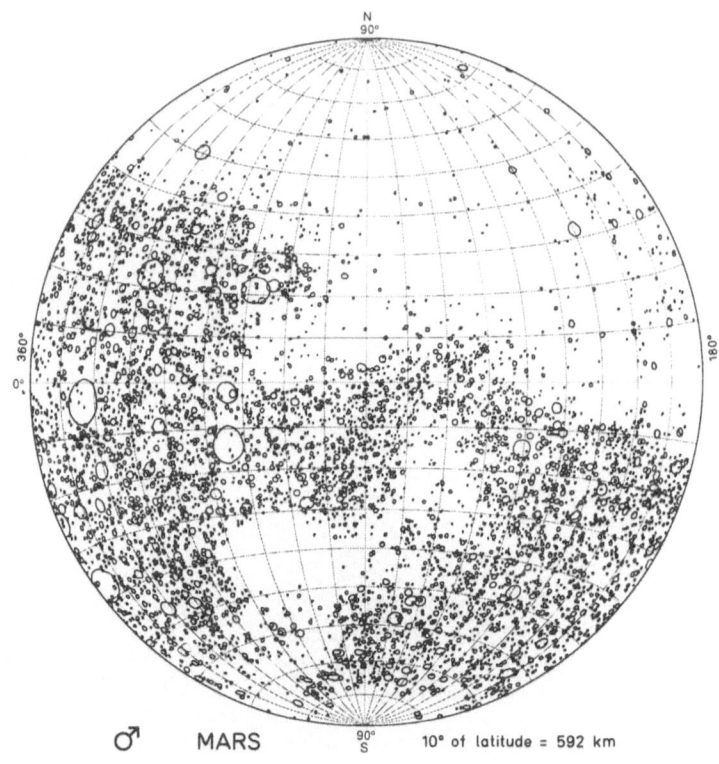

σ'   MARS   90°   10° of latitude = 592 km
                        S

Fig. 95. Distribution of exposed martian impact craters larger than 15 km in diameter in the hemisphere between Long. 180° and 360°W (Mutch et al. 1976, fig. 4.46)

identify and characterize the obliteration event. He also proposed that regional variations in densities of degraded craters are the result of variation in the rate of obliteration. According to Chapman (1974), if the cratering rates for Mars and the Moon were the same the erosional event must have occurred 3.5 Ga ago, and if the rates were 5 to 1 the event would have occurred 1.5 Ga ago. Woronow (1977) also reported that the craters in 12 regions of Mars differ significantly from the planet-wide crater population average, a difference that he ascribed to sedimentation processes, with some differences being influenced by the presence of subsurface ice and by evolution of the impact bodies with time. Soderblom et al. (1974) noted that the densities of 4- to 10-km-diameter craters on the 3.5-Ga-old lunar maria and the ancient cratered terrain on Mars were the same, suggesting that both sets of features were formed at the same time on planets that were undergoing similar impact histories.

    Assuming that lunar and martian fluxes were essentially identical, Soderblom et al. (1974) used the densities of 4- to 10-km-diameter craters to define four stratigraphic units: (1) the heavily cratered surfaces (200–250 craters/10^6 km^2) are the oldest at 3.0–3.5 Ga; (2) the equatorial cratered plains and polar mottled cratered plains (130–180 craters/10^6 km^2) are 1.8–2.4 Ga old; (3) the Elysium volcanics (60–120 craters/10^6 km^2) were extruded 0.8–1.6 Ga ago; and (4) the Tharsis volcanic plain (10–70 craters/10^6 km^2) is the youngest, having been emplaced during the past 0.8 Ga. Soderblom et al. (1974) also ascribed the abrupt drop in density of 0.6- to 1.2-km-diameter craters at latitudes higher than Lats. 35°S and 35°N to burial by a debris blanket, and the variation of distribution in the equatorial region was ascribed to local topographic control of dust deposition and mantling. In contrast, Oberbeck et al. (1977) noted that the mercurian and martian uplands together with lunar highlands far from basins and their associated secondary craters are deficient in craters smaller than 50 km in diameter; and they ascribed these deficiencies not to erosional effects but to production statistics. Application of crater statistical analyses also led Barlow (1990) to propose that the downturn in crater size-distribution curves at diameters generally < 70 km is a reflection of the size frequency of the impacting bodies rather than of removal by crater erosion. Laboratory experiments (Croft et al. 1979) also suggested that terrains composed of ice or ice-saturated soil may have greater count ages than other geological units having identical influx histories. The correction required for this difference in material strength effect may be as great as the correction needed for impact velocity and surface gravity. Such corrections must be considered when comparing interplanetary chronologies. In addition, if the size of craters in ice and ice saturated terrain is related to the tensile strength of these materials, then crater characteristics also must vary with latitude.

8.4.4 Endogenic

8.4.4.1 Densely Cratered Terrains

The martian surface is dominated by endogenic densely cratered and sparsely cratered terrains, flow-and-ridged plains, and high-latitude plains (Figs. 92 and 93), all produced by resurfacing volcanic flows. Other martian features such as volcanoes, impact structures, and channels are superimposed on these two provinces (Carr 1984). The densely cratered terrain is mainly a plateau, which probably includes Mars' oldest surfaces and superficially resembles the lunar highlands and intercrater plains. Martian highlands differ from those on the Moon, however, in that they are dissected by fluvial erosion, and the impact craters are more subdued and have a different size distribution than those on the Moon's highlands (Fig. 96). They also differ in having much more plains material of depositional or volcanic origin than the older terrains on the Moon (Wilhelms 1974). These differences may reflect a more intense crater obliteration early in Mars' history (Carr 1984). The heavily cratered terrain is the result of a period of high rates of impact of bombardment early in the planet's history and a subsequent period when impact rates declined rapidly to the present level. As on the Moon, this change in the rate of bombardment may have occurred around 3.9 Ga ago (Table 8). Crater obliteration rates also were high at the time when bombardment was greatest (Hartmann

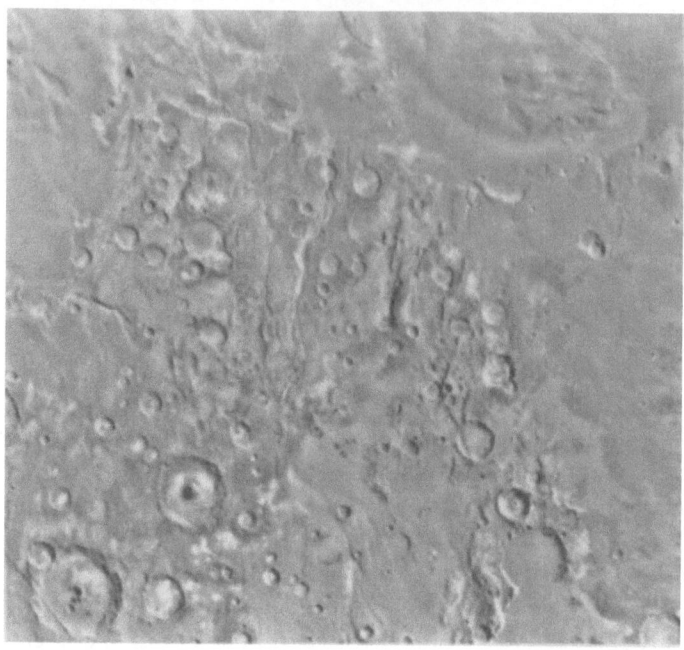

Fig. 96. The contact between the cratered uplands of the southern hemisphere and the northern plains
of Mars near Mangala Vallis. The contact between these two morphological features is an escarpment.
Albee et al. (1992) stated that the rolling plains at the northwest may consist of ignimbrite deposits.
Viking mosaic photograph at Lat. 6°S, Long. 150°W, courtesy of A. L. Albee of California Institute of
Technology and the American Geophysical Union

1973). Contributing to this high rate of crater removal was a combination of the
craters destroying each other, high rates of volcanism (an activity enhanced by
brecciation and heating caused by impacts and greater heat production during the
early history of Mars), and higher rates of fluvial and eolian erosion (rates
enhanced by the thicker atmosphere that existed at that time). The obliteration
processes also affected the fluvial system, leading to a wide range in the degree of
preservation (Carr 1984). Tempe Terra, the northern large block of heavily cratered
terrain that rises 2–3 km above the Acidalium Planitia to the north and south
(Fig. 92), and 4–5 km below the high volcanic Alba, appears to have undergone
three major resurfacing events of Cratered Plateau, Lunae Planum, and Vasitas
Borealis ages (Frey and Grant 1990). The first event is similar to the intercrater-
plains-resurfacing event during the late Noachian. The second event occurred at
the time of emplacement of the ridged plains and is found not only in the heavily
cratered terrain but also in the ridged and knobby plains; it may have been a
planet-wide event at the late Noachian/early Hesperian boundary. The final event
in late Hesperian time that produced the Vastitas Borealis Formation involved
not only volcanic activity but also periglacial processes. Possibly even younger

resurfacing events may be represented in the knobby and mottled plains. Geo-logical mapping in the Amenthes and Tyrrhenas region of the southern cratered highlands indicates that resurfacing terminated during late Noachian to early Hesperian time (Craddock and Maxwell 1990).

A thickness of between 0.5 and 1.5 km of material was removed from the highlands, with some of the material accumulating locally and the remainder being deposited in Isidis Basin (Fig. 92) and other lows in the northern plains. Subsequent eolian erosion (during which as much as 1 km of sediment was removed) exhumed some of the smaller craters and reduced the density of the valley network. In the heavily cratered terrain (Lats. ~ 20° to 30°N, Longs. ~ 285° to 315°W; Fig. 95) of northeastern Arabia the region is overlain by a partly eroded several-hundred-meter-thick horizontally bedded deposit (Moore 1990). Large craters in the region are obscured and those smaller than 10 km are shallow rimless lows. That the layer subdues but does not mask the larger morphologic features and buries the smaller features led Moore (1990) to postulate that the unit was deposited from atmo-spheric suspension or saltation and may be a welded tuff or differentially com-pacted eolian dust. Crater density indicates that the sequence was deposited immediately after the decline of heavy bombardment. Deposition of the layer was episodic, allowing for the development of a succession of channel formation, infilling, erosion, and subsequent new channeling of the Auqakuh Vallis. High rates of volcanism persisted after the decline of cratering rates in the Tharsis and Elysium lows of the northern hemisphere, but declined in the cratered terrains of the southern hemisphere, because the intercrater plains in this hemisphere are more heavily cratered than those in the northern one.

The intercrater plains that partly buried the heavily cratered terrain are superimposed on most of the craters (the rims of old craters protrude through the plains), covering the ejecta deposits and preserving only those craters formed since the time of enhanced obliteration (Carr 1984; Greeley 1985, p. 158). Many inter-crater plains have wrinkle ridges that resemble the volcanic wrinkle ridges on the Moon. The wrinkle-ridged plains cover 36% of the heavily cratered terrain (Greeley and Spudis 1978). Whereas the heavily cratered terrain is dissected by a fluvial system displaying varying degrees of preservation from barely discernable to relatively fresh, the intercrater plains are less dissected and the fluvial system is well preserved. Fluvial erosion apparently continued beyond the date of decline in impact rate, but at lower rates; this accounts for the undissected nature of the intercrater plains (Carr 1984). The intercrater plains contain horizontal layers with some of the layers being unconsolidated (Malin in Carr 1984). Many of the large craters in these plains have level floors, some of which are saturated with craters as small as tens of kilometers in diameter, while others are filled with partially eroded layered deposits, and still others contain wrinkle ridges.

8.4.4.2 Sparsely Cratered Plains

Sparsely cratered plains were formed after the decline in early high rates of cratering (Fig. 97; Carr 1984). The oldest of these plains, which have greater crater

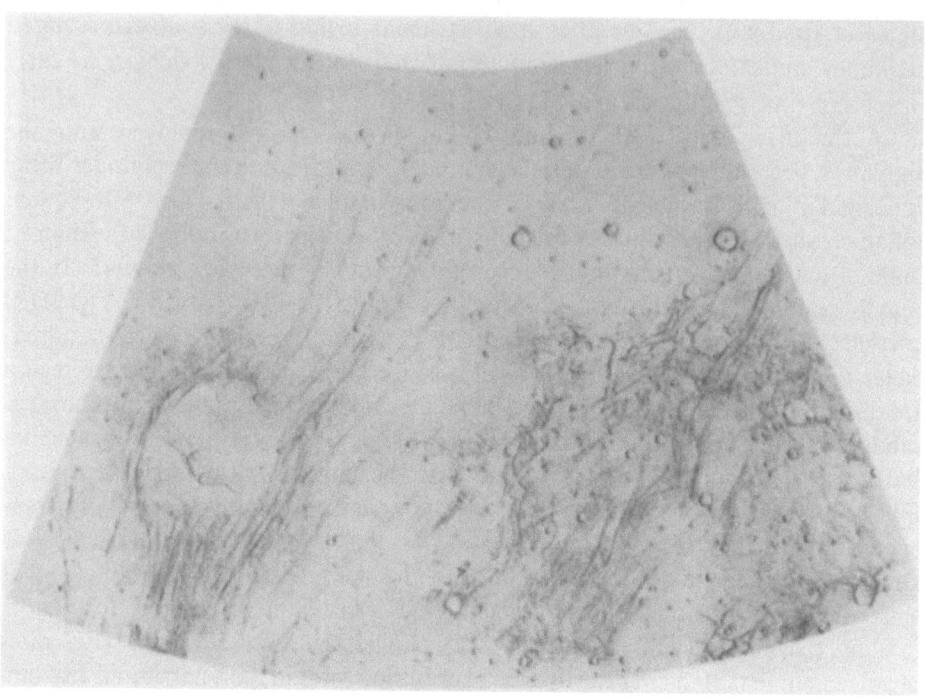

Fig. 97. Sparsely cratered plain in the vicinity of Alba Patera on Mars. Note the ridged plain at the right of Alba Patera, a volcanic structure, and the fractures and grabens associated with Alba. Shaded airbrush relief map (Batson et al. 1979, p. 20)

densities than most lunar maria, probably date from the decline in cratering rate about 3.9 Ga ago, and the youngest plains (those in the vicinity of the Tharsis shields) may be only a few hundred million years old. Most of the sparsely cratered plains are in the northern hemisphere, but a few large plains at the south end of the Tharsis bulge extend into the southern hemisphere. Additional plains are in Hesperia Planum and the floors of Argyre (see Figs. 92 and 111) and Hellas basins. South of Lat. 30°–40°N the plains are topographically well defined and are young looking. Lava-flow fronts, wrinkle ridges, craters, grabens, and channels are recognizable, and albedo patterns are simple and mostly confined to crater-related streaks (Carr 1984). At higher latitudes plains topography is more complex, and the albedo patterns are irregular and produce some of the most confusing morphology on Mars. The plains in the equatorial region appear to be endogenic (volcanic) in origin, but some of them have been modified by fluvial and eolian processes. The high-latitude plains also may be volcanic, but if so they have been so extensively modified that their volcanic features largely have been destroyed. Frey and Schultz (1990) believed that the character of the northern plains in eastern Mars is the result of meteoritic bombardment and early and rapid relaxation accompanied by prolonged volcanic flooding. This volcanic flooding was especially pronounced in regions of basin overlap where the lithosphere was severely thinned and fractured. The load of these Hesperian and younger volcanics and the loss of impact-induced heat led later to thermal subsidence of the basins. Carr (1984) also believed that ice has had a major role in modification of these northern lowlands.

8.4.4.3 Flow and Ridged Plains

The plains within 30° to 40° latitude of the equator can be classified as of two types: flow plains composed of mappable volcanic flow units superimposed one on another, and ridged plains whose only features consist of craters and wrinkles. The plains in southern Amazonis, however, resemble those at higher latitudes, and those of Isidis Planitia contain lines of lava cones having summit vents. Most of the flow plains are on the flanks of the Tharsis and Elysium bulges, with the flows of Tharsis merging with the flow field around Alba Patera (Fig. 97). The three large volcanoes in central Tharsis (see Fig. 93 and 105) appear to have been the source of fan-shaped flows toward the northeast and southwest of each of the shields. As described by Carr (1984), flow morphology ranges from narrow finger shapes to broad tabular types, and individual flows may be as much as hundreds of kilometers long and 10- to 80-m thick, with most being in the 30–60 m range. The narrow flows, which are only a few kilometers wide, commonly contain a central levee channel. The presence of leveed channels and the great lengths of the flows suggest that the martian lavas are basalt. Based on the number of superimposed craters, the flow plains of Tharsis range from the time when the impact cratering rates declined, about 3.9 Ga ago, to plains adjacent to Olympus Mons that are only a few hundred millon years old; thus, formation of the Tharsis flow plains spans most of the history of Mars (Schaber et al. 1978). Carr (1984) believed that volcanic activity in the region may continue, and the plains still are being formed.

Ridged plains occur in a 1000-km-wide north-south trending belt on the eastern flank of the Tharsis bulge (Figs. 90 and 92). The ridges display either a linear parallel trend or a concentric pattern and are regularly spaced 25 to 50 km apart (Zuber and Aist 1990). Included in the belt is Lunae Planum and the eastern margin of Solis Planum and Sinai Planum. Ridged plains also occur in Syrtis Major Planitia, Elysium Planitia, Hesperia Planum, and locally in the densely cratered terrain. Why there is no evidence of flow structures in these plains has yet to be resolved. The wrinkle ridges are tectonic in origin, with those east of Tharsis being clearly related to the bulge itself; they may be penecontemporaneous or postdate the bulge. Ridges in the plains of Hesperia Planum, Syrtis Major Planitia, and the patches in the heavily cratered terrain may have formed in response to local stresses. Zuber and Aist (1990) stated that the regular spacing of the ridges is controlled by contrasts in internal strength and by thickness ratios of competent and incompetent layers within the shallow lithosphere. They proposed a model where horizontal compression prior to ridge-related faulting of a lithosphere consisting of a strong surface plains unit underlain by a weak megaregolith atop a strong lithospheric basement, explains the ridge spacing. Watters (1991) proposed that the wrinkle ridges in the Tharsis bulge region are folds caused by buckling followed by reverse to thrust faulting. Neukum and Hiller (1981) estimated that the age of Lunae Planum is 4.0 Ga, and that Chryse Planitia with 2100 craters larger than 1 km/10^6 km^2 and Elysium Planitia with a crater count of 2000 are slightly younger. Crater counts indicate that the ridge plains of Syrtis Planum, Hesperia Planum, and Lunae Planum are 3.6, 3.4, and 3.2 Ga old, respectively (Hartmann et al., 1981). The best estimated ages based on crater counts given by Carr (1984, table 7.3) are: Hesperia Planum, 3.9 Ga; Hellas Planitia, 3.8 Ga; Lunae Planum, 3.5 Ga; Chryse Planitia, 3.0 Ga; Syrtis Planitia, 2.9 Ga; Amazonias Planitia, 2.8 Ga; Noachis, 2.5 Ga; Utopia Planitia, 1.8 Ga; Sinai Planum, 1.4 Ga; and Maria Acidalium, 1.2 Ga.

8.4.4.4 High-Latitude Plains

Extensive areas of the northern plains (Maria Acidalium and Utopia Planitia; Fig. 92) are disrupted by cracks that form polygons a few to 10 km across (Fig. 98). Most blocks are angular, but where they are superimposed on buried craters their

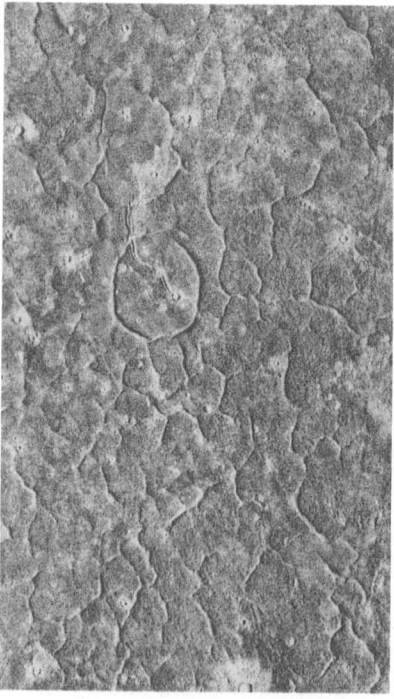

Fig. 98. Ice-wedge polygon-like features on the northern plains of Mars. These features are orders of magnitude larger than the ones on Earth. Photograph 032A18, courtesy of Viking Orbiter Experiment Team Leader M. H. Carr and NASA

cracks tend to form a circle. In some instances the features extend from the adjacent plains across the troughs as though the floors of the troughs have been down faulted (Pechmann 1980). The pattern of cracks also becomes coarser toward the boundary of the heavily cratered terrain. The crack pattern indicates uniform tension in all directions. On Earth such features are due to cooling of lava sheets or occur in permanently frozen ground, but the ones on Earth are much smaller than those on Mars (Squyres 1984). As a result of this scaling problem, Pechmann (1980) proposed that the martian structures were caused by tension associated with large-scale warping of the surface and McGill and Hills (1992) proposed that they were caused by differential compaction over buried topography.

The only plains of any extent at high southern latitudes are in and south of Hellas Planitia and within Argyre Planitia (Fig. 92). The plains of Hellas tend to be obscured by frost, clouds, or haze, and they have an etched appearance as though their surfaces are partly eroded (Carr 1984). The southern edge of these plains also is characterized by ridges and valleys that originate from several volcanic centers beyond the basin. In Argyre Planitia are some long interconnected ridges almost parallel with the southern margin of the plain and which maintain a uniform width of 1 to 2 km for several hundred kilometers. They may represent volcanic intrusions that attained their relief by preferential erosion of surrounding sediment. Carr (1984) also suggested that the ridges may represent eskers (ridges of debris deposited by subglacial rivers) left standing after the ice cover was removed.

8.4.4.5 Volcanoes

In single-plate bodies, such as the Moon, Mercury, Venus, and Mars, interior warming, volumetric expansion, and resultant thermal stresses lead to global extensional tectonics and a lithospheric stress field that tends to enhance widespread volcanism. As the interior cools, expansion changes to contraction so that compressional tectonics and surface volcanism cease. The date that the change in tectonic regime occurs in a planet's history is a function of the initial heating of the planet when it aggregated and the balance between subsequent heat production and heat loss (Solomon 1978). Tensional features such as faults, grabens, and large troughs (or chasms) from such thermal expansion are quite extensive on Mars, and volcanic activity has been widespread throughout much of its history. This activity appears to have modified extensively the cratered-terrain hemisphere that comprises about 55% of the martian surface, mostly in the southern hemisphere. The volcanic units that have modified this cratered terrain consist of: (1) paterae (0.3% of the cratered terrain) that consist of large low-profile volcanoes, older shields, and other unique types; (2) plains volcanoes (comprising 2.9% of the hemisphere) which are low-volume eruptions that formed cones, low shields, and other smaller features; (3) flood volcanics (4.7% of the hemisphere) are high-volume eruptions having lava fronts and wrinkle ridges that postdate the heavy bombardment; and (4) plateau plains consisting of intercrater plains, wrinkle ridges, and floor-fractured craters that comprise 36% of the cratered terrain hemisphere (Greeley

and Spudis 1978). Volcanic processes apparently have been important in obliter-
ating impact craters smaller than 10 km in diameter on Mars. Greeley and Spudis
speculated that as much as 44% of the surface rocks in the cratered terrain may be
volcanic in origin.

Carr (1984) summarized the volcanic history on Mars. The oldest recognizable
volcanic event (which took place shortly after the impact rate declined 3.8 Ga ago)
is represented by the ridged plains of the highly cratered terrain. The next oldest
event is represented by 3.5- to 2.5-Ga-old volcanics. Included in this sequence are
the volcanic features of Hellas, Hesperia Planum, and Tyrrhenia Patera within the
heavily cratered terrain, the slightly younger volcanic features in the ridged plains
of Lunae Planum, Solis Planum, Isidis Planitia, and Syrtis Planitia within the
slightly cratered hemisphere, some of the heavily cratered plains peripheral to
Tharsis and most of the small Tharsis edifices, and the still younger volcanics in the
Elysium, Amazonis, and Noachis plains, Alba Patera, and the Elysium volcanoes.
The youngest units are in Tharsis, the plains of Mare Acidalium and Utopia, and
possibly at Apollinaris Patera. Volcanism in the plains of Tharsis ranges in age
from 3.0 Ga ago to the present, with volcanism on the surface of the large Tharsis
shields and Olympus Mons being only a few hundred millon years or younger in
age. These shields have been sites of volcanism for billons of years, with activity
terminating earlier on the flanks than on the Tharsis bulge proper. Volcanism in
the Sinai-Solis plains peripheral to the bulge is approximately 3.0 to 1.5 Ga old.

Volcanism on Mars initially was quite extensive, and between 4.0 and 2.5 Ga
ago it formed most of the sparsely cratered plains of the northern hemisphere and
the ridged plains of the heavily cratered terrain. Beginning about 2.5 Ga ago
volcanic activity was contained within the periphery of the Tharsis bulge, and
during the past billion years it was confined to near the crest and northwestern
flank of that bulge. Geissler et al. (1990) stated that volcanism in post-Noachian
time (Table 8) was dominantly effusive with some localized pyroclastic activity
documented by cinder cones. The magnitude of pyroclastic volcanism on Mars has
yet to be resolved, with some investigators interpreting the ancient highland
paterae as pyroclastic (Reimers and Komar 1979; Greeley and Spudis 1978), while
others (Francis and Wood 1982) contend that the structures are too large to have
been produced by interaction of magma with groundwater or ice. On the basis of
spectral reflectance, erosional morphology, and the tendency for eolian erosion,
Geissler et al. (1990) interpreted a thick and extensive regional deposit on the walls
of Juventae Chasma and Coprates as a mafic ash. If that interpretation is correct,
pyroclastic eruptions must have occurred at the surface of the planet and later were
buried. These pyroclastic deposits, which are quite old, are analogs with the
basaltic glass deposits associated with older basalts beneath the mare lavas on the
Moon and possibly with deposits on Mercury.

Volcanism on Mars is most evident north of the equator and in the western
hemisphere (Fig. 93C–E). The largest volcanic constructional features occur at
Tharsis, Elysium, and the Hellas region, with the youngest volcanoes on Tharsis
resembling terrestrial basaltic shield volcanoes. The Tharsis province contains
more volcanoes than the other provinces, most of them concentrated near the

summit and around the northwestern flank of the 5500-km-wide and 7-km-high Tharsis bulge that straddles the plains/upland boundary. It is also near this province that the largest volcanoes are to be found on Mars. Arsia Mons, Pavonis Mons (Fig. 99), and Ascraeus Mons in the central Tharsis bulge are 700 km apart and appear to be aligned atop a northeasterly fracture zone northwest of the bulge (Fig. 92). They are 350 to 400 km in diameter with a summit caldera 27 km above

Fig. 99. Pavonis Mons, a volcanic structure on Mars, and associated features. Photograph of mosaic 211-5170, courtesy of Viking Orbiter Experiment Team Leader M. H. Carr and NASA

the planetary datum. The caldera of Arsia Mons is 100 km in diameter and that of Ascraeus Mons is 3.6 km deep, much larger than similar structures on Earth, where the caldera on Mauna Loa is only 2.7 km at its widest point and little more than 200 m deep (Carr 1984). The flanks of the martian shields have a fine radial fabric, produced by long thin flows and channels and wider-spaced concentric lineations caused by terraces, grabens, and pits. Embayments on the northeastern and southwestern flanks of the volcanoes were the source of extensive lava flows that flowed downslope into the adjacent plains, burying the flanks of the volcanic highs and forming an array of fan-shaped flows at the distal ends. All three volcanoes have lobe-shaped features which start at elevations of 10 to 11 km and terminate at about 6 km. Zimbelman and Edgett (1992) believed that the features resulted from gravity tectonics followed by effusive and possibly pyroclastic volcanism. Carr et al. (1977) stated that the lobe-shaped features extending about 350 km from the base of Arsia Minor may be debris flows that formed when the flank of the volcano failed. Ice, which covered the surface of the adjacent plain at the time of slope failure, may have facilitated the lateral transport of the flow. All three structures appear to have experienced the same development: construction of the edifice by eruptions at the summit and along concentric fissures on the flanks, and a second stage where extrusion was concentrated along the northeast-southwest fracture belt causing repeated collapse of the shield, with the lava flows spreading over the adjacent plains to form the flow plains. Carr (1984) stated that building of the central edifice terminated earlier in Arsia Mons, but that activities within the central calderas and embayments in all three structures still may continue.

Alba Patera (Figs. 92 and 97) at the northern edge of the Tharsis bulge is more than 1600 km across and less than 6 km high. Its 100-km-wide caldera has numerous north-south and northeast-southwest fractures forming a 600-km-wide ring. On its flanks are several flow features including 400-km-long 8-km-wide sharp-crested ridges built by channels along their crests that are less than 500 m wide or along lines of pits suggestive of collapsed lava tubes. Some channels leave the crests of the ridges and continue between topographic highs. Other flows display tabular forms with steep fronts, and a third type (also found in Olympus Mons and the shields of Tharsis; Fig. 98) consist of short flows less than 2 km wide with a central channel. These last flows occur mostly near the summit of the volcano. From terrestrial lava analogs, morphological data from the lava flow fields of Alba Patera show that at least for basaltic-andesitic compositions the conditions of eruption on Mars (eruption duration, average discharge rates, average velocities) were similar to those on Earth (Lopes and Kilburn 1990). Geomorphic and stratigraphic analysis of Alba Patera by Schneeberger and Pieri (1991) indicates that the edifice underwent four volcanic phases: extensive flood-like flows from fissures, deposition of the eroded smooth unit that may be a pyroclastic deposit, tabular flows that originated from a central vent or from fissures associated with a circumferential graben system, effusion of tabular flows and levee-like flows followed by collapse of the summit calderas, and a final tectonic episode of graben formation.

Olympus Mons 1600 km northwest of the Tharsis shield volcanoes has a diameter of 600 km with a 65-km-wide caldera at the summit (Fig. 100; Carr 1973, 1984). Within this caldera is a complex crater that exhibits collapse pits at several different centers of activity. Surrounding the shield is an outward-facing cliff 6 km high. Normal and thrust faulting, ice-lubricated gravity sliding, and erosional processes have been proposed for the origin of this cliff (Thomas et al. 1990). Borgia et al. (1990) stated that the basal scarp of Olympus Mons is morphologically similar to the central Costa Rican volcanic range. This similarity suggests that, like the volcanic range, the scarp was formed by a combination of crustal failure caused by volcanic loading and subsequent spreading of the edifice along thrust faults. Beyond the scarp is the 700-km-wide Olympus Mons aureole, a series of deposits consisting of huge, partly overlapping lobes. Models proposed for this unique feature include flows of lava, erosion of subglacial deposits, ash flows, gravity-assisted thrust sheets from the base of Olympus Mons, landslides, and gravity spreading. Surrounding the central caldera is a series of terraces that Thomas et al.

Fig. 100. Olympus Mons and surrounding aureole on Mars. This is a 65-km wide caldera containing collapse pits and small impact craters and surrounded by an outward-facing cliff. Photograph of mosaic 211-5276, courtesy of Viking Orbiter Experiment Team Leader M. H. Carr and NASA

(1990) believed may represent thrust faults caused by compressional failure of the cone. Crater counts indicate an age of a few hundred million years for Olympus Mons (Borgia et al. 1990), an age suggesting that the volcano still may be active (Carr 1984).

Volcanic structures in the Elysium province (Fig. 92) also are concentrated in a crustal bulge, but this bulge is much smaller than the Tharsis one, being only 2000 km across and 5 km high. Elysium Mons has a relief of 9 km with a 400-km-wide fracture ring centered on the summit caldera. The ring probably resulted from flexure of the crust under the load of the edifice (Carr 1984). Outside the ring are numerous east-southeast troughs resembling grabens that are roughly parallel to the Cerberus Rupes fractures southwest of Elysium and that change northwest-ward to a system of fluvial channels extending several hundred kilometers across the plains to the northwest.

The smaller shield volcanoes in Tharsis and Elysium differ from the larger structures not only in size, but also in having proportionally larger summit calderas (Carr 1984). Three of the smaller structures, the 60-km-wide Uranius Tholus, the 120-km-wide Ceraunius Tholos in Tharsis, and the 180-km-wide Elysium, have steeper flanks than the larger structures and have numerous fine radial channels on their flanks. However, these channels lack levees, and they are not atop ridges. One channel that is 2 km wide starts at the summit caldera and terminates at an impact crater in the adjacent plain, but there is no deposit at the end of the channel within the crater. The resemblence of these channels to channels on the flanks of the Barcelo Volcano in Mexico led Reimers and Komar (1979) to suggest that these channels were carved by volcanic density currents composed of mixtures of gas and particulate debris moving radially outward from the volcanic center (similar to neues ardantes at Martinique and St. Vincent in the West Indies during 1903; Agar et al. 1929). Such a current probably was produced by explosive activity or collapse of a gas and ash column.

Many of the large shield volcanoes appear to have accumulation records that may be billions of years long. Crater densities, for example, indicate that Arsia Mons may have been a site of igneous activity for more than 3.0 Ga with average accumulation rates of only 0.0005 km^3/yr (Blasius and Cutts 1976); in comparison, Mauna Loa has an accumulation rate of 0.01 km3/yr. The shapes of the shields are suggestive of fluid lava eruption with little ash, and the complex caldera geometry indicates that the structures have experienced cycles of inflation and deflation. The large size of the martian structures must be related not only to the size of the magma chamber but to lack of plate movement, thereby allowing a structure to remain fixed above its source. Thus as described by Carr (1973; 1984), a volcano under those conditions can continue to grow as long as the lithostatic pressure in the magma chamber exceeds that of the pressure exerted by the column of magma between the chamber and the summit of the volcano. Carr calculated that the depth to melting in Olympus Mons is about 160 km, so topographic relief of the volcanic shields would indicate that the chamber below the crest of Tharsis is shallower than the one below Olympus Mons. Evidently the lithosphere on Mars must be much thicker than the lithosphere on Earth, with its thickness dependent on proximity to the main volcanic provinces (Solomon and Head 1982b). In contrast to Hawaiian volcanoes, martian shields have few pits and vents related to lava tubes. Flows on Mars tend to be longer and wider than their counterparts in Hawaii, supporting the concept that effusion rates are higher on Mars (Carr et al. 1977).

In the area of Hellas are old low-relief volcanoes such as Hadriaca Patera and Amphitrites Patera with circular depressions at their centers and numerous radial ridges comparable with those in Alba Patera. These edifices appear to have been the sources of flows in adjacent plains (Carr 1984). Tyrrhenum Patera, 800 km northwest of Hadriaca Patera, has a 12-km-wide circular caldera surrounded by a 300-km-wide layered unit. In places this unit has been eroded into irregular escarpments, flat-floored radial valleys, and mesa-like highs. Its massive erosion indicates that the unit probably consists of easily eroded material such as ash. From terrestrial pyroclastic flow analogs the flows at Tyrrhenum Patera are gravity-driven ash flows generated by hydromagmatic or magmatic explosive eruptions. Such an origin would require large volumes of water (27.5×10^{16} kg) delivered at flow rates of about 10^5 to 10^6 m^3/s (Greeley and Crown 1990). Apollinaris Patera at Lat. 9°S, Long. 186°W has had a history comparable with that of Arsia Mons – an initial stage during which the main edifice was built, and a later stage when flank eruptions constructed a broad ridge adjacent to and overlapping the volcanic shield.

Igneous activity on Mars was not restricted to the large shield volcanoes and plains described above, but it occurs in different morphologic settings and in features that are smaller than 5 km high. Deep-seated fractures associated with the largest impact craters, for example, served as pathways for igneous activity similar to basaltic flooding in the lunar maria (Schultz and Glicken 1979). Analogs to terrestrial subglacial volcanoes have been recognized on the northern plains and near the south polar region (Allen 1979). The relief of the landforms indicates that the ice-rich volcanic unit originally was 100 to 1200 m thick. Hodges and Moore (1979) described conical mounds on steep sided plateaus and table mountains in the region between Lat. 40°N and the margin of the north polar plains; they believed that these mounds resemble features in Iceland formed by subglacial eruptions. They proposed that both Olympus Mons and its aureole also had a subglacial beginning beneath an ice cap several kilometers thick. Pseudocraters, produced when lava flowed over water-logged ground, are believed to be present in the Cydonia region of southern Acidalium Planitia (Frey et al. 1979), and features resembling spatter cones and ramparts have been reported from Isidis Planitia. The domes in Isidis are aligned in long strings, probably along faults (Carr 1984). Some crater highs in the knobby plains resemble stratocones, and the conical highs with slot-like vents in the Tempe region also are volcanic in origin. As described by Carr (1984), the highs in the Tempe region tend to lie astride grabens and merge outward with the surrounding volcanic plains. The line of craters in the Tempe grabens also may have had a volcanic origin.

Magma production not only creates volcanic morphology but also includes a larger volume of intrusive bodies, some of which change the surface levels of the crust and thereby yield broad morphological effects. On Earth the volume ratios of intrusive to extrusive rocks range from 2 to 12, perhaps averaging 8.5, according to Greeley and Schneid (1991 and Table 5). On Mars, 70% of the total magma production (654×10^6 km^3) appears to have occurred during Hesperian through to early Amazonian times. Greeley and Schneid found that the average rate of magma production in km^3/yr for Mars (0.17) is much less than for Earth (26 to 34) and Venus (< 20), but much more than for the Moon (0.025).

8.4.4.6 Tectonism

The most prominent topographic feature of Mars is the Tharsis bulge, a 6000-km diameter, roughly circular high with a relief of 11 km that occupies about 25% of the planet's surface (Figs. 92, 93, and 101; Wise et al. 1979). Other first-order tectonic provinces of the planet are a slightly cratered terrain in the middle and upper latitudes of the northern hemisphere, covering about one-third of the surface of Mars, and a heavily cratered terrain that occupies about two-thirds of the planet, mostly in the south, and stands about 3 km above the slightly cratered terrain (Fig. 96). The cratered units are separated by a belt of transition that is about 700 km wide and that has receded southward for several hundred kilometers (Mutch et al. 1976. p. 56; Wilhelms and Squyres 1984). This transitional belt, a great circle inclined 35° to the equator contains landforms that are unknown on other planets. Both the lower, slightly cratered terrain and the highly cratered terrain are

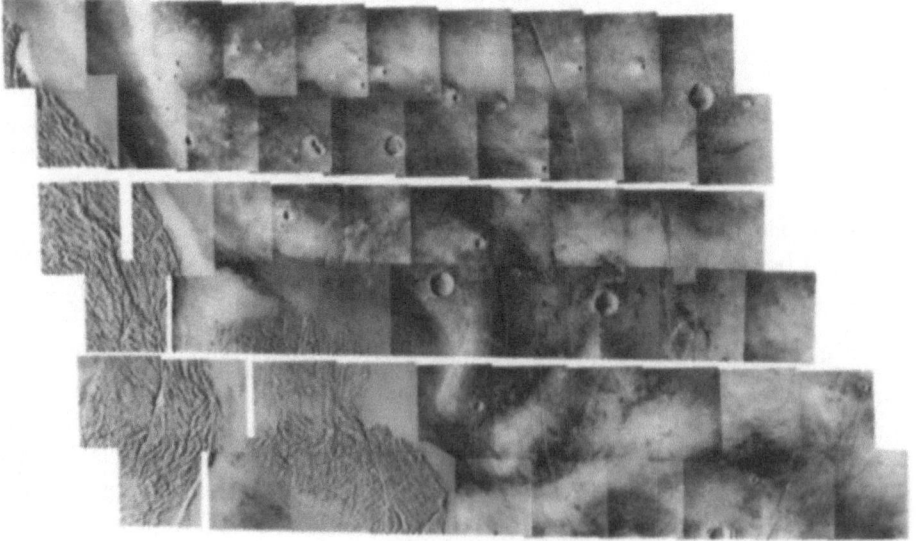

Fig. 101. A segment of the Tharsis bulge of Mars east of Olympus Mons aureole. Note warp, grabens, and fractures at the *right* of the aureole. Photograph of mosaic 211-5771, courtesy of Viking Orbiter Experiment Team Leader M. H. Carr and NASA

in isostatic equilibrium, indicating that they are underlain by different types of crust. Wise et al. (1979) stated that the crustal dichotomy of Mars is the result of subcrustal erosion, thinning, crustal foundering, and lateral migration of lighter materials of the mantle into other parts of the mantle, whereas other workers have proposed that the low region in the northern hemisphere is the result of either the formation of a giant impact basin early in Mars' history (Wilhelm and Squyres 1984) or multiple overlapping large (not giant) impacts (Frey and Schultz 1988). McGill and Dimitriou (1990) postulated that this marked dichotomy between the northern lowland and the heavily cratered southern upland resulted from a tectonic event near the Noachian/Hesperian time boundary that was characterized by enhanced heat loss, extensive fracturing, and formation of faults along much of the highland/lowland boundary. Such a tectonic sequence might have allowed the crust of the northern lowland to be formed by a large convection cell or plume. Yet, according to Barlow (1990), the cratering record on the oldest units in the northern hemisphere indicates that the crustal hemispheric dichotomy took place prior to the heavy-bombardment-intermediate epoch (middle to late Noachian). This cratering record also is not compatible with the multiple-impact model for the origin of the hemispheric dichotomy.

Located on the flanks of the Tharsis bulge are Mars' youngest and largest volcanoes as well as a system of equatorial canyons (Figs. 92 and 101). At the center of the high is an extensive set of fractures oriented roughly radially to the volcanic center (Fig. 93). This radial fault system is most pronounced at the northeastern and southwestern extensions, where the distal ends are 10 000 km apart and cover a large part of the planet. Displacement on the fractures is normal and displays no evidence of lateral slip. The structures form two opposing fan-like geometries that converge on the line of the Tharsis shields. The northeastern fan includes the structures around Alba Patera (Fig. 97), the Mareotis and Tempe fossae, some fractures north of Olympus Mons, and possibly those parallel to Kasei Valles. The less developed southern fan consists of Sirenum and Memnonia fossae (Fig. 92) southwest of the Tharsis shields, Claritas Fossae south of Tharsis, and the fractures in the area of Labyrinthus Noctis. Whereas the peripheral areas of the fracture system display recognizable patterns, those in the general region of Tharsis and Labyrinthus Noctis are less coherent (Carr 1974). A morphotectonic analysis of a north-south graben cut into the Hesperian- and Amazonian-age volcanic plains at Lat. 35°N accommodated about 8 km of post-Amazonian east-west extension, with the graben dimensions suggesting a regional discontinuity at a depth of about 2 km. This discontinuity may represent the base of the megaregolith or a strength contrast within the megaregolith, and it is depressed almost 7.5 km by volcanic loading between Longs. 105° and 115°W (Plescia 1991). Development of the faults may have been caused by superposition of stresses from Tharsis and the volcanic load of Alba Patera. Syria Planum (Fig. 92) in the Tharsis bulge is a rather unique martian high, a rise with radial fractures and a well-developed volcanic center. The Tharsis bulge appears to separate two distinct volcanic types. The prerise volcanics are lower than the postrise ones, indicating a shallower or more fluid magma. The postrise volcanic centers are large and have high relief shields, a morphology implying a deep magma source and thus a thick (about 200 km) and rigid lithosphere.

Carr (1974) postulated that this complex pattern of fractures in the Tharsis bulge is the result of asymmetrical epirogenic upwarping of the martian crust, that the upwarping was centered in the Labyrinthus Noctis region, and that the uplift probably was the result of mantle convection. He further suggested that formation of the bulge occurred before construction of the shields whose locations were controlled by these fractures. Some fractures may have been reactivated as a result of this volcanism. This tectonic event appears to have terminated about 1 Ga ago. Wise et al. (1979) suggested that development of the bulge was controlled by the following events: (1) during the early high-temperature convective phases masses of the core were dispersed throughout the mantle as well as forming lighter eroded subcrustal material; (2) cooling by convection lowered the thermal gradients and major overturn ceased; (3) core material coalesced; (4) core masses fell to the center of the planet, resulting in large-scale mantle motions and overturn; and (5) isolated lighter masses in the mantle became concentrated above the descending core, underplating and thickening the crust which rose isostatically to form the Tharsis bulge. Fractures on the uplifted region resulted from gravity creep of the high topography, with the thermal effects of the underplated mass and the adiabatically heated lower mantle providing the energy for Tharsis volcanism (see also Fig. 93). More recently, Tanaka et al. (1991) proposed that the structural history of the Tharsis bulge region was controlled by a lithosphere beneath the bulge consisting of a thin elastic crustal cap mechanically detached from the strong upper mantle by a hot and weak lower crust; these layers merged laterally into a single cooler strong lithospheric layer. Such a crustal setting could be responsible for late Noachian/ early Hesperian plains volcanism in the surrounding highlands and on the Tharsis bulge, an extensive field of normal faulting on the bulge extending outward at Valles Marineris and Tempe Terra, wrinkle ridges along the edge of the bulge, and Tharsis volcanism on the crest of the bulge from compressional stresses caused by planetary cooling and contraction during late Hesperian/Amazonian time.

8.4.5 Exogenic

8.4.5.1 Erosional Features

The high latitude, sparsely cratered plains are dominated by erosional rather than depositional features. Geological features reported from the northern plains include mottling caused by sharp contrast in albedo between ejecta and plains sediments, regions of numerous closely-spaced mounds and depressions that commonly are aligned, stripes resulting from linear ridges and troughs, closed depressions, knobby hills, flat-topped mesas, cratered domes, pedestal craters, craters surrounded by inward-facing escarpments, and giant polygonal fractures (Schaefer 1990). Guest et al. (1977) described these northern plains as an intricate mixture of young rock units, surfaces cut into old rocks by wind, exposed permafrost, and old exhumed surfaces comparable to the southern cratered highlands. The linear features are described jointly as thumbprint or curvilinear terrains. Possibly the main difference between low- and high-latitude plains is that the high-latitude structures are covered with a blanket of eolian sediment tens of meters to kilometers thick, whereas the low-latitude areas have largely been swept free of debris. As the area swept free of debris moves north and south in response to the 50 000-year precessional cycle and the climates of the northern and southern hemispheres alternate, the boundary between swept equatorial and unswept high-latitude regions must have undergone repeated depositional and erosional cycles (Soderblom et al., 1973a, b). The pattern of a relatively smooth unit with gently undulating topography and a stippled or dimpled unit at a lower elevation, within 500 km of the northern boundary of the old cratered terrain, may be the result of these repeated episodes of deposition and erosion (Arvidson et al. 1976; Guest et al. 1977).

Under the present climatic regime water is stable at high latitudes, but it is unstable in the equatorial region, so that the surface in the latter region is permanently dehydrated. This latitudinal difference in the stability of water also may account for differences between high- and low-latitude plains. Plains north of Lat. 50°N have a mottled appearance due to differences in albedo between the ejecta around craters and the much darker adjacent plains. Differences in albedo may be caused by variations in the degree of entrapment of fine-grained, wind-blown detritus by the ejecta (Carr 1984). West of the Viking 2 landing site, between Lats. 40° to 50°N and Longs. 240° to 290°W, the plains at the termination of channels from the Elysium dome are characterized by a rough topography of a few-hundred-meters relief across hills and hollows separated by smooth areas and local concentrations of closely spaced low linear ridges less than 1 km wide. The channels in the region are indistinct dark linear features. Carr (1984) believed that this change in plains morphology must be related in some way to changes in the character of the channels; if the channels were formed by water, possibly a modification of the ice/fluvial deposits may have caused the change in morphology.

Schaefer (1990) proposed that some of the enigmatic morphologic features in the northern plains may be due to karstic weathering of a former ocean floor. In this scenario an ocean may have come into being early in Mars' history when abundant liquid water was available; that body of water was located primarily in the northern hemisphere lowlands and was in equilibrium with a dense carbon dioxide atmosphere. Such an atmosphere forced the ocean to become quite acid and led to chemical weathering of volcanic material. Weathering also was supplemented by groundwater and surface runoff on any terrains that rose above the surface of the primeval ocean. Calcium and other cations added to the primeval ocean were precipitated as calcium carbonate and other minerals. Weathering of the regolith and precipitation of carbonates in the ocean would have led to the slow removal of carbon dioxide from the atmosphere in a time span as short as 100 Ma if the precipitated carbonates were not reworked, and 1 Ga if they were reworked (Schaefer 1990, and references therein). When the carbon dioxide was removed from the atmosphere and reduced to less than 0.75 to 5.0 bar, the surface of the ocean froze first at the pole and then slowly toward the equator. In time, the entire ocean froze, with the frozen water overlying a layer of salts or concentrated brines atop a layer of carbonate. After the ocean froze, sublimation and ablation processes gradually removed the water–ice until only a residue was left at the poles. Schaefer (1990) proposed that if the entire martian carbon dioxide budget of 1 to 10 bars were used to form the carbonate layer, the carbonates would be 100 to 1000 m thick. In this model the stratigraphic section in the northern lowlands should consist of a highly altered regolith capped by several hundred meters of carbonate that is sulfate- and salt-rich toward the top. Near the highland/lowland boundary (the last region to freeze over and the first one to have its ice cover removed by ablation and sublimation) differential solution by thermokarstification and true karstification produced the unusual topographic features. Schaefer (1990) believed that the karst morphology on Mars probably is comparable to the arid karst terrain in the Nullarbor Plain of Australia – a topography carved in flat-lying carbonate layers during a regime of limited water.

8.4.5.2 Chaotic, Fretted, Grooved, and Knobby Terrains

Chaotic terrains consisting of disordered, jumbled blocks that may be the result of extensive mass wasting, have been described from the eastern part of Valles Marineris, the western boundary of Lunae Planum, south of Apollinaris Patera, and north of Hecate Tholus (Figs. 92 and 102; Greeley 1985, p. 168). This type of terrain is believed by Squyres (1984) to be the result of geothermal warming of subsurface ice in the manner described by McCauley et al. (1972) and Sharp and Malin (1975). Fretted terrain along the northern lowland plains and the cratered

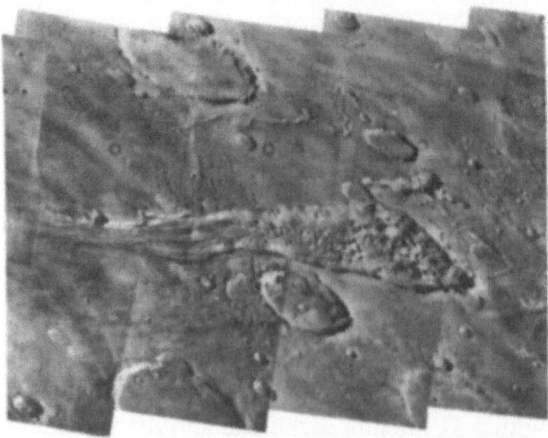

Fig. 102. Chaotic terrain (Aromatum Chaos) on Mars at the head of Ravi Vallis and a graben terrain in a 20-km-wide outlet zone at the *left* of Aromatum. Photograph P16983, courtesy of Viking Orbiter Experiment Team Leader M. H. Carr and NASA

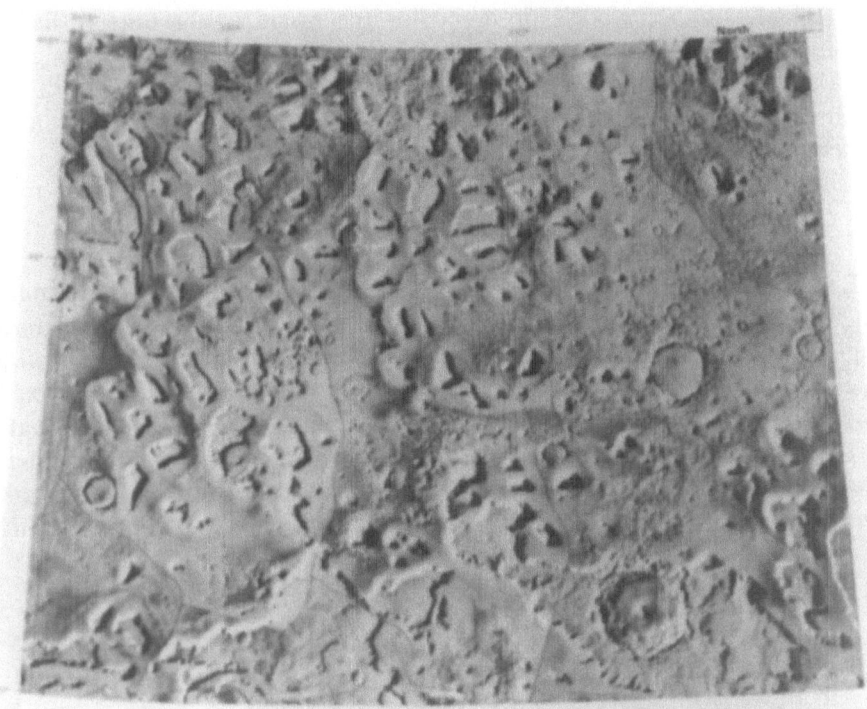

Fig. 103. Fretted terrain in the northern lowlands of Mars, perhaps produced by melting or other loss of buried ice. (US Geological Survey 1985)

highland boundary also may have resulted from the removal of subsurface ice (Fig. 103). This terrain, first described by Sharp (1973b), is characterized by smooth, flat, low areas separated from a cratered upland by scarps displaying complex morphologies. The landforms range from isolated buttes, mesas, and plateaus separated by wide flat-floored valleys and fretted channels. They appear to have been formed by a combination of scarp retreat and breakup of the uplands. Retreat of the scarps may have been caused by evaporation of exposed ground-ice or the emergence of groundwater (Squyres 1984).

Extending for several hundred kilometers from the base of Olympus Mons is a grooved terrain which also is known as the Olympus Mons aureole (Figs. 100 and 101). It consists of four partly overlapping lobes, suggestive of four independent events (Francis and Wade 1983). The lowest lobe occurs on all sides of the volcano, with the other lobes originating from the northern half of the volcano and involving only half of the outward-facing scarp surrounding Olympus Mons. Whereas the oldest lobe represents a single radially directed spreading sheet, the younger lobes were formed by spreading from volcano flanks that already had scarps. Lateral spreading of the material is believed to have been caused by a major plutonic intrusion within the volcano, and the intrusion supplied radial compressive forces. Other models proposed for the origin of the aureole include pyroclastic and lava flows, landslides, erosion of ash sheets, eruptions through ice sheets, and mass wasting (Francis and Wade 1983; Greeley 1985, p. 172, and references therein). The knobby terrain east of Elysium (Fig. 92), within the northern plains and along the cratered uplands/northern lowland plains boundary, consists of knobs up to 10 km in diameter. Some of the knobs appear to be erosional remnants, but others having small summit craters may be small volcanoes.

Exogenic terrains formed by fluvial, glacial, and eolian agents or a combination of endogenic/exogenic processes are quite extensive on Mars. The equatorial troughs, the Valles Marineris, are endogenic in origin but have been modified by exogenic processes Figs. 104–107). The system is about 5000 km long, 700 to 150 km wide, extends from about Long. 20°W to Long. 100°W between the equator and Lat. 15°S, and some of the segments have as much as 9 km of relief (Sharp 1973a; Baker 1982, p. 27). Imagery shows that landslides, avalanches, faulting, and both eolian and fluvial processes have had roles in forming this immense trough

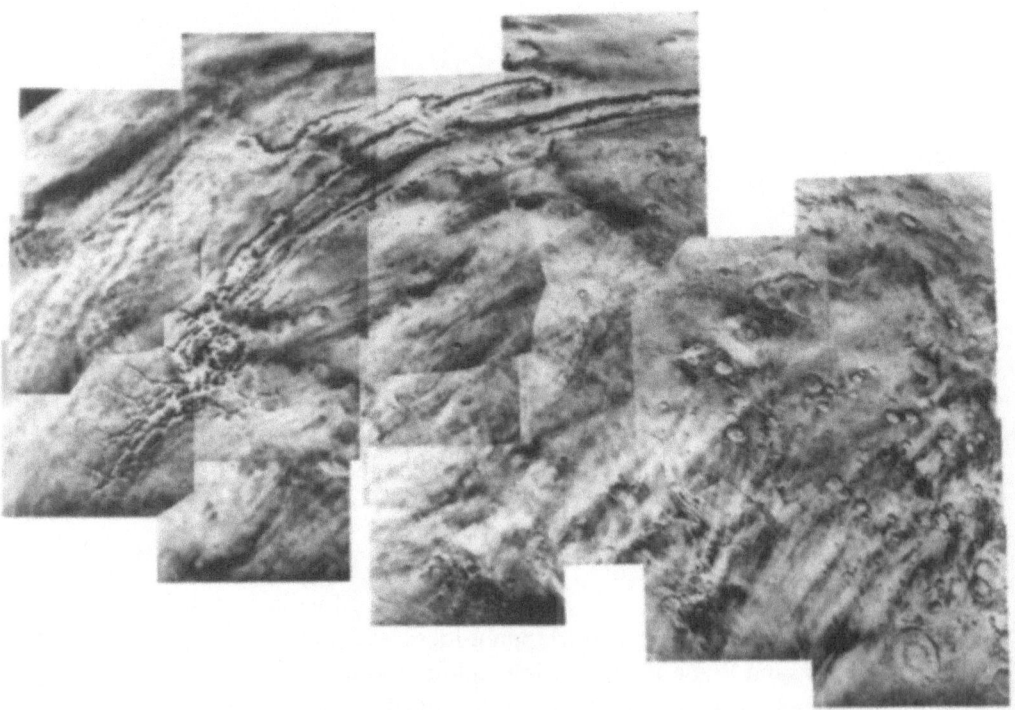

Fig. 104. A segment of Valles Marineris extending eastward from the Tharsis bulge on Mars. Note ridged plains southeast of the bulge. Photomosaic 211-5048, courtesy of Viking Orbiter Experiment Team Leader M. H. Carr and NASA

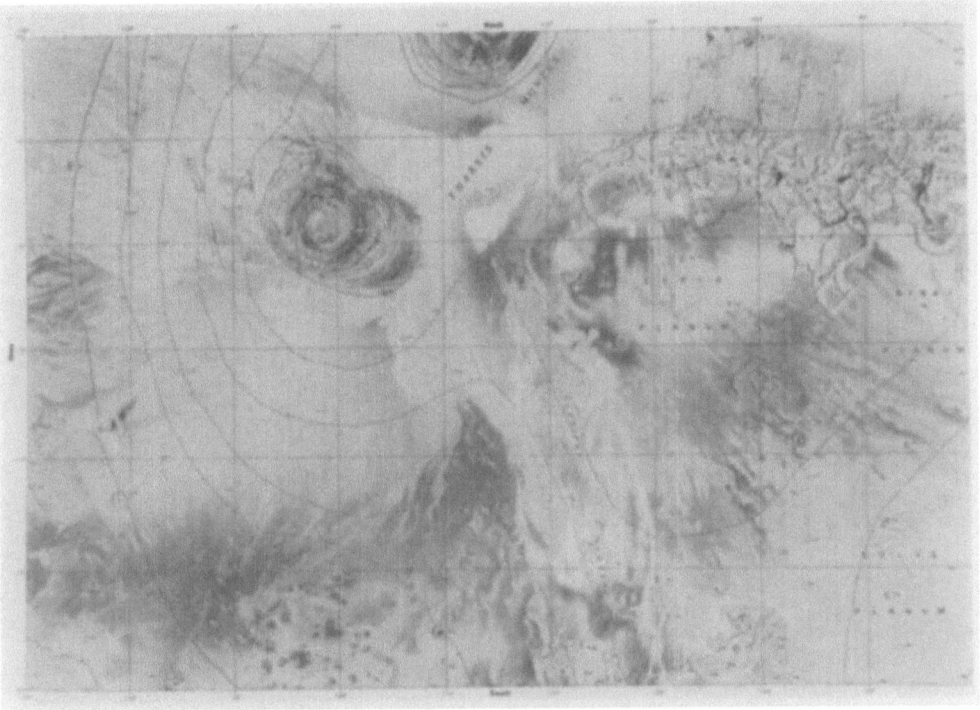

Fig. 105. Western end of Valles Marineris at the eastern flank of the Tharsis bulge, a 4000-km-wide plateau on Mars with its apex at about Lat. 0°, Long. 114°W (Fig. 92). The complex morphology of Noctis Labyrinthus is caused by fracturing associated with uplift of the bulge. Arsia Mons, Pavonis Mons, and Ascraeus Mons (beyond the map) are aligned along the crest of the bulge. Olympus Mons (see Fig. 100) is located on the northwest flank of the bulge. Other points of interest are the ridged plains north of Noctis Labyrinthus. Shaded airbrush relief map (Batson et al. 1979, p. 56)

system. Faulting (which has acted over a considerable length of time) has created the linear lows, whereas the exogenic processes have tended to fill them. The sections exposed along the canyon walls consist of an upper (one third to one fourth of the exposed section) well-stratified unit, which is 1 to 2 km thick and may be a sequence of basaltic flows, and a lower more homogenous unit. The lower unit is associated locally with huge lobate flows (Baker 1982, p. 28). Canyon fill consists of bedded sediments that have been attributed to cyclic sedimentation – a depositional regime probably related to changing planetary climates (Blasius et al. 1977). The system shows a progressive change in morphology from west to east. At the western end is a region of intersecting graben-like features of fault origin and known as the Noctis Labyrinthus (Figs. 104 and 105). East of it are long narrow canyons, Tithonium Chasma and Ius Chasma, choked with landslide debris. Ius displays tributaries with elongate light and dark markings indicating down-canyon transport. A concentration of these tributaries along the south wall of Ius may have resulted from regional slope enhancement by spring sapping (Sharp and Malin

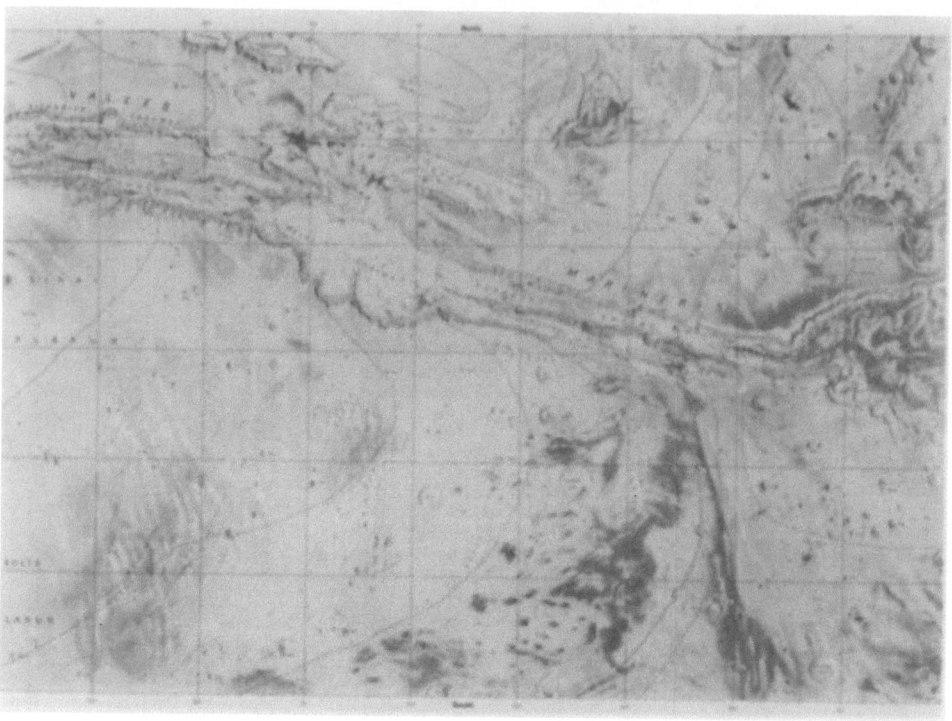

Fig. 106. The central canyon area of Valles Marineris, a region of deep and wide chasms on Mars. Shaded airbrush relief map (Batson et al. 1979, p. 58)

1975). As described by Higgins (1984 and references therein) headward growth of these tributaries was due to: (1) a rising thermal front produced by regional heating under Tharsis or local injection of magma into the Valles Marineris rift caused the subsurface ice to thaw; (2) the meltwater together with suspended ash, volcanic clasts, and other particles began to flow in the direction of Valles Marineris; (3) where the strata were thick or coherent this sapping led to slumping, block slides, and debris flows; and (4) where the strata dipped toward the trough concentration of the subsurface flow along structural weaknesses led to headward growth of tributary valleys. Higgins stated that the time required to form the tributaries depended on the rate of thaw and the nature and degree of consolidation of the material. It is these processes, subsequent sublimination of ice ('dry' sapping), and talus accumulation that created the tributaries along the margins of Valles Marineris.

The central part of Valles Marineris is dominated by the broad Melas and Coprates chasmata characterized by straight low scarps along the bases of their walls (Fig. 106). These scarps, commonly parallel to fractures on the heavily cratered upland, represent traces of normal faults that locally truncate spurs and gullies to form hanging valleys. Atop these scarps also are landslide lobes. Collapse

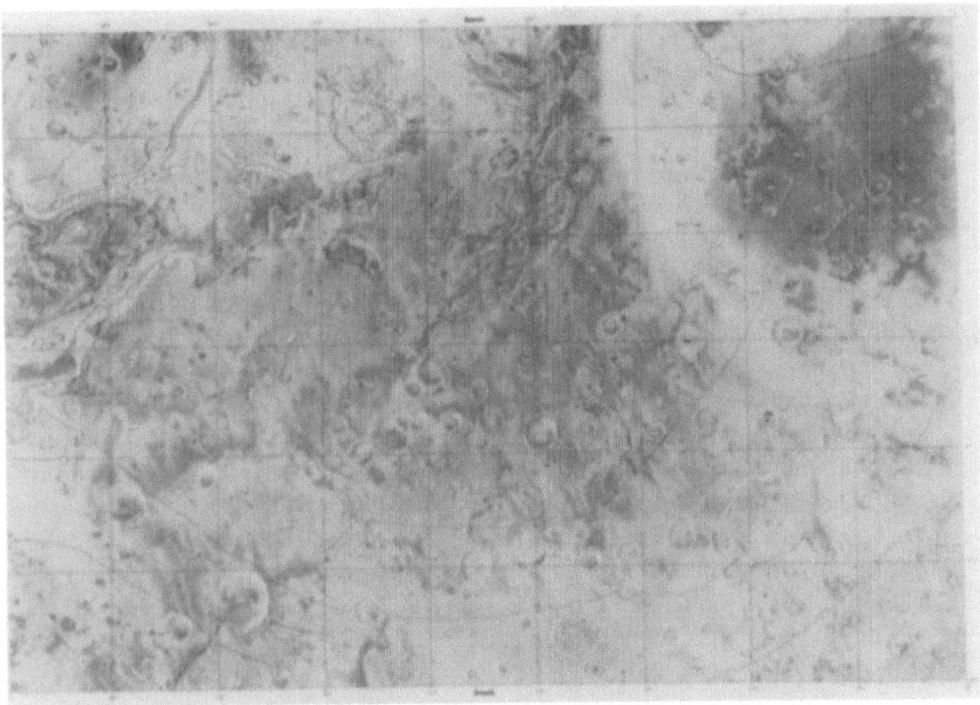

Fig. 107. Eastern end of Valles Marineris, a system of outflow channels and chaotic terrains on Mars. Shaded airbrush relief map (Batson et al. 1979, p. 60)

of the Coprates Chasma graben was caused by volcanic tectonic activity in Tharsis Montes during early late Hesperian time (Schultz 1992). Faulting in Ophir Planum is related to faulting in the chasmas. At the eastern end of Valles Marineris, which includes Capri Chasma transitional to the chaotic terrain, is a region of slump and collapse blocks in lowland depressions bounded by steep walls (Fig. 107; Sharp 1973b). North of Valles Marineris are Echus and Juventae chasmas containing areas of chaotic terrain. These chasmata are the source of the channel systems in Kasei and Maja valles.

The canyons that form Valles Marineris are eroded in strata deposited before or during the end of the heavy bombardment, because the plains in which they are entrained are late Noachian or Hesperian in age (Spencer and Fanale 1990). the sparsity of craters in the canyon fill indicate that their interiors are Amazonian in age, with canyon erosion possibly extending to the present (Scott and Tanaka 1986). Although the evidence for tectonic control in the trend of the canyons is strong (they parallel a trend radial to the Tharsis bulge followed by a graben swarm on the plateau peripheral to the canyon system, and fault scarps are present at the base of some canyon walls (Blasius et al. 1977; Spencer 1984), most canyons tend to terminate abruptly and bluntly along the strike. Closed depressions are common throughout the canyon system, the most impressive being Hebes Chasma. The smaller depressions are chains of pits along small grabens on the adjacent plains. According to Spencer and Fanale (1990), there is a continuum between pits and canyons, with some canyons having been formed by coalesence and enlargement of pits. Spencer and Fanale (1990) proposed several models to explain the closed depressions, which appear to have been formed by collapse. In one model the depressions were formed by decay of ice-rich bodies in nearly sediment-filled grabens that predate the overlying cratered and ridged plains. In a second model the lows were formed by removal of an equatorial carbonate layer that was deposited during the early history of Mars. Solution of the carbonates could have been caused by atmospheric carbonic acid (which would have required extensive recycling of the water supply), through solution by groundwater acids (which may have been derived from the Tharsis magmas and required less water), or through decarbonation by early heat flow or Tharsis-related heat pulses. Other linear features of endogenic/exogenic origin include linear troughs in younger volcanic terrains that have been modified by subsequent erosion. Fractures also guided the development of channels by fretting and sapping. Other endogenic linear features are open at both ends, are relatively straight, and have steep sides and flat floors. According to Sharp and Malin (1975), they occur in groups that display parallel, en echelon, or intersecting geometric patterns.

Martian channels of fluvial origin can be classified in two groups: outflow and runoff channels (Sharp and Malin 1975). Outflow channels, which are most common along the northern lowland/southern highland boundary, originate in the chaotic terrains and may extend for hundreds of kilometers (Figs. 96 and 108). Some channels are narrow and deeply incised, whereas others are broad and anastomosing (Squyres 1984), resembling channels on Earth that were formed by catastrophic floods such as the flood features of the Channeled Scabland of eastern Washington (Bretz 1923), which were the result of catastrophic breakouts of ice-dammed Lake Missoula and overflow of Lake Bonneville (Bretz 1923; Baker 1973; Baker and Kochel 1979; Baker 1982, pp. 140–174). Recent calculations indicate that flood discharge from Lake Missoula may have exceeded 17 ± 3 (10^6) m^3/s with the flow duration being about several days (O'Connor and Baker 1992). The features on Mars resemble their supposed terrestrial counterparts in many ways, but they differ in that their components are several times larger than their supposed terrestrial counterparts (Squyres 1984). Impact-crater density indicates that the channels are younger than the heavily cratered highland from which they originate. Using the Soderblom et al. (1974) influx model and Mariner 9 images, Malin (1976) came to the conclusion that the channels date from the period of heavy bombardment 4 Ga ago. An influx model by Neukum and Wise (1976) suggested the age of the channels to be 4.0–3.5 Ga (Squyres 1984). From Viking images and the Soderblom et al. model, Masursky et al. (1977) estimated that the age of the channels is only 2.5 to 1.0 Ga.

According to Baker (1982, p. 174) and Baker and Kochel (1979), fluvial processes involved in the large-scale floods that shaped the martian landscape included ice covers formed as a result of low pressures (7 to 10 mbar) and low temperatures (190° to 250°K) on the surface of Mars, macroturbulence, stream-

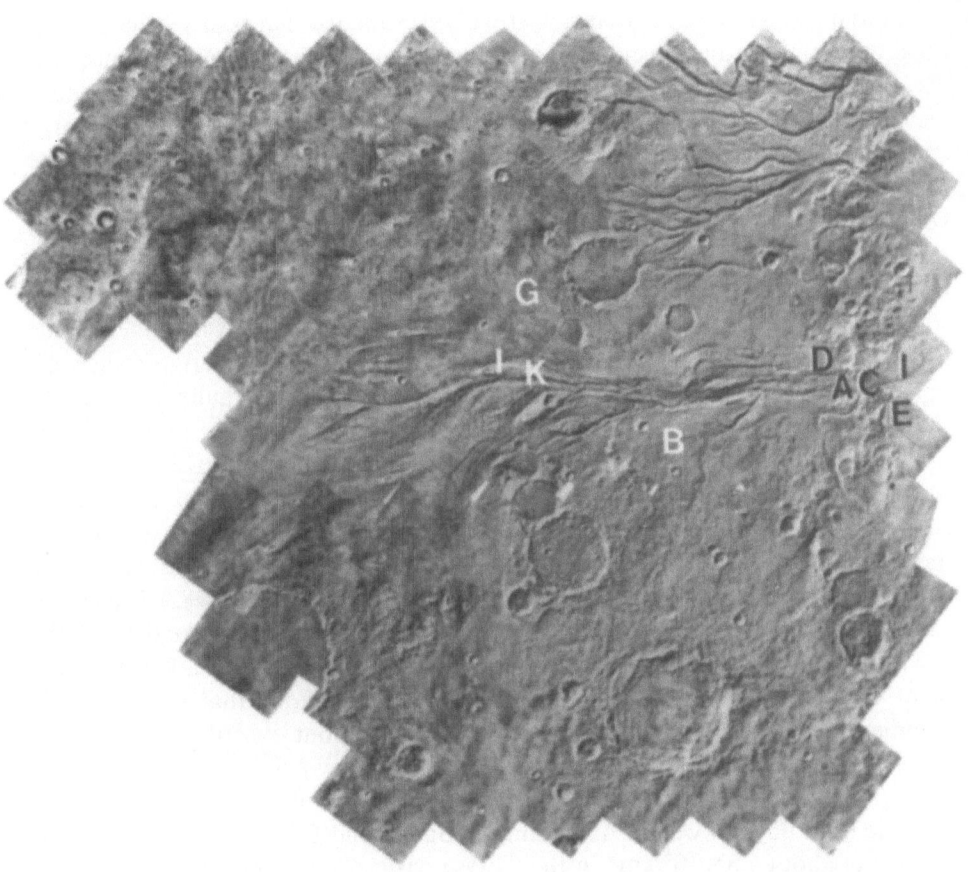

Fig. 108. Maja Canyon segment of the Maja Vallis system of Mars. The canyon, an outflow channel, is incised on the heavily cratered terrain. Baker (1982, pp. 124–125) recognized the following features in the canyon area: **A** lobate deposits, **B** possible thermokarst topography, **C** constricted channel, **D** inner channel cataracts, **E** possible mid-channel bars, **G** overflow area connecting Maja flows with flows from channels at the north, **I** terrace-like benches along the side of the channel, and **K** slump blocks. The channels north of Maja Canyon may have been produced by spring sapping. Sediments deposited from these channels have filled the flat-floored 30-km-wide crater. Photograph VI 211-5190, courtesy of Viking Orbiter Experiment Team Leader M. H. Carr and NASA

lining, and cavitation. Turbulence would have tended to break up the ice cover formed by evaporative cooling, with the ice blocks contributing to a variety of river ice processes, channel enlargement by bank scour and incision, ice jamming where ice scour would be enhanced, backflooding by waters imponded by ice jams, and catastrophic floods when the ice dams broke up. An example of such an out-flow valley is the Maja Vallis system that originates at Juventae Chasma, a 250 × 100 km depression. The southern part of the system consists of 1000-km-long northward-trending channelized terrains that converge on canyons (Maja Canyon, Maumee Vallis, and Vedra Vallis) incised on the heavily cratered terrain. The canyons debouch onto a series of channels on Chryse Planitia (Fig. 92) at the terminus of the Maja Vallis system. Carr (1979) estimated that peak discharge for a flood that originated in Juventae Chasma and extended across Lunae Planum was 7×10^6 to 10×10^8 m^3/s. Mechanisms proposed for triggering such large floods include volcanic eruptions beneath glaciers, pressure release, dissociation of large amounts of subsurface clathrate, and geothermal warming of subsurface ice in the chaotic terrain (Squyres 1984, and references therein). According to Carr (1979), the volume of water that could be removed from the chaotic terrain was insufficient to provide the flood discharge needed to erode the outflow channels. Thus, he proposed that the water was released at high pressure from an extensive ground-water system, a pressure that was due to capping of the aquifer by an impermeable frost layer or by tectonic warping of the surface. The channels formed when the pore pressure exceeded the lithostatic pressure, leading to hydraulic breakout. In the scenario by Tanaka and Chapman (1990) there were two major coeval periods of erosion and faulting in the Mangala Valles and Memnonia Fossae regions of Mars during the late Hesperian and late Hesperian/early Amazonian. Both sets of channels originated from the same graben system, and the faulting correlates with the Pavonis I and II episodes of radial faulting in the Tharsis bulge when lavas were extruded in and east of the channel system (Plescia and Saunders 1982). Erosion of Mangala Valles is supposed to have been caused by faulting that broke through the cap layer and tapped the confined aquifer in the higher regions east and west of the point of discharge. Tanaka and Chapman (1990) estimated that the minimum volume of water required to erode the valleys was about 5×10^{12} m^3, a volume released through two floods and representing only a few percent of the volume of the aquifer. They also estimated that the peak of the floods may have lasted anytime from days to weeks. According to Gulick and Baker (1990), a discharge volume of about 7.5×10^6 km^3 in the Chryse basin region could have formed an ocean in the northern plains, an ocean that had an average depth of 660 m and a surface area of 1.1×10^7 km^2, or about 7% of the surface area of Mars.

Another model that has been proposed for the origin of the outflow channels is subsurface liquifaction of clay sediments in the chaotic terrain and release of debris flows, a generation that was enhanced by high water pressures (Squyres 1984, and references therein). Erosion by ice streams comparable in dimensions to those in Antarctica originating in the chaotic terrain is another mechanism that has been suggested for the channels (Lucchita et al. 1981), but, according to Squyres (1984), the chaotic terrain is too small to be the source for streams as large as those peripheral to the Antarctic ice sheet. Cutts and Blasius (1981) proposed that the channels were formed by sand blasting and deflation of the surface by sand-sized particles weathered out of the chaotic terrain and transported by wind across the adjacent plateau surfaces. They believed that eolian erosion formed the outflow channels in about 1 Ma; however, such massive erosion could have occurred only early in the planet's history. Although eolian erosion may have helped to modify the channels after they were formed, it probably was not responsible for their formation.

Some of the linear erosional features known as runoff channels or valleys are narrow and commonly sinuous, they tend to increase in size and depth distally, and their headwaters usually have tributaries (Figs. 96 and 108; Sharp and Malin 1975; Theilig and Greeley 1979). According to Sharp and Malin (1975), crustal structures may influence the configuration of these runoff channels. They differ from fluvial terrestrial features in that they have steep sides with talus slopes, flat floors, and tributaries that do not display a dendritic pattern but tend to have amphitheater-like terminations. There also is no evidence for fluvial erosion on intervalley surfaces (Squyres 1984). The morphology of the martian valleys is more indicative of spring sapping and runoff of subsurface fluid than of rainfall. Pieri (1979), who named these features valleys to distinguished them from features caused by fluvial action, believed that they were the result of erosion by water flowing on the surface, or in the subsurface, from lithospheric sources. He stated that there is no evidence to indicate that they were formed by rainfall-supplied water, that they were eroded early in martian history, or that they predate the major plains units. According to Pieri, at similar orbital resolution there are no terrestrial features comparable with the martian valley network. Not all these channels, however, may be exogenic in origin. Such features as Nergal (with its long main valley having a few short stubby tributaries) are more suggestive of a collapsed lava tube or a graben. Others are due to faulting and subsidence along linear fracture zones caused by ground-ice deterioration or magma withdrawal (Sharp and Malin 1975). Such observations led Schumm (1974) to propose that the large martian runoff channels are essentially endogenic features. Brakenridge (1990) reported that the valleys in the Aeolis quadrangle display the same orientation as the thrust faults in the region, suggesting that they are partly controlled by preexisting faults and related fracture systems. He proposed that the valleys were formed by headward sapping of thermal springs, thermokarst subsidence, and limited downslope fluid flow. The water supposedly was derived from outgassed water stored as frost, snow, and ice within the cratered terrains during the time of heavy bombardment and deposition of ejecta, volcanic ash, and eolian sediment. Later effusive volcanism and sill injection melted the subsurface ice, and the water rose to the surface via faults and fractures.

A few of the valleys on Mars do display very complete dissection and lack broad intervalley surfaces; these valleys more probably were carved by rainfall (Squyres 1984). This channel system is in heavily cratered terrain and appears to be concentrated in the equatorial region (Pieri 1980a). Theilig and Greeley (1979) reported that both outflow and runoff channels occur in Lunae Planum, an upland centered at Lat. 15°N, Long. 65°W and at Chryse Planitia, a depression northeast of Lunae Planum located at Lat. 24°N, Long. 45°W. These features denote four periods of channeling, the first two of which date from early in the history of Mars and are represented by narrow channels in the heavily cratered terrain, and the third and fourth are represented by larger channels (Vedra, Maumee, Bahram, and Maja valles) formed by two catastrophic floods. During the first flood a sediment wedge was deposited by Vedra, Maumee, and Bahram valles on the western slope of Chryse Planitia, and during the second flood Maja Valle eroded the wedge. The crater density indicates that the catastrophic flood channels are as old as the cratered highlands, with their formation being limited to a time prior to 4.0 Ga ago. As pointed out by Milton (1973), degradation by running water was important in molding the martian landscape some time during the past. Although water might flow for a considerable distance beneath a thick ice layer in Mars at present (Wallace and Sagan 1979), the valleys probably were not formed under climatic conditions similar to those of the present. Now the tributaries probably would become choked by ice, cutting off the flow from the main trunk (Squyres 1984). The channels must have been eroded at a time very early in Mars' history when the surface temperature and pressure were much higher. As these channels are absent from terrains younger than 4.0–3.5 Ga, such ideal conditions for fluvial processes must have existed only during Mars' earliest history.

Other features ascribed to a combination of fluvial and sapping erosion are channels/valleys on the sides of moderate-sized martian volcanoes (Gulick and Baker 1990). Those on Tyrrhena Patera are wide, flat-floored, monofilament amphitheater-headed channels resembling valleys on Earth carved by spring sapping. Along the channel walls are horizontal layers that may be ash flows. On Hadriaca Patera the easily eroded volcanic materials contain long, highly degraded, trough-shaped channels having a radial pattern that originates near the central caldera. Apollinaris Patera also has radial, degraded, trough-shaped channels, and these can be traced to the rim of the central caldera. Channels in the eastern flank of Ceraunius Tholus consist of pristine features having steep walls and degraded valleys with eroded, less steep sides. Both types have tributaries that originate from sources just below the central caldera. This type of terrain is similar to the morphology of the northeastern flank of Kohala volcano on Hawaii and formed by runoff and sapping. The southwestern flank of Ceraunius Tholus is entrained by chains of pits that radiate out from the caldera to the base of the volcano, and on the northern flank are three large channels that originate at or near the caldera and can be traced down the side of the structure. Pristine and degraded channels also occur on Hecates Tholus, and on the northern flank of Alba Patera are gully-like forms, which are linear and display a pitted or discontinuous texture along the axis of what may be subdued lava flows, and pristine valleys imposed on gully-like branching forms. This distinct channel form of a combination of two different morphologies has been termed 'enigmatic'.

These volcanic channels/valleys, whose relative ages are indicated in Table 8, are believed by Gulick and Baker (1990) to have been formed by fluvial processes, with lava flows and possibly volcanic density flows also having significant roles. The formation of valleys on Ceraunius Tholus, Hecates Tholus, and the modified valleys on Apollinaris, Hadriaca, and Tyrrhena are an indication that surface and near-surface water was widespread on Mars during the Noachian, and the presence of valleys on Alba Patera (Fig. 97) shows that such conditions persisted locally until the middle to late Amazonian; thus, fluvial erosion persisted for more than 3 Ga. Channels/valleys that formed during the heavy bombardment exhibit a less developed tributary system than those on Alba Patera; apparently more water was available during formation of the Alba valleys. According to Gulick and Baker, the lack of spring sapping at Alba also suggests that fluvial activity in this region was short lived, and its presence on Alba indicates that erosion can occur under the present climate. Possibly erosion was active under a protective ice layer, as described by Wallace and Sagan (1979). Increased precipitation could occur during high obliquity of Mars' ecliptic followed by melting of snow or ice during periods of low obliquity. Possibly the water that carved the valleys could have been introduced into the surface system by hydrothermal activity, a process that led to formation of local warmer and moister environments, which in turn would have enhanced precipitation. Precipitation also may have been enhanced on Alba Patera by northerly winds that delivered saturated air from the ocean that existed in the polar region. The local ash mantle on Alba would tend to inhibit filtration and increase surface runoff. Evidence for such an ocean is the nature of the overlap contact between the gradational and fretted terrains along the lowland/upland boundaries in West Deuteronilus Mensae at Lat. 45°N, Long. 345°W. According to Parker et al. (1989), the morphology along the overlap requires an unfrozen condition that allowed wave erosion and redeposition of sediment along the boundary. Of all the channels on Mars those on Alba most closely reflect an atmospheric hydrologic cycle, and they apparently were formed late in martian history when climatic conditions were similar to those at present. These observations led Gulick and Baker (1990) to postulate that channels on Mars do not necessarily indicate a former warmer climate and denser atmosphere, but may denote local hydrologic cycles resulting from atmospheric aberrations or endogenic processes such as hydrothermal circulation.

Other types of channels that may owe their morphology in part to fluid erosion and deposition include fretted and excavated channels and modified endogenic features. Fretted channels are steep walled with smooth floors and in plan view they have irregular indented walls, integrated craters, and they appear to be structurally controlled (Sharp and Malin 1975). Isolated buttes or mesa-like features are common, and locally there is extensive integration of adjacent channels. Excavated channels form a series of linear depressions. Although these features may be endogenic in origin, their presence now may be the result of removal of cover by eolian deflation or evaporation. Sharp and Malin expressed the notion that some of the linear lows formed by endogenic processes may have undergone secondary modification by exogenic processes, such as collapse features that subsequently were modified by erosion. Detailed mapping of the fretted terrain of Ismenius Lacus led Kochel and Peake (1984) to propose that it was produced by mass wasting of old cratered terrain. They interpreted the island mesas as extensive debris aprons whose flow was facilitated by interstitial ice.

Some morphologic features on Mars reflect the presence of ice now on the planet (Figs. 109 and 110; Squyres 1984). If such ice is present at latitudes lower than 40° it must be buried, because ice cannot exist in equilibrium with the present atmosphere (Fanale 1976). According to Smoluchowsky (1968), however, a sediment layer as thin as a few tens of meters may act as an effective barrier, preventing sublimation of the water ice. A feature that may owe its origin to ice is the chaotic terrain described by Sharp (1973b) as a region of jumbled slumps and collapsed

Fig. 109. North polar region of Mars. Mutch et al. (1976, p. 88) ascribe the *spiring dark lines* to frost-free cuesta-like slopes. Photograph 211-5359, courtesy of Viking Orbiter Experiment Team Leader M. H. Carr and NASA

blocks bounded by steep walls (Fig. 102). It was formed by collapse of a cratered terrain through removal of subsurface ice or magma. Sharp believed that collapse caused by evacuation of magma may be more feasible because of the extensive volcanism in the northern hemisphere. After the collapse, sapping by ground-ice led to local slumping and modification of the collapsed blocks. According to Sharp,

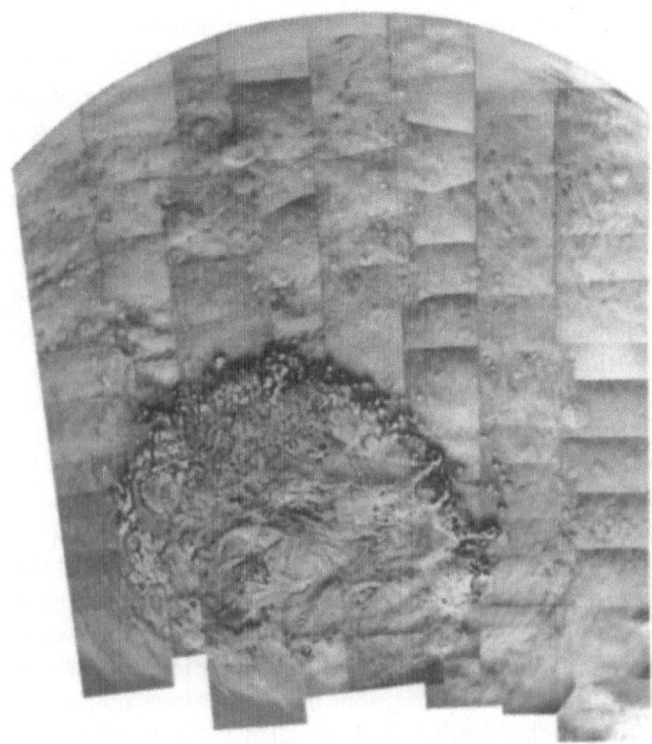

Fig. 110. South polar region of Mars. Like the north pole, this region is a quasi-linear swirl-like feature. Photograph 211-5393, courtesy of Viking Orbiter Experiment Team Leader M. H. Carr and NASA

some segments of the chaotic terrain have undergone smoothing and now resemble a fretted terrain.

Another area whose morphology may have been influenced by ice is the fretted terrain – a low smooth region mainly along the northern lowland/southern highland boundary and separated from cratered upland by complex scarps (Fig. 103; Sharp 1973b). Retreat of the scarp that produced the smooth lowland could have been caused by evaporation of exposed ground-ice or emergence of groundwater. Material left behind when the scarp retreated was removed by subsequent eolian activity (Squyres 1984). Lobate aprons at the bases of the scarps at Lats. 30° to 50° are other examples of terrains formed by the action of ice. They appear to be the result of interstitial ice flow by creep and are perhaps comparable to rock glaciers on Earth, according to Carr (1984). Small-scale collapse regions characterized by tablelands with scalloped edges and small depressions (oases) in the southern part of Chryse Planitia (Fig. 92) are believed by Carr and Schaber (1977) to be similar to thermokarst features on Earth formed by melting of ice at high latitudes. In the heavily cratered terrain in the Nilosyrtis Mensae are numerous flat-floored lows some of which have semicircular terminations at their upper ends. Most of the valleys contain numerous longitudinal ridges that tend to merge with similar ridges in other lows at valley intersections somewhat like the merging of glacial medial moraines. Carr and Schaber interpreted these features as having been formed by gelifluction, a slow flow from higher to lower altitude of masses saturated with ground-ice. Sediment lobes associated with rampart impact craters (also known as fluidized craters) may be the result of melting of ground-ice by impacts. Lucchitta et al. (1981) proposed that the outflow channels on Mars were carved by ice streams, but, as pointed out by Squyres (1984), the ice sheet in the chaotic terrain was too small to provide the large ice streams needed to erode such an extensive channel system; a catastrophic flood appears to be a more reasonable origin for these features.

Kargel and Strom (1992) ascribed to glaciation the long sinuous ridges that are particularly abundant at middle high latitudes, features on Argyre impact basin at Lat. 51°S, 42°W that they interpret as glacial features such as esker systems, tunnel valleys, glacial outwash, glaciolacustrine plains, arêtes, cirques, and horns produced by glacial erosion, rock glaciers, kettles, kames, hills sculptured by ice and/or meltwater, and glacial grooves and ridges, deeply scoured features in Hellas Impact Basin (Fig. 92) at Lat. 45°S, Long. 290°W, Dorsa Argentea, which is a complex sinuous ridge system at Lat. 78°S, Long. 40°W, and the thumbprint terrain (a complex of arcuate and cuspate ridges) in the northern plains (Fig. 111). Kargel and Strom proposed that these features indicate that vast austral and boreal ice sheets were active late in martian history. If this is true, then Mars would have had a relatively warm moist climate and a dense atmosphere in recent geological history.

Volcanic features formed by eruptions beneath ice have been recognized in the northern plains and polar regions of Mars (Allen 1979); possibly Olympus Mons and the aureole deposits around its base (Figs. 100 and 101) may have had such subglacial births (Hodges and Moore 1979), but the 7 to 10×10^6 km^3 ice needed for this to happen appears to be excessive. Another feature related to ice is a network of polygons in the younger northern plains and which resemble some of the structures of the high latitude patterned ground on Earth (Fig. 98). There, they are formed by repeated diurnal and seasonal freezing and thawing of ice-rich soil – solifluction. The network of polygonal fractures on Mars, however, are one to two orders of magnitude larger than those on Earth. As discussed by Carr (1984), the major problem with a frost origin is that the diurnal or seasonal thawing needed to form such polygons requires warmer temperatures than are possible now or during the recent past. Thus, the large size of the polygons and the absence of the optimum climatic conditions needed for their formation suggest that they are structural in origin, the result of extensional tectonics or cooling fractures in lavas or compaction over an irregular buried topography (Pechmann 1980; McGill and Hills 1992). Possibly, their large size may be the result of longer-period climatic fluctuations rather than diurnal or seasonal temperature changes (Helfenstein in Squyres 1984). Other features of possible glacial origin are curvilinear terrains that may represent solifluction lobes, or ice-cored ridges. An alternate origin is that they are nonglacial in origin and were formed by scrap retreat (Rossbacher and Judson 1981).

Eolian processes also have had an important role in sculpturing the surface of Mars (Figs. 112–115; Cutts 1973a, b; Cutts and Smith, 1973). As pointed out by

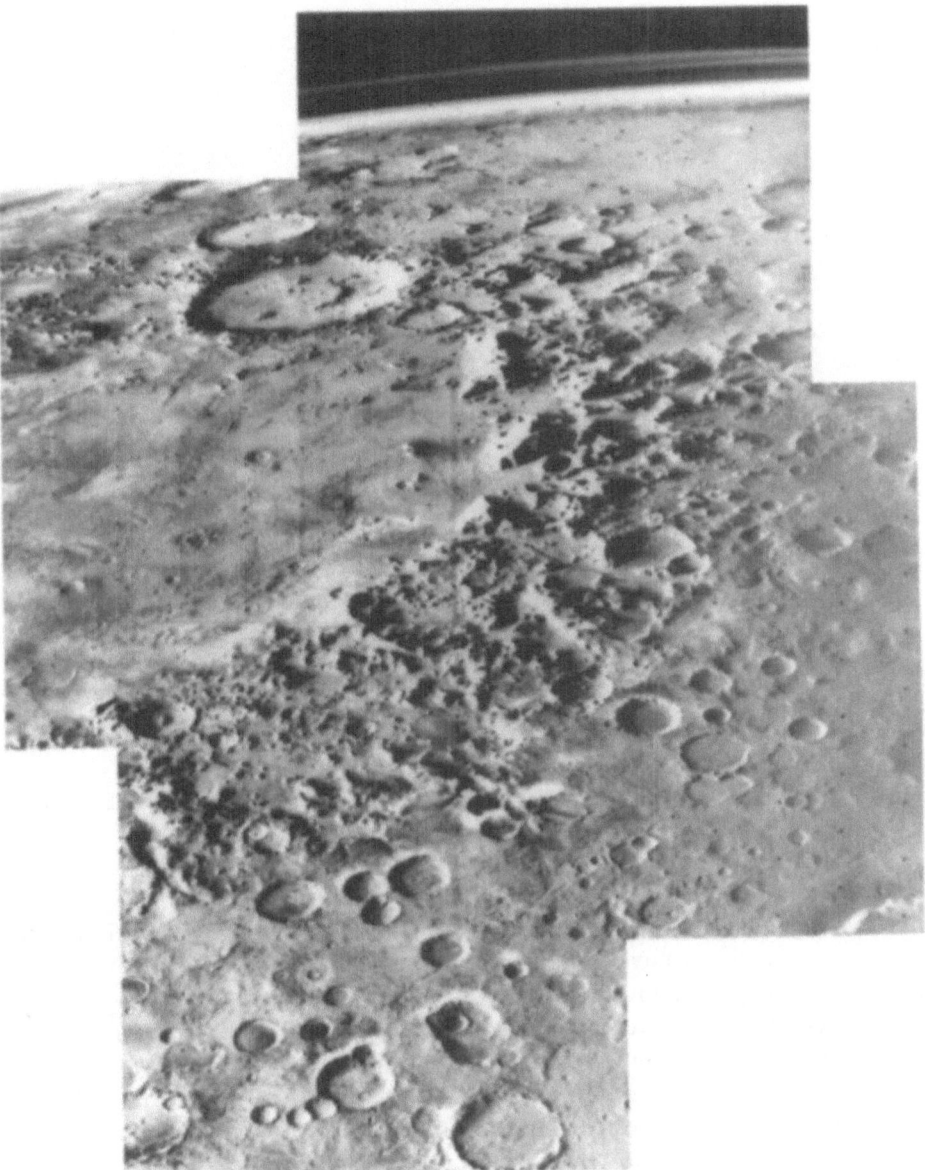

Fig. 111. Oblique photograph of Argyre Basin of Mars. Some features in this basin have been interpreted by Kargel and Strom (1992) as glaciofluvial in origin. Photograph 22A92, courtesy of Viking Orbiter Experiment Team Leader M. H. Carr and NASA

Sagan (1973), because the martian mean atmospheric pressure is less than that on Earth by two orders of magnitude, the wind velocity needed to transport sediment on Mars has to be much greater. Eolian erosion rates on Mars may be in the order of > 3 km/Ma. As eolian erosion is meteorologically dependent, the erosion rate

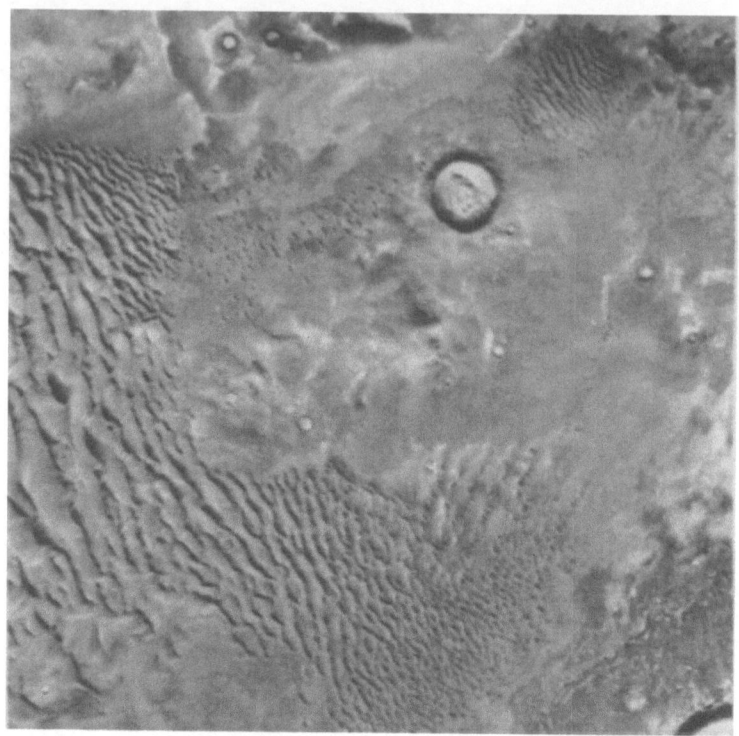

Fig. 112. Transverse and barchan dunes on Mars. Photograph 575B60, courtesy of Viking Orbiter Experiment Team Leader M. H. Carr and NASA

must vary geographically. Although this rate is much higher than on Earth, the absence of significant fluvial weathering and erosion must cause the total eolian erosion on Mars to be much less than that on Earth (Sagan 1973). On the other hand, McCauley (1973) reported that the processes of saltation and suspension transport are more significant planetwide on Mars than on Earth. Features associated with wind action on Mars include modified crater rims, irregular pits and hollows, streamlined ridges, linear grooves, fluted cliffs, reticulated ridges, and subdued and sand-mantled terrain (McCauley 1973). Thermal inertia, calculated from mid-infrared emission data, indicate that martian dunes are much coarser-grained than those on Earth, having an average particle size of $500 \pm 100 \ \mu m$, in the coarse to medium-sand range, whereas most dune sands on Earth are in the medium-size range. Calculations based on particle size, determined from grain trajectories and particle-size transitions from suspension to saltation yielded similar results (Edgett and Christensen 1992). Apparently, wind that is pervasive throughout the planet is the principal agent responsible for sculpturing the planet's surface. Such activity occurs both prior to and during global storms and it explains the color changes noted at both Viking lander sites during a martian year when an

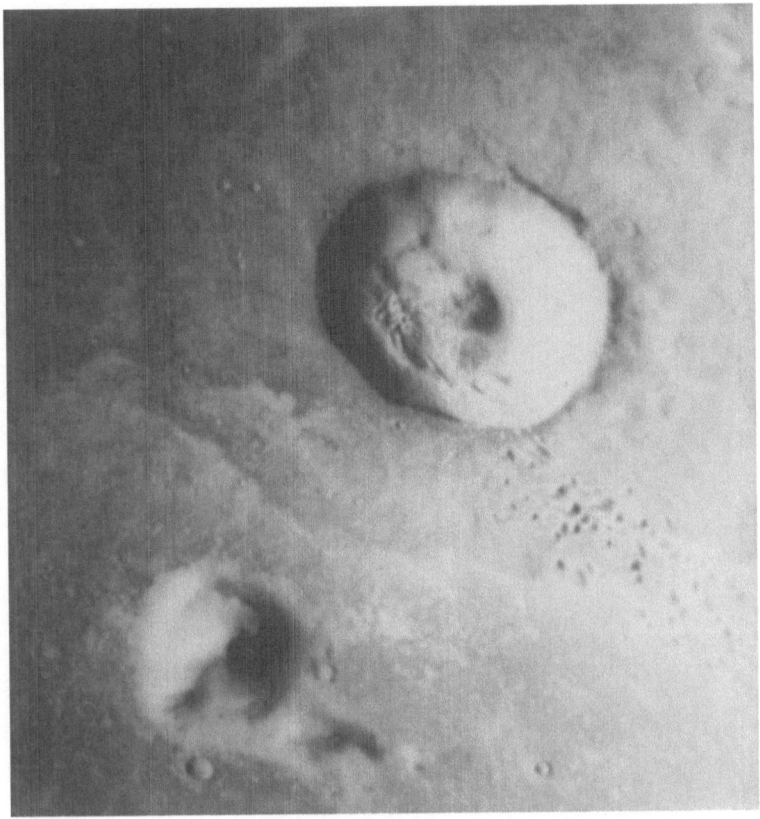

Fig. 113. Climbing dunes on Mars. Greeley (1987, p. 181) believed that these dunes are climbing over the rim of the 160-km-wide impact crater. Photograph 571B53, courtesy of Viking Orbiter Experiment Team Leader M. H. Carr and NASA

increase in the red to blue ratio, loss of contrast between soil units, and a general brightening was noted (Guinness et al. 1979). These changes are ascribed to deposition of a thin layer of eolian red dust. Condensation of CO_2 and H_2O around the dust nuclei led to accumulation of dust on Viking Lander 2. Apparently, the bright areas on Mars are produced by the accumulation of red dust, and dark areas by the stripping of this dust by wind erosion. Physical properties suggest that the soil beneath the sediment drift at both Viking lander sites should be relatively stable, with erosion occurring only when wind speeds are high (Moore et al. 1979). The intensity of erosion varies with the rock type, erosion having caused the plains to be lowered by tens of meters and the material removed from equatorial regions to be accumulated in polar regions. According to McCauley (1973), the landforms in the rainless desert of central Peru are excellent analogs to the eolian features on Mars. The Western Desert of Egypt, one of the most arid regions on Earth, also provides excellent analogs of such eolian features as sand

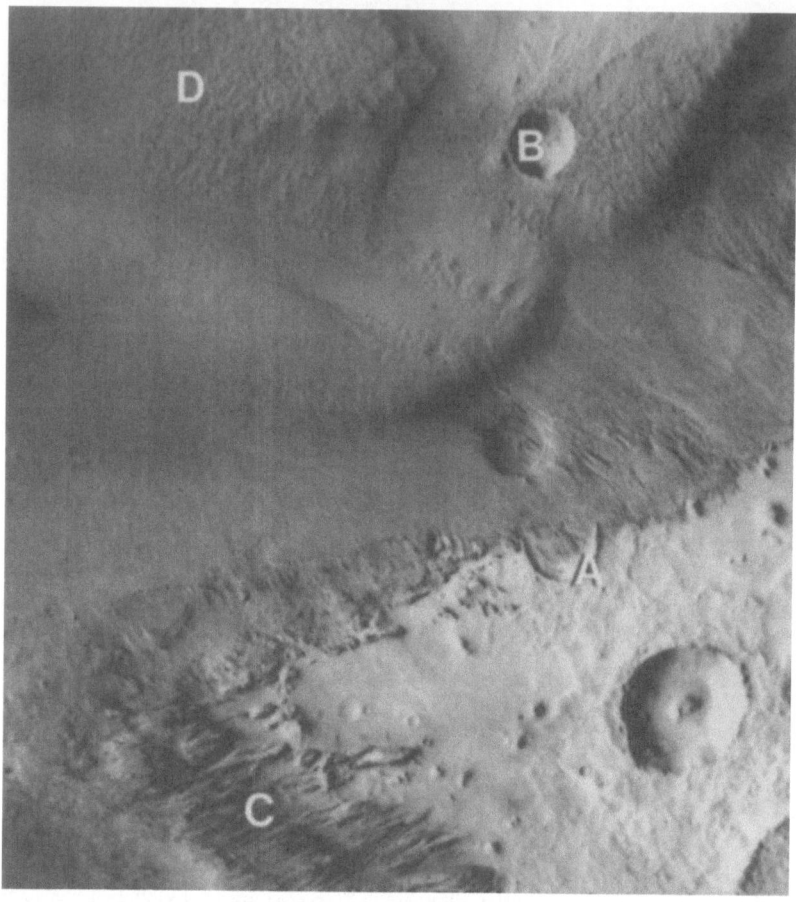

Fig. 114. Eolian features on Mars. **A** Partly exhumed impact crater, **B** a young crater superimposed on the eolian deposits, **C** yardangs, and **D** sand dunes. Photograph 438S01, courtesy of Viking Orbiter Experiment Team Leader M. H. Carr and NASA

dunes, alternating light and dark streaks, knob 'shadows', yardangs, and fluted and pitted rocks (El-Baz et al. 1979; McCauley et al. 1979).

Yardangs, streamlined erosional features that resemble inverted ship hulls and that are aligned in the direction of the wind, are best developed in the equatorial region of Mars (Fig. 114; W.R. Ward 1979). In the Amazonis region these features are tens of kilometers long and separated by valleys 1 km wide. These yardangs appear to be relatively young and were formed probably from such easily eroded deposits as ignimbrites, mudflows, or lithified regoliths. According to Ward, similar features also occur in Ares Valles and in the Aerolis and Iapygia regions. A light-colored plateau (White Rock) inside a crater was interpreted by Ward as a yardang cluster formed by erosion of a pyroclastic deposit.

Fig. 115. Dark wind streaks on eolian sediments of Mars. Photograph 056A20, courtesy of Viking Orbiter Experiment Team Leader M. H. Carr and NASA

8.4.5.3 Depositional Features

One of the largest eolian exogenic terrains on Mars is more than 2000 km long in an easterly direction near the Tharsis volcanoes (Muhleman et al. 1991). This feature, known as Stealth and imaged using the 70-m antenna of the Deep Space Network at Goldstone, California, has been interpreted as an eolian deposit of ash or dust blown westward from the Tharsis Montes. The deposit has a density of less than about 0.5 g/cm^3, is free of clasts larger than 1 cm in diameter, and is at least several meters thick. Its wind streaks with typical lengths of tens of kilometers are the most numerous eolian features on Mars (Fig. 115). Bright streaks tend to be long and narrow, whereas dark streaks are shorter and broader; most emanate from craters, but a few are from positive features. Splotches are irregular in outline, with smaller structures being concentrated inside craters and larger ones extending beyond the rim of the craters onto adjacent plains. Sagan et al. (1973) stated that the dark streaks were formed by the removal of surficial bright sand and dust, and that generation of streaks and albedo changes required threshold wind velocities of only about 2 m/s for a day. They also proposed that the dark color that follows the north polar cap in its retreat is caused by scouring of the bright surface material by winds driven by temperature differences between the frosted and unfrosted regions. Wind streaks on the Tharsis bulge and its central volcanoes appear to be related to slope winds on Tharsis and by global circulation in Elysium (Lee et al. 1982). Differences in streak geometry may be related to the efficiency of the slope winds, which are controlled by the gradient of the slope, its surface roughness, and thermal inertia.

According to Thomas and Veverka (1979), most bright streaks, inferred to consist of dust-storm deposition in the lee of obstacles, change very little in form or orientation during a martian year (2.2 Earth years). Some other bright features experience no effective eolian action, but those few streaks that are subjected to both global and topographic winds change rapidly during dust storms. Erosional dark streaks also are quite stable between storms, but the dark streaks in Oxia Palus, which consist of material deflated from the dunes in the craters, have lengthened secularly since 1972. Thomas and Veverka (1979) stated that changes in the streaks occur from late southern spring to early southern fall, and that these changes reflect the present north-south asymmetry of the martian seasons. Soderblom et al. (1973b) also reported that many topographic features of the polar regions have been molded by wind action; such features include the elongate basins and grooves in the pitted terrain, small-scale flutes in laminated terrain beyond the perennial southpole ice cap, and deposition and erosion of the central polar layered deposits.

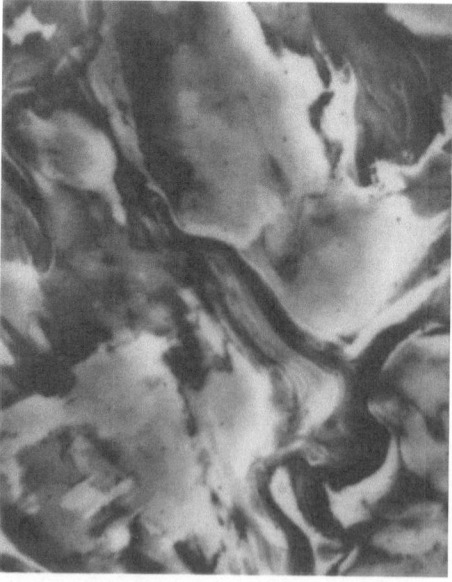

Fig. 116. Layered terrain in the north polar region of Mars. Photograph 056B84, courtesy of Viking Orbiter Experiment Team Leader M. H. Carr and NASA

The layered terrain of the complex sequences of layered deposits in the polar regions (Figs. 116–118), which are believed to be of eolian origin, cover 1.1×10^6 km^2 in the planet's northern hemisphere and 1.5×10^6 km^2 in the southern hemisphere and are several kilometers thick (Cutts 1973b). They are confined to the regions poleward of Lat. 70°N and S, occupy most of the area poleward of Lat. 80°, and underlie the perennial ice cap. The source of this dust is believed to be the depressed and eroded equatorial terrains, and gravitational deposition was enhanced by precipitation of volatiles on the dust grains. The accumulation time represented by the deposits is 500 Ma at a rate of 30 m/2.2 Ma. Analysis by Cutts (1973b) indicated that as much water as dust is accumulating near the poles. Except on the frost cap with its high albedo, both dry ice and water–ice are sublimated into the atmosphere. The regularity of stratification is believed to result from climatic changes caused by periodic changes in obliquity of the ecliptic, eccentricity of the orbit, and precession of the equinoxes. Wind action has eroded the edges of the layered terrain beyond the ice of the perennial frost to form subdued uplands dissected by curving and sinuous scarps – an erosional episode indicative of some

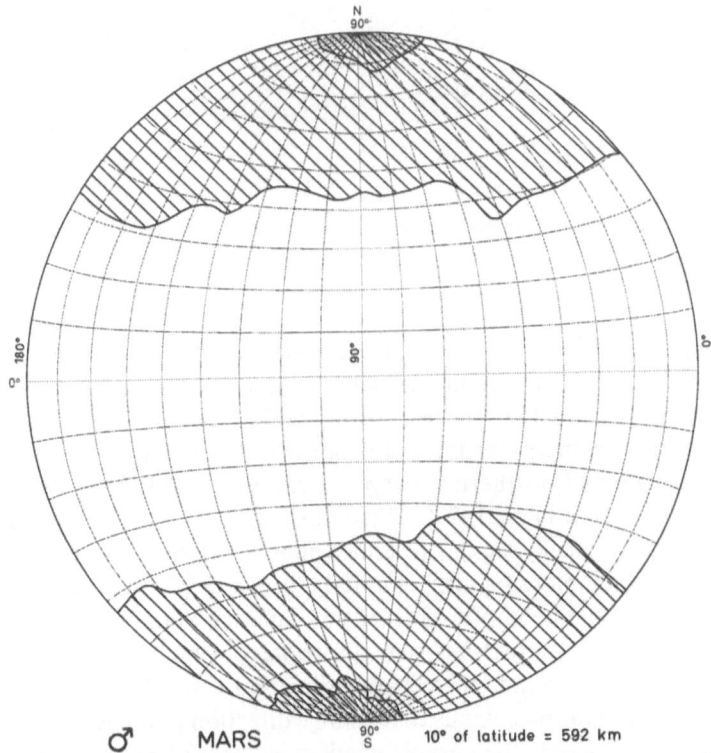

Fig. 117. Widespread relatively recent sediments on the western hemisphere of Mars (between Long. 0° to 180°W). Layered deposits in polar regions (*close diagonal line pattern*), mantled terrain (*wide diagonal line pattern*), and terrain containing only thin or no mantle (*blank areas*). (After Murray et al. 1972; Cutts 1973b; Soderblom et al. 1973a, b)

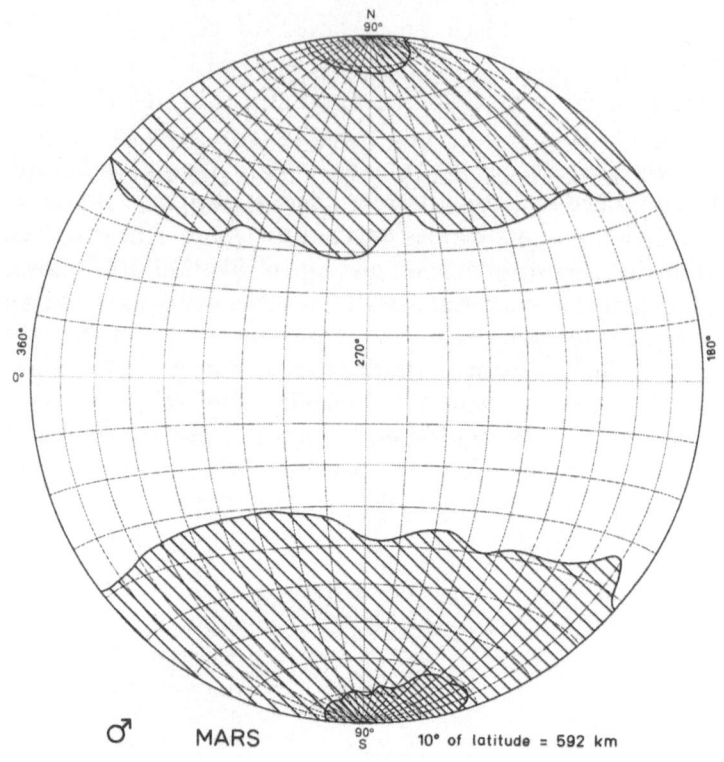

Fig. 118. Widespread relatively recent sediments on eastern hemisphere of Mars (Long. 180° to 360°W). *Symbols* as for Fig. 117

major secular change in the martian environment. The partly eroded mantles, distributed symmetrically about the polar regions and extending poleward from Lat. 30° (Figs. 117 and 118) are believed by Soderblom et al. (1973a) to be eolian debris derived from the polar layered deposits, a transfer that may still continue. Around the perennial northern ice cap is a sand sea (erg) that covers an area of $> 5 \times 10^5$ km^2 (Tsoar et al. 1979). These dunes (Fig. 112) and those of the crater floor are morphologically similar to dunes in desert-basin ergs and dune fields on Earth; thus the dynamics of dune formation must be similar on both planets (Breed et al. 1979). This eolian complex of transverse and barchan dunes is the result of a wind regime having more than one direction: during the summer the major directions are off-pole winds that change direction toward the east as a result of Coriolis forces, and on-pole winds that change direction to the west; and during the winter and/or spring only the on-pole winds exist. According to Tsoar et al. (1979), the strongest winds in the region occur during summer over the transverse dunes between Longs. 110° and 220°W.

In addition to the polar ice caps previously described, the polar regions contain two mappable morphologic units, layered, etched, rippled plains, and a

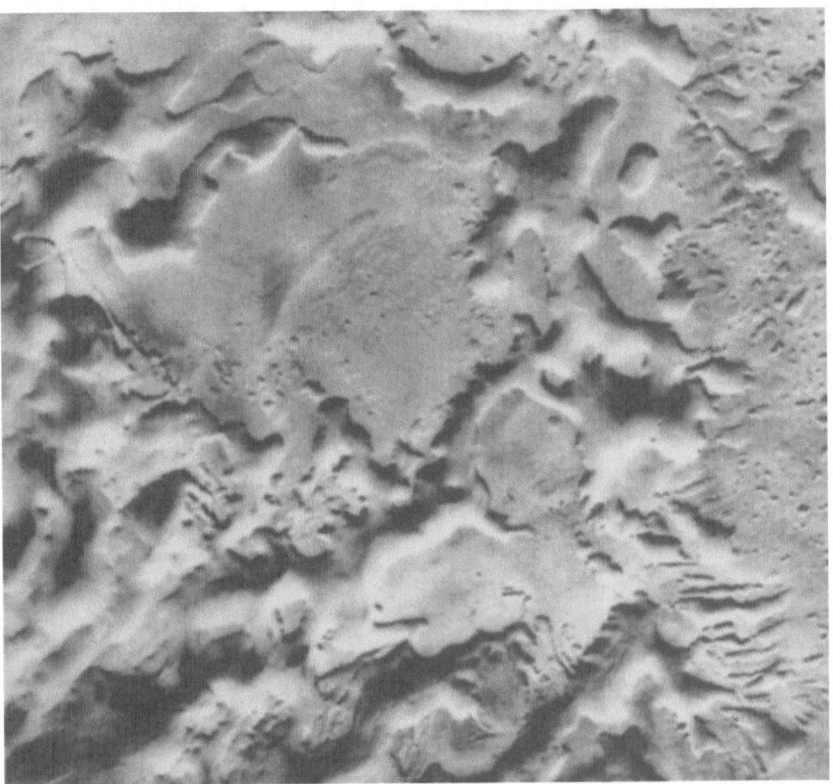

Fig. 119. Etched terrain in the south polar region of Mars. Photograph 390B90, courtesy of Viking Orbiter Experiment Team Leader M. H. Carr and NASA

debris mantle (Fig. 119; Soderblom et al. 1973b; Greeley 1985, p. 165). Near the north pole the layered terrain (Fig. 116) rests on a mottled cratered plains unit, and in the southern region it rests on a heavily cratered terrain. The unit near the north pole consists of fine bright and dark bands about 10 to 50 m thick that can be traced for hundreds of kilometers. The sedimentary sequence appears to be disrupted by discontinuities. Application of new photoclinometric techniques on high-resolution Mariner 9 images indicates that slopes as steep as 10° to 20° occur in the exposed layered deposits of the south polar region (Herkenhoff and Murray 1990). Albedo variations indicate that frost is present on level areas and that there is evidence for temporal variations of the frost. Sublimation of water-ice would leave only loose dust that would slump or be blown away, exposing more water-ice to the sun and allowing it in turn to be sublimated – a process leading to rapid erosion of the layered sediments. The presence of steep slopes suggests that they must be protected from such erosion by some sort of weathered surface. Herkenhoff and Murray (1990) postulated that this protective cover is a layer of dark self-cementing sublimation residue particles. The present consensus is that the polar layered unit consists of a combination of eolian dust, water-ice, and carbon-dioxide ice (Cutts and Lewis 1982) that has accumulated in annual or longer-term cycles, and that the number of layers indicate that the unit is less than 1 Ma old. The etched terrain (Sharp 1973c) has pits and valleys that are 0.5 km to tens of kilometers wide, have a maximum relief of 400 m, and are formed by erosion of the layered unit, possibly by wind action.

Mass wasting also molded the surface of Mars (Fig. 120). The unusual landforms in two regions of fretted terrain (Nilosyrtis Mensae and Protonilus Mensae; Fig. 92, northern hemisphere) were ascribed by Squyres (1979) to slow creep or flow of erosional debris. Flow of the debris is believed to be the result of small amounts of water frost deposited on the surface of the debris each winter. Continued loading of fresh debris causes the debris apron to deform and flow away from the scarp. These displaced materials commonly formed lobate debris aprons or linear valley fill when confined to narrow valleys. Although the mode of their displacement has yet to be resolved, the aureole deposits around Olympus Mons (Fig. 98) also have been ascribed to mass wasting. Flows may be concentrated in two latitudinal bands about 25° wide and centered at Lats. 40°N and 45°S. The band in the northern hemisphere is associated with old highland surfaces, and that

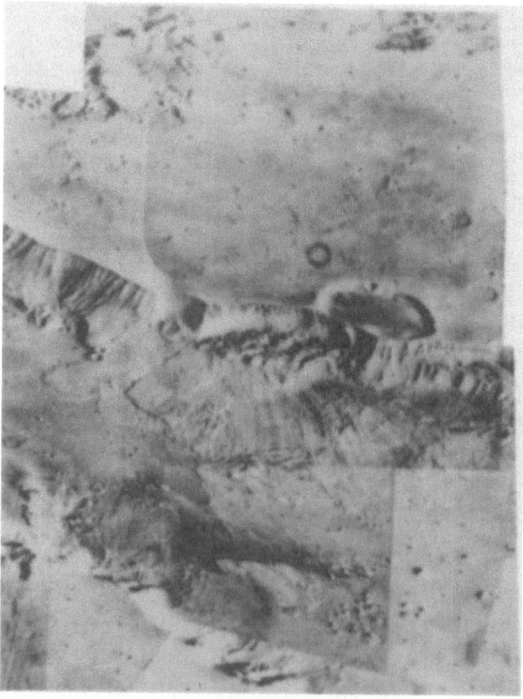

Fig. 120. Massive landslide in Gangis Chasma of Mars (Fig. 92). Photograph P12952, courtesy of Viking Orbiter Experiment Team Leader M. H. Carr and NASA

in the southern hemisphere is associated with the major impact basins Argyre and Hellas. This distribution suggests that the northern highlands are more susceptible to mass wasting, whereas, except for the regions surrounding the two large impact basins, the southern highlands are more resistant to mass wasting (Squyres 1979). Apparently, there may be a hemispheric asymmetry in mass wasting comparable to the asymmetry of surface geology. Landslides also have been imaged in Valles Marineris near the equator (Lucchitta 1979). Many of these structures probably resulted from a major faulting episode, with some slides being coeval with late igneous activity in the Tharsis Montes. A bright cloud, probably composed of dust, was noted on 10 September 1978 in Valles Marineris on a cliff at the west side of Baetis Mensa in central Candor Chasma at Lat. 5°40′S, Long. 71°35′W. Lucchitta and Ferguson (1992) suggested that the cloud came from a landslide, and that such slides may be common; thus, the steep slopes of Valles Marineris probably are still being eroded by gravitational processes.

8.4.6 Summary

Mars' somewhat similar characteristics to Earth makes it a potential target for planetary engineering that would transform it to a planet suitable for habitation by plants and possibly humans (McKay et al. 1991). Such a transformation would begin by the introduction of chlorofluorocarbons to initiate a greenhouse effect that would warm the surface of the planet by about 20 °K. At that temperature carbon dioxide would be released from the polar regions and regolith into the atmosphere. McKay et al. estimated that 100 to 100 000 years may be long enough to build up enough carbon dioxide in the atmosphere for plants to survive. If the reduced carbon could be stored, the plants could slowly produce an oxygen-rich atmosphere in about 100 000 years. If enough nitrogen could be released from the soils and the atmospheric content of CO_2 could be kept low, an atmosphere amiable to human habitation could be created. To determine whether such a scenario is feasible, more data are needed from Mars.

Mars, like the Moon, Mercury, and Venus, has a single lithospheric shell or plate. Interior warming during the early history of such planets led to extensional tectonics, producing a stress system that enhanced widespread volcanism. When the planet's interior cooled, a compressional regime was established and surface volcanism ceased. On the basis of surface tectonics, Solomon (1978) speculated that the degree of early heating and the age of the youngest volcanic rocks decreased in the order: Mercury, Moon, and Mars. On a one-plate planet the volcanic structures tend to remain fixed over their magma sources, and they continue to grow as long as magma is available and the pressure is sufficient to pump it to the surface. Volcanoes can grow to dimensions measured in hundreds of kilometers and develop summit calderas as much as 100 km across and several kilometers deep.

Like the other rocky planets, Mars' earliest history is one of massive impact bombardment, outgassing, and crustal differentiation. Deep-seated structures produced during the heavy bombardment served as pathways for subsequent extrusive activity, as on Mercury and the Moon. Initially, this volcanism may have been pyroclastic as a result of interaction with groundwater and ground-ice producing ash plains and patera, with volcanism becoming more effusive with time. Effusive volcanism via central calderas during the second phase led to the construction of Alba Patera, shields of the Tharsis bulge, and extrusion of plateau basalts via fissures that flooded the northern plains (Greeley 1985, p. 186). During the final phases of the heavy bombardment the crustal hemispheric dichotomy was established, the initial uplift of the Tharsis bulge in the northern hemisphere occurred, and segments of the topography sculpted by the heavy bombardment were obliterated by endogenic/exogenic processes. Outgassing of the planet generated an atmosphere that was much thicker than at present and allowed the presence of surface water on the planet and possibly even an ocean at the north. About 90% of the volume of water became ground-ice and groundwater (Rossbacher and Judson 1981). The rest was chemically locked in silicates, physically absorbed in the regolith, trapped as ice in the polar deposits, and lost to the atmosphere (the maximum amount of water present in the martian atmosphere at any time during the year currently is about 1.3 km^3) and to space (probably in the order of 250 g/cm^2; Squyres 1984).

Although some of the erosional/depositional features do appear to be the result of glacio-fluvial, pluvial, and nearshore marine processes, the outflow channels (which are concentrated along the northern lowland/southern highland boundary and arise from the highlands and terminate on the lowlands) are the result of catastrophic floods during the distant past. The water was released from extensive groundwater reservoirs. Valleys (also known as runoff channels) with their steep sides, flat floors, and amphitheater beginnings of their tributaries were formed by spring sapping. In many instances the configuration of these linear lows was influenced by local structures (for example, the Coprates Chasma from the Tharsis bulge; Hartmann 1974) that are primarily endogenic in origin with secondary exogenic modifications (Valles Marineris; Sharp and Malin 1975).

McKinnon and Tanaka (1989) proposed a model where a heavily bombarded martian crust, generally consisting of a 10-km thick layer of highly permeable fractured basement overlain by a 1- to 2-km-thick impermeable ejecta layer, best explains erosional and mechanical discontinuities at depths of 1 to 3 km, noted over some broad areas on Mars, and initiation of the processes (fluvial, debris flows, and mass wasting) that have formed various martian landforms. They also suggested that rises and fracturing of the Tharsis Montes and Valles Marineris regions may have led to high pore-water pressures in the fractured crust that ultimately led to catastrophic floods. Features of ice origin appear to be the result of glaciofluvial processes, sublimation of ice along scarps, removal of subsurface ice, and magmatic extrusion beneath an ice cover. Eolian processes apparently have operated throughout much of martian history and still are active on Mars (Greeley 1987, p. 186). It is these exotic, endogenic, and exogenic processes that are responsible for Mars' unique surface morphology.

To this point the discussion about Mars has described the planet's morphological terrains, described their causative agents and processes, and provided some comparisons with those aspects of other planets. With the discussion at this level of understanding completed, it is now appropriate to make broader generalizations about the distribution of exotic, endogenic, and exogenic morphologies on Mars, using compilations in map (Figs. 121 and 122) and tabular (Table 8) formats. As shown by the maps, the endogenic morphology (volcanic shields, domes, craters, and larger associated lava plains) is more abundant and largest in the western hemisphere centering around the Tharsis bulge. The exogenic morphology mainly occupies the polar regions of lower topography shown in Figs. 121 and 122.

All three morphologies developed throughout the geohistory of Mars, each with different epochs of dominance. The distribution exhibited in Figs. 121 and 122 represents merely the present palimpsest pattern at Mars' surface, although the early igneous activity may well have forced the pattern to be rather similar to the present one during most of the later history of the planet. As shown by the total areas listed in Table 4, the basic endogenic morphology presently occupies only about 19% of the total surface, even though in Mars' earliest history it may have

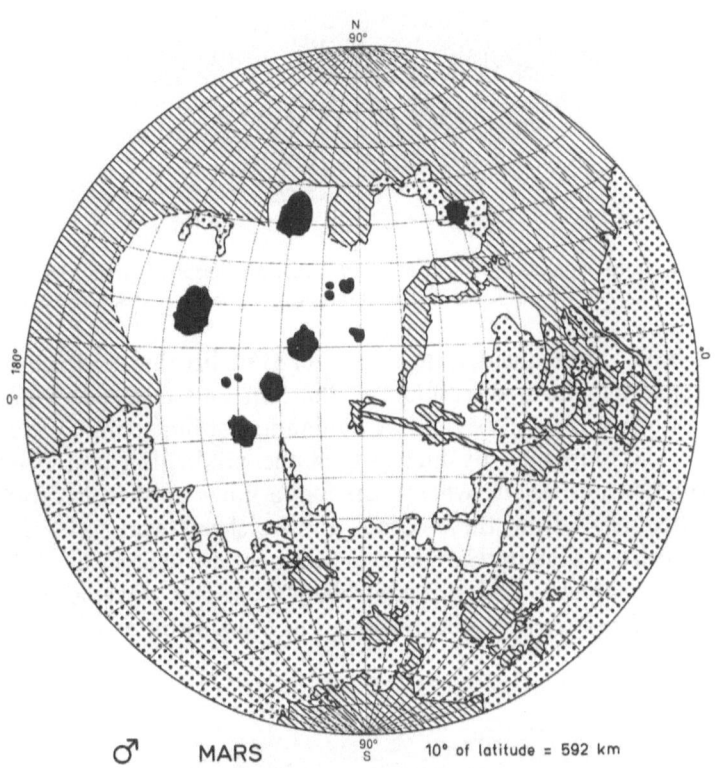

Fig. 121. Present morphological provinces of Mars – western hemisphere (Long. 0° is on *right*, Long. 90° is in the *middle*, Long. 180° is on the *left*). Endogenic: lava plains (*blank areas*) and volcanic shields, domes, and craters (*black*); exogenic: sediments deposited by wind, streams, ice (*lined*); exotic: impact craters (*stippled areas*). Boundaries are from Murray et al. (1972), Carr et al. (1973), and Soderblom et al. (1973b). (After Mutch et al. 1976, fig. 3.2)

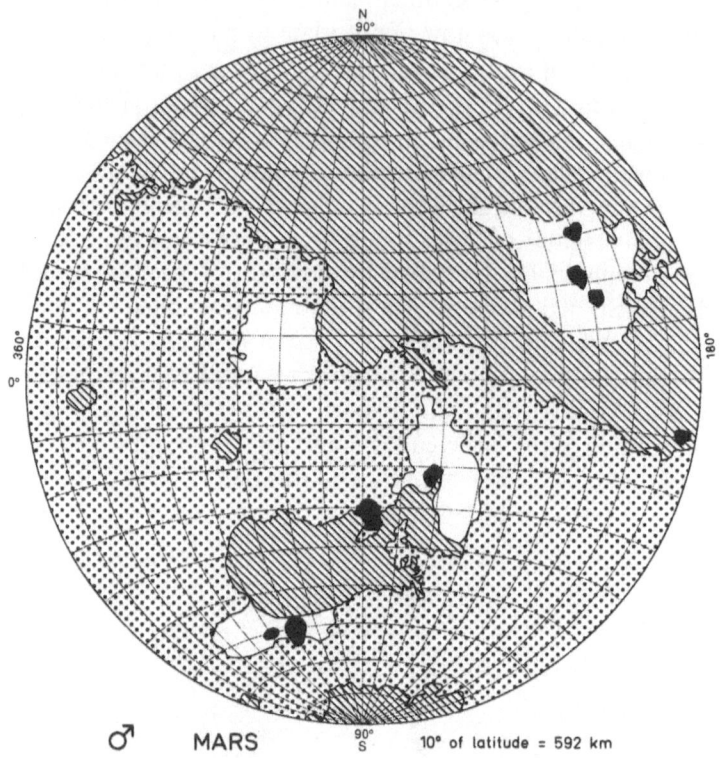

Fig. 122. Present morphological provinces of Mars – eastern hemisphere. Long. 180°W is on *right*, Long. 270°W is in the *middle*, and Long. 360°W is on the *left*. *Symbols* and sources as for Fig. 121

been nearly 100%. Subsequently, the surface of Mars became modified by impacts of bodies remaining from disruption of the Sun and perhaps of debris from elsewhere in our galaxy and beyond. A diminution of impacting exotic materials generally was accompanied by increases in the production and deposition of exogenic sediments, especially when Mars had a now-vanished hydrosphere and cryosphere. Sediments from these oceans and ice sheets were followed by widespread deposits from wind that continue to be moved during epochs of erosion and redeposition at different latitudes, as controlled by periodic and episodic changes of climate produced largely by cyclical changes in Mars' orbit around the Sun.

9 Outer Planets and Satellites

9.1 Exploration

In contrast with the rocky planets, the gaseous planets beyond the Asteroid Belt consist of a rocky core mantled by thick ices of water, methane, and ammonia, liquid metallic hydrogen, and liquid molecular hydrogen (Hartmann 1983, p. 294). Jupiter may be experiencing an anomalously slow cooling because of condensation and settling of helium-rich droplets from a hydrogen-helium mixture; a similar process also may be occurring on Saturn (Klepeis et al. 1991). The satellites orbiting the gaseous planets have various origins ranging from accretion from a disk of gas from a previous larger satellite, to fragmentation, and reconsolidation (a process that may have occurred more than once), and capture. The internal structures displayed by these satellites include a total molten stage model except for a rigid thin solid crust on Io. However, Schubert et al. (1986) believed that there are problems with this molten model of Io and have proposed instead a model consisting of an Fe-S core, a solid mantle, a molten or partly molten asthenosphere, and a rigid crust. Models proposed by Schubert et al. for the other satellites include mostly rock (hydrated or dry silicates) covered by water and an ice crust (Europa), and differentiated bodies having rock cores covered by ice mantles and undifferentiated structures of homogenous ice rock mixtures. Thus, the tectonics and internal dynamics of these satellites are controlled by the rheology and viscosity of high-pressure and low-temperature forms of ice (Poirier 1982).

Information about the morphologies of Jupiter, Saturn, Uranus, and Neptune has come from observations from Earth and images recorded by man-made satellites. Knowledge of Pluto and its satellite Charon is based solely on observations from Earth. Satellite exploration of Jupiter, Saturn, Uranus, and Neptune was begun by a Pioneer 10 flyby, whose closest approach to Jupiter during 3 December 1973 was 130 000 km, with the vehicle exiting the Solar System on 14 June 1983. The Pioneer 11 Jupiter flyby occurred on 2 December 1974, and its closest approach was 34 000 km (Greeley 1987, p. 189). Voyager 1 and Voyager 2 had encounters with Jupiter on 5 March and 9 July 1979, and with Saturn on 12 November 1980 and 26 August 1981 (Stone and Miner 1981). The Voyager 2 encounter with Uranus was from 4 November 1985 to 25 February 1986, and with Neptune from 5 June 1989 to 2 October 1989 (Stone and Miner 1986, 1989).

9.2 Jupiter and Satellites

9.2.1 Planetary Setting

Jupiter, the largest of the planets with its diameter of 143 884 km, has more mass than all the other planets together and, as described by Hartmann (1983, p. 31), dynamically the Solar System has two main bodies – Sun and Jupiter. Jupiter has a density of 1.33 g/cm^3, a rotation of 9.93 h, a revolution of 4332 days (11.86 Earth years), a magnetic dipole tilted 11° to the rotation axis, an equivalent equatorial surface magnetic field of 402 000 mT (microteslas; Merrill and McElhinny 1983, table 12.1), and an internal structure consisting of a silicate core, a lower mantle of liquid metallic hydrogen, and an upper mantle of liquid molecular hydrogen (Hartmann 1983, p. 194). The jovian planet has an enormous magnetic tail, about 300 to 400 R_j (R_j is the radius of Jupiter, 71 942 km) in length in the lee of the solar wind (Ness et al. 1979a, b). Its dense atmosphere, which obscures the planet's surface consists of molecular hydrogen, helium, water vapor, methane, and ammonia. The atmosphere has semipermanent bands of clouds having various colors (brown, tan, yellow, and red) parallel to the equator and displaying large irregularities. The largest of its features is the Great Red Spot, a giant eddy of variable size that can reach a diameter four times the size of Earth.

Surrounding Jupiter is a ring system the main part of which consists of an 800-km-wide bright segment surrounded by a somewhat dimmer and broader 5200-km wide outer segment. The main ring extends from 1.72 to 1.81 R_j. An inner, much dimmer segment, which may reach the top of Jupiter's atmosphere, extends from 1.0 to 1.72 R_j (Smith et al. 1979b; Jewitt 1982). The rings are circular and equatorial with the typical ring particle size being about 5 μm (Morrison 1982; Jewitt 1982). Its dust may be generated by the erosion of parent bodies, having dimensions of 1 m to 1 km within the ring, while old particles are swept out of the ring system by electromagnetic and gravitational forces. Associated with the ring system are the Inner Satellites: Andrasta, Metis, Amalthea, and Thebe. Beyond Thebe are the Galilean or Jovian satellites Io, Europa, Ganymede, and Callisto, and beyond them are the Outer Satellites: Himalia, Elara, Lysithea, Leda, Pasiphae, Sinope, and Carne (Morrison 1982). Of all these satellites only Andrasta, Metis, Amalthea, Thebe, and the Galilean satellites were imaged by Voyager (Jewitt et al. 1979; Smith et al. 1979a, b; Synnott 1980, 1981).

9.2.2 Inner Satellites

9.2.2.1 General

The dark-colored Andrasta and Metis at the outer edge of the ring system at a distance of 1.80 R_j have orbiting periods of 7 h and 5 min and diameters of tens of kilometers (Morrison 1982). Amalthea, at a distance of 2.55 R_j is a 270×165 ×150 km irregular satellite with its long axis pointing toward Jupiter. Its surface is heavily cratered, and it has an orbital period of about 12 h (Thomas and Veverka 1982). At 2.55 R_j from Jupiter, Amalthea is deep within the gravity field of Jupiter and in the very intense part of its magnetosphere. The satellite has a disk-averaged temperature of 180 ± 5 K, which is an unusually high temperature because solar models give a maximum surface temperature of 164 K. This thermal state cannot be due solely to solar heating, but must result from a combination of solar radiation, charged particles, and inductive Joule heating. However, Amalthea's low velocity with respect to Jupiter's magnetic field causes Joule heating to be probably insignificant. Although protons and electrons do deposit energy within the surface layers, it is not clear that they can account for the temperature discrepancy that exists. Thebe, beyond Amalthea, at a distance of 3.11 R_j, has an orbital period of 16 h and 11 min, and a diameter of 75 km.

9.2.2.2 Amalthea

Amalthea, the only Inner Satellite to have been mapped, has a very dark red aspect (the reddest known object of the Solar System) with the trailing edge being slightly redder than the leading one (Thomas and Veverka 1982). At least two circular depressions (Pan and Gaea) probably are impact structures, and five other possible impact sites have been mapped on the surface of the satellite (Fig. 123). Between Pan and Gaea near Lat. 0°, Long. 60°W is a complex of ridges and troughs or possibly of large nested noncircular depressions tens of kilometers in length and well over 20 km wide. The anti-Jupiter point lies near a 40- to 50-km-long ridge whose ends are marked by bright patches about 15 km across and known as Ida and Lyctos. Similar bright areas also occur inside Gaea and on the eastern rim of Pan (Thomas and Veverka 1982). Gradie et al. (1980) stated that the very dark red color of the satellite surface is due to alteration by charged particle radiation, high-velocity impacts, sulfur allotropes (polymorphism in an element; i.e. sulfur as orthorhombic and monoclinic), and sulfur-compound contaminants from Io. Greenish bright areas on relatively steep slopes may represent recently exposed regions containing less sulfur-rich glass (Thomas and Veverka 1982).

The two large impact craters formed on the irregular surface have diameters of 75 and 90 km and irregular rims, with Pan having a relief of 8 km and Gaea a relief of at least 10–20 km. These bowl-shaped craters are among the largest in the Solar System. As a result of the focusing effect of Jupiter's gravity, crater production on Amalthea has been at least 10 to 40 times that of Callisto, located 26.6 R_j from Jupiter, and probably 4 to 20 times that of the Moon. However, because Callisto is made of ice, whereas Amalthea is rocky, a given impact energy produces a larger crater on Callisto than on Amalthea. According to Thomas and Veverka (1982), the presence of at least two craters having diameters of more than 50 km on Amalthea is compatible with the cratering record on Callisto, a satellite that is nearly saturated with craters larger than 20 km (Smith et al., 1979b). Such cratering indicates that Amalthea did not have its present shape during the past 4.5 Ga and must be a result of collisional fragmentation of a larger satellite. Similarly, such an

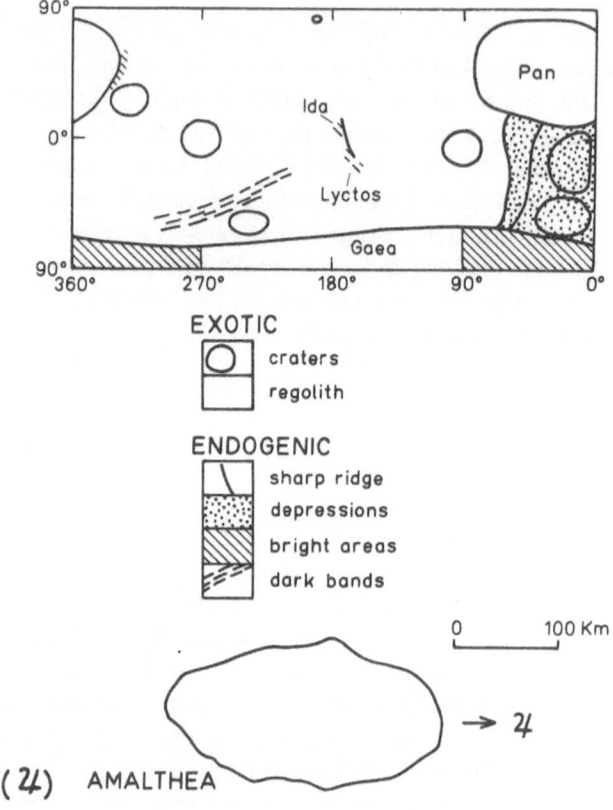

Fig. 123. Silhouette and morphology of the inner satellite of Jupiter, Amalthea. (After Veverka et al. 1981, figs. 2 and 7)

impact history must have produced a considerable amount of regolith on the satellite; some of this material must have been ejected from Amalthea whose gravity probably is about 10 cm/s² (assuming a density of 3 g/cm³). Some of the lost material, however, probably later was swept up again. Thomas and Veverka (1982) estimated nearly complete excavation of material to a depth of 2 km.

Lunar and terrestrial analogs suggest that blocks as large as 300 m in diameter may have been produced during formation of Gaea. Such massive impacts also must have led to severe fracturing near the craters, landslides on their upper slopes, and severe antipodal effects in the regolith. Pan, which is antipodal to Gaea, probably was smoothed during the formation of Gaea. The rough topography between the two craters may have been produced by mobilization of the regolith during impacts. Other exotic processes that modeled the surface of Amalthea include micrometeoritic impacts and charged particle bombardment. The heavy ions and dust, largely of sulfur, oxygen, and sodium from Io, probably produced abundant melt within the ejecta, with glass forming a significant component of the regolith (Thomas and Veverka 1982). However, the relative amount of this glass, whose color was controlled not only by its cooling history but also by the amount of sulfur present cannot be ascertained from existing data.

9.2.3 Galilean Satellites

9.2.3.1 General

The Galilean satellites of Jupiter (the four that were viewed by Galileo in 1610) consist of Io, Europa, Ganymede, and Callisto. They are the best known satellites of Jupiter because of their much larger size than the other satellites. All have a synchronous revolution-rotation (Table 1), because of the effect on them of Jupiter's large gravitational attraction. The same attraction has had important but different tectonic effects on each of the four satellites.

9.2.3.2 Io

Io, about the same size and density as the Earth's Moon, has an orbital radius of 422 000 km, a diameter of 3630 ± 10 km, a density of 3.55 g/cm^3, and a rotation and revolution period of 1.8 Earth days. Its internal structure consists of a possible Fe-S core about 700 km thick, a solid silicate 900-km-thick lower mantle, an asthenosphere 50 to 200 km thick and is partly or wholy molten, and a lithosphere that is 5 to 100 km thick and includes a sulfur and salt crust (Greeley 1987, p. 190; Gaskell et al. 1988). The surface of Io is largely covered by sulfur allotropes, SO_2 frost, and the sulfide salts Na_2S and K_2S (Fanale et al., 1982; Sill and Clark, 1982). Its atmosphere is mainly SO_2 with small amounts of H_2O, CO_2, NH_3, and CH_4 and with the photochemistry of SO_2 possibly leading to O_2 as the major gas on the night side (Kumar and Hunten 1982). The 35% of the surface of Io (most of it in the equatorial and south polar regions; Greeley 1987, p. 190) imaged by Voyagers 1 and 2 has a variety of yellow, orange, red, brown, black, and white colors due to the presence of various allotropes of sulfur at volcanic vents and fissures (Figs. 124 and 125). Other endogenic processes influencing the surface morphology include nine active volcanic plumes. Such processes have completely obliterated the impact structures that except on Europa and Earth, dominate the planets, satellites, and asteroids of the Solar System (Schaber 1982).

Endogenic terrains on Io are the result of tidal heating caused by their proximity to Jupiter – a tidal energy that is dissipated not within a thin elastic lithosphere but mainly within a molten or partly molten asthenosphere. The remaining one-third of the heat is dissipated deep within the mantle (Gaskell et al. 1988; Ross et al. 1990). However, this heat pattern does not agree with the equatorial topography which indicates that Io's lithology (or Io, itself) recently has undergone a rotation of about 25°. The endogenic terrains on Io can be classified in four groups: mountains, plains, vent materials, and linear features (Schaber 1980). The mountain unit, the oldest mapped terrain, comprises 1.9% of the mapped area,

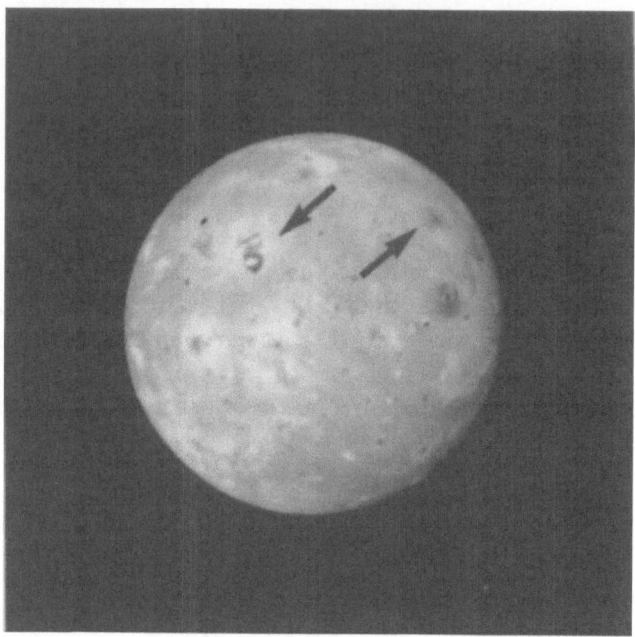

Fig. 124. Image of Io, a Galilean satellite of Jupiter. According to Greeley (1987, fig. 8.3), the diffuse areas (*arrows*) are active volcanic plumes. Photograph 42J1 + 000, courtesy of Voyager Experiment Team Leader B. A. Smith and NASA

has a relief of about 9 km, and is topographically rugged; the roughness is caused by tectonic stresses that produced at least one set of orthogonal fractures. Slumping along the mountain fronts has defaced the steeper mountain fronts with large-scale terracing. The relief of the mountain unit suggests that it consists mainly of silicate volcanic rock containing significant amounts of sulfur, because sulfur alone could not support such elevations. Within the imaged area the mountain unit appears to be evenly distributed relative to latitude and longitude. At Lat. 6°S, Long. 267°W large peripheral massifs associated with a 172-km-diameter vent, represent rim deposits or individual volcanic constructional features. Lobate scarps north of Creidne Patera, along which mountain material crops out, resemble large land-slides or terrestrial lava fronts. The surfaces of these supposed flow fronts are fretted, indicating some form of erosional or flow structure. The grooved terrain terminates abruptly on a lobate front that Schaber (1982) believed is the result of faulting or a fissure-vent source of the flow apron.

Grabens and lineaments on Io exhibit a northwesterly and northeasterly planetary grid pattern that may be the result of tidal flexing. Erosional and tectonic scarps, lineaments, and grabens are present at all latitudes. At Lat. 8°S, Long. 284°W a system of grabens connects several volcanic vents and outcrops of mountain material. Although normal faulting appears to be dominant in this region, there also is some evidence of transcurrent motion. South of Lat. 46°S the

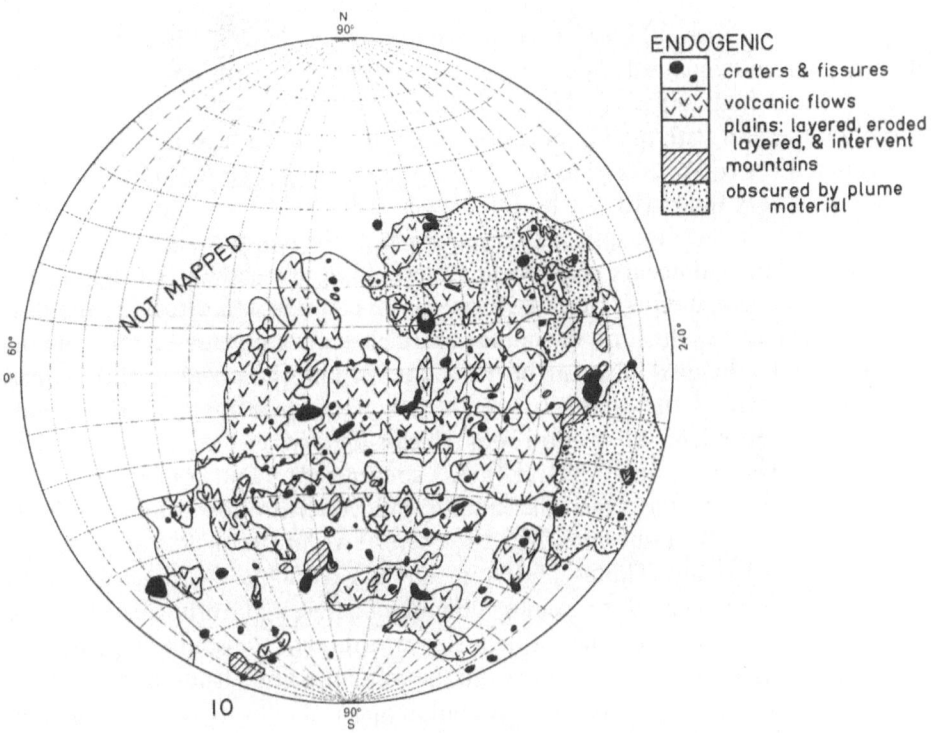

Fig. 125. Morphology of a segment of Io, a Galilean satellite of Jupiter (Fig. 124). The craters and fissured terrain consists of crater cones, walls and floors of pit craters, and fissures, and the volcanic flows unit consists of pit craters, shield craters, and fissure flows (After Schaber 1982, pl. 6, on a Lambert equal-area projection)

main tectonic features are grabens concentrated on the layered plains. The presence of scarps with reliefs of 200 to 600 m in a satellite having an average resurfacing rate of 10^{-2} cm/yr indicates that erosional and tectonic processes may dominate over volcanism, particularly in the south polar region where no active plumes were observed and where volcanic terrains are less abundant (Schaber 1982).

The vent terrains have been subdivided by Schaber (1982) into five units: vent floors and walls, pit-crater flows, shield-crater flows, fissure flows, and crater cones. The youngest volcanic unit is represented by thin fumarolic haloes around vents and shield craters, and their deposits are restricted to equatorial and midlatitudes (Schaber 1982). The vent wall and floor unit is 4.0% of the mapped area, occurring on the walls and floors of pit craters, shield craters, and fissures; it varies in color from white to orange, yellowish orange, brown, and black. The unit displays multiple-stage floor and wall structures indicative of repeated eruptions from individual vents. The high relief of the features and absence of gravitational structures are evidence of considerable wall strength, indicating that the features are not just sulfur but that some other material of greater strength, such as silica glass, must be present.

The pit-crater flow unit comprises 20.2% of the mapped region and occurs generally on the sides of pit-crater vents as a massive coalescing flow having colors that range from white, yellowish orange, and orange to brownish red, brown, and black – colors that indicate relative abundance of various allotropes of sulfur. Although the flows are mainly of sulfur, some silicate lavas also may be present. Individual flows can be traced as far as 700 km, reflecting high reproduction rates, extrusion of very fluid lavas, or both (Schaber 1982).

The shield-crater flow unit that covers 9.7% of the mapped area is associated with crater rims and occurs near shield craters. Although similar in origin to the pit-crater flow unit, it differs from the former in its composition, viscosity, and flow history. The unit has the form of sinuous channels that originated on the summits of shields, and individual lobes can be traced for 300 km. Color and albedo changes along the length of the channels are the result of quenching of the various temperature-dependent sulfur allotropes. Shields and associated flows are limited to the region between Lat. 30°N and 45°S, a concentration that coincides with the most active volcanic plumes near the equator. This concentration also may indicate a more siliceous volcanism here than elsewhere. Two of the shields at Lat. 13°S, Long. 350°W (Inachus Tholus and Apis Tholus) resemble Olympus Mons on Mars – as they have circular outlines, summit calderas, and abrupt basal scarps. No flows were observed on these shield slopes, although the south slope of Apis Tholus is overlapped by shield flows from the west, and flows from the north have been diverted by the basal scarp of Apis Tholus. Spectral reflectivity data from their surfaces indicate that the flows are sulfur and sulfur compounds.

The fissure flow unit constitutes 1.2% of the mapped area and is associated with elongate fissure vents that may be fault controlled. West of Mazda Catena at Lat. 4°S, Long. 320°W is a raised 92-km-long dike-like feature with thick flows emanating from both sides of the fissure. The flows are grooved along the flow directions and have steep, high, lobate fronts. A similar flow emanating from a 100-km-wide front north of these flows is disrupted by a circular pit crater. The morphology of these fissure flows suggests that they have higher viscosity and yield strength than most of the pit-crater and shield-crater flows. This high viscosity may be caused by extrusion of sulfur at temperatures between 433 and 463 K or of materials having a high silica content. Their high-relief flow fronts also resemble flow edges on Mars that are believed to have a high silica content. The flows and surroundings of the fissure dike are orange at the north and whitish orange at the south. These colors probably reflect only a surface coating of sulfur and sulfur dioxide and probably do not indicate the composition of the flows themselves.

The crater cone unit that forms the raised rims around pit and shield craters and comprises 0.1% of the mapped area resembles terrestrial pyroclastic cones of silicate ash and cinders. The best example of this unit occurs in the pit-crater vent Amaterasu Patera at Lat. 36°N, Long. 306°W, a crater with a terraced wall that may represent former strandlines from ash, sulfur, or silicate lava lakes. Other examples of the crater cone unit are the crater cone atop Inachus Tholus and a ring-shaped feature on the intervening plains at Lat. 45°S, Long. 330°W (Schaber 1982). Although these crater cones resemble terrestrial pyroclastic cones, they could have been formed by low-energy fire-fountains of mostly sulfur and sulfur dioxide.

The mapped area on Io contains 170 vents having diameters larger than 14 km. They are more randomly distributed than on the Earth, Moon, and Mars. A slight increase in number of vents per unit area in the equatorial belt and their deficiency in the polar region may be related to the concentration of plumes and hot spots near the equator. A lack of structural control suggests the presence of some anomalous thermal conditions near the surface associated with a high degree of interior melting. The dynamic nature of the surface is believed by Cassen et al. (1982) to indicate intense heating by dissipation of tidal strain energy. According to Pearl and Sinton (1982), however, the disk-averaged heat flow from the interior of Io ranges from 1.5 to 23.0 W/m^2 and is higher than can be accounted for by tidal dissipation alone. High surface temperatures on Io are at spots on a much cooler background. Spots having temperatures of < 400 K are associated with dark calderas and flows, a small area of 650 K is associated with the source of the Pele plume, and a third type, observed from Earth as a sudden possibly explosive event is characterized by a 600 K color temperature (Pearl and Sinton 1982). During the Voyager 1 encounter over a period of 6.5 days nine erupting plumes were observed on Io. Four months later during the Voyager 2 encounter eight of these plumes still were active. The plumes ranged in height from about 300 to 600 km with ejection velocities of 0.5 to 1.0 km/s. According to Strom and Schneider (1982), the plumes are within level plains, either on fissures or calderas, rather than on topographic highs. This volcanism is so intense that resurfacing rates in recent geological time are of the order of 10^{-3} to 10 cm/yr (Johnson and Soderblom 1982)

The planar surfaces on Io were subdivided by Schaber (1982) into intervent, layered, and eroded layered plains. The intervent plains cover 39.6% of the mapped area, are relatively smooth, and vary in color from black to white to reddish orange to orange, yellow, and orange-brown (Masursky et al. 1979). These plains are disrupted by straight to sinuous scarps that are < 10 to 100 m high. This type of plain is believed to consist of stratified material that fell from volcanic plumes. Imbedded with these plume strata composed of sulfur and its compounds are local fumarole sediments and flow deposits from pit and shield craters, and fissure vents. The layered plains are smooth and have boundary scarps that are 150 to 1700 m high with grabens either parallel to or transverse to the scarps; this suggests a composition other than of pure sulfur, possibly a mixture of sulfur and silicate ash or lava. However, devolatilization of the layered plains into eroded layered plains is evidence that the plains materials have a low silicate content. The layered plains, which cover 9.3% of the mapped area, are believed to have the same origin as the intervent plains. They differ only in the presence of boundary scarps on the former. Some scarps represent erosional limits controlled by normal faults and grabens, and others reflect erosion that has progressed beyond the fault structures. Scarp erosion is believed to be due to sapping by SO_2 similar to the spring sapping processes that formed the box-canyon network on Mars. Semicircular outcrops of layered plains are believed by Schaber (1982) to be controlled by arcuate scarps, grabens, or outward erosion from a central point (a center of deposition or of high heat flow). The eroded layered plains comprising 0.7% of the mapped area, represent isolated remnants of volatile collapse of the layered plain unit and are restricted to the south polar region. This polar distribution may indicate that in this region erosional processes are faster than resurfacing by volcanic deposits.

9.2.3.3 Europa

Europa has a diameter of 3138 $\pm$ 20 km, a revolution and rotation period of 3.55 Earth days, and an orbital radius of about 671 000 km (Lucchitta and Soderblom 1982). According to Greeley (1987, p. 190), the satellite has a density of 3.04 g/cm^3, a 25-km-thick crust of ice, a 1200-km-thick silicate mantle, and an 850-km-thick Fe-S core. Squyres and Croft (1986) described Europa as having a silicate interior overlain by a liquid water ocean that in turn is capped by a relatively thin ice shell. This water layer may be more than 100 km thick if the satellite is completely differentiated (Squires et al, 1983a). On the other hand, Lucchitta and Soderblom (1982) considered that surface features on Europa are more compatible with a crust that is about 25 km thick and firmly anchored to a hydrated brittle silicate lithosphere 200 to 300 km thick, rather than with a 100-km ice shell atop a dry silicate asthenosphere. Crustal development may be the result of planetary expansion because of formation of hydrated minerals, possibly modified by internal convective processes in the silicate interior. As the present surface of the satellite postdates the heavy bombardment 4 Ga ago, because the lineations could not have survived such a bombardment. The present tectonic fabric does not appear to have been influenced by this bombardment, both the ice and the silicate crustal layers must have continued to develop after the end of the bombardment. How old is this surface? Shoemaker and Wolfe (1982) believed that the flux of impact bodies producing craters > 10 km in diameter probably was several times higher for Europa than for the Moon during recent time, yet the frequency of Class 1 craters on Europa is several times lower than that of similar size craters on the lunar maria. The apparent low crater density suggests either that the surface is about 0.1 Ga old, or that the surface is much older with resurfacing or collapse of the craters by cold flow and annealing of the ice crust during this time period (Lucchitta and Soderblom 1982). The complex pattern of fractures lacing the surface of this satellite is the result of stresses on this thin icy crust (Figs. 126 and 127). The relief on the surface of Europa is low, with a variation of less than a few hundred meters. Features identified on the satellite include craters, dark-halo craters, crater palimpsests, brown mottled and gray mottled terrains, plains, bright, dark, and fractured plains, brown spots and bands, and light and gray bands.

Exotic features recognized on Europa consist of impacts of a few kilometers to a few tens of kilometers in diameter, sharp-rimmed bowl-shaped craters, structures with bright rims, diffuse ejecta, bright central peaks, a multiring crater, and a bright crater having a dark halo, all of which Lucchitta and Soderblom (1982) termed class 1. Class 2 features are large flat brown circular features which are 100 km or

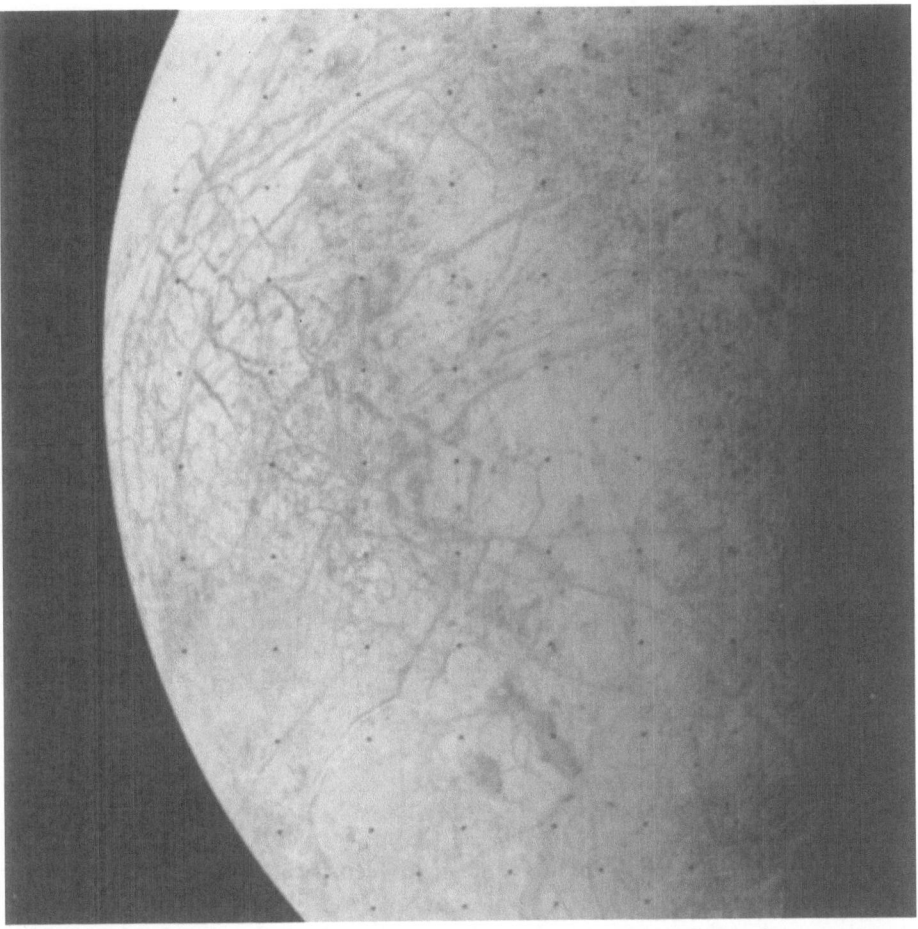

Fig. 126. Voyager 2 image of Europa, a Galilean satellite of Jupiter. On the *right* is a brown mottled terrain, in the *center* are plains and gray mottled terrain, and on the *left* are fractured plains. Photograph C2065022, courtesy of Voyager Experiment Team Leader B. A. Smith and NASA

larger in diameter. Tyre Macula, a flat, dark, circular patch > 100 km in diameter on the northern part of the mapped area, is an excellent example of a class 2 structure. It is similar to structures on Ganymede that Smith et al. (1979a) and Shoemaker et al. (1982) named crater palimpsests, craters that have been erased by subsequent crustal relaxation or other processes. The origin of the class 2 structures is uncertain, but if they were formed by impact they are much older than class 1 structures. Lucchitta and Soderblom (1982) ascribed their low relief to impact on an early crust that was soft, or because either the long time span since they were formed or their large size permitted relaxation. Possibly, the craters penetrated the silicate mantle and subsequently were buried by a new ice shell. They ascribed the brown stain on the structures to upward mixing of mantle silicate because of the

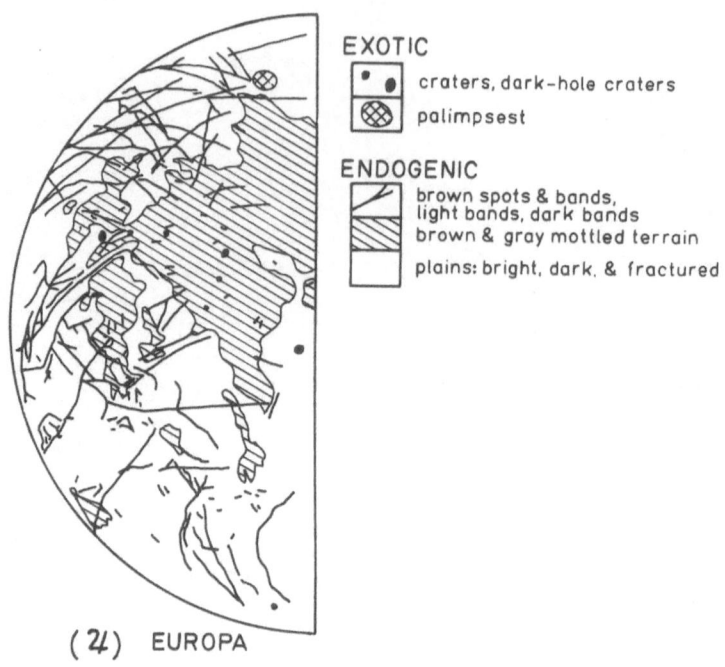

EXOTIC

• • craters, dark–hole craters

⊗ palimpsest

ENDOGENIC

brown spots & bands,
light bands, dark bands

brown & gray mottled terrain

plains: bright, dark, & fractured

(♃) EUROPA

Fig. 127. Morphology of a segment of the jovian satellite, Europa (After Lucchitta and Soderblom 1982, p. 5)

impact, or a result of disruption of the overlying ice crust and injection of dirty ice. According to Shoemaker and Wolfe (1982), the crater retention age of the surface of Europa is short with the satellite-wide mean density of about 1.5 craters per $10^6 \, km^2$ corresponding with an age of about 30 Ma. Even when allowing for an overestimation of the cratering rate and for additional craters near the threshold of recognition, Shoemaker and Wolfe estimated that the surface observed on Europa is unlikely to be older than 0.1 or 0.2 Ga. In contrast, the oldest preserved surface of Ganymede is 3.9 Ga, and the oldest surface near the antapex (the point from which a celestial body is moving at a given time) of Callisto is 4.4 Ga. Squyres et al. (1983a) proposed that the recent resurfacing of Europa is the result of endogenic extrusion of water from a liquid layer.

Brown mottled terrain occurs mainly in the north-central part of the mapped area, and the gray mottled terrain is in the west-central part. This terrain is rough textured, hummocky, and lineations tend to break up and disappear where they extend into this unit. The gray mottled terrain consists of plains materials with isolated dark patches. Bright plains in the southern part of the mapped area are in sharp contact with the dark plains, and are superposed by straight or crossing broad bands to faint streaks that are more orange than the lineations in the equatorial area. The dark plains occur in the southwest and surround the northern part of the gray mottled terrain. Stripes, similar to those on the bright plains, but grayer, are superposed on the dark plains. The fractured plains are interrupted by a fine network of gray curved and short streaks that terminate against one another to form a system of pentagons and hexagons. Darkening in the equatorial region may be the result of bombardment by ions trapped in the magnetosphere that contaminated the surface, removal of surface ice by sputtering to leave a lag deposit of silicates, or a lag that was left after sublimation of the ice (Lanzerotti et al. 1978; Haff and Watson 1979; Squyres in Lucchitta and Soderblom 1982). Brightness in the polar regions may be caused by poleward migration and deposition of ice, a mechanism proposed by Purves and Pilcher (1980) for Ganymede. The orange color of stripes near the poles may be the result of brightening by frost deposits in the stripes; elsewhere on the satellite the stripes appear brown. The low albedo of the dark plains unit could be due to the presence of a dense network of fractures or to processes described above.

The linear and curved marks superposed on the plains and mottled terrains range from dark wedge-shaped bands to triple bands or gray bands and ridges (Smith et al. 1979a). These features, some of which are global in extent, vary in color, size, and pattern. All appear to be controlled by fractures that opened along old lines of weakness (Finnerty et al. 1980; Pieri 1980b). Their uniform albedo and continous color suggest that they were the result of a single episode. Lucchitta et al. (in Lucchitta and Soderblom 1982) suggested that although fractures formed from a late single stress field, the fracturing occurred along old lines of weakness established in response to older variable stress fields. Dark wedge-shaped bands occur in a southeast trending belt across the south equatorial region of the mapped area and they cross most of the other linear features. Their wedge shape and apparent displacement of crossing lineations led Schenk and Seyfert (1980) and Lucchitta and Soderblom (1982) to suggest that incipient plate tectonics led to opening of fractures and endogenic infilling by darker contaminated water or ice from below. The curvilinear bands characterized by pentagons and hexagons in the western part of the mapped area may denote an isotropic stress field, and the straight bands in the northeastern part characterized by tetrahedrons may be caused by oriented stress fields (Pieri 1980b). Helfenstein and Parmentier (1980) proposed that the stresses that produced the fractures were from tidal forces from Jupiter. Finnerty et al. (1980) and Ransford et al. (1980), who believed that tidal forces are too weak to rupture the satellite's surface, suggested that tensional stresses were caused by upwelling in the silicate interior. This upwelling, influenced by Europa's tidal lock with Jupiter, led to fracturing of the antijovian region.

The triple bands consist of pairs of dark bands flanking a central narrow bright strip that locally merges with single brown streaks. The central part of the bands appears to be ridges, suggesting that the central ridges seen only in the terminator areas (see below) are the central stripes of triple bands (Malin 1980, in Lucchita and Soderblom 1982). Parts of the triple bands cut through other streaks, and parts of the triple bands are cut by other bands. This geometry suggests that they were formed throughout an extended time period. Lucchitta and Soderblom stated that most triple bands disappear where they enter the brown mottled terrain, indicating that either the bands are older than the brown mottled terrain, or the nature of the brown mottled terrain is such that it prevents the fractures from propagating into the region. Finnerty et al. (1980) and Ransford et al. (1980) proposed that the triple bands are the result of global expansion with the fractures aligned along tidally induced lines of weakness. The bands trend northwestward north of the equator and southwestward south of the equator, a change in trend that led Lucchitta and Soderblom (1982) to speculate that global stresses were responsible for their formation. Helfenstein and Parmentier (1980) believed that they are the result of conjugate shear fractures from stresses in cyclical deformation caused by orbital eccentricity. According to Finnerty et al. (1980), the cracks subsequently became filled with aqueous solutions containing silicate xenoliths that rose from the silicate mantle to form breccia dikes of dark material, and the bright central dike-like ridges were formed by clean water forced out of the fractures by an increase in volume due to freezing.

The gray curvilinear bands in the southern part of the mapped area are relatively old and are cut by other lineations and ridges. They appear to have been produced by local processes such as impacts, volcanic structures, or structures centered on former poles of tidal deformation (Lucchitta and Soderblom 1982). Brown to dark patches occur in the various bands, in aligned spots in the brown mottled terrain, and where bands intersect. Brown material excavated from the subsurface also accumulated in the rim of a small crater. Finnerty et al. (1980) proposed that the brown material flanking the central bright strip in the triple bands is from dike-like intrusions of water containing silicate material. Other brown patches may be be due to emplacement of the crust by contaminated water. Lucchitta and Soderblom considered the last origin less likely because of the difficulty of extruding the higher-density muddy water.

The youngest endogenic features on Europa are the numerous ridges along the terminator and south polar region that cut all other lineations. Such a youthful age also is suggested by control of their radial pattern around a relatively young crater at the terminator. The ridge configuration ranges from straight segments in the northern and central parts of the mapped area to a cycloidal pattern in the south. The ridges, which trend northwesterly in the northern part of the mapped area and southwesterly in the south, were formed probably by convection of an anhydrous interior, a convection that became oriented by an interaction with Jupiter that produced upwelling at the antijovian point (Finnerty et al. 1980). This model calls for compressional forces due to radially spreading and downward moving convection currents. Lucchitta and Soderblom (1982) proposed that the ridges, like triple bands, were formed by the intrusion of clean ice into the fractures, a model that they believed is supported by the observation that some central stripes merge with ridges and that some ridges and triple bands have the same trends. If the ridges were formed by intrusion into fractures, they were the result of tension and not compression, as suggested by Finnerty et al.

The dominant process on Europa is endogenic, a process of tectonic disruption of units already in place (Lucchitta and Soderblom 1982). The geological history of the satellite can be divided into four stages. (1) Development of the bright plains unit; all other units were derived through disruption of this unit. (2) Development of the dark plains, the fractured plains, and possibly the gray mottled plains. The gray albedo features of these units formed as a result of fracturing, intrusion of gray material, or alteration or aging of other material – an aging caused by sputtering or reworking through meteoritic plowing that altered the optical properties of the material with time. (3) Intrusion of the brown material (the major component of the triple bands) and wedge shaped dark bands into preexisting plains, cracks, and fractures. Brown material also was excavated from the subsurface during impacts and deposited on the surface. Near the end of this stage or soon after the mottled brown terrain was disrupted, brown material was added to the mottled terrain. (4) Emplacement of the ridges as bright dike intrusions.

Like the morphologic units classified by color and albedo, the tectonic features also were grouped by Lucchitta and Soderblom (1982) into a time sequence. The oldest tectonic unit is the curvilinear and concentric gray bands. Triple bands formed throughout Europa's history; some of them formed along fractures created by impacts, whereas others are from expansion of Europa or hydration of its core and mantle. Relatively late crustal expansion combined with rotation of plates led to opening of fractures south of the antijovian area to form the dark wedge-shaped bands. Narrow ridges and central stripes of some triple bands were emplaced. These tectonic features followed preexisting tectonic weaknesses to form the pentagons, hexagons, and curved traces in the southern half of the mapped area, and the tetrahedrons and straight fractures in the northern half.

9.2.3.4 Ganymede

Ganymede, the largest satellite in the Solar System, has a diameter of 5262 $\pm$ 20 km, an orbital radius of 1 070 000 km, a rotation and revolution period of 7.1 days, and a density of 1.93 g/cm^3 (Morrison 1982; Table 1). Its low density indicates that the satellite is about half water ice. The satellite may have a 500-km ice lithosphere over a 900-km ice-silicate mantle, and its silicate core has a radius of 1250 km (Greeley 1987, p. 190). Shoemaker et al. (1982) stated that the surface morphology and geological record of Ganymede closely resembles those of the Moon and some of the rocky planets. Nearly 40% of the surface of the satellite is occupied by a dark heavily cratered terrain that resembles the cratered highlands of the Moon, Mercury, and Mars (Smith et al. 1979a). Various endogenic forms of topography (grooved and reticulated terrains, and smooth plains) comprise the rest of the satellite's surface (Figs. 128 and 129; Greeley 1987, pp. 205–206). The dark heavily cratered terrain with its low albedo is represented on the surface of Ganymede as a series of discrete polygonal, rounded, and irregular areas. Shoemaker et al. (1982) interpreted these regions as fragments of ancient crust within a matrix of younger grooved terrain. The cratered terrain contains impact basins ranging in diameter from the limit of resolution to about 50 km; craters larger than 50 km are rare. Crater shapes range from bowl-like to smooth floored, central peak and central pit structures, and palimpsests. According to Shoemaker et al., the densities of craters larger than 10 km are about 200 to 300 structures per 10^6 km^2,

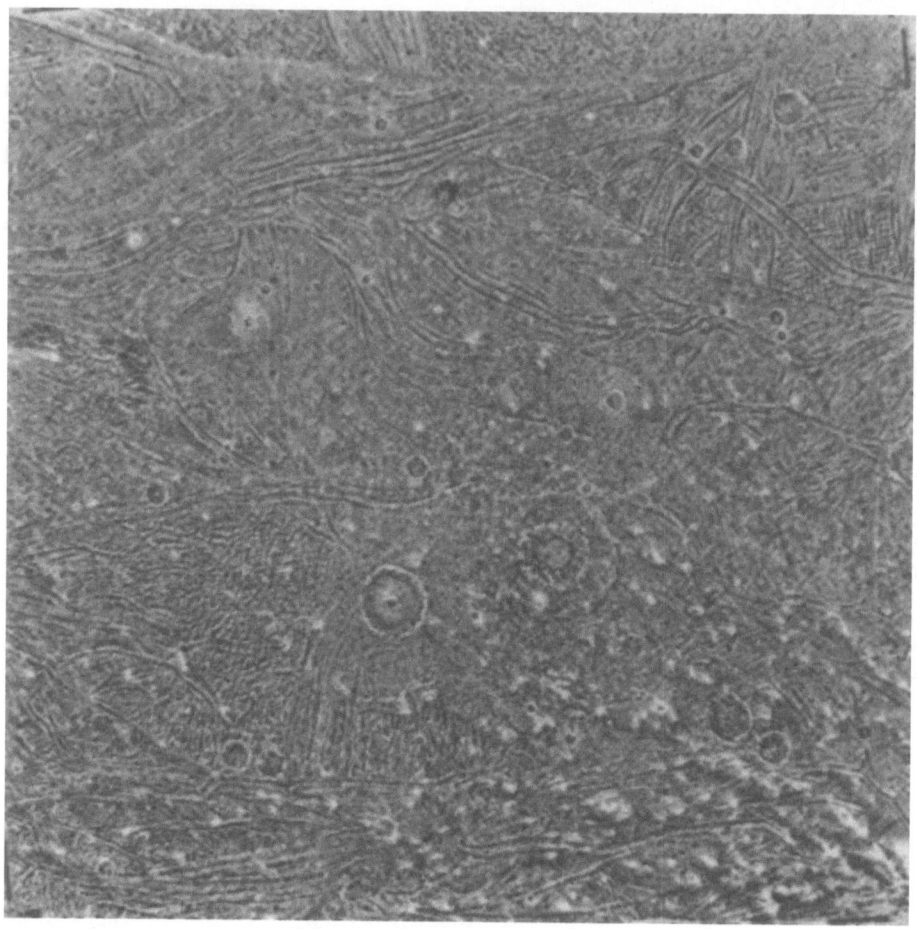

Fig. 128. Reticulate terrain on Ganymede, the largest Galilean satellite of Jupiter. The terrain consists of orthogonally intersecting grooves. Photograph 0494J2-001, courtesy of Voyager Experiment Team Leader B. A. Smith and NASA

with local densities as high as 400 per 10^6 km^2. Most craters have a very low relief. As predicted by Johnson and McGetchin (1972) and verified by Passey and Shoemaker (1982), the low relief of craters is the result of plastic or viscous flow of the satellite's icy crust. Such crustal relaxation has removed some of the larger structures. The sites of these former craters or palimpsests are indicated by bright patches having albedos similar to those of the grooved terrain. The palimpsests constitute about 25% of the heavily cratered terrain and are circular, oval, or irregular in outline with diameters ranging from about 50 to 400 km. The bright patches represent ejecta deposits from destroyed craters, and the moderate albedo ejecta is believed to represent clean ice excavated from an ice layer from which

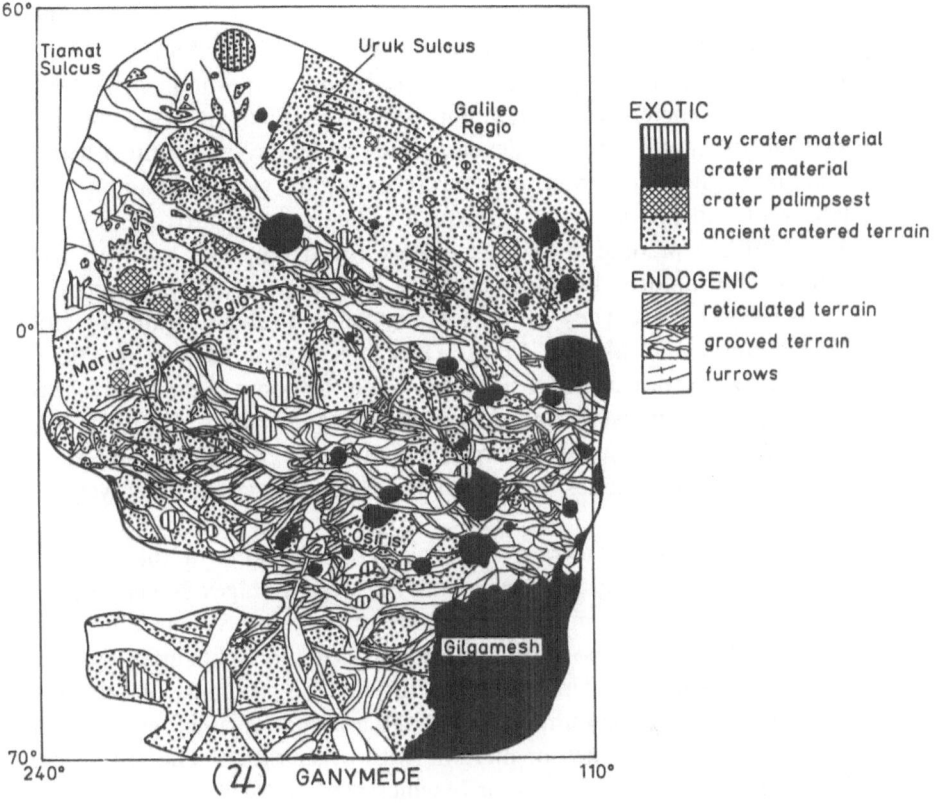

Fig. 129. Morphology of a segment of the jovian satellite Ganymede (After Shoemaker et al. 1982, p. 3)

silicates have been removed by processes of differentiation. Modeling of the relaxation processes by Passey and Shoemaker suggested that the lithosphere was about 10 km thick at the time the structures were formed. This thickness at the time of impact also is suggested by the absence of bright rim deposits on craters having diameters smaller than 50 km (Shoemaker et al. 1982). Greeley et al. (1982) and Croft (1983) proposed that the palimpsests and some of the shallower craters were formed during the final stages of the heavy bombardment as the crust solidified, and that they were the result of impact into ice that melted and deformed. Also present on Ganymede are features that are transitional in morphology between craters and palimpsests; these Passey and Shoemaker termed penepalimpsests. These structures have diverse topographic and albedo characteristics, and their vestiges of crater rims or other topographic features prove that they have not quite disappeared by viscous relaxation. Included within this group are two types: extremely relaxed craters, and complex structures having low central domes surrounded by low relief complex annuli.

The dark rocky material on the satellite's heavily cratered terrain is believed by Shoemaker et al. (1982) to represent debris from bombardment by silicate plane-tesimals prior to formation of the palimpsests. As described by Shoemaker et al., the heavily cratered terrain is disrupted locally by linear or sublinear structures several hundred kilometers long. These furrows are irregular in shape, and some are offset from others in a poorly developed *en echelon* system. Still others are broadly arcuate with the center of curvature of one system being near the antijovian point and just south of the equator. The rimmed furrows of Galileo Regio (Fig. 129) that predate most large craters and palimpsests, are 5 to 10 km wide, 50 to several hundred kilometers long, a few hundred meters in relief, and spaced 25 to 100 km apart. They are part of a system of grabens created by global stresses (McKinnon and Melosh 1980). Their raised rims have reliefs formed probably during relaxation of the initial relief, when the floors of the lows rose and the side scarps rotated and were carried upward (Shoemaker et al. 1982). The origin of the linear to slightly arcuate features on Galileo Regio, which intersect the rimmed furrows at large angles and are older than most craters, has yet to be resolved.

The grooved endogenic terrain that covers nearly 60% of the surface of Ganymede is characterized by grooves having intervening ridges. No close analog to this terrain has been seen on any other body of the Solar System (Shoemaker et al. 1982). The features are about 100 km long, spaced 3 to 10 km apart, and have reliefs of 300 to 400 m with a maximum of 700 m. The gradients of side slopes average 5° and reach a maximum of 20°. Ridge termination is by either abrupt truncation, gradual loss of relief, or merging with other ridges. The grooves and ridges are grouped into cells or sets some of which are hundreds of kilometers long and tens of kilometers wide, but others are short and as wide as 100 km. The cells have straight or curvilinear boundaries, smooth centers, and are bounded by deep grooves. The ridges and grooves in other cells roughly parallel the long axis, while others are orthogonal to this axis, and still others are fan-shaped. The boundaries of cells tend to truncate adjacent cells. Some grooves also occur outside the cells as solitary short and stubby or long gashes. These straight, curvilinear, or irregular gashes are circular or semicircular in plan view, as though controlled by a preexisting crater, and they occur in both grooved and cratered terrains, locally transecting the boundaries of these terrains. According to Shoemaker et al. (1982), the contact between grooved and cratered terrains in most places is sharp and defined by a single groove or groove set. Because the grooved terrain transects the cratered terrain, it must be younger. In some areas the grooved terrain cuts the smooth plains, but elsewhere it is overlapped by smooth plains.

Squyres (1980) and Parmentier et al. (1982) proposed that the grooves are the result of internal differentiation from one ice phase to another, which led to global expansion that produced the extension tectonics exhibited by the grooved terrain. According to Shoemaker et al. (1982), however, only a fraction of this expansion of 7% may have occurred during formation of the grooved terrain. That the expansion was much less is suggested by the preservation of Galileo Regio, a cap-like remnant of cratered terrain (Fig. 129). Extension of the magnitude proposed by

Squyres and Parmentier et al. would tend to cause tensional hoop (as for a barrel) stresses in the cap – leading to the breakup and spreading of the outer parts of Galileo Regio. Shoemaker et al. (1982) proposed a process whereby a preexisting crust was displaced downward. After early differentiation, the density of the lithosphere was increased by planetesimal and cometary debris, and thermal contraction occurred as the lithosphere thickened and cooled. A slight expansion of the core during thermal evolution of the satellite led to fracturing and foundering of the crust and to filling of the depressions by extruded water as the blocks sank. Analyses of relaxation craters suggest that the lithosphere was about 35 km thick at the beginning of formation of the grooved terrain. The vertical dimensions of the subsiding blocks probably corresponded to the full thickness of the lithosphere. The grooves are 3 to 10 km wide, so individual blocks were deep narrow slabs. The dimensions of the global driving mechanisms probably are reflected in the dimensions of individual cells, with 2000- to 3000-km grooved terrains indicating a satellite-wide system of deep convection within the icy mantle, and with structural cells about 100 km in dimension showing a convection system restricted to a shallow layer immediately beneath the lithosphere (Shoemaker et al. 1982).

Closely associated with the grooved terrain are smooth plains. They occur generally in small patches as a facies of a grooved unit in areas where grooves become shallow or indistinct, or as thin stripes parallel to adjacent grooves. Shoemaker et al. (1982) also reported that smooth plains occur as large discrete areas resembling lunar plains. These plains of endogenic origin apparently were formed by surface flows of very fluid material. Reticulate terrains, also of endogenic origin, form where grooves on the heavily cratered terrain are superposed one on another (Fig. 129). Hummocky terrains formed where the grooves are less regularly superposed are transitional between dark heavily cratered terrain and grooved terrain.

In addition to the questions of origin of exotic and endogenic topography on Ganymede, the interrelations of these two kinds of topography provide considerable information on the relative dates of morphologic development. Crater densities on the grooved terrain for structures having diameters down to 10 km range from 25 to 100 structures per 10^6 km^2 with crater densities as high as 200 per 10^6 km^2 in the south polar region, a density comparable with that on the least heavily cratered parts of the heavily-cratered terrain. Thus the grooved terrain is older in the south polar region than at low latitudes. Although most craters on the grooved terrain exhibit topographic relaxation, they have not been erased by flow of the icy lithosphere. The structures also have greater depth-to-diameter ratios than on the heavily-cratered terrain, a preservation permitted by their overall younger age. Possibly the craters are better preserved not only as a result of their younger age, but because of greater thickness and stiffness of the lithosphere at the time they were formed. Their albedo characteristics suggest that the material in which the craters were excavated is similar to the surface material in the grooved terrain. The largest of the impact structures is Gilgamesh, located at Lat. 50°S, Long. 128°W (Fig. 129) with a 150-km diameter, a relief of 1.0 to 1.5 km, and a central depression having a radius of about 75 km surrounded by mountainous regions 250 to 300 km wide; beyond the mountains are incomplete escarpments out to about 400 km. Beyond 400–500 km are secondary craters that form subradial chains extending for nearly 1000 km from the basin. Associated with the secondary craters are ejecta in the form of herringbone and subradial ridges. These structures apparently were emplaced shortly after the grooved terrain in the region was formed. A flattened basin, the Western Equatorial Basin, with a well defined 200-km-diameter rim and a relief of about 600 m, is centered at Lat. 7°S, Long. 115°W in the grooved terrain. It has hummocky rim deposits that extend outward from the rim crest for a distance of 265 km, with secondary craters present within 150 to 450 km from the basin's center. Like the Gilgamesh Basin, the Western Equatorial Basin probably is very near the age of the grooved terrain in which it is found. Its low relief suggests that at the time of formation the lithosphere was warmer and thinner near the equator than at high latitudes where Gilgamesh was formed, allowing greater relaxation than in the polar regions (Shoemaker et al. 1982).

Bright structures also may change abruptly from high- to low-albedo material with a bright crater having a dark floor, a bright ray system changing outwardly to a dark one, a bright crater being surrounded by dark rays, or a dark crater being surrounded by dark rays. These changes do not appear to be related to the nature of the substrate, and they occur in both the cratered and grooved terrain. Bright ray craters range in diameter from less than a few kilometers to 155 km. Osiris Crater, in a polygonal cratered terrain at Lat. 39°S, Long. 161°W (Fig. 129), is surrounded by a bright ray system that formed probably rather recently. Shoemaker et al. (1982) stated that this structure was caused by impact of a comet nucleus having a 10-km diameter that struck the region about 0.1 Ga ago. Near the apex of orbital motion (facing the direction of motion) ray craters constitute less than 10% within the post-grooved structures, increasing linearly away from the apex. The density of rayed craters is 40% higher on grooved terrain than on heavily cratered terrain. Recent impacts (mainly by comet nuclei) should be denser at the apex, so the present distribution must result from preferential erasure of bright rays near the apex. Passey and Shoemaker (1982) believed that this erasure was caused by the impact of small meteroids in nearly circular heliocentric orbits, which would tend to be concentrated at the apex of motion of the satellite. This distribution also implies that most ray craters on the leading hemisphere are younger than on the trailing hemisphere, and that at a given distance from the apex of orbital motion the average age of craters on the impacted terrain is 40% less than on the grooved terrain.

About 40 craters with dark rays have been identified on Ganymede. Like the bright-rayed structures, they increase from the apex of orbital motion toward the antapex (facing away from the direction of motion). Conca, in Shoemaker et al. (1982), ascribed the dark rays to ablation of surface strata through sputtering by high energy ions in the jovian atmosphere. Shoemaker et al., noting the latitudinal dependence of the frequency of dark rays, suggested that sublimation of ice due to insolation may be more important than sputtering. Dark material could be from contamination by the impacting bodies or from rocky material in the crust of Ganymede.

Intense impact throughout its history must have modified the surface of Ganymede. Shoemaker et al. (1982) stated that impact plowing by meteorites that formed a regolith was the dominant surface process. Other mechanisms that may have aided in regolith development included contamination of the surface by infalling meteoritic debris, sublimation at low latitudes with possible redeposition at midlatitudes, and sputtering and collision of high-energy ions in the jovian atmosphere. Shoemaker et al. estimated that the regolith was about 20 m thick near the apex of orbitial motion, 10 m thick on the polar grooved terrain, and about 1 m thick near the antapex. It has been inferred that most of the craters and much of this regolith were produced during an interval of a few tenths of a Ga after the grooved terrain was formed during the waning stages of the heavy bombardment. Others factors that have affected the micromorphology of the surface of the satellite include latitudinal thermal migration of water and surface erosion by sputtering or ion erosion.

Using the cratering scales constructed for the Galilean satellites by Shoemaker and Wolfe (1982), Shoemaker et al. (1982) estimated the ages of the heavily cratered terrain and grooved terrain. They noted that the average density on the heavily cratered terrain is below that of Callisto and the lunar highlands, a distribution ascribed to loss by viscous relaxation. Since impacts on Ganymede would have been much higher than on Callisto, because of gravitational focusing by Jupiter, such a low crater density implies that cooling and thickening of the lithosphere on Ganymede occurred later. This delay may be a result of Ganymede's larger size and possibly of its greater collisional heating during the last stages of accretion. The grooved terrain in the south polar region is estimated to have an age of 3.8 Ga. The mean age in the leading hemisphere is 3.4 Ga and in the trailing one 3.6 Ga. Apparently, the grooved terrain (which formed during the waning stages of the heavy bombardment) developed first in the polar region (where it is more extensive), then on the trailing edge, and finally on the leading edge; for comparison, a retention age of 3.4 Ga is near the time that the Imbrium Basin was formed on the Moon (Shoemaker and Wolfe 1982). The preservation of craters on the grooved terrain relative to the heavily cratered terrain indicates that the grooved terrain is younger, and formed during a period of cooling and thickening of the lithosphere. The grooved terrain apparently formed first in the polar regions where lithospheric cooling was more advanced at any given time, and in the trailing hemisphere where cooling preceded that in the leading hemisphere. At the time that the Gilgamesh and Western Equatorial basins (Fig. 129) formed near the end of heavy bombardment in the leading hemisphere the lithosphere must have been much thicker and stiffer so as to be able to support such large structures. From the density of small craters on the rim of Gilgamesh, this basin and the Western Equatorial Basin are estimated to be about 3.5 to 3.2 Ga old. Thus in the relatively short time of 0.3 to 0.6 Ga the lithosphere of Ganymede thickened enough to support basins of the dimensions of Gilgamesh and the Western Equatorial Basin. The ray craters are the youngest features on Ganymede, with the oldest craters near the apex being formed 0.5 Ga ago and the older ones in the trailing hemisphere about 1.1 Ga ago (Passey and Shoemaker 1982).

9.2.3.5 Callisto

Callisto, the darkest of the Galilean moons, has a diameter of 4800 ± 20 km, an orbital radius of 1 880 000 km, a period of 16.69 days, a revolution and rotation period of 16.7 Earth Days and a density of 1.83 g/cm³ (Table 1). Its interior consists of an ice lithosphere approximately 250 km thick, a 1300-km-thick ice-silicate mantle, and a silicate core having a radius of 1200 km (Greeley 1987, p. 190). Callisto is comparable in size to Mercury and has a density similar to that of Ganymede, but it lacks the endogenic tectonic pattern that is dominant on Ganymede. The surface morphology of these two satellites is so different that it may indicate that, although perhaps similar originally, they have followed different evolutionary pathways. Possibly Callisto's lack of an endogenic terrain (Fig. 130) is an indication that the satellite is either undifferentiated or less differentiated than Ganymede. Assuming that accretion on both satellites was homogeneous, McKinnon and Parmentier (1986) proposed six scenarios to explain their morphologic differences, including runaway differentiation on Ganymede triggered by early tidal heating by Jupiter, or Ganymede being partially differentiated during accretion with further differentiation triggered accretionally. In a third scenario both Ganymede and Callisto underwent differentiation, but heavy bombardment masked endogenic processes on Callisto, and in a fourth scenario Callisto was more differentiated. A fifth scenario calls for both satellites to be deeply differentiated during accretion, but with Callisto evolving faster; in the final scenario Ganymede was deeply differentiated during accretion, but Callisto was only partially melted. Greeley (1987, p. 214) stated that differences between Callisto and Ganymede came from much less heating on Callisto from various sources (impact cratering, radionuclides, tidal stresses). This allowed the lithosphere to thicken faster on Callisto than on Ganymede, leading to the preservation of an earlier cratering record on Callisto and preventing the formation of the endogenic terrains that dominate the surface of Ganymede.

Although the gravitational focusing effect of Jupiter should result in more cratering on Ganymede than on Callisto, Callisto is more heavily cratered than even the heaviest cratered surviving terrain on Ganymede (Greeley 1987, p. 210). Craters produced during the recent past are due to nuclei of long-period comets, including the parabolic and nearly parabolic comets that account for one-eighth to one-third of the present production of craters, and of short-period comets (the Jupiter family comets) that produce roughly half of the present impact craters (Shoemaker and Wolfe 1982). However, most craters on Callisto, as well as those on Ganymede, probably date from a late part of the heavy bombardment early in the Solar System history prior to 3.8 Ga ago. The oldest surviving surface on Callisto occurs near the antapex, and it has a retention age of 4.4 Ga. The Asgard

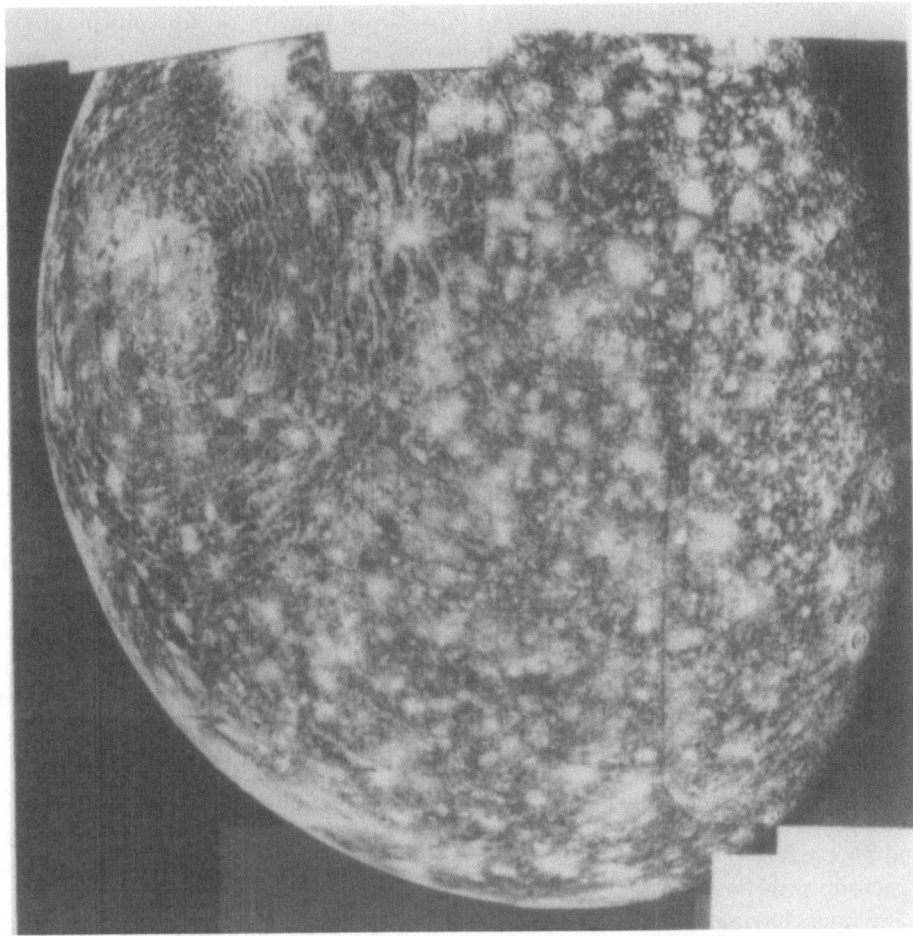

Fig. 130. Photomosaic of Callisto (the outermost Galilean satellite of Jupiter) showing Valhalla Basin (bright spot on the *left*) and adjacent concentric rings. In contrast to Ganymede, the surface of Callisto is dominated by impact structures. Photograph P212821, courtesy of Voyager Experiment Team Leader B. A. Smith and NASA

multiring structure is slightly older than the oldest craters on Ganymede, and the Valhalla multiring structure is somewhat younger, near the age of the Imbrium Basin on the Moon (Shoemaker and Wolfe 1982). This bombardment was due to the late-arriving Neptune and Uranus planetesimals that also may have been responsible for the late heavy bombardment of the rocky planets. A consequence of this concept is that any evidence of impact by Jupiter planetesimals or objects displaced from the outer Asteroid Belt by Jupiter's perturbations was destroyed by the Neptune and Uranus planetesimal bombardment. However, Woronow et al. (1982) stated that a different population cratered the surfaces of the outer planets and cautioned against extrapolating time scales from the Moon record to the Galilean satellites.

As on Ganymede, impact features on Callisto display various morphologies including bowl-shaped craters, smooth-floored craters, and craters having central peaks and central pits. Long chains of craters (catena) to 700 km and containing as many as 27 craters have been recognized on both satellites. The chain craters generally are circular and some of them overlap considerably. Those with diameters of about 20 km have central peaks, and craters with 30- to 35-km diameters have a slight central depression. These structures probably represent secondary craters associated with large craters that have been erased either by prompt collapse or by slow relaxation (Passey and Shoemaker 1982). Those at Lat. 50°N, Long. 350°W and Lat. 12°S, Long. 14°W may be related to Valhalla (Fig. 130). Fifteen palimpsests and possible palimpsests have been recognized on Callisto. Albedo contrasts on the Callisto structures are less than those on Ganymede, and their outer boundaries are diffuse and irregular. Eight multi-ring structures also have been mapped on Callisto. The largest of these structures, Valhalla at Lat. 11°N, Long. 57°W has in its center a palimpsest structure that is 600 km wide. Surrounding the palimpsest is a concentric ridge system which extends outward for 2000 km from Valhalla's center. The multi-ring system has three parts, an inner 200- to 300-km-wide zone with a fairly continuous band of sinuous and scalloped ridges, a middle 200- to 400-km-wide zone of discontinuous ridges, and an outer more than 500-km-wide zone of narrow sinuous troughs (Greeley 1987, p. 211). Variation in ring morphology is believed to be the result of inward flow of the asthenosphere when the impact cavity collapsed. This inward flow produced a distinct fault pattern in the lithosphere away from the point of impact (Melosh 1982). Asgard, the second largest multiring structure on Callisto at Lat. 30°N, Long. 39°W also has a 230-km-wide palimpsest in its center. According to Passey and Shoemaker (1982), the ridges on Asgard extend 800 km from the center of the structure, with the innermost ridge about 80 km from its center.

Craters having bright rim deposits or rays are more abundant on Callisto than on Ganymede. As reported by Passey and Shoemaker (1982), the maximum diameter of bright-rim and bright ray craters on Callisto is about 120 km and the mean frequency of structures having diameters greater than 25 km is about 4 craters per 10^6 km². On both Ganymede and Callisto bright ray and rim structures display the least topographic relaxation, indicating that they are the youngest impacts. Cassen et al. (1980) and Passey and Shoemaker (1982) summarized the history of Callisto as follows:

1. Heating of the lithosphere during the heavy bombardment by impacts and from the interior prevented the retention of any craters. This heating, augmented by heating from radionuclides and tidal stresses, led to crustal differentiation and formation of a siliceous core. Heating from these sources was less on Callisto than on Ganymede, as a result of which a lithosphere was formed much earlier on Callisto – allowing preservation of a longer cratering record than on Ganymede.

2. When the bombardment rate and the heat transfer from the interior waned, the upper part of the lithosphere cooled to the point where craters could be retained by the crust.

3. The oldest craters retained by Callisto (as well as Ganymede) have diameters less than 10 km and occur either in the polar regions or near the antapex.

4. Crater retention spread from these two areas toward the apex, and at any given distance from the apex progressively larger craters were retained by the crust with the passage of time.

9.2.4 Outer Satellites

The outer group of satellites beyond the Galilean satellites consists of four known bodies moving prograde in eccentric orbits inclined about 27° to Jupiter's equator (Leda with a diameter of about 10 km, Himalia with a diameter of about 185 km, Lysithea with a diameter of about 20 km, and Elara with a diameter of 80 km) at orbital radii of 11 110, 11 470, 11 710 and $11\,740 \times 10^3$ km, respectively. Four other satellites move retrogradely (Ananke with a diameter of about 20 km, Carne with a diameter of 30 km, Pasiphae with a diameter of 40 km, and Sinope with a diameter of 30 km) at orbital radii of 20 700, 22 350, 23 300, and $23\,700 \times 10^3$ km, respectively. These satellites have periods in days, Leda 240, Himalia 251, Lysithea 260, Elara 260, Ananka 617, Carne 314, Pasiphae 735, and Sinope 758. Little is known about these small objects. Himalia and Elara closely resemble C-type asteroids in the outer Asteroid Belt and Pasiphae and Sinope in the retrograde group resemble the dark red asteroids. According to Cruikshank et al. (1982), these two satellite groups are comparable with the two compositional classes of Trojan asteroids and other asteroids in the main belt.

9.2.5 Summary

The Jovian System can be considered an analog of the Solar System with Jupiter representing the Sun, and its 16 known satellites representing the planets (Pollack and Fanale 1982). About 95% of the mass of Jupiter is a gaseous liquid envelope of H_2 and He over a core consisting of silicates and perhaps water and other low-temperature condensates. The regularity of orbits of the eight inner satellites may indicate their formation within a jovian nebula, just as the rocky planets were formed with variation of density away from the Sun, reflecting the temperature gradient within the nebula. The outer eight smaller asteroid-sized satellites have irregular orbits, indicating that they were captured, probably as a result of gas drag by Jupiter's outer atmosphere. This jovian system, however, differs in some important respects from the inner Solar System. For example, none of the rocky planets has the bulk composition of Ganymede or Callisto, and, although Io and Europa have densities comparable to those of the rocky planets, Europa is almost entirely covered by water ice, and Io has a level of volcanism higher than on the rocky planets.

The inner satellites and the three satellites discovered by Voyager (Andrasta, Metis, and Thebe) probably formed in the outer regions of a proto-Jupiter. As Jupiter decreased in size, a flattened disk of gas and dust was left behind. The satellites were formed from this disk by accretionary processes, a formation that probably occurred about 0.1 to 1.0 Ma after the end of the collapse phase. Differentiation led to concentration of rock components toward the center and the ice fraction toward the surface. Water ice blankets all the surface of Europa, some of Ganymede, and a little of Callisto. The rest of the surfaces of Ganymede and Callisto have a rocky component of exotic origin. Sulfur compounds, elemental sulfur, and sulfur dioxide cover the surface of Io. Endogenic processes had a major role in sculpturing the surface of these four major satellites. On Io this process has been so pervasive that all exotic terrains that ever were present have been destroyed.

9.3 Saturn and Satellites

9.3.1 Planetary Setting

Saturn has a rotation period of 10.22 h, a diameter of 120 536 km, a density of 0.71 g/cm^3, and a composition by weight of 67% hydrogen and helium, 12% water, ammonia, and methane, and 21% iron, silicates, and oxides (Moore 1988, p. 51). As described by Hartmann (1983, pp. 293–294), a massive fluid-hydrogen ocean overlies a liquid metallic-hydrogen region over a massive silicate metallic core. Ices of water, ammonia, and methane may form the surface layers of the core. Saturn's low density is the result of its hydrogen-rich and distended outer layer. Its magnetic field, a centered dipole tilted approximately 0.8° from the rotation axis, is relatively smooth and displays no evidence of azimuthal asymmetry or magnetic anomaly in the planetary field (Ness et al. 1982). The planet's atmosphere is dominated by hydrogen (88% by mass of H_2) followed by 11% of helium, and traces of ammonia, methane, phosphine (PH_3), ethane, and acetylene that total less than 1% of the mass. This dense atmosphere with dark belts and bright zones obscures the surface of the planet (Hartmann 1983, p. 436). Orbiting the planet are seven rings (designated D, the inner ring; C, B, A, F, G, and E, the outer ring) with the inner edge of D Ring being at a distance of about 67 000 km from Saturn (1.11 R_s; R_s is the radius of Saturn, 60 268 km) and the inner edge of E Ring being at a distance of about 210 000 km (3.48 R_s) and its outer edge at about 300 000 km (4.98 R_s; Stone and Miner 1981). Ring D consists of numerous narrow features, the Ring C contains a number of dense ringlets and particle size of 2 m, and Ring B has numerous narrow ringlets that may have been formed by the action of moonlets within the ring. Ring B also has sporadic radial markings or spokes supposedly produced by levitation of particles above the plane of the ring. The Cassini division, separating rings B and A, contains five broad rings having particle diameters of 8 m, and Ring A with a particle size of 10 m is marked by narrow features corresponding to orbital resonances with satellites 1980S1 and 1980S3. The Encke division within Ring A is characterized by satellite-driven density waves. Beyond Ring A are three additional rings, F, G, and E. Ring F is very narrow, has local concentrations of debris, is multicolored, and is shepherded by satellites 1980S26 and 1980S27 (Stone and Miner 1981).

Although by the early twentieth century telescopic observations had revealed that Saturn had nine satellites, the largest known regular satellite system (Table 1), and that one of the satellites, Titan, had an atmosphere. Most of what is known about these nine large satellites and eight smaller ones is the result of Voyager 1 and Voyager 2 flybys near the planet (Smith et al. 1981, 1982; Morrison et al. 1986). As described by Greeley (1987, p. 217), the densities of the satellites indicate that they are mixtures of rock and water ice (up to 70% ice by mass); Titan, which has a density of 1.88 g/cm^3, consists of about half rock and half ice. Whereas the densities of the satellites in the Jupiter system decrease with radial distance from the planet, those of Saturn are randomly distributed. The surfaces of the satellites appear to be blanketed by ice frost, a cover that makes Enceladus, one of the Saturn satellites, the brightest object in the Solar System. Except for Phoebe, the satellites of Saturn, like those of Jupiter, rotate synchronously with their orbital period, and thus the same hemisphere of each satellite always faces Saturn.

9.3.2 Satellites

Mimas, with a diameter of 394 $\pm$ 6 km and a density of 1.2 g/cm^3, is the smallest and innermost of the regular satellites of Saturn. Its surface is heavily cratered and its albedo is relatively uniform. Although the cratering is rather dense, it is not uniform. Mimas' most noticeable impact structure is Herschel Crater that is centered on the leading hemisphere (Table 1, Fig. 131). According to Morrison et al. (1986) and Greeley (1987, p. 218), the crater has a diameter of 130 km and its walls are about 5 km high. Parts of the floor of the depression are as much as 10 km deep, and a 20- to 30-km wide central peak has a relief of 6 km. Greeley stated that the impactor at Herschel probably was the maximum size that Mimas could have experienced and still remained intact. Most craters are less than 50 km in diameter, with size frequency distributions suggesting a mixing of two populations, which Morrison et al. (1986) postulated may be the result of resurfacing. Other features that disrupt the surface of Mimas are grooves 90 km long, 10 km wide, and 1 to 2 km deep, which Greeley (1987, p. 218) stated may be related to the formation of Herschel Crater or of some other impact crater. In addition, the trailing hemisphere of the satellite displays a cluster of hills 5 to 10 km across and about 1 km high that may represent ejecta deposits.

The surface of Enceladus, outboard of and slightly larger than Mimas, displays a much more complex geological history (Figs. 132 and 133). This high-albedo satellite, having a diameter of 502 $\pm$ 10 km, a density of 1.20 g/cm^3, and a

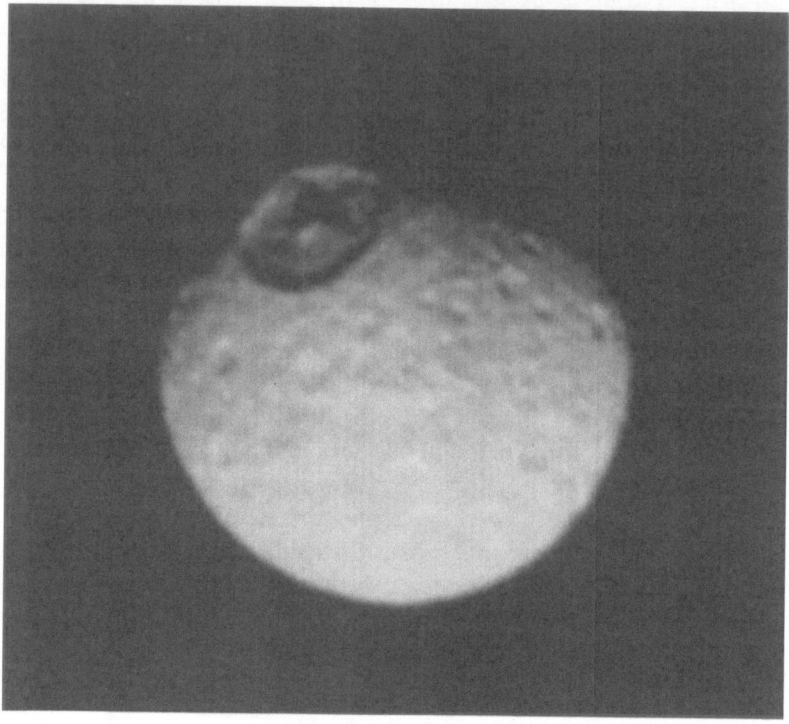

Fig. 131. Image of Mimas, the innermost and smallest of the regular satellites of Saturn, showing the impact crater Herschel. Voyager 1 JPL-23210 photograph, courtesy of Voyager Experiment Team Leader B. A. Smith and NASA

lithosphere consisting of ammonia-ice and water-ice (Passey 1983; Table 1), displays a wide diversity of terrains including highly cratered smooth plains with a rectilinear groove pattern, ridged plains, cratered terrains, and crater-free smooth plains. The cratered terrains include regions of flattened craters 10 to 20 km in diameter, a region of well-preserved craters 10 to 20 km in diameter, and a terrain having bowl-shaped craters 5 to 10 km in diameter (Morrison et al. 1986). Ridge spacing ranges from 7 to 15 km and ridge relief from a few hundred meters to 1.5 km (Passey 1983). Morrison et al. postulated that the crater-free terrains are younger than 1 Ga, and possibly even younger than 10 Ma. The denser part of Saturn's Ring E is coincident with the orbit of Enceladus; this ring consists of spherical, micrometer-sized particles with a life time of less than 10^4 years (Morrison et al. 1986, and references therein). Possibly the same event that produced the uniform, uncontaminated optical surface of Enceladus also sprayed out the ice particles that form the E ring, an event that may have occurred within the past few thousand years. Geological processes documented by the terrains on Enceladus include replacement of the older units, formation and subsequent flattening of craters on the new surface, and emplacement of a ridged terrain. The

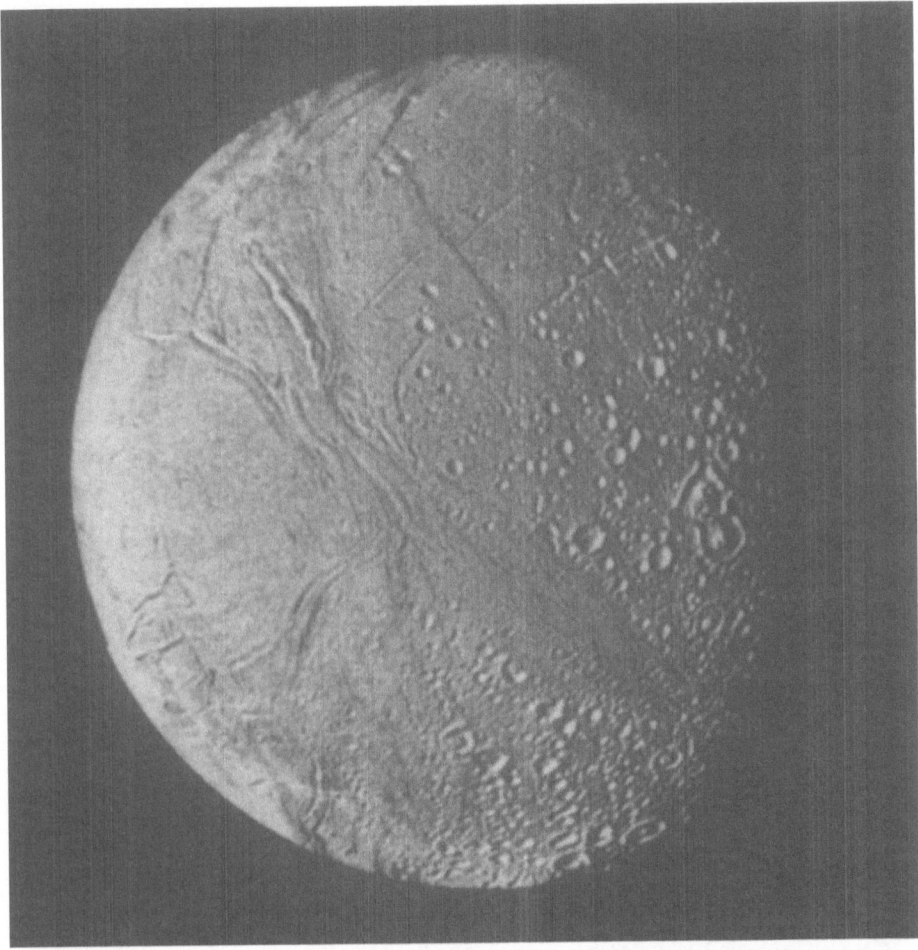

Fig. 132. Image of the saturnian satellite Enceladus showing cratered, grooved, and smooth terrains and Samarkand Sulci, the graben-like structure that transects the satellite. Note that the fracture to the *right* of Samarkand Sulci displays evidence of right-lateral transcurrent motion. Photograph P23956, courtesy of Voyager Experiment Team Leader B. A. Smith and NASA

grooves are interpreted as grabens produced by extension and brittle failure of the lithosphere, and the ridges are thought to be produced by compression associated with upwelling convection from the interior, or from expansion of freezing water that had intruded into fractures (Greeley 1987, p. 222). The tectonic deformation and the various resurfacing episodes that involved extrusion of water or water-ice slurries may have resulted from tidal heating. Enceladus' present eccentricities would not allow the melting of an initially frozen body, particularly if pure water is involved, but an inclusion of hydrates of ammonia or methane clathrate would lower the melting temperature and enhance melting of the frozen Enceladus (Squyres et al. 1983b).

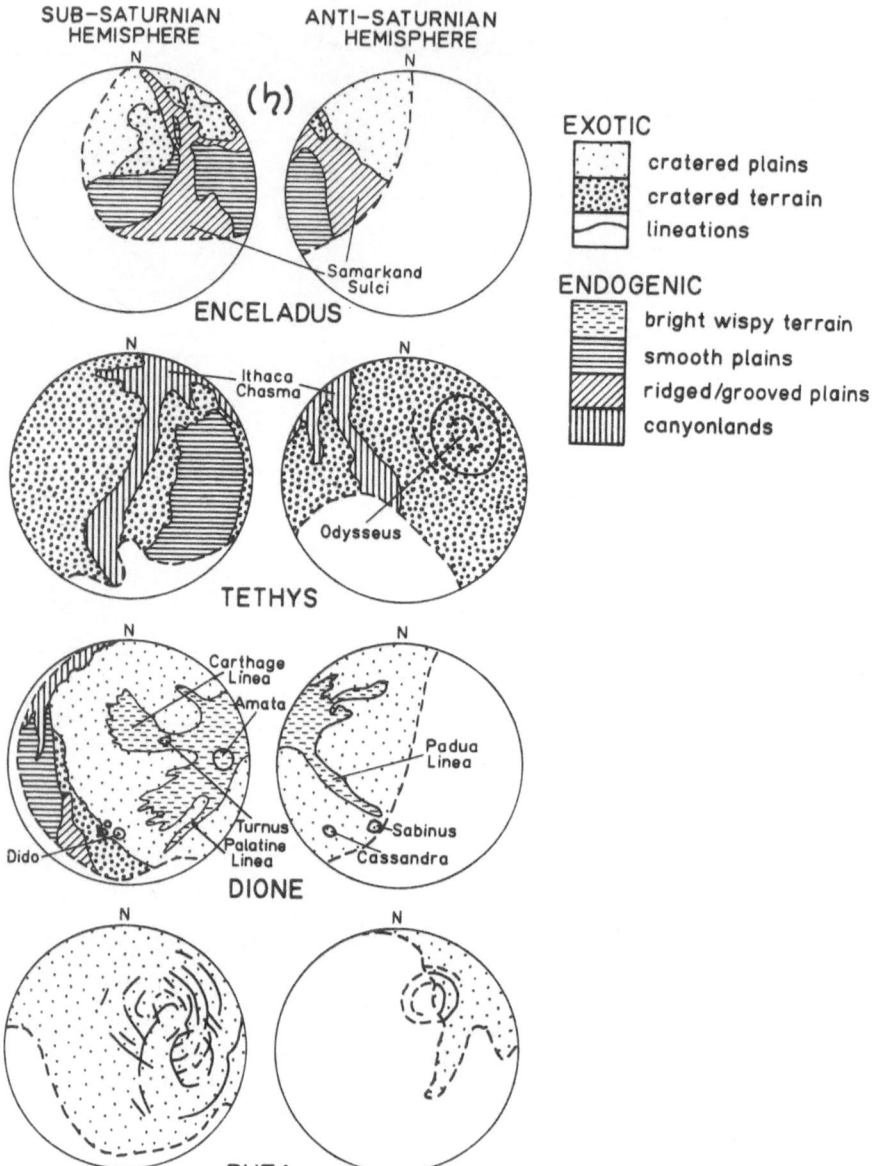

Fig. 133. Morphology of the saturnian satellites Enceladus, Tethys, Dione, and Rhea (After Greeley 1987, figs. 9.9, 9.10, 9.15, and 9.25). Greeley compiled these maps from Passey (1983) for Enceladus, Moore and Ahern (1983) for Tethys, Plescia (1983) for Dione, and Moore et al. (1985) for Rhea. *Blank areas* are unmapped

Outboard from Enceladus is Tethys (Table 1) with a diameter of 1060 ± 20 km (more than twice that of Enceladus), a density of 1.21 g/cm³, and an albedo of 0.8 (Morrison et al. 1986). The major topographic features of this satellite are Odysseus Crater with a diameter of 400 km (~ 40% of the diameter of Tethys; Fig. 133) and a trench or valley of global proportions, Ithaca Chasma (Figs. 133 and 134). The crater in the leading hemisphere centered at Lat. 30°N, Long. 130°W has rebounded tens of kilometers, producing its present convex face; this rebound was enhanced by creep or viscous flow of its icy lithosphere (Smith et al. 1982; Passey 1983). Impact occurred probably when the interior of Tethys was partly liquid or

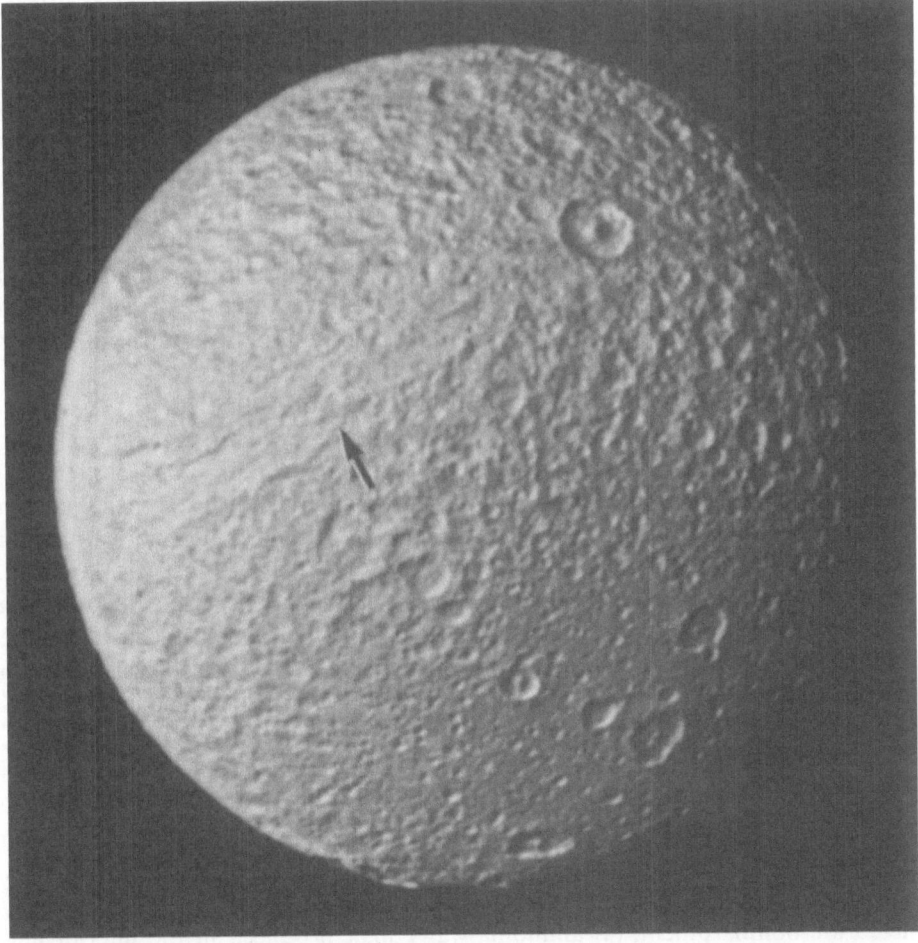

Fig. 134. Image of the saturnian satellite Tethys displaying impact craters and Ithaca Chasm (*arrow*), plains unit and cratered terrain. Photograph P24065, courtesy of Voyager Experiment Team Leader B. A. Smith and NASA

soft ice, and less than 2% of the ejecta escaped the satellite and was placed in orbit around Saturn (Soderblom and Johnson 1983). The 1000-km long, 100-km wide, and up to 4-km deep Ithaca Chasma, whose rim rises 500 m above the surrounding terrain, may be a lithospheric response to the impact that formed Odysseus Crater (Moore and Ahern 1983), or to the freezing of water whose volume would expand by about 10%, leading to a 7% increase in surface area (Greeley 1987, p. 229).

Dione with a diameter of 1120 km, displays a wide range of albedo patterns and has the highest density (1.43 ± 0.04 g/cm^3) among the inner satellites, indicating a higher proportion of rocky material than most of the other satellites, whether smaller or slightly larger. Morphologic features of the satellite include a rough, heavily cratered terrain with numerous craters having diameters wider than 20 km, cratered plains with a lower frequency of large craters than the heavily cratered terrain, and smooth plains having few craters or other topographic features (Figs. 133 and 135; Greeley 1987, p. 229). The largest craters with diameters as large as 200 km occur on the trailing hemisphere, and craters with diameters of more than 100 km are common on the heavily cratered terrain. Most larger impact structures have terraced walls and central peaks, with smaller structures being simple bowl-shaped depressions. In general, craters on Dione tend to be shallower than those on Tethys, and except for crater rays, clearly defined ejecta deposits on this and other saturnian satellites are unrecognizable (Greeley 1987, pp. 234–235). Statistical studies by Plescia and Boyce (1982) indicate that even the heavily cratered terrain has a cumulative density that is significantly less than the least cratered terrain on Rhea. Distribution curves also suggest that the heavily cratered and cratered plains terrains may reflect different populations of impact bodies as well as different ages of the plains.

On the trailing hemisphere of Dione is a network of bright stripes superimposed on a dark background. In the center of this network is an elliptical area, about 240 km in diameter, that may represent an impact scar. Smith et al. (1981) interpreted the stripes as deposits of frost-like material formed by the sudden release of volatiles from the interior via fractures. Other endogenic features on Dione include ridges that are 50- to 100-km long and less than 0.5-km high, interpreted by Plescia (1983) as lobate deposits similar to ejecta flow deposits on Mars; scarps as long as 100 km on the heavily cratered terrain that may be the result of faulting or mass wasting; troughs that generally are 30 to 100 km long with some more than 500 km long and a pit crater at one or both ends; and crater chains that are as long as 100 km, 10 km wide and may be secondary impact craters, volcanic craters, or collapse features. In spite of its small size, Dione's level of endogenic activity exceeded in intensity and duration that of the larger Rhea. Moore (1984) summarized the geological history of Dione as follows: (1) formation of a brittle lithosphere after accretion; 2) global expansion from heat generated by radionuclides and tidal stresses that produced global linear patterns; (3) heavy meteoritic bombardment; (4) ammonia-water melt produced in the interior and extruded onto the satellite surface to form the plains; (5) cooling of the interior and/or a phase change leading to compression of the surface and formation of ridges by thrust or high angle reverse faults; and (6) light impact cratering mainly by cometary objects.

Fig. 135. Leading hemisphere of the saturnian satellite Dione displaying impact craters and a trough network. Photograph P23113, courtesy of Voyager Experiment Team Leader B. A. Smith and NASA

Rhea has a diameter of 1530 km, a density of 1.24 g/cm³, and an albedo of 0.6. According to Morrison et al. (1986), it may represent the archetype of an icy satellite of the outer Solar System. Its trailing (facing away from the direction of orbit) hemisphere is dark and has bright wispy markings, and, except for a very bright feature that may be an ejecta deposit, the leading hemisphere is uniformly bright (Smith et al. 1981). The surface morphology of Rhea is dominated by impact craters that resemble the highland provinces of the Moon and Mercury (Figs. 133 and 136). They differ from the flattened features common on Ganymede and Callisto, reflecting the lower surface gravity on Rhea and the presence of a stiffer ice

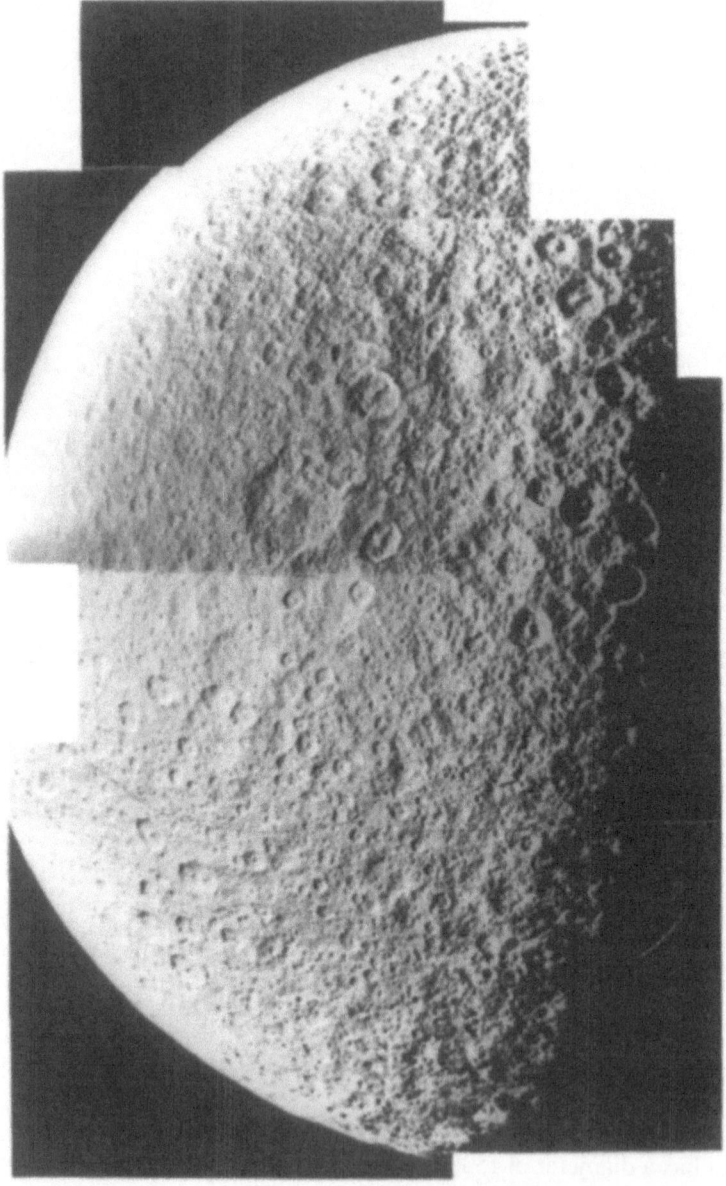

Fig. 136. Image of the heavily cratered Rhea, a satellite of Saturn. Photograph of mosaic P23177, courtesy of Voyager Experiment Team Leader B. A. Smith and NASA

crust. Ejecta blankets are missing and some of the craters display bright patches on their walls, which Smith et al. (1981) interpreted as fresh ice deposits or other fresh material exposed by slumping. Moore and Horner (1984) described a ~ 450-km multiring basin, and Pike and Spudis (in Greeley 1987, p. 235) discovered a second, much larger, very degraded multiring structure centered at about Lat. 30°N, Long. 316°W. This structure in the resurfaced terrain has been subdued by subsequent impacts. Plescia and Boyce (1982) noted that whereas larger craters have been retained in the polar regions, they are absent in the equatorial region near Long. 30°W with the line of demarcation being very near sub-Saturn Long. 0°. They postulated that this demarcation boundary is the edge of a major resurfacing event that occurred near the end of heavy bombardment of the satellite. Moore et al. (1985) proposed that ejecta associated with two multi-ring basins mantles part of the surface of Rhea. They also proposed two distinct cratering episodes that were disrupted by periods of endogenic resurfacing. According to Greeley (1987, p. 240), some of the resurfaced terrains, particularly those centered on Lat. 25°N, Long. 310°W, may indicate flooding by water 'magmas'. On the resurfaced terrain from Long. 270°W to Long. 360°W, particularly near the crater Pedn, are N30°E-striking ridges or scarps several hundred kilometers long, about 5 km wide, and 0.25 km high. Some craters, about 20 km in diameter, are superimposed on the ridges, and other structures are transected by them. Features of endogenic origin on Rhea also may be represented by parallel linear troughs and coalescing pits that are 5 to 10 km wide, as much as 100 km long, less than 0.5 km deep, and oriented N45°E in the old cratered terrain between Lats. 0° and 60°N and Longs. 60° and 120°W (Greeley 1987, p. 240). These features are superimposed on the older impact structure and cannot be traced into the younger cratered terrain. They continued to form after the end of the heavy bombardment but prior to resurfacing. Evidently they were produced by extensional tectonics and volcanism between the end of the heavy bombardment and the resurfacing event.

Titan has a diameter of 5120 km (exceeding that of Mercury) and a density of 1.88 g/cm^3, suggesting that it consists of 45% water ice and 55% rocky material (Greeley 1987, p. 240). It has an atmosphere predominantly of nitrogen with ethane, acetylene, ethylene, hydrogen cyanide, and various carbon-nitrogen components as minor constituents. An extensive cloud cover obscures Titan's surface (Fig. 137). The surface temperature and pressure are 94 ± 2 K and 1.5 bar, respectively (Morrison et al. 1986). When Titan accreted, ammonia and methane ices were incorporated into its mass and later released by heating and satellite differentiation. The ammonia reacted with sunlight causing the release of nitrogen. Photolysis of methane led to the production of ethane and acetylene, which then condensed and rained down onto the surface of Titan to form an ocean one to several kilometers deep. If this ocean were in equilibrium with an atmosphere of 3% methane, its composition would have been 70% ethane, 25% methane, and 5% nitrogen (Lunine et al. 1983). Sagan and Dermott (1982) proposed that large tides (about 10 m) produced by the present satellite eccentricity would have readily eroded any surface that rose above its surface. Such an ocean would have obscured all earlier geological features so that if volcanism, tectonism, and cratering had left landforms, they cannot be identified now.

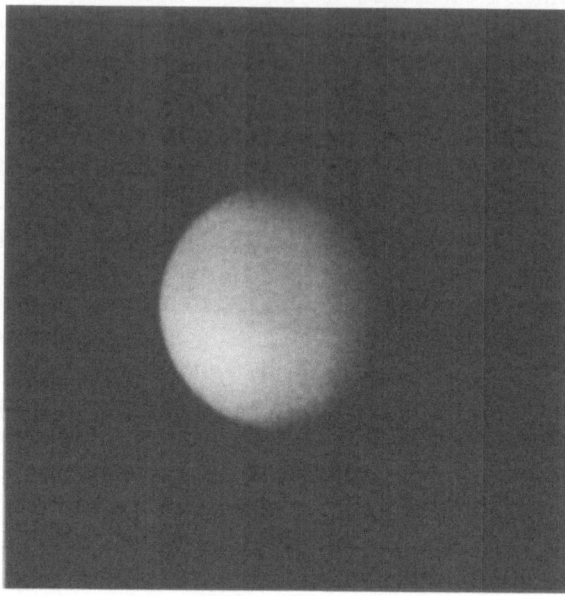

Fig. 137. The cloud covered Titan, largest satellite of Saturn. Photograph P23076, courtesy of Voyager Experiment Team Leader B. A. Smith and NASA

Hyperion, whose orbit lies just outside Titan, is a small irregular object 350 ×235×200 km (Greeley 1987, p. 242). This, the darkest object of the inner saturnian satellites, is believed to consist of a mixture of ice and rock and is heavily cratered. Its irregular shape, heavily impacted surface, and a possible impact scarp nearly half its size suggest that the satellite is a remnant of a larger object.

Iapetus, outboard from Hyperion, has a diameter of 1437 ± 36 km and a density of 1.16 g/cm³ (Greeley 1987, p. 16). The leading hemisphere of this satellite is very dark, and the trailing one is much brighter with a well-defined boundary separating the two regions (Fig. 138). Although no details can be discerned in the dark area, both the north polar region and the bright side are heavily cratered. The largest structures display central peaks and well defined rims (Greeley 1987, p. 242). The floors of the craters along the bright/dark boundary are covered with dark material that has the same spectral characteristics as the dark terrain. Spectrophotometric data obtained from the Mauna Kea Observatory suggest that the dark area may be mantled by a mixture of meteoritic organic polymers (10% and hydrated silicates 90%; Bell et al. 1985). Four origins have been proposed for the dark material:

1. Impacts on the leading edge eroded the ice and left behind a lag deposit of dark rocky material (Cook and Franklin 1970); however, Greeley (1987, p. 243) believed that the low density of the satellite is not compatible with such an origin for the dark material.

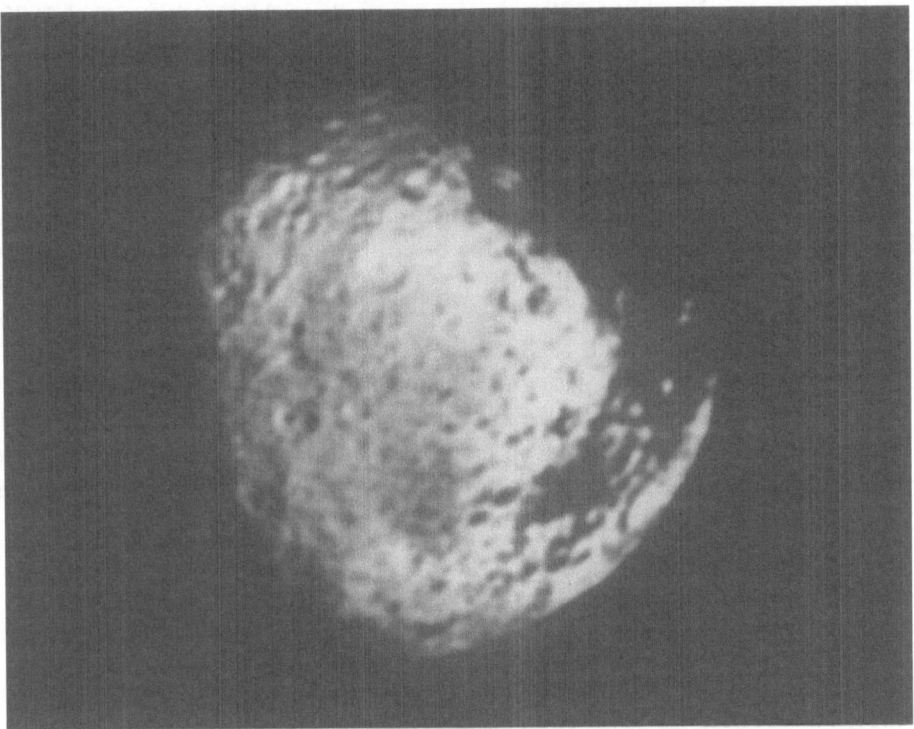

Fig. 138. Image showing the bright and dark hemispheres of the polar region of Iapetus, a saturnian satellite. Photograph P23920, courtesy of Voyager Experiment Team Leader B. A. Smith and NASA

2. The dark material was derived from Phoebe; impact cratering supposedly led to escape of some of the ejecta that fell toward Saturn and may have landed on Iapetus (Soter in Greeley 1987, p. 243), but, according to Greeley, the spectral characteristics of Phoebe and the dark material are different.

3. Bell et al. (1985) proposed that the dark material is a primitive component of the original condensate that formed Iapetus and has been concentrated on the leading hemisphere by exotic processes. This implies that impacts have been greater on the leading side. Yet, according to Lissaur (1985), the distribution of craters on these satellites is symmetrical and uniform, and they are not more abundant on the leading hemisphere (as theoretical analysis predicts) than on the trailing one.

4. The material was erupted from the interior onto the surface of the satellite, possibly in the form of a carbonaceous-rich or organic-rich ice slurry (Smith et al. 1982). Morrison et al. (1986) stated that although the origin of the dark material remains problematical, while its distribution with respect to the orbital direction calls for some external control, its topographic relationship on the bright side, however, suggests an internal origin.

The outermost saturnian satellite is Phoebe with a diameter of approximately 220 km, a nonsynchronous rotation having a period of 9.4 h (Morrison et al. 1986), and a retrograde inclined orbit. Its spectral characteristics suggest a surface blanketed by carbonaceous material with circular markings and indentations probably indicating craters (Greeley 1987, p. 243). Morrison et al. stated that its dark surface and retrograde orbit indicate that the satellite is a captured object having a primitive surface composition.

Of the eight known small satellites of Saturn having dimensions smaller than Hyperion or Phoebe, five are within the orbit of Mimas. The largest of these are Janus, 220×190×160 km, and Epimetheus, 140×115×100 km. They are coorbital, within 50 km of each other, and may be remnants of a larger satellite (Greeley 1987, p. 243). Satellites 1980S26 and 1980S27 are F ring shepherds and confine the orbits of the particles of the ring. Atlas, with a diameter of 30 km, is a shepherd of the A ring. Calypso, Telesco, and 1980S6 are known as the Lagrangian satellites (small objects that have the same orbital speed as a larger object and maintain 60° of arc behind or ahead of a larger satellite); Calypso, Telesco, and two other small objects are associated with Tethys, and 1980S6 is associated with Dione.

9.3.3 Summary

The satellite system of Saturn (except for Phoebe, which probably was captured) is believed by Morrison et al. (1986) to have a common origin in the circum-saturnian protonebula, and the presence of water-ice in the rings and satellites indicates that temperatures in the nebula were low enough to permit the condensation of water. The composition of the atmosphere on Titan indicates that ammonia and methane were incorporated on the outer part of the saturnian system. Satellite masses show that they contain 60 to 70% water-ice, with the remainder being chondritic. This uniformity in composition means that either the saturnian protonebula was uniform in composition, or during accretion the nebula became homogenized. The larger satellites were heated and melted by accretional energy, tidal despinning, or radioactive energy from their chondritic component, and this led to differentiation. The composition of the saturnian satellites and rings can be related to the following sequence: (1) an early stage lasting 10^4 to 10^5 years when Saturn was several hundred times larger than now; (2) a hydrodynamic collapse phase during which proto-Saturn shrank to about five times its present size in about 1 year; and (3) another hydrostatic stage during the past 4.6 Ga when Saturn slowly contracted to its present size. Smith et al. (1982) proposed that the ring and regular satellites came from a flattened disk of gas and dust that formed near the end of the hydrodynamic stage when the outer portions of proto-Saturn were not able to follow the contraction of the rest of the protoplanet and its rapid rotation.

For satellites as small as the saturnian moons, cratering rates were influenced by the gravitational attraction of the planet. On synchronous satellites one would expect that cratering would be enhanced on the leading hemisphere, but as this is not true it may indicate that the satellites were not in synchronous rotation during the time of heavy bombardment or that the impact bodies were in planetocentric rather than heliocentric orbits (Lissauer 1985; Morrison et al. 1986, and references therein). However, if Rhea, Dione, and possibly Tethys developed a global wispy terrain during an early episode of igneous activity the wispy terrain's survival on the trailing hemisphere would suggest that these satellites underwent more massive bombardment on their leading hemisphere. Cratering rates due to long-period Saturn family comets yield an age of less than 1 Ga, but such rates probably are too low, because at such an estimated rate 1000 Ga would have been required to produce the existing crater density on Iapetus. If the crater density on the bright side of Iapetus were produced by projectiles external to Saturn, then the inner satellites would have been cratered more than they show, indicating saturation or resurfacing after stabilization of Iapetus' surface (Smith et al. 1982). However, if the impact history of Iapetus were due to debris associated with Saturn, catastrophic impacts on the other satellites may have been less common. Such a rate also implies impact disruption for the satellites inboard of Rhea; thus, Enceladus and Mimas may have been disrupted and reaccreted four or five times during the period that Iapetus developed its present surface. The E ring may have been formed by condensation from a liquid, with the water originating from an impact melt associated with a cratering event on Enceladus (McKinnon 1983).

The cratering record of the satellites appears to reflect two populations, the older group representing the tailing off of postaccretion bombardment, and the second group being impacts by debris generated by collisions within the system. Population 1 is represented by Rhea, the older parts of Dione, and the heavily cratered terrain of Tethys, and population 2 is represented by the younger terrains of these three satellites and most of the surface of Mimas and all of Enceladus.

Among the endogenic features of the satellites is the Ithaca Chasma on Tethys produced by freezing and expansion of a liquid inner layer covered by a thin frozen crust. However, there is a problem with such an origin because one would expect expansion to be documented by multiple global fractures rather than being concentrated in a single narrow belt (Morrison et al. 1986). Outgassing along fractures may have produced the wispy terrain on Rhea and Dione. Other features of endogenic origin on the satellites are presented by resurfacing events, plastic deformation, curvilinear grooves, and flooding of the satellites' surfaces with water-ammonia magma which, on Enceladus, reached the surface via extension fractures in the relatively brittle crust (Stevenson 1982; Squyres et al. 1983b).

9.4 Uranus and Satellites

9.4.1 Planetary Setting

Uranus is rather unusual in that its axis of rotation is not approximately perpendicular to the plane of the Solar System, as approximated by the other planets, but is near the plane of the Solar System so that sometimes its north pole points toward the Sun and half a revolution later it faces away from the Sun (Hartmann 1983, p. 45). The rotation of the planet is approximately 17.3 h. The planet has an atmosphere of hydrogen and helium, but a trace of methane in the upper atmosphere gives the planet its blue-green appearance (Stone and Milner, 1986). This dense atmosphere obscures the surface of the planet, as on Jupiter and Saturn. Like the other gaseous planets Uranus, with a density of 1.20 g/cm^3, has a core of silicate rocks and metals, a lower mantle of ices of water, methane, and ammonia, and an upper mantle of liquid molecular hydrogen (Hartmann 1983, p. 294; Moore 1988, p. 51). All the models that have been proposed to-date consist of a small rock core, a central layer with a density intermediate between rock and hydrogen-helium, and a low density outer layer (Padolak et al. 1991). Uranus, like Neptune, has less hydrogen and helium than Jupiter and Saturn but much more water, ammonia, and methane. It also has a strong magnetic field with its magnetic axis being offset 60° from its rotational pole (Ness et al. 1986). The large tilt of the magnetic dipole indicates that the dynamo generation is not in the center of the planet but is in a fluid convecting "ice mantle" of water, ammonia, and methane (Ness et al. 1989). These authors also believe that the large tilt is not caused by Uranus' large orbital obliquity or that the dynamo is in the process of a field reversal.

Uranus has 11 rings, of which all but one are eccentric, and all but three are inclined to the equator (Smith et al. 1986; French et al. 1991). The innermost ring, 1986U2R, is 37 000–39 500 km from the planet, and the outer ring, Epsilon, is 51 160 km distant. The width of the rings ranges from about 2500 km for the 1986U2R inner ring to 0 to 2 km for the seventh ring, Eta. In contrast with the bright and reddish surfaces of the particles of Jupiter and Saturn, the spectral reflectance of the particles in the uranian rings is low (less than 5%) and one ring (Epsilon) is gray. Dust concentration in the uranian rings is low, being in the order of 0.1 to 0.01%, whereas the thick parts of the saturnian rings contain several percent of dust; Jupiter has comparable optical depths due to dust and larger particles (Smith et al. 1986). Two satellites, Cordelia and Ophelia, shepherd the Epsilon ring and are on opposite sides of it. The other rings also may have shepherds that were not imaged by Voyager because they are too small and black like the ring particles. Stone and Miner (1986) stated that longitudinal variability in the main rings and possible presence of numerous adjacent ring arcs indicate that the rings are dynamic and may be younger than Uranus The dust of the rings may have various origins. It may have originated from collisions or spiral-wave damping or shepherding by nearby small satellites, collision among present bodies in the interring regions, impact of meteorites on an as yet undetected satellite only a few kilometers across, or debris trails maintained by such satellites in "horseshoe orbits".

Orbiting Uranus are 15 satellites, ten of which were discovered by the Voyager vehicle (Brown et al. 1991; Pollack et al. 1991; Veverka et al. 1991). The ten new satellites orbit between Uranus and Miranda at distances from the center of Uranus ranging from R_u (R_u is the radius of Uranus: 15 343 km) 1.94 for Cordelia to 3.36 for Puck. All 15 satellites are nearly circular in orbit, and all except Miranda orbit Uranus in the equatorial plane. The five major satellites (Miranda, Ariel, Umbriel, Titania, and Oberon) have synchronous rotation, with one side always facing Uranus (Stone and Miner 1986). The small satellites have diameters 154 ± 3 km for Puck, 108 ± 12 for Portia, 84 ± 10 for Juliet, 66 ± 6 for Belinda, 62 ± 8 for Cressida, 59 ± 4 for Rosalind, 54 ± 3 for Desdemona, 42 ± 6 for Bianca, 30 ± 4 for Ophelia, and 26 ± 4 for Cordelia, listed in order of distance from Uranus (Veverka et al. 1991, table I).

9.4.2 Satellites

9.4.2.1 General

The five largest uranian satellites are Miranda (the inner satellite) at a distance of 129 900 km and with a diameter of 484 km and an orbital period of 33.9 h; Ariel at a distance of 190 900 km and with a diameter of 1160 km and an orbital period of 60.5 h, Umbriel at a distance of 266 000 km and with a diameter of 1190 km and an orbital period of 99.5 h; Titania at a distance of 436 300 km and with a diameter of 1610 km and an orbital period of 208.9 h; and Oberon at a distance of 583 400 km and with a diameter of 1550 km and orbital period of 323.1 h. The densities, in g/cm^3, are: Miranda 1.35, Ariel 1.66, Umbriel 1.51; Titania 1.68; and Oberon 1.58 (Table 1; Pollack et al. 1991). These densities indicate large fractions of rock, suggesting that radiogenic heating may have had a larger influence on these bodies than on the saturnian satellites (Smith et al. 1986). The smaller satellites inboard of Miranda are uniformly dark and the larger bodies, except for Phoebe and the darker areas of Iapetus, tend to be darker than the saturnian satellites (Smith et al. 1986).

Initial studies led investigators to propose that the impact crater population on Oberon and Umbriel is similar to that of the lunar highlands and the oldest parts of the Saturnian satellites, a crater population referred to by Smith et al. (1986) as population I. This population is believed to date back to the early sweep-up of the postaccretional debris 4 Ga ago. The crater populations on Titania and Ariel have a size-frequency population generated by impact of secondary ejecta from large primary ejecta. This population, also recognized on some saturnian satellites, was designated population II. Craters of this population have been interpreted by Smith et al. to have been produced by secondary debris generated by collisions within the uranian system. As Titania and Ariel do not display abundant population I craters, their present surfaces must postdate this period. Miranda's impacts are typical of population I craters with their number exceeding that of Oberon, Umbriel, and the Moon by a factor of 2. In addition to old and worn impacts is another population of young and fresh structures. Greenberg et al. (1991) recognized that this contrasting impact morphology indicates that Miranda's impactor source changed with time. Cumulative size-frequency plots of the surface of Miranda also indicate population II craters. In a recent investigation McKinnon et al. (1991) proposed that populations I and II are equivalent and were produced by heliocentric comets that were responsible for much of the bombardment of Jupiter and Saturn. They also suggested that a bombardment component by secondary debris from cratered or disrupted uranian satellites also may be present. The present production of craters on the satellites results from collisions by short-period comets (population III). Because of the low orbital velocities of these short-period comets at the position of Uranus in the Solar System and the proximity of the satellites to the planet, the cratering rate of population III increases inward, a concentration related to an increase of comet flux in the uranian gravity field.

9.4.2.2 Puck

Puck, the only small satellite (108 km in diameter) to be imaged at a close enough range to resolve its surface morphology, is quite spherical for such a small body. Its surface is heavily scarred by craters, one of which (Bogle) has a diameter of 47 km. Two lanes, about 9 km wide and with midlines 25 km apart, also have been identified on the surface of the satellite. Croft and Soderblom (1991) suggested that these lanes may represent either fractures comparable with the grooves on Phobos (Fig. 87, Mars) or alignments of impact craters near the limit of resolution. Puck's surface has a low albedo, a darkening that may be due to irradiation of surface methane, sweep-up of coorbiting dark particles, or preferential removal of ice from the surface leaving behind a residue of uniformly dark material (Bergstralh and Miner 1991; Croft and Soderblom 1991). The available data are insufficient to determine whether Puck is a product of multiple disruptions or an original accretionary body. Croft and Soderblom stated that its surface probably is a primitive one dating to the end of the heavy bombardment in the uranian system, but the dark lanes are not obscured by impacts so they must have been formed late in the bombardment.

9.4.2.3 Miranda

Miranda displays two contrasting terrains, an older heavily cratered rolling terrain having a uniform albedo, and a younger terrain dominated by three coronae of subparallel ridges and scarps that display a trapezoidal to ovoid geometry (Figs. 139 and 140). The trapezoidal region near the south pole (in the center of the figures) is about 200 km on a side, and its outer boundary and internal patterns of ridges and bands of contrasting albedo have numerous sharp corners. The center of this unit consists of intersecting ridges and grooves surrounded by an outer band of ridges and grooves that sharply truncate the structures of the center. The complex, banded, ovoid region on the right side of Fig. 140 (leading hemisphere) is about 300 km in the direction of the equator. Its outer margin has rounded corners with the dark bands parallel to the boundary curving smoothly around the rectangular core. The dark bands along the outer margin correspond to outward-facing scarps. The ridged ovoid terrain on the left side of Fig. 140 (trailing hemisphere) has a complex central set of intersecting ridges and troughs truncated by the concentric

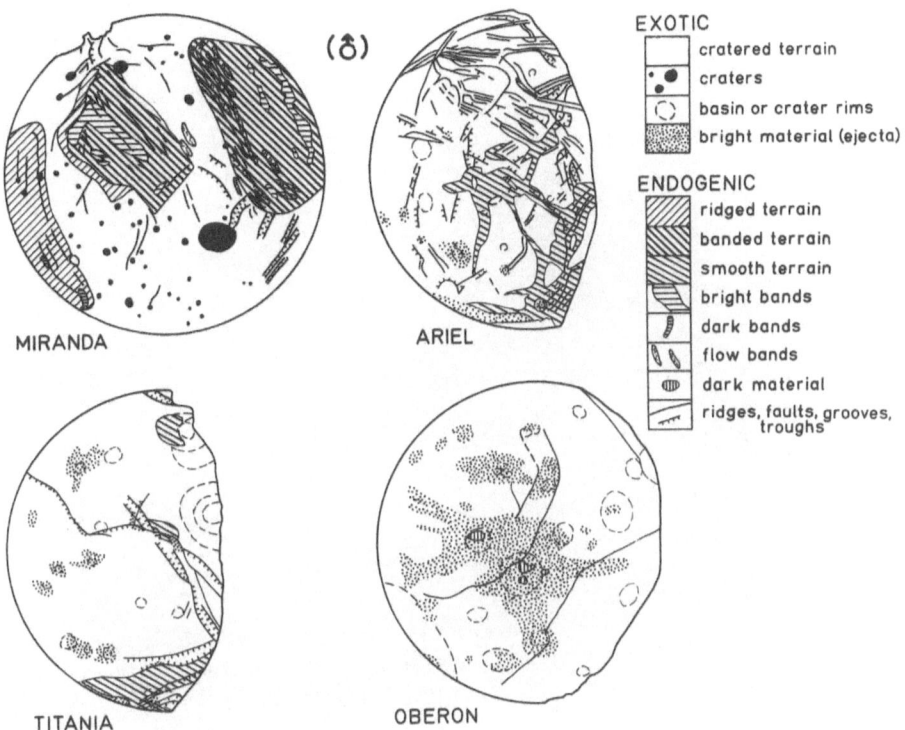

Fig. 139. Morphology of the uranian satellites Miranda, Ariel, Titania, and Oberon (After Smith et al. 1986, figs. 27, 25, 22, and 23)

linear ridges and troughs of the outer belt. Included in this outer belt are small segments of undulating cratered terrain. The belt appears to have been formed by faulting and possibly by intrusion and extrusion of fluid or plastic material. Both the ridged and banded ovoid terrains, which display similar morphologies, were formed probably by the same processes, with the banded unit possibly undergoing a late-stage doming and subsequent collapse. The trapezoid terrain was created probably by the same geological processes that formed the ridged ovoid terrain.

Other features of endogenic origin on Miranda are fault scarps that cross the globe; some are older and others younger than the trapezoid and ovoids. The most striking of the fault structures extends along the margin of the trapezoid and over the limb, forming a deep gorge on the outer margin of the banded ovoid (Smith et al. 1986). A graben complex also occurs near the terminator and bright limb. The highest fault scarps have widths of about 20 km, and a graben on the bright limb is 10 to 15 km deep.

Bright material is exposed on the upper parts of the scarps near the banded ovoid and trapezoid, and in places it extends 11 km beneath the rolling cratered surface. Most of the brightest material occurs in impact craters within the trapezoid and ovoid terrains, in the outer margin of the ridged ovoid, in the center of the

Fig. 140. Shaded relief map of Miranda (After US Geological Survey 1988)

banded ovoid, and in many places in the trapezoid. This may indicate that the subsurface ice in these regions is brighter and cleaner than ice beneath the cratered plains. Dark material is exposed in one location next to the banded ovoid, in craters, and along scarps in the cratered terrain. Very dark material occurs mainly where the trapezoid and banded ovoid have been breached by faults that expose the subsurface strata.

    Why is Miranda so different from other satellites? Is its morphology the product of accretion of heterogeneous blocks, and if so why was this heterogeneous material not fragmented and redistributed during accretion? Is it possible that the coronae, which are so unique to Miranda, may not necessarily be related to accretion, but may have come from relaxation of topographic highs, lithospheric stresses driven by density anomalies in the asthenosphere, or large-scale volcanic extrusion through preexisting fractures? Previously, Smith et al. (1986) proposed that the morphology of Miranda was caused by several meteoritic impacts, catastrophic disruptions, and reaccretions. More recently, Greenberg et al. (1991) ascribed the unique morphology of Miranda to the availability of heat after cessation of the heavy bombardment to drive the satellite's tectonism and volcanism in the presence of an unusual ice composition.

9.4.2.4 Ariel

The oldest terrain of Ariel is a widespread cratered unit displaying chiefly population II craters, the largest of which is 60 km wide. Most population I craters have been destroyed by either viscous relaxation or extensive extrusions. The largest population I crater is flattened and has a gently domed floor partly encompassed by a shallow trough – evidence of viscous relaxation of the topography. The cratered terrain is disrupted by a global system of fractures and faults, the youngest of which can be traced to the limb of the satellite at midlatitudes in the leading hemisphere (Figs. 139 and 141). This fault system consists of normal faults developed in response to crustal extension. Some population II craters are atop scarps whereas other scarps are nearly crater free, indicating that extension spanned the episode of population II bombardment. The youngest terrains on Ariel are bright-rimmed craters and bright deposits resulting from the impact of comets during the past 3 to 4 Ga.

A unit of smooth material on the floors of most grabens in the sub-Uranus hemisphere of Ariel and in the extensive plain at high latitudes was emplaced probably as a sequence of flows that overlapped and partly buried older craters. This smooth material with its superimposed population II craters was emplaced during population II bombardment. The form of the smooth unit within fault valleys is bounded by troughs, suggesting that it was emplaced as a highly viscous material, possibly as an ice or clathrate at temperatures near the minimum melting point. This material upwelled along the axis of the valleys, with the median ridges representing late stage renewal extrusion activity. Extrusion of the viscous material in the grabens probably is related to expansion of the crust during the late-stage freezing of the interior that led to extrusion of warm subcrustal material via fractures formed during the expansion.

The surface flows of Ariel and the other satellites probably consist of ammonia-water mixtures with a eutectic melting point of 173 K. Schenk (1991) believed that ammonia-water-ice is a reasonable composition for the flows on both saturnian and uranian satellites. Methane or carbon monoxide clathrates may have been important in low-temperature resurfacing. Volcanic forms on Ariel and Miranda consist of linear ridges with crest grooves, flatter continuous flow bands, and broad flooded plains. Morphological differences between Miranda and Ariel may reflect subtle changes in composition or extrusion history. In contrast with the broad flows on the saturnian and on the jovian Galilean satellites (Dione, Enceladus, Ganymede), those on the uranian satellites are quite restricted, suggesting that

Fig. 141. Photograph of mosaic of the uranian satellite Ariel displaying a network of fractures and craters. Photograph P29520, courtesy of Voyager Experiment Team Leader B. A. Smith and NASA

these flows were more viscous. Rapid cooling due to volatile exsolution and incorporation of greater amounts of impurities in volatile ice phases is believed to have led to the high viscosities and yield strengths of the uranian flows.

9.4.2.5 Umbriel

Umbriel is much darker than the other larger uranian satellites, and it has a weaker water-ice spectral signature (Fig. 142; Smith et al. 1986). Its uniform albedo suggests that the surface of the satellite is young and undergoing resurfacing by some process that is erasing the impact structures. Yet, its surface is marked by dense population I craters, so the surface of Umbriel, along with that of Oberon, is really one of the oldest of the uranian system.

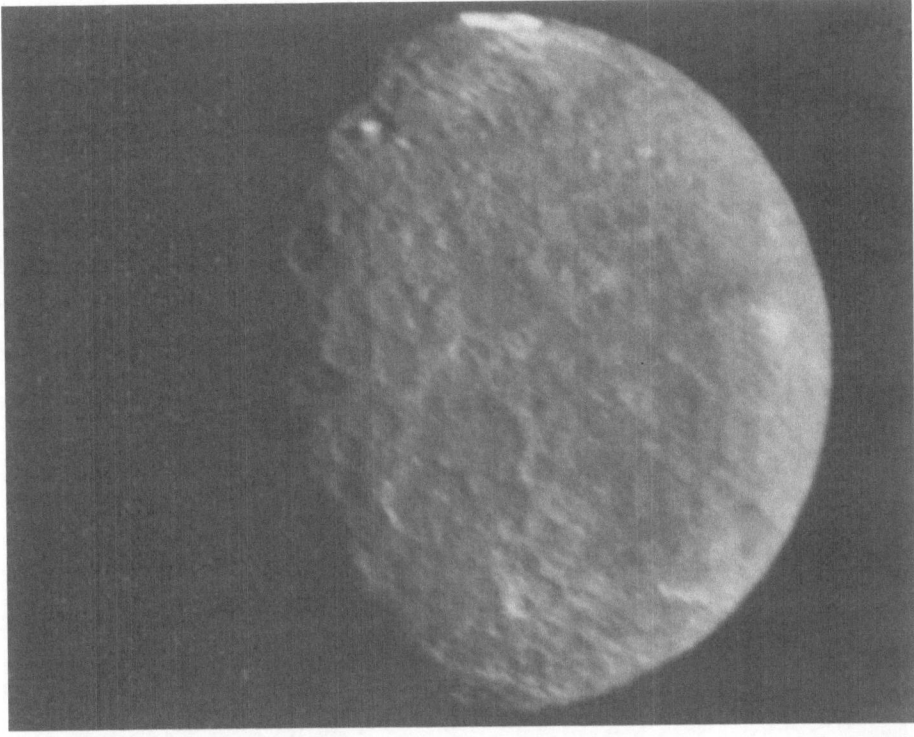

Fig. 142. Image of Umbriel, darkest of the uranian satellites. Its morphology is dominated by impact craters. Photograph P29521, courtesy of Voyager Experiment Team Leader B. A. Smith and NASA

The origin of the uniform blanket of dark material that erased older surface features and gave Umbriel an albedo pattern is not known. Possibly, like Enceladus in the saturnian system, Umbriel has been blanketed recently by dark material orbiting the satellite. One source of this material could have been Umbriel itself, from which material from a major impact or an explosive eruption escaped the satellite into space, allowing it to become well mixed to give its dark color, and then it fell back onto the satellite. Alternately, the satellite never differentiated and it developed at an early stage a deep dark layer that became excavated during the heavy bombardment. A 10-km crater, for example, could produce a 1-cm layer over the entire satellite. Such an impact probably is not the answer, because formation rates of smaller craters are much higher than the formation rate of craters the size of the 10-km hypothetical crater, and these small impacts would rapidly have destroyed the 1-cm layer (Smith et al. 1986). Collision by a large bolide that remained in orbit and later collided with Umbriel in very late geological history also is considered by Smith et al. (1986) to be highly unlikely. Possibly the dark material was ejected into orbit by explosive volcanism caused by methane or dissociation of carbon monoxide clathrates, but Smith et al. (1986) also stated that such volcanic activity is inconsistent with Umbriel's ancient surface. The surface of the satellite may be so dark because Umbriel was never differentiated and developed a deep dark layer before the end of the heavy bombardment, but Smith et al. (1986) also rejected such a scenario, believing it to be inconsistent with the bright ring and other local bright patches on Umbriel.

9.4.2.6 Titania

Titania has abundant population II craters scattered over its surface with several patches of smoother terrains indicating early periods of resurfacing. Features of endogenic origin include an extensive network of normal faults (Figs. 139 and 143).

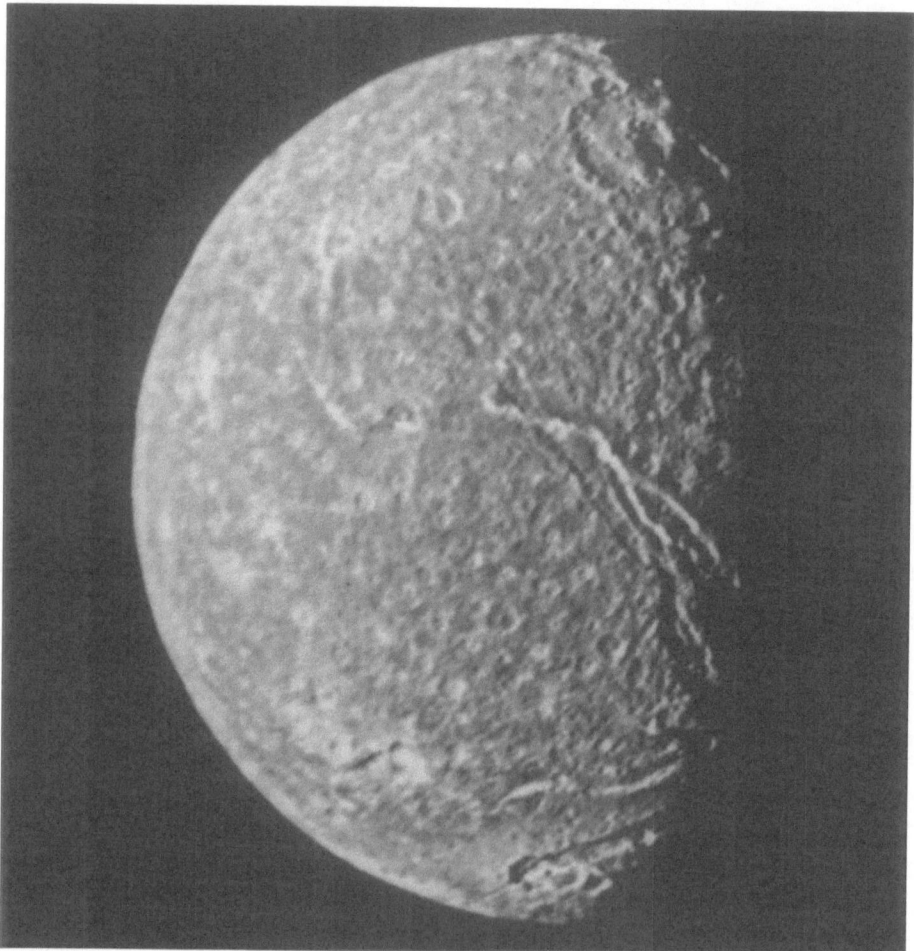

Fig. 143. The largest uranian satellite Titania, whose surface morphology is controlled by impact craters, grabens, and multiring basins. Photograph P29522, courtesy of Voyager Experiment Team Leader B. A. Smith and NASA

Exposed along several of these fault scarps is brighter material. Grabens bound by these fault scarps are 20 to 50 km wide with the flanking scarps having reliefs of about 2 km. This global faulting is believed to have resulted from global expansion of the crust caused by the last stages of freezing of water-ice in the interior of Titania. These faults cut across the large craters and are not strongly modified by population II craters, so they are among the youngest geological features on Titania. Smith et al. (1986) proposed the following geological history for the satellite: (1) formation of population I craters; (2) resurfacing by extrusion of material onto the surface, viscous relaxation, or impact degradation that largely obliterated population I craters; (3) generation of population II debris by objects colliding with Titania or with one another in orbit and the debris being swept up by the satellite; (4) resurfacing by another extrusion of material; (5) rupture of Titania by tensional faulting that produced normal faults and grabens; and (6) impact by comets (population III) during the past 3 to 4 Ga, producing the bright-ray craters.

9.4.2.7 Oberon

Oberon, the outermost satellite, is similar in diameter, density, color, and albedo to Titania, but its cratering record is different (Figs. 139 and 144; Smith et al. 1986). A possible center peak of an impact crater several hundred kilometers in diameter rises more than 20 km above Oberon's bright limb. Isolated patches of dark material, resembling those on the trailing edge of the saturnian satellite Iapetus, also occur on the floors of a few large craters. This material was deposited after the craters were formed, possibly a result of eruption of some dark material. Alternatively, the extruded material acquired its color after its surface exposure. Erosion of the material may have been triggered by the bombardment, or the material was deposited long after bombardment ceased.

9.4.3 Summary

Some workers have proposed that craters on the heavily cratered terrains of the uranian satellites are the result of impacts during the early history of the Solar System about 4.0 Ga ago by either uranian or neptunian planetesimals (population I), by small objects orbiting within the Uranus system (population II), or by a combination of the two groups (Smith et al. 1986). If the craters on Oberon (the outermost satellite) were produced by the impact of population I projectiles, then all or some of the other satellites inboard of Oberon probably underwent intense collisions at that time. As a result of these collisions some, and perhaps all, of these large satellites were disrupted during the early history of the Solar System. Miranda probably was reaccreted several times, Ariel was reaccreted at least once, Umbriel may have been disrupted once, and possibly Titania was disrupted. Population II craters on Titania could reflect the last stages of reaccretion from a prior disruption, from a breakup of another satellite in the orbit of Titania, or from a large impact on Titania. Other workers have questioned this scenario and

Fig. 144. Image of the uranian satellite Oberon, showing its morphology consisting of impact craters and ejecta deposits. Photograph P29501, courtesy of Voyager Experiment Team Leader B. A. Smith and NASA

proposed that both populations are equivalent and the result of a bombardment by heliocentric comets, bodies that also were responsible for cratering on satellites of Jupiter and Saturn (McKinnon et al. 1991). Whether the inner planets were catastrophically fragmented depends on the extent of the poorly documented large-scale (100- to 200-km) impacts.

Catastrophic disruption of the smaller satellites inboard of Puck due to comet impact occurred about 2.0 Ga ago or later, and the life time of the satellites inside the radius of ring Epsilon is 1.0 Ga. Smith et al. (1986) postulated that the small satellites may have accreted from fragments derived from catastrophic disruption of Puck, with most of the satellites having been broken and reaccreted several times since then. The rings themselves may be continuously produced by fragmentation of satellites that in turn were formed by fragmentation of larger bodies. Shepherd satellites associated with the rings probably have collisional lifetimes of a few hundred Ma with the meter-size particles having collisional lifetimes of only a few million years.

The satellites of the outer Solar System are a mixture of rock and ice, a composition that allows them and their host planets to be related to the chemical-equilibrium condensation of the solar nebula. These models indicate that material formed at the orbit of Jupiter and beyond contained approximately equal amounts of silica and water ice with more volatiles, ammonia hydrate, and methane clathrate being present at greater distances, and still farther out pure methane. The presence of ammonia and methane compounds in water ice would tend to lower the eutectic melting point, leading to possible geological activity. In addition, methane clathrates tend to darken under the influence of various types of irradiation to form the dark material on the surfaces of some satellites such as Iapetus and the dark uranian rings.

How much methane and ammonia was incorporated into the uranian satellites is a reflection of the conditions that existed in the nebular gas from which they condensed. As described by Smith et al. (1986), the major carbon and nitrogen species may have been in the form of carbon monoxide and molecular nitrogen, not methane and ammonia. A satellite of rock, water, and carbon and nitrogen clathrate should have a density of 1.7 to 1.8 g/cm^3, whereas a satellite formed from rock, water ice, ammonia, and methane hydrate should have an uncompressed density of about 1.25 g/cm^3. The densities in the uranian system are too high to be modeled as an equilibrium constant condensation of a mixture of rock, water, ammonia, and methane, and too low where the carbon is in the form of carbon monoxide. However, considering the errors involved in estimates of the density calculations, the values are compatible with either system. As pointed out by Smith et al. (1986), if the satellites are related to a single equilibrium assemblage, they would seem to be closer to material rich in carbon monoxide but poor in water, a composition that also would affect the composition of Uranus. Is it possible that the nature of the satellites is related to a large impact that created the large obliquity of Uranus? Such an impact may have thrown out shock material poor in water.

The bulk density of the uranian satellites may imply a larger rock fraction which in turn implies that radiogenic heating may have had a larger role than in the saturnian bodies. A lack of ammonia hydrate would make the materials more difficult to melt. That such melting may not have occurred is suggested by the absence of extensive geologically-recent resurfacing on Umbriel, Titania, and Oberon, perhaps because of the low abundance of ammonia and methane. Titania, the densest of the satellites, however, does display evidence of resurfacing comparable with that on Dione in the saturnian system. Resurfacing on Ariel and Miranda is quite extensive, indicating a greater abundance of low melting-point ammonia. An alternative explanation for the resurfacing is that these satellites were subjected to other sources of heat that triggered this activity. Their relatively large eccentricities and Miranda's inclination may be the consequence of complex orbital evolution, a change that involved a large amount of tidal heating.

The origin of the extensive dark material in the outer Solar System has yet to be resolved. Possibly it represents radiation-darkened methane in the clathate form, implying that satellite composition may be far from an equilibrium mixture of silicate and ice; or it is mainly primordial dark material, an origin more in accord with the density distribution, but also implying a deficiency in ammonia hydrate that would enhance geological activity (Smith et al. 1986).

9.5 Neptune and Satellites

9.5.1 Planetary Setting

Like Uranus, Neptune has a rocky core, a lower mantle of water, methane, and ammonia ices, and an upper mantle of liquid molecular hydrogen. It has a diameter of 50 538 km, a rotation period of 17.8 h, and a density of 1.64 g/cm^3 (Hartmann 1983, p. 294; Moore 1988, pp. 46, 52; Stone and Milner 1989). Its orbital period is 165 Earth years. According to Ness et al. (1989), the planetary magnetic field is due to a dipole that is inclined 47° with respect to the rotation axis, and the magnetic field intensity ranges from < 0.1 gauss in the northern hemisphere to > 1.0 gauss in the southern hemisphere. The large offsets at both Neptune and Uranus suggest that dynamo generation is in a fluid and convecting "ice mantle" composed of water, ammonia, and methane, and not in the core. As with the other gaseous planets, hydrogen is the dominant atmospheric constituent, followed by helium. Methane, which is more abundant in Neptune's upper atmosphere than at Uranus, helps give this planet a blue color, from which the name Neptune (Roman god of the sea) was given. Also detected in the atmosphere are acetylene and ammonia. This dense atmosphere with its dark spots including the Earth-size "Great Dark Spot", obscures the surface of the planet. Orbiting Neptune are four prograde, equatorial, circular rings at distances ranging from 41 900 (1989N3R) to 62 900 km (1989N1R). The outermost ring includes three optically thicker arcs 4°, 4°, and 10° ring longitude with azimuthal lengths of features responsible for the ring occultation events seen from Earth. In the course of analyses of the rings, Porco (1991) discovered five additional arcs on the outer ring at Longs. 285°, 277°, 265°, 263°, and 252°W. These ring arcs are believed to be radial distortions with an amplitude of about 30 km traveling through the arcs, a perturbation due to satellite Galatea at a distance of 62 000 km from Neptune.

9.5.2 Satellites

9.5.2.1 General

All six satellites newly discovered during the Voyager missions orbit Neptune in prograde circular orbits of low inclination, with five of them orbiting within 1° of the planet's equatorial plane and the sixth having an inclination of 5°. In contrast, the two larger previously known satellites, Triton and Nereid (the outermost one), have inclinations of 157° and 29°, respectively. Satellites 1989N6 to 1989N1 are at distances of 48, 50, 52.5, 62, 73.6 and 117.6 × 10^3 km from Neptune, have periods of 7.1, 7.5, 8.0, 10.3, 13.3, and 26.9 h, have diameters of 54, 80, 180, 150, 190, and 400 km, respectively, and they orbit probably in synchronous rotation.

The only recently discovered satellites that were imaged by Voyager with sufficient resolution to define their surface morphologies were the irregular-shaped 1989N1 and 1989N2. The 400-km slightly elongated N1 has a topographic feature with a relief of about 20 km on its limb. It also displays craters, with one near the terminator having a diameter of 150 km (Smith et al. 1989). The two larger satellites that can be seen from Earth are Triton (named from a pre-Greek merman) and Nereid (named for primitive mermaids). Triton orbits in a synchronous rotation, is at a distance of 354 800 km from the planet, has an orbital period of 141 h and has a diameter of 2705 $\pm$ 6 km (Table 1). Its large inclination to Neptune's equatorial plane suggests that Triton is a regular satellite formed around the planet and was later disturbed by a passing planetesimal, or it was formed in the nebula and later captured by Neptune (Lunine et al. 1989, and references therein). The rotation of the highly eccentric Nereid, at a distance of 5 513 400 km and with a diameter of 340 $\pm$ 50 km, has yet to be determined. The low-resolution Voyager images of this satellite provide only the albedo of the body and yield no information about its surface morphology (Stone and Miner 1989).

9.5.2.2 Triton

Triton with a surface temperature of 38 $\pm$ 4 K has a density of 2.08 g/cm^3, much more than those of the saturnian and uranian satellites, but similar to that of Pluto (Table 1; Smith et al. 1989). Evidently its silicate fraction is higher than that of the large icy satellites of Jupiter, Saturn, and Uranus, but similar to that of the Pluto/Charon system. Triton is large enough that radiogenic and accretional heat may have led to differentiation to form a silicate core, approximately 1000 m in radius, and an ice-I mantle approximately 350 km thick. As the interior cooled, a layer of ice-II may have developed. According to Smith et al., water ice is the primary component of the near-surface crustal material and it is overlain by thin coatings of nitrogen- and methane-ices and their derivatives. Triton's reddish color is due to polymerization of the methane at the surface and in the atmosphere by cosmic rays, ultraviolet photolysis, charged particles, or all three. The albedo patterns on Triton consist of two units: a brighter polar unit blanketing most of the southern hemisphere, and a somewhat darker and redder plains unit extending roughly northward from the equator (Fig. 145). As these albedo patterns do not correlate with topographic features, they must represent a cover on the terrain units (Smith et al. 1989). The brighter polar unit is ascribed to the presence of bright N_2 and CH_4 frost, but the northern hemisphere appears much darker because these bright elements are concentrated in lows so small as to be beyond the resolution of Voyager, giving it a darker appearance (Yelle 1992).

Thompson and Sagan (1990) used cluster analysis to identify six major spectral units on the surface of Triton. Color/albedo classification is based upon five narrow-angle images of Triton at a phase angle of 39° filtered in the ultraviolet, violet, blue, green, and orange ranges. These units differ mostly in the quantity of admixed chromophore or the overlying bright frost or both. Class I in the polar cap and equatorial band is due to redistribution of old chromophores, classes II (thick frost at the cap edge) and IV (frost at the leading edge) constitute transient seasonal frost; class III (equatorial region) is suggestive of the presence of methane or other

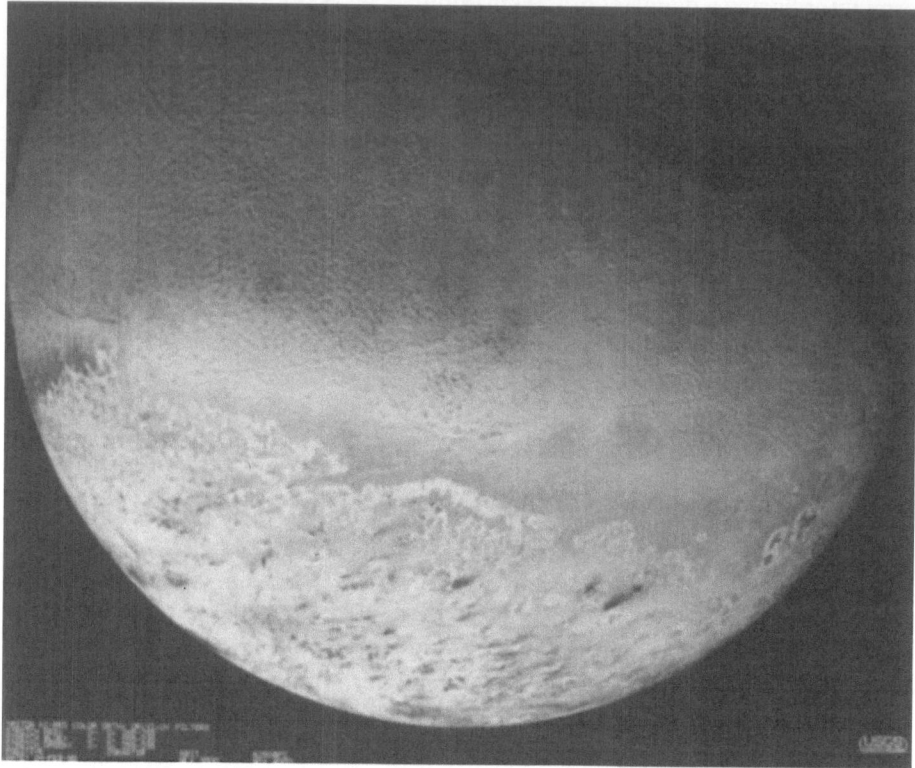

Fig. 145. Photograph of mosaic of the largest neptunian satellite Triton showing cantaloupe terrain and the south polar cap. Photograph provided by the US Geological Survey and courtesy of Voyager Experiment Team Leader B. A. Smith and NASA

hydrocarbons with similar infrared spectral properties at the surface, and class V (polar cap bright area, orange-absorbing units) and VI (polar cap bright area, ultraviolet-absorbing units) have colors which may be due to fine-grained ammonia frost which is subject to more rapid processing. Associated with these albedo classes are dark features having smooth-rimmed irregular margins and surrounding aureoles. A thermal inversion may occur in the lower 5 km of the atmosphere, and a tropopause has been postulated at an elevation of 25 to 50 km. The clouds observed in the tropopause may be caused by condensation or surface eruptions. The atmosphere is dominated by nitrogen, although methane is present in the lower atmosphere, and the surface pressure is 16 ± 3 μbar. A cover of seasonal ice (probably nitrogen) blankets the polar regions south of Lat. 15°S. This ice has a slight reddish tint due to the presence of organic compounds produced from methane and nitrogen by photolysis. A thin layer of what may be nitrogen frost, and is faintly blue. also blankets most of the equatorial region (Stone and Miner 1989).

The less than 40% of the surface of Triton that was imaged by Voyager 2 appears to be relatively young, and the heavily cratered terrain dating back to the early postaccretional bombardment is absent (Fig. 146). Impact structures having diameters to about 12 km display sharp rims and bowl-shaped interiors, and larger ones, ranging up to 27 km, are complex with flat floors and central peaks. The transition from simple to complex craters relative to size is the same as observed on other satellites of the outer Solar System having similar compositions. No ejecta was discerned (possibly because of the low resolution), and the craters also lack rays. Smith et al. (1989) ascribed this absence to masking veneers of mobile surface volatiles. Comparison of crater statistics from the most heavily cratered terrain in the leading hemisphere, the least cratered terrain, and the ice cap of Triton with the lunar highlands, lunar maria, and Miranda's rolling cratered plains indicates that the most cratered terrain in Triton is similar to that of the lunar mare. The decrease in flux between the most heavily cratered and least cratered terrains must reflect a difference in absolute age, even after taking into account Triton's retrograde rotation which would lead to preferential cratering on the leading hemisphere.

Smith et al. (1989) and Strom et al. (1990a) speculated that bombardment by comets during the last 3.5 Ga probably accounts for the craters on Triton and for the origin of most small satellites by collisional fragmentation. The sharp craters on the 'canteloupe' terrain may not be the result of impact, but may represent volcanic explosion structures. Extrapolation from the observed Jupiter-family short-period comets to Neptune suggests that the scarcely cratered terrain on Triton could have been generated during the past Ga or so. These conservative estimates also suggest that the smaller satellites 1989N5 and 1989N6 could not have survived intact

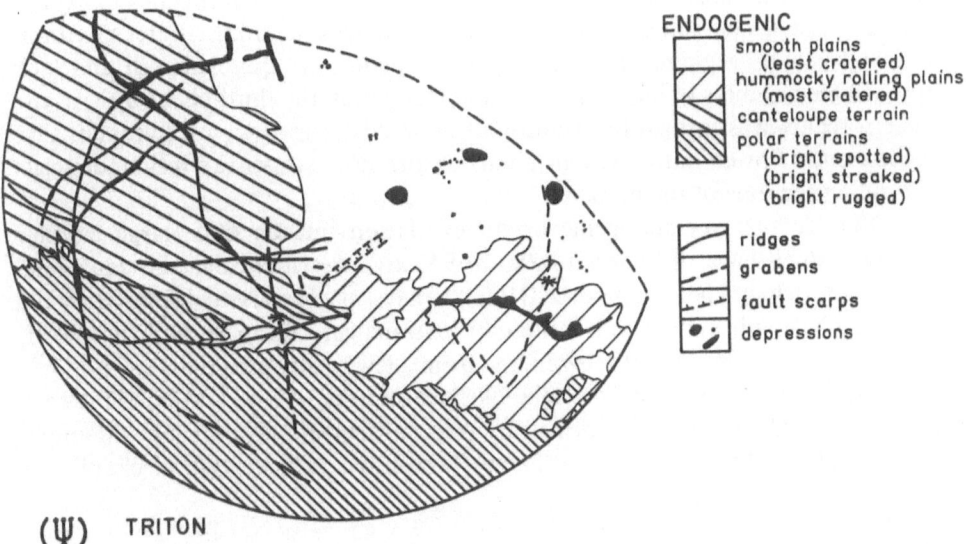

Fig. 146. Endogenic morphology of the neptunian satellite, Triton (After Smith et al. 1989, fig. 31)

during the past 3.5 Ga, and probably they represent fragments of a larger body that was disrupted during the past 2 Ga. Although 1989N2, 1989N3, and 1989N4 may have escaped destruction by the recent comet bombardment, they would have been destroyed during the early heavy bombardment. Thus these bodies probably are destruction products of a satellite the size of 1989N1 and were formed near the end of the heavy bombardment (Smith et al. 1989)

Morphological features of endogenic origin on Triton (Fig. 146) are an extensive terrain on the western part of the equatorial region termed the "cantaloupe" terrain, smooth plains, and hummocky plains. The cantaloupe terrain is characterized by pits and dimples criss-crossed by ridges of viscous materials erupted into grabens of global scale that extend onto other terrains farther east and south (Smith et al. 1989). Most of the dimples, which belong to two size classes of 5- and 25-km diameter, are on the western part of the terrain and are organized into linear, equally spaced units. Smith et al. ascribed this morphology to a combination of viscous flow, collapse, and deterioration of the primary features by sublimation of surface materials.

The smooth eastern plains contain three units. One unit, described by Smith et al. (1989) as "lake-like", is several hundred kilometers across and is near the terminator. The surface of this unit is flat, and it embays scarp-rimmed margins and onlaps hills and knobs that protrude through the plains. The floors of the unit are terraced, suggesting the presence of several episodes of emplacement. Each terrace has a cluster of irregular pits surrounding a central pit, probably an eruptive vent. The presence and survival of an impact crater within the "lake-like" plains unit indicates that the unit consists of rigid material. Another plain, a high-standing plain, is south of "lake-like" depressions having floors at two levels and appears to be the result of an eruption from large circular depressions. These plains are comparable with those on Ariel that are believed to have been formed by extrusion of highly viscous materials, probably partly crystallized ammonia-water mixtures. The high standing plains, like the "lake-like" plains, are quite smooth and lack appreciable numbers of impacts. The hummocky unit, the third plains unit, in the eastern region was formed by eruption of material along sections of grabens that subsequently flowed onto adjacent plains to form the hummocky rolling deposits and filling sections of the grabens.

Other features of endogenic origin on Triton include three large circular features, which have diameters of 280 to 935 km, are situated in the equatorial region, and which Helfenstein et al. (1992) believe may be cryovolcanic in origin; and a global fault system that is mainly tensional in origin. Smith et al. (1989) reported strike faults with offsets ranging from a few to 30 km in the transition belt between the cantaloupe and the "lake-like" features. A 100- to 120-km depression also is believed to have been formed by two transcurrent motions having amplitudes of about 30 km. When the feature is restored it has a typical cantaloupe depression morphology.

During Voyager 2 imaging of Triton two active geyser-like plumes were identified and several others were suspected to be present. All are within the complex central zone in the south polar region. The western plume is at Lat. 50°S,

Long. 334°E with emission angles that range from 37° to 75°. The plume, a tall dark stem, rose from a dark surface and terminated in a small dark dense cloud from which a more diffuse cloud rose in the form of a narrow 5 km band extended 150 km westward. The eastern plume at Lat. 57°S, Long. 12°E, with emission angles of 53°, 72°, 76°, and 77°, rose 8 km and terminated in a dense black cloud that diverged as it moved westward (Smith et al. 1989). Stratification of the atmosphere probably controlled the shape of the plume, with the 8-km altitude reflecting an inversion at the tropopause. Its westward displacement also suggests that wind speed increases abruptly at this elevation. The dark streaks on the south polar region are related to these active plumes. A mechanism to drive the plume, described by Smith et al. (1989), involves venting of the gas from the surface, entraining of fine particles that are carried upward by a combination of ballistic and buoyant forces, and downstream transport by sublimation winds. The driving gas probably is nitrogen, or possibly concentrated methane. Other energy sources include insolation and geothermal heat. That plumes occur near the present subsolar latitude argues for a solar-driving mechanism (Smith et al. 1989; Brown et al. 1990). In their model they proposed that a dark layer beneath a clear nitrogen layer on the surface of Triton absorbs radiation until the thermal gradient reaches a point where excess heat is conducted and radiated back. With increasing temperature the vapor pressure on the surface nitrogen ice increases. As this pressure increases it fills permeable subsurface reservoirs, and if the seal becomes ruptured by pressure or the gas migrates laterally to an open vent, the gas decompresses rapidly and launches a plume of nitrogen gas and ice. This plume entrains dark particles encountered in the exit area and carries them upward to elevations of 8 km, where they are dispersed globally by winds. In the model described by Kirk et al. (1990), plume lifetimes of several years can be explained as the result of thermalized solar energy and nitrogen gas flowing laterally from surrounding areas. Ingersoll and Tryka (1990) proposed that the plumes are an atmospheric rather than a surface phenomenon. Like the dust devils on Earth they are atmospheric vortices originating in the unstable atmospheric layer close to the ground. This instability is the result of patches of unfrosted ground, near the subpolar point, that became heated relative to the surrounding frost. Convection currents produced in this manner could then raise particles from the plume region into the atmosphere, provided that they are less cohesive, by a factor of 10^3 to 10^4, than those on Earth.

Soderblom et al. (1990) estimated that the mass flux in the trailing clouds accumulates as much as 10 kg of fine dark particles per second, about 5 kg of nitrogen ice per second, and several hundred kilograms of nitrogen gas per second. They calculated that each eruption may last a year or more during which as much as 0.1 km³ of ice become sublimated. It is these plumes that are the source of the long (10 to about 100 km) streaks scattered over the interior of the south polar ice cap. These streaks, which surficially resemble the martian wind streaks, probably are only a few thousand years old, because if they were older they would have migrated deep into the ice deposits by now. In addition to this material from active plumes being carried downwind, other eolian exogenic features imaged on Triton include wind streaks, caused by other injections of material into the atmosphere and subsequent settling or saltation of the sediment grains (Hansen et al. 1990). These streaks are most abundant from Lats. 15° to 45°S with orientations of 40° and 80° measured clockwise from the north.

9.6 Pluto and Charon

Pluto, the outermost planet of the Solar System with a diameter of 2300 km and a density of 2.1 g/cm^3, is the smallest planet in the Solar System, being about the size of our Moon (Fig. 147). Its companion, Charon, has a diameter of 1200 km, about one-ninth the volume of Pluto (Lunine et al. 1989; Stern, 1993). Both Pluto and Charon have rotation periods of 6.4 days, and a 248-year orbit; they appear to be locked in synchronous revolution/rotation (Burns 1986a; Cruikshank and Brown 1986). Burns (1986b) proposed that Pluto/Charon, like Triton, are leftover planetesimals, but, whereas Triton was captured by gas drag into orbit around Neptune, Pluto and Charon remained in solar orbit as their 3:2 resonance prevented them from approaching Neptune. Some of these planetesimals, which Weidenschilling (1991) called plutons and which were within 20 AU of the Sun were ejected to the Kuiper Disk (orbiting beyond 35 AU) and the Oort Cloud (at 10^4 AU), and others were forced into the inner Solar System (Stern 1991) where they may have had a role in the evolution of the rocky planets. Stern also proposed that the tilts of Uranus and Neptune may be the result of collision with these planetesimals.

Infrared measurements suggest that the surface of Pluto consists of methane, nitrogen, and CO and is nonuniformly distributed across the surface of the planet. Infrared signatures indicate that methane is absent from the surface of Charon, but water ice may be present (Marcialis et al., 1987; Stern, 1993). Whereas Charon appears to have a uniform albedo, on Pluto the albedo ranges from a very bright zone in the south polar region to a dark zone at mid-southern latitudes, and from a bright band at mid-northern latitudes to a dark band at high northern latitudes (Lunine et al. 1989; Binzel and Young 1992). According to Binzel and Young, this albedo distribution is the result of Pluto's south pole receiving less insolation than the north pole over time spans that are much longer than a single orbit. Pluto's

Fig. 147. Artist's view of Pluto (foreground), Charon and the Sun (background). Photograph 83-HC-219. Courtesy of NASA

reddish color in the photovisual spectral region suggests that methane is being converted to higher hydrocarbons by sunlight or charged particles. This color probably inspired the name Pluto (the Greek god in charge of Hades) and Charon (the ferryman who carried, for a fee, the dead across the river Styx to Hades). Fink et al. (1980) reported that the planet has a methane atmosphere with a surface pressure of 0.15 mbar. Adiabatic expansion of this escaping atmosphere could create a cold trap which would tend to reduce the escape rate to values consistent with the presence of methane-ice on the surface (Hunten and Watson 1982). To date no information is available on the morphology of the Pluto/Charon double planet system.

10 Summary and Conclusions

10.1 Introductory Statement

Intellectually, this grand cruise across the Solar System has been one of the most stimulating voyages we have taken. One could not have asked for more informed guides than the geologists who have been involved in the many National Aeronautical and Space Administration projects on the Solar System. Their numerous reports have provided us with concise views rather than a maze of unrelated images of that part of the cosmos which we inhabit. It is obvious from their descriptions that the cosmos is not only well organized, but also it is beautiful. In the early eighteenth century the *Dictionnaire de l'Academie Française* defined the literary form 'romance' as a light verse that tells an ancient tale. This short book is our version of a romance of the Solar System. It is a tale that roams within the vastness of the cosmos with our gaze concentrating on one aspect of the cosmos, then turning abruptly to another phenomenon. Our discourse seems to show a cosmos that is accumulating tension, leading to some unknown climax as it expands ever outward since the Big Bang. Sometimes it is difficult to avoid the feeling that our study of the Solar System and its place in the Universe is rather superficial as though an understanding of this immense canvas is beyond our comprehension. To rephrase the words that Fuentes (1992, p. 53) used to describe the great mosque at Cordoba, the cosmos is a vision of the infinite where the creation searches for the creator to complete the task. We believe that this task is to make the unknown known. It is an image of the cosmos where different events are occurring at different places at the same time, and yet we believe that all these alterations form part of the same event. This tale of ours must not be looked upon as the ultimate explanation of the Solar System. On reading this essay errors will be found, errors of omission and/or commission, and there will be some readers who will strongly disagree with what we say. So be it. We hope that those who read this book will find reading it as enjoyable as we found writing it, and we hope that we have given proper credit to all.

The tale that we present here is a tale of what is past, passing, or yet to come, like that which the poet sang to the lords and ladies of Byzantium in the *Sailing of Byzantium* by W. B. Yeats. It is akin to a Mahler symphony of images of a vast rotating nebula that collapsed to form the Sun, a condensation and accretion to form the planets instead of a second sun, a solar wind that cleansed the Solar System of its excess dust, transporting it to the outer fringes, planets that underwent massive bombardments early in their histories, capture and loss of satellites, a bombardment that led to the creation of the Moon, the hemispheric crustal dichotomy of Mars, removal of the mercurian lithosphere, the tilts of Uranus and Neptune, degassing of the planets to form their present atmospheres and hydrospheres (many of which later were partly or completely lost), and creation of a biosphere that gave Earth its unique ambience. It is a tale constructed from a chronology known only on the Moon, and from the supposition that processes that produced some diagnostic features on Earth also produced similar ones on other planets, a concept that may be not only bad science but also presumptuous on the part of Earth-dwellers.

Following true geological procedure, we believe that our summary of the evolution of the history of the various members of the Solar System includes many aspects that can be applied to increased understanding of our Earth to ascertain parts of its history that are not accessible on Earth itself. This belief follows from recognition that the Earth is an integral member of the Solar System so that events or processes that importantly affected several members probably also affected other members as well. In a sense, the agents that shaped other rocky members of our Solar System may have shaped the Earth during the ancient past, but the products of their work on Earth have largely been concealed by subsequent tectonism, erosion, or deposition on this – the most active of the rocky members of our extended family because of the presence on Earth of an unique associated atmosphere, hydrosphere, and biosphere.

10.2 Properties Inherited from the Origin of the Solar System

After attempting to summarize the current knowledge about the composition and structure of the various individual members of the Solar System and the relationships of these members to our galaxy and to the Universe, we think it appropriate to attempt to consider in this concluding chapter the similarities and differences of these properties in the various bodies relative to events in approximate chronological sequence. The prime objective of this scientific recapitulation is to compare the morphologies of the Earth and the other rocky planets, so these concluding remarks are largely restricted to the planets and Moon inboard of the Asteroid Belt. Most of our information regarding planetary properties comes from morphology–the study of the shapes of the surfaces of the rocky members in terms of irregularities in relief, both broad-scale and local in extent. Some of the morphology of the planets, moons, and other smaller bodies is related to subsurface crustal processes (from mantle convection) that are grouped as endogenic forms, whereas other morphologies result from action of surface agents (wind, water, and ice) and are grouped as exogenic forms. A third kind of morphology, consisting of both large- and small-scale effects of bombardment by debris from within and beyond the Solar System is termed exotic. Variations in distribution of these three groups of surface morphology upon a given planet or satellite and their differences between the bodies provide clues as to the similarities and differences in events and histories of the various rocky members of the Solar System. Information obtained from a study of morphology is strongly supplemented by other information about mineral and chemical compositions obtained by remote measurements from the Earth and from satellites, especially from landers, and most valuably from astronauts who to date have landed only on our Moon.

Knowledge about the origin and composition of the Earth and its associated rocky members in the Solar System and of our galaxy has evolved from earlier beliefs in animistic and divine interventions, sometimes coupled with reaction to perceived immoral acts by humans. Modern observations provide much new information, the accuracy and extent of which are highly indebted to the transfer and use of instrumentation and technology from other fields of science to astronomy, and particularly to planetology. The recent development of the scientific approach to planetology is exemplified by the long and vigorous controversy, which ended only a few decades ago about whether the craters on the Moon were produced by endogenic (volcanic) or exotic (meteorite impact) processes. As the flow of information continues to increase during the near future we expect that some of the conclusions drawn herein will become obsolete. It is this questioning and thinking by humans that ultimately may lead to unravelling the origin of the Universe and our place in this vast cosmos. The fact that religious beliefs for the origin of the Solar System remained unchallenged for millennia indicates more the failure to make new observations and to think about their meaning than it supports the accuracy of early religious beliefs. In fact, the rapid changes in scientific interpretations during recent decades reflect the advent of new data and not an irresponsible change of unbacked opinion. In contrast, if we had learned nothing new, there would be no reason to change our ideas about the distant past and of the future. Is it not intellectually worthwhile to investigate why we are on the only member of the Solar System habitable to us? If it is worthwhile, we should consider whether the reasons why this particular planet is habitable might be applied to determine whether similar habitability may be present on planets that orbit other stars of our galaxy or stars of more distant galaxies in the Universe. Although many traditional religious explanations for natural phenomena are implausible, a scientific explanation for the origin of the Solar System is not an argument against belief in a Creator. The scientific approach has limits. Although science can determine the physical attributes and evolution of an object, its essence is beyond scientific comprehension; a few lines by the poet Robert Burns have done a better job of capturing the nature of a field mouse than all the learned scientific tomes that have and ever will be written.

The oldest date that is significant for the history of the Solar System and its member Earth is that of the Big Bang 10 to 20 Ga ago, although many questions remain about conditions and even time prior to the Big Bang. Some writers even question this concept of a solitary big bang, believing that it reflects a search by some cosmologists for a Creator and for a beginning. They propose instead that creation is continuous and is occurring in a series of little big bangs. In the single big bang concept an estimated 10^{11} glaxies began to aggregate 1 Ga after the bang from singularities (irregularities) in the distribution of atoms, molecules, and dust resulting from the explosion. These galaxies are distributed throughout an enormous volume of space, with the ones most distant from the center of the big bang traveling outward at the greatest speed. Attempts to improve the precision of the date of the big bang have led to the building of larger telescopes and ones with reduced disturbance from motions of Earth's atmosphere (which produce the interfering twinkle of stars). Thus, one of the objectives of the Hubble telescope mounted on a satellite was to see farther into space and to measure the largest red shift (Doppler effect) and the intensity of light from stars in the most distant galaxies to determine the speed, distance, and age of the farthest galaxies, and thereby to learn more precisely the date of the big bang and perhaps whether there is an alternative to the Big Bang. Galaxies have various shapes, but ours appears to be fairly typical – a spiral with arms that rotate around the nucleus. Within this galaxy are about 10^{11} stars, of which our Sun is a younger, smaller, and cooler member that is located near the side of one of the galaxy arms (where new stars are born). Comparison with other stars in our galaxy indicates that the Sun has a life expectancy of about 10 Ga, about twice its present age. Stars ten times the size of the Sun have a life expectancy of only a few Ma before they explode into supernovas. Thus, we are fortunate that our Sun started at a size that allowed it to last long enough to evolve a complex planetary system whose members exhibit a wide range of properties including those shown by the Earth.

The Solar System may have begun as a binary pair of stars. The companion may have been larger, hotter, and thus shorter lived. When it exploded as a supernova, much of its material escaped outward, but our proto-sun may have been able to capture some of its debris to increase its percentage of heavier elements and to provide substance for its family of planets and associated smaller bodies. This explosion and the capture of some material by the proto-sun occurred about 4.6 Ga ago. Soon thereafter some of the captured material accumulated in the planets, but most of it was blown away by solar (stellar) radiation pressure – the solar wind. Alternately, the proto-sun was alone and attracted nearby interstellar material, possibly produced by an unrelated supernova. Another view is that both the Sun and planets formed from a single vast rotating cloud. Central condensation and gravitational contraction led to formation of the Sun and development of a cocoon nebula. Dissipation of the cocoon nebula by the solar wind eventually exposed the proto-sun, allowing it to change character and reach the T-Tauri stage when it increased its rotation rate, magnetic-field strength, and changed into a flat disk-like mass. The loss of gas to the solar wind and effects of the magnetic field slowed the proto-sun's rotation to its present rate. The finely dispersed material within the rotating disk-shaped solar nebula in approximate equilibrium (inward gravity forces balanced by gas pressure and outward-directed centrifugal forces) gradually accumulated into planets that continued to revolve around the Sun or into another star to form a binary system.

The solar wind had an additional effect that was responsible for much of the compositional differences between planets. Temperatures within the original nebular cloud decreased markedly outward from about 2000 K near the proto-sun to less than 200 K near the outer limit of the cloud. This temperature gradient allowed the solar wind to preferentially remove uncondensed gases but permitted condensed materials to remain in the proto-planets and become part of the final planets. As a result, the planets have compositions that are related to their relative distances from the Sun. Thus, Mercury is refractory-metal rich, Venus and Earth have lower density silica-rich compositions, Mars has still lower density oxidized-iron compounds, and the outer planets largely consist of gases and their moons are rich in ices of water, ammonia, and methane. These compositional differences between the planets produced differences in their densities, gravities, and thicknesses of cores, mantles, and lithospheres – and thus in their surface morphologies.

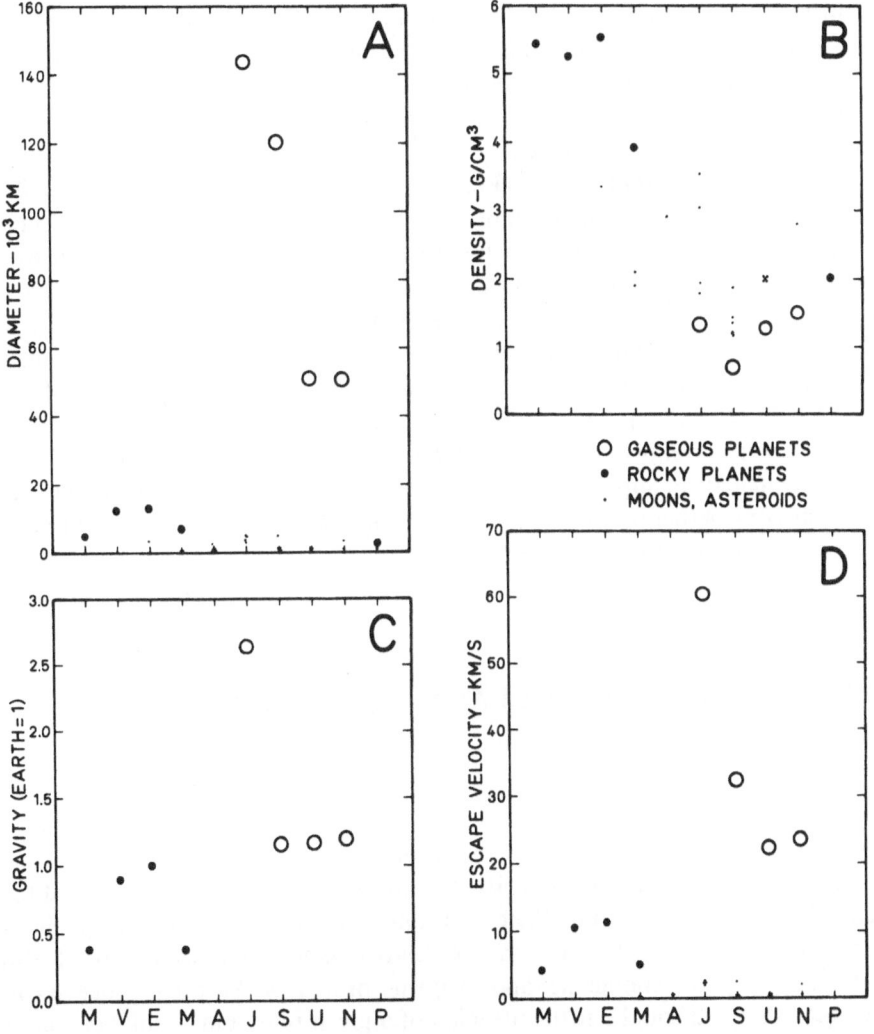

Fig. 148. Relationships of diameters (**A**), densities (**B**), gravities (**C**), and escape velocities (**D**) among planets, moons, and asteroids of the Solar System. Data from Table 1

In the present Solar System, planets and other smaller bodies revolve around the Sun along orbits that lie near the plane through the Sun's equator, the ecliptic. Most rotate in the same direction as their movement in orbit. In general, the larger the body the more circular the orbit. The inner planets are smaller, denser, and rotate more slowly than the outer planets. Planet-moon systems are analogous to that of the Sun-planet system. Many moons nearest their host planets have identical periods of revolution and rotation because of the gravitational control of nonspherical moons by the planets. Some moons have axes of rotation that are far from perpendicular to their orbits of revolution; these probably are asteroids or other bodies captured from other planets. Several planets (Venus, Uranus, and Pluto) even have retrograde rotation and peculiarities that imply capture or alteration by the impact of large bodies early in their histories. Such impacts produced some moons (the Earth's Moon), as indicated by compositions and movements which differ from those expected of moons obtained by capture.

Some of the similarities and differences between planets and moons are listed in Tables 1 and 2, but they are more easily visualized in the plots of Figure 148. These plots clearly show the presence of three groups: large (> 50 000 km) gaseous planets, moderate (13 000 to 6000 km) rocky planets, and the small (< 6000 km) planets Mercury and Pluto, plus moons and asteroids. These groups are effectively distinguished from each other on the basis of size, density, gravity, and escape velocity. Note that the moons and asteroids are more similar to each other than are the planets. The small diameters of some planets are partly compensated by their denser compositions and slower rotations so that some smaller bodies (Venus and Earth) have nearly as much gravity as have some much larger bodies (Saturn, Uranus, and Neptune). This gravitational strength in turn partly controls the escape velocity of gases and solid materials that may have been lifted above the surface by meteorite impacts or by volcanic eruptions. Thus, the escape velocity in turn controls whether or not the planet grows larger and whether it might retain an atmosphere and/or a hydrosphere.

10.3 Morphologies Developed After Separation of the Planets

10.3.1 Exotic

Once the Solar System, with its planets, moons, and smaller bodies, was established the individual bodies began to evolve and to differ from each other according to their distance from the Sun, their size, and their composition. These changes eventually produced differences in the relative kinds of morphology – features produced by exotic, endogenic, and exogenic processes. All three processes may have been active throughout the life span of a given body but usually exotic ones were more active during the earlier stages and exogenic ones predominated during the later stages of the history of the body.

. Exotic morphology represents a continuation of the processes that produced the planets and moons rather than of processes that began after initiation of the planets or moons. There can be no sharp boundary between the period of original accumulation of materials that formed planets and moons and the gradually diminishing accumulation of debris that temporarily remained in transit. However, severe impacts by large bodies on planets, moons, and other bodies produced episodic increases in interplanetary debris, and thus episodic changes in rates of production of exotic morphology on other planets and moons. Similarly, capture of moons by planets was episodic and could have occurred long after the large planets and smaller moons-to-be were formed. This record of episodicity of exotic morphology is well represented by the chronology first established on the Moon and which then was extended to Mercury, Venus, Mars, and the satellites of the gaseous planets. The characteristics of exotic terrain on any given body were controlled by the presence and density of an atmosphere and the composition of the impacted material. Gravitational focusing by a nearby planet also led to extensive development of exotic features on satellites. Records of chronology were not retained during the earliest history of once-molten planets, on planets or satellites that still are molten, or on planets or satellites whose surfaces still are concealed by opaque atmospheres. Also, the ancient record of exotic morphology has partly been lost where surfaces have been renewed extensively by plate tectonics, as on Earth, or by widespread volcanism, as on Venus and other bodies. As a result, the total area that consists of exotic morphology on the rocky planets ranges from less than 1% (Earth and Venus) to at least 84% (Moon) – an average of 14%. In general, the larger bodies can be expected to have lower percentages of exotic morphology because of obliteration by later endogenic or exogenic activities.

10.3.2 Endogenic

Endogenic morphology includes mid-ocean ridges, subduction complexes, volcanoes, lava fields, folded and faulted mountains, plains, plateaus and broad uplifts, broad basins, fault blocks, and long strike-slip and transform faults. Where these features are present, subsurface structures and compositions must be involved. The nature, distribution, and abundance of morphological features differ on various rocky members of the Solar System, so we can be sure that subsurface structures also must differ. Thus many inferences about morphology/structure/composition may be checked by independent geophysical measurements of gravity, magnetic fields, heat flow, seismicity, and by seismic surveys. Some of these measurements can be made from unmanned satellites or landers, but most of them require human involvement and more or less elaborate equipment on the ground. Thus, most information by far is that which has been gained about Earth during centuries of studies. Similar knowledge about other large rocky bodies is scant but is led by results for the Moon, is least for Mercury, and is impeded on Venus (Table 9) because of its very hot and inhospitable surface conditions. Subsurface structural and compositional knowledge of the moons of gaseous planets is low because of poor accessibility resulting from their great distance from Earth.

Table 9. Depth (km) to bottom of each concentric shell of planets (From text discussions)

Shell	Earth	Moon	Mercury	Venus	Mars
Crust	10–40	48–74	0	10–80	8–77
Lithosphere	70	1000	800	300	1900
Asthenosphere	150	1600–1800			
Upper mantle	400				
Transition	1000				
Lower mantle	2900			3300	(1400–1900)
Outer core	5080				
Inner core	6378	1738	2439	6052	3397

The internal structure of the Earth probably is more complex than that of other planets and the Moon (Table 9). However, this concept of greater complexity is biased by the greater accessibility and enormously longer time that has been devoted to measurement and study of the interior of the Earth than of other bodies. As shown in Table 4, the dominant morphology of the Earth is endogenic, but many of the endogenic features of the Earth differ in origin from those of other bodies in the Solar System. These unique features are ones that developed because of movements of lithospheric plates – plate tectonics. Cellular movements within the mantle result from internal radiogenic heating and heat flow from the core, and the tops of these cells within the mantle are able to rift and drag the overlying lithosphere with them. These lithospheric movements (divergence, convergence, and translation) enhance extensive durable convection currents on Earth. Divergent movements rift continental plates and continue as spreading belts within oceanic plates. The lateral separation of lithospheric plates provides avenues for magma from the asthenosphere and upper mantle to rise to the surface and form mid-ocean ridges. This global network of mid-ocean ridges is not contiguous but is segmented by transform faults, overlapping spreading belts, microplates, bends in the ridge's rift valley, gaps between volcanic chains, and changes in axial linearity. Each ridge segment is dominated by axial magma chambers with magma outflow occurring both in the direction of sea-floor spreading and parallel to the ridge's axis. As the two parts of the rifted original lithosphere (mainly continental) continue to drift apart at a rate of a few centimeters per year, the gap becomes filled with oceanic crust produced by the igneous activity along mid-ocean ridges. Thereby, the belt of active rifting moves laterally away from the continental plateaus, and the ocean floor gains width and slowly subsides and evolves into abyssal hills that eventually become buried beneath sediments of prograding continental margins. Thus, the belt of sea-floor spreading is marked by a long wide oceanic ridge that carries many individual volcanic mountains (including seamounts and guyots) on its back and sides.

Plate convergence occurs where the plates atop the asthenosphere and upper mantle collide with a plate that is relatively stationary: continent-to-oceanic, continent-to-continent, or oceanic-to-oceanic, in order of decreasing frequency. The continent-to-oceanic convergences are the most spectacular, marked by linear deep-ocean trenches bordering a continent (as for western South America), volcanic fronts 200 to 300 km inland from the trench axes, and accretion of allochthonous terrains as in western and eastern North America. Andesite magma rises within the edge of the continent to produce a long line of andesitic volcanoes, as exemplified by the Andes of western South America, because of partial melting of a mantle source, fractional crystallization, contamination from subducted sediments and melting of overlying continental crust, and magma mixing. The cores of these volcanic arc systems are diorite to granite plutons, such as the Sierra Nevada in California. Continent-to-continent convergence marks the end of an ocean after a length of oceanic plate has been completely subducted and the once separate continental plates collide, as for the Appalachian, Ural, Zagros, Alps, and Himalaya mountains. These and many other mountain ranges consist chiefly of older sedimentary and igneous rocks that have partly metamorphosed during the process of convergence and collision and have formed a long series of high folded mountain ranges having a sinuous pattern and containing many overthrust faults. Calc-alkaline granitic plutons are produced during the formation of these collisional mountain fronts. Lastly, oceanic-to-oceanic plate convergence produces long deep-ocean trenches bordered on one side (the convex side) by a belt of volcanic mountains many of which rise above sea level as islands. Examples are the island arcs of the Marianas, Aleutians, the East Indies, the Leeward Islands of the West Indies, and the South Sandwich Islands. Igneous rocks associated with these oceanic convergent belts are basaltic andesites (immature oceanic arcs) and andesites (evolved oceanic arcs).

Pure plate translation is less important in producing morphological forms than are plate divergence and convergence, because few avenues are provided for the escape of magma and movements are lateral rather than opposed. Nevertheless, long straight fault scarps are produced during the lateral movement. Examples are the West African Gulf of Guinea, northern Brazil, southeastern Africa, and the Dead Sea Rift.

Another product that is formed probably mainly by plate movements is widespread lava plateaus from fissure flows, such as those of the Permian in Africa, South America, and Antarctica, the Late Cretaceous and Paleogene Deccan basalts of India, and the Cenozoic Columbia River basalts of the northwestern United States. These widespread lava fields, emplaced by superplumes, resulted from plate configurations that led to the breakup of Gondwana and the convergence of various plates. Additional volcanic endogenic morphology on Earth is formed by magma elevated by mantle plumes over a very long time span from deep-origin sites near the contact of the lower mantle and the outer core. These localities, known as hot spots, form chains of volcanoes whose direction and length provides evidence of the direction and speed of a moving lithospheric plate through which the magma penetrates to the surface. Probably the best known example is the Emperor-Hawaiian Island Chain.

Although the endogenic morphologies produced by plate tectonics are the most abundant ones on Earth, these features are unknown on single-plate planets and the Moon. We must attribute their presence on only the Earth to a thinner lithosphere on Earth than elsewhere, so that it can be rifted, subducted, and penetrated by magma more easily than the much thicker lithospheres of other rocky members of the Solar System. The Earth's lithosphere ranges from less than 25 km thick in oceanic basins to 100 to 300 km beneath the continents, in contrast with 300 to 1900 km for the lithospheres of Venus, Mercury, Moon, and Mars. Moreover, some of the other planets may contain no asthenosphere (or mobile layer) that would allow lateral movement of a lithospheric plate. This stiffening of the lithosphere has led to a style of mantle convection on Venus and Mars that inhibits the breakup of their lithospheres; thus the major features on these two planets are caused by cylindrical mantle plumes.

The Moon's surface is dominated by exotic impact craters with only 16% of it consisting of features attributed to endogenic causes (Table 4). No features related to plate tectonics have been recognized, and volcanoes are rare, possibly consisting only of cinder cones and associated domes. By far the most extensive features of the endogenic area are maria (Figs. 49 and 50), extensive lava plains whose magma sources are believed to be within the upper lithosphere. These flows occurred after most of the impact topography was produced; most of them formed 3.9 to 3.1 Ga ago, although some are as young as 1.0 Ga. Atop the maria are many rilles (Fig. 52) considered to be collapsed lava tubes. There also are wrinkle ridges or folds atop the maria and some of them present evidence of simultaneous lava extrusion, both features caused by the pressure of lava beneath thin crusts. The widespread maria and rarity of volcanoes accord with the very thick lithosphere that probably made it difficult for small magma vents to penetrate. The absence of large long-time-span volcanoes (unlike on Mars) suggests the absence of hot spots, in agreement with the Moon's small core and low heat flow.

Mercury, in contrast with the Moon, has a morphology that is dominated by endogenic features; 76% of the slightly less than half of the total surface area that has been imaged so far consists of endogenic features. Almost all are widespread volcanic plains formed during several episodes of lava extrusion through fissures (Figs. 58 and 60). Perhaps the first and most extensive of these fissure flows resulted from early expansion of the planet caused by settling of iron to form the relatively large diameter core. This widespread flow obliterated most of the impact craters from the original heavy bombardment, and it ceased when cooling of the planet led to compression that closed the conduits. Later impacts of planetesimals left successions of craters atop the various lava plains, providing a chronology that appears to be closely similar to that of the Moon. Many of these impacts melted the basaltic regolithic rocks and produced small impact-lava plains mainly within the craters. Atop the broader intercrater lava plains are patterns of intersecting grid-like lines (tesserae) that are believed to owe their origin to the slowed rotation of the planet and to collapse of an early equatorial bulge. Some lines of this grid are sources for lava extrusion, implying compression akin to that which produced wrinkle ridges on the Moon's maria. No evidence of plate tectonics is evident on Mercury, probably because the lithosphere is too thick (between 150 and 800 km), and if an asthenosphere once was present it has cooled to become part of the lithosphere. In fact, there appears to be no present tectonic activity and no active production of endogenic morphology, so that both Mercury and the Moon are dead bodies orbiting within the Solar System.

In contrast to the Moon and Mercury, Venus may have a limited morphology produced by plate tectonics, and, unlike them, it has only a small area (about 2%) of impact craters (Table 4). The rarity of impact craters probably resulted from repeated resurfacing by igneous activity. This resurfacing has been so efficient that very little terrain older than 1 Ga has survived. As a result of extensive fissure flows and accompanying volcanoes of various shapes and sizes the crust and lithosphere formed early during the hot planet's history and soon became too thick to be rifted and drifted by movement of circulation cells within the mantle, but it could be penetrated by mantle plumes (the chief source of the magma). The lithosphere became thicker and the mantle more viscous than on Earth because of the absence of much water (most carried away by early solar winds and the remainder lost by the extreme greenhouse environment). The largest mantle plumes evidently produced the largest mountain regions on Venus: Aphrodite Terra, Ishtar Terra, and Lakshmi Plateau (Figs. 65–68, 71, 74, and 78). Arching and rupturing of the lithosphere by the largest plumes allowed local lateral transport and broadening of these large volcanic features; this led to the belief by some investigators of local plate-tectonic activity. Smaller mantle plumes uplifted the lithosphere to form coronae, novae, and arachnoids – patterns nearly unique to Venus (Fig. 82). Many differences between Venus and Earth may be the result of the failure of Venus to be struck by a large planetesimal such as the one which carried away much of the Earth to form its Moon and melted much of the Earth that remained.

Mars displays a crustal dichotomy with a heavily cratered upland in the southern hemisphere and a less heavily cratered 3-km-lower lowland in the northern hemisphere (Figs. 90 and 91). Both terrains are in isostatic equilibrium, indicating that they are underlain by different types of crust. This crustal dichotomy is not caused by sea-floor spreading, but is the result of some other endogenic process or the impact by a giant body early in martian history. Mars' present endogenic morphology covers about 19% of the planet's surface, although during its middle history, before becoming broadly buried under eolian sediments, this value was as much as 50%. Just as for the Moon, Mercury, and Venus, Mars' lithosphere has been rigid and thick since early history. Therefore, plate tectonics had no role in forming the endogenic morphology. Thick cylindrical mantle plumes that rose from the lower mantle and penetrated the lithosphere built the huge Tharsis and Elysium volcanic edifaces to heights of as much as 27 km above Mars' datum level (Figs. 90 and 91). If plate movements had been present, the extruded magma would have been distributed along chains of smaller volcanoes, as on Earth, where the lithosphere drifted across long-lived plumes. In addition to the two major volcanic features, there are many smaller volcanoes on Mars that appear to have been inactive for several Ga. The volcanoes are associated with volcanic plains that were produced along and beside extensive fissures caused by the upward pressure of mantle plumes especially in the Tharsis region (Fig. 90). The broadest morphological features on Mars are crustal basins (Fig. 91), parts of which lie more than 3 km below the planetary datum level in Vastitas Borealis of the northern hemisphere and Hellas Planitia of the southern hemisphere. Elsewhere, particularly south of Lat. 20°S, are large regions less than 1 km below the datum. The pattern of the low regions seems to be controlled more by the distribution of major volcanic edifices than by the variable thickness of the lithosphere or by other broad structures in the interior of Mars. The hypsometric curve for Mars (Fig. 5), obtained from the more recent harmonic analysis, has a slight flattening that reveals the presence of these broad basins.

10.3.3 Exogenic

All the rocky members of the Solar System undergo minor erosion through continuous impact by small meteorites that dislodge fragments of rock and leave small pits in the rocky surface. The topographic effect of these impacts is too minor to justify their assignment as exogenic morphology, but the debris that is produced may be transported and redeposited to form thin films to thick layers of sediment, mainly very local in extent. Where the sediment deposit is quite thin, as on the essentially airless and/or waterless bodies (Moon, Mercury, and Venus), it hardly can be considered a morphological feature. These three bodies also have mass movements of slopes made oversteep, as the perimeters of planetesimal impact craters (Fig. 46), the craters and sides of volcanoes, the fronts of lava flows, and the slopes of fault scarps. The resulting landslides and terraces are truly exogenic in origin, but they are so local as to be incapable of being mapped as exogenic morphology on a planet-wide scale.

On Mars and Earth, each of which has air and water, considerable work of erosion, transportation, and redeposition has been, or is being, done by wind, water, and polar ice caps, so that the effects of erosion and deposition are planet-wide and produce widespread and distinctive morphological features that must be considered exogenic in origin. On Mars the presence of ancient glacial and karst terrains (Fig. 103), channels formed by catastrophic floods (Fig. 108), and valleys representing drainage routes from areas where melted permafrost has created amphitheater-shaped spring-sapping heads document a time when the planet was warmer and wetter than at present. No present standing water, as lakes or seas, occurs now, but such may once have existed because the equivalent of an average 10-m thick surface layer of water is believed to have been expressed from the martian crust during the ancient past. In other words, the exogenic-modified endogenic morphology of Mars is transitional between that of the waterless planets and that of the water-rich Earth. Further progress toward Earth-like features of erosion and deposition was arrested by a lack of sufficient water on Mars' surface. On Earth there are abundant features of both erosion and deposition, the typical subjects of geomorphology textbooks. These erosional features formed by water range from rainwash through stream valleys and peneplains to seacliffs. Glacial erosional features include whalebacks, roche moutonne, grooves, rock basins, troughs, drumlins, cirques, and horns. Accompanying depositional features formed by water are alluvial fans, stream-valley fills, deltas, coastal plains, carbonate buildups (reefs, atolls, banks, platforms, and sediment aprons), and terrigenous continental margin deposits on the floors of deep basins. Glacial constructional features include outwash plains, drumlins, kames, eskers, tail ridges, and various types of moraines, some of which were produced by ice thrust. Eolian terrains of depositional and erosional origin are extensive on Earth and Mars (Fig. 112). This process also appears to have had a role in sculpturing the surface of Venus. Most features of ground-water erosion on both Earth and Mars are concealed below the ground surface, except for local spring-sapping and karst topography, and ground-water deposits are mostly thin or local.

10.3.4 Summary

The preceding text indicates that the morphology of each of the different rocky members of the Solar System was formed by the same processes (exotic, endogenic, and exogenic), but they acted at different times for the different bodies. A summary diagram (Fig. 149) is modified from Head and Solomon (1981, fig. 12). About 80% of the present surface of the Moon and Mercury was formed during the first 600 Ma of their origin and the remaining 20% was formed between 4.0 and 3.0 Ga ago. These surfaces have undergone very limited subsequent modification. Most of the martian surface was formed prior to 3.0 Ga ago by a combination of endogenic and exogenic processes, and it also has undergone limited modification since then. The surface of Venus is more recent in age, having been developed by endogenic processes during the past 1 Ga. Endogenic processes may or may not be active on

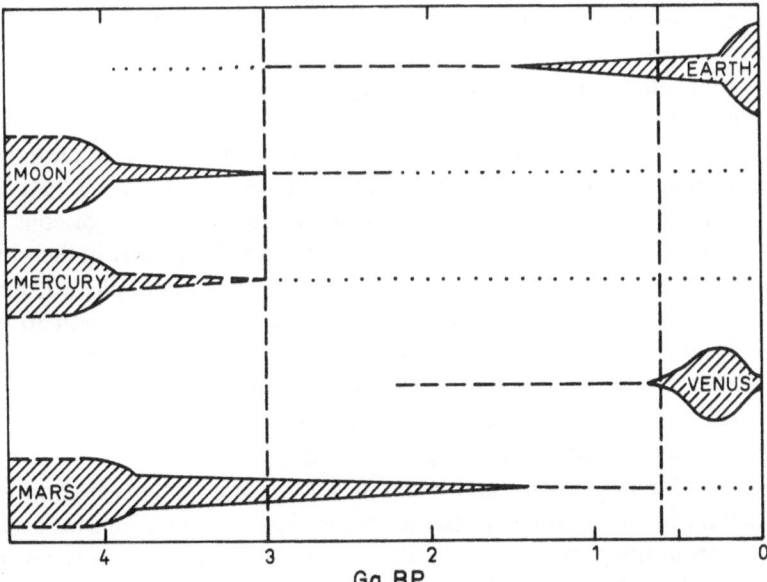

Fig. 149. Ages of production and relative surface areas of the rocky planets and the Moon since the origin of the planetary system. *Hatching* Areas of morphology most certainly formed by exotic, endogenic, and exogenic processes; *dashed line* indicates small contributions to areas by these same processes, *dotted line* possible contributions by these processes (After Head and Solomon 1981, fig. 12)

these planets today. The present dynamic Earth has a surface that was formed mainly by endogenic processes during the past 200 Ma, as documented by the wide extent of its ocean basins. Remaining surfaces, formed by endogenic processes on Earth prior to 200 Ma ago, are limited in extent. The efficiency of Earth's exogenic regime has caused these ancient primary surfaces to become so extensively modified that much of Earth's surface morphology has an age of only a few thousand years. This combination of planet-wide endogenic/exogenic processes has given Earth its youthful morphology.

10.4 Oceans of Magma and Other Non-Water Liquids

As summarized above, the planets in the Solar System display their own unique characteristics, features that are related to the way that they were formed and the way that they have evolved since they were created. Beyond the Asteroid Belt are the gaseous planets which, with their many moons, resemble miniature solar systems. Sunward of the belt are the rocky planets that have different internal and external characteristics. This planetary individuality also is reflected in the oceans that cover or have covered some of their surfaces. Among the gaseous planets Saturn has a massive fluid-hydrogen ocean resting on a liquid metallic-hydrogen mantle. One of its moons, Titan, has an ocean several kilometers deep formed by condensation and precipitation of ethane and acetylene, produced by photolysis of methane in the atmosphere. This ocean, if in equilibrium with an atmosphere containing 3% methane, consists of ethane (70%), methane (25%), and nitrogen (5%). Large tides of about 10 m, produced by the satellite's eccentricity would have destroyed all geological features that formed before the ocean was created. Possibly, endogenic processes are producing features on the floor of the ocean, but our present technology does not yet allow us to image them.

The jovian satellite Europa has a liquid water ocean capped by a thin ice shell. Stresses on this shell have produced a complex pattern of fractures that lace the surface of the satellite. Possibly these and other satellites of the gaseous planets may have had 'magma oceans,' similar to those on the Moon early in their histories if impact and radioactively induced heating was high enough. Such oceans, however, have not been documented to date. Even the jovian moon Io with its intense volcanic activity caused by its proximity to Jupiter cannot be said to possess a 'magma' ocean. The composition of the satellites may preclude extensive heating from radiogenic sources and heavy bombardment of such small bodies would have tended to break them up rather than melt them.

Possibly Mercury, Venus, and Mars may have had 'magma oceans' early in their planetary histories. Some investigators also believe that Venus may have had a water ocean in the past. If such an ocean once did exist, a runaway greenhouse effect removed it along with the rest of the hydrosphere, with the hydrogen escaping into space and the residual oxygen reacting with surface lavas. Venus has had, and is still experiencing extensive igneous resurfacing, and local seas comparable with those during the 1969 Mauna Ulu eruption of Kilauea, Hawaii, may have existed. Degassing on Mars led to the production of an atmosphere, which was much thicker than the one of today and possibly even an ocean tens to hundreds of meters deep that flooded the lowland in the northern hemisphere. Coastal-like features formed by littoral processes have been identified along the boundary between the lowlands of the northern hemisphere and the uplands of the southern hemisphere. The bulk of this early water, about 90%, became ground ice and groundwater and the rest was chemically locked in silicates, absorbed into the regolith, trapped as ice in the polar regions, and lost into the atmosphere and space. Occasional catastrophic release of some of this groundwater by melting of the ice produced massive floods and carving of outflow channels along the northern lowland/southern highland boundary. Spring sapping also is respsonsible for numerous valleys on Mars.

The Moon never had a water ocean, but early in its history its surface may have been blanketed by a 'magma ocean,' with the lunar uplands representing flotation layers of a magma cumulate series during solidification. Within 150 Ma after formation of the Moon 4.55 Ga ago this ocean solidified to form a felspathic crust with ferro-anorthosites at the top grading downward to norites at the base of the crust. The Earth, like the Moon, may have had a 'magma' ocean early in its history. This ocean was the result of the large amount of heat available on the accreting Earth from the presence of a primitive solar type (H_2–He) atmosphere and from meteoritic impacts. As the primitive atmosphere escaped into space, the 'magma ocean' began to solidifiy, cooling to a mush within 1000 years. Extensive degassing during and after accretion led to the formation of the present water ocean and a carbon dioxide and nitrogen atmosphere within 50 Ma after accretion ended. This presence of a water ocean permitted the establishment of a plate tectonic regime that produced the present two-tier morphology of the Earth. Formation of a water ocean also permitted the development of life. This biosphere, in conjunction with weathering of the ocean floor and precipitation of carbonates, prevented a runaway greenhouse effect and destruction of the oceans and the rest of the hydrosphere as probably happened on Venus. With continental growth, subaerial erosion allowed further reduction of the atmospheric carbon dioxide to its present level. It is this biosphere, which created the present oxygen-rich atmosphere by photosynthesis, that is the ultimate creator of the morphology of Earth.

10.5 Ocean Basins and Water Oceans

The only member of the Solar System that has basins that contain oceans at present is the Earth. This morphology of continental plateau and oceanic basins was formed after a hydroplanet stage when no continents existed. Continental crust was initially formed along belts of plate convergence by crustal delamination, partial melting and crustal differentiation, and intrusion of large volumes of alkaline magmas. Continental growth was augmented by additions of magma, thrusting and stacking of crustal rocks, continental and arc collisions, and accretion of sediment wedges. Various episodes of rifting, dispersion of continental plateaus, closing of oceanic basins, and continental collisions have created the present terrain morphology. The lifetime of an individual ocean basin (the intervals between times of extensive orogenies) ranges from 500 to 210 Ma; therefore, the positions and shapes of the Earth's ocean basins have by no means been constant. This continuous change provided a mechanism for repeatedly reworking the crust of the Earth, thereby separating and recombining the materials of both continental and oceanic crusts. The hydrosphere of the Earth is essentially the only one in the Solar System. If spread evenly over the entire surface of the planet the water would have a thickness of 2680 m. Most of this water was expressed from the mantle during the first 50 Ma after accretion of the planet, and after it was impacted by a large planetesimal that extracted the Moon and excavated deep into the mantle. The volume of water in the basins has remained approximately constant for about 4.4 Ga, and especially so during the most recent and best-known Wilson ocean-basin/continent cycle that began about 220 Ma ago. This volume of water was sufficient to fill the ocean basins (Fig. 5), allowing narrow to broad straits to exist between the basins and smoothing the brims of the basins by controlling the level of sediment deposition in the form of coastal plains and continental shelves and slopes. The main cause of variations in the volume of water during this and earlier Wilson cycles was temporary transfer of water onto the continents in the form of glaciers.

Mars has had no plate tectonics and so its deep basins cannot be so changeable with time and place as the Earth's basins have been. In fact, the positions of Mars' basins appear to have been fixed by the long-term growth of the two major volcanic edifaces (Tharsis and Elysium). These edifaces certainly did not move around, and so the basins between them cannot have moved either. There is no geomorphic evidence that any of the basins ever were filled to overflowing with water. The total volume of water expressed from within Mars may have been only about 10 m times the total surface area of the planet (145×10^6 km^2), or 1.5×10^6 km^3. If all this water had been present at one time and was concentrated within the basins, it would occupy only the deepest parts. The basin floors are not at a uniform level, so if the water filled just the deepest parts of the present basins, the seas would have a total area of 3×10^6 km^2 (2% of the total area of Mars and 7% of the area of all regions below the planetary base level). The average water depth would have been 0.5 km and the maximum water depth about 1.3 km. The largest of these seas would have been in the Hellas Planitia and the Acidalium Planitia (the latter centering at about Lat. 60°N, Long. 40°W in the eastern Vastitas Borealis), each about half the area of the Earth's Mediterranean Sea – too small to be considered an ocean, almost just large lakes. On the other hand, more water than this may have been expressed and available for oceans when carbon dioxide was more abundant in the martian atmosphere, before it became deposited as calcium carbonate by combination with calcium derived from the weathering of calcitic igneous rocks. At that time, the more abundant carbon dioxide would have provided a higher than present air temperature that would have allowed more water to be in a liquid form then than now. Although oceans are absent on Mars, water is present. Its most obvious form is as polar ice caps that alternate with the seasons, with the larger ice cap being at Mars' south pole. Associated with the water ice is carbon dioxide frost in polar regions, especially during the winters. In addition, Mars' atmosphere contains a small concentration of water vapor (perhaps 0.04%). Probably most of the water on Mars now is immobilized and is out of view as ground ice or permafrost at high latitudes where the temperature is low enough to prevent it from melting and escaping. In Mars' early history, when carbon dioxide was more abundant in the atmosphere much of the ground-ice melted and its water formed amphitheater-like heads of valleys through spring sapping, flowing thence down slopes to carve stream valleys and deposit chaotic terrain and related exogenic morphology. Some of the ground-ice also appears to have been melted by volcanic intrusions, thereby forming episodic catastrophic floods with consequent flood deposits of sediments.

Although Venus may have very limited plate tectonics, it has no oceanic basins akin to those on Earth. Moreover, the very high temperature (the highest of any rocky member of the Solar System) prevents there being any water ice, water, or much water vapor (the latter being only about 0.1% of the atmospheric gases below the clouds and less above them). The high temperaure is caused by the at least 95% concentration of carbon dioxide in the atmosphere – a runaway greenhouse effect.

The Moon and Mercury are at an extreme in connection with ocean basins, having neither large basins nor water in any free (non-mineral) form.

10.6 Atmospheres

The light elements, hydrogen and helium, comprise more than 99% of the mass of the Sun and probably of the entire Universe as well, so we should not be surprised to find them present as major constituents of the Sun's surface. Also present, but much less abundant, are methane and ammonia. The ratios are $H/He = 9$; H/C (C of CH_4) = 3000; and H/N (N of NH_3) = 13 000. The abundances indicate that the elements must have been inherited from the nebula that formed the Sun and its planetary system. Indeed, these gases are the chief components of the atmospheres present in the large outer gaseous planets – Jupiter, Saturn, Uranus, and Neptune (Table 2). They must also have been the main components of the primitive atmospheres of the small inner rocky planets – Mercury, Venus, Earth, and Mars. Evolution of the Sun into its T-Tauri phase about 50 Ma after its formation, however, produced solar winds that were able to strip the primitive atmospheres from these inner planets because of their close proximity to the Sun. The replacement atmospheres had to be derived from degassing of the host planets. Mercury and the Moon, however, had gravities too low to be able to hold atmospheres, and thus they have no atmospheres now (Table 2).

Venus, Earth, and Mars are large enough to have renewed their atmospheres by degassing from the planets and to have retained some to most of the gases afterward, but the nature of degassing and retention differs from planet to planet. Mars is 3.75 times as distant from the Sun as Mercury is, thus it received only 1/14th as much solar flux. It also has only 10 mbar of atmospheric pressure (far more than Mercury and the Moon have, but only 1/100th that of the Earth's). Carbon dioxide comprises 95% of its atmosphere, but this partial pressure of carbon dioxide is far too low to produce a greenhouse warming effect (estimated to require 2 bar carbon dioxide to permit an average martian surface temperature above freezing). In fact, the surface temperature of Mars is so cold (150 to 280 K; Table 2) that nearly all the existing water is present as ground ice or permafrost, and the rest is in the form of polar ground frost, fog, and a little water vapor. However, during the time of heavy bombardment, 3.5 Ga ago, some melting of the rocks occurred because of heat liberated by the impacts. At that time the climate became warm enough to allow spring sapping by melted ground ice, and the meltwater that was produced cut sinuous stream valleys downslope from the resulting amphitheater-shaped tributary heads, and deposited stream-bed sediments on the floors of the valleys and at their mouths. A major component of the martian atmosphere is suspended dust; it is abundant because of the absence of much water. Although the atmosphere is thin, the wind velocity is sufficient to pick up surface dust that was formed by the impact of small meteorites, and perhaps by ancient weathering, and to transport and deposit it as a widespread blanket atop the rocky topography, especially at high latitudes (Figs. 117 and 118). Dust storms commonly are so extensive that they obscure much of Mars' surface from the view of observers on Earth and of satellites, and they give rise to the red color of that planet.

Venus is unique among the rocky planets and moons in that it has an atmosphere with such a high albedo from its thick cloud cover that nothing of the planet's surface can be seen from the Earth. This albedo and Venus' nearness to Earth (closest planet when Earth and Venus are on the same side of the Sun during their orbits) are the causes of its brightness and designation as the Morning Star and the Evening Star. Carbon dioxide comprises 95% of Venus' atmosphere, distantly followed by nitrogen, water, and other gases. The barometric pressure is 90 bar, enormously more than for any other rocky planet, and this gives rise to questions of why carbon dioxide is so abundant and water so scarce in comparison with the concentrations in the atmospheres of its neighbors (Mars and Earth). The carbon dioxide and water were released along with other gases during the outgassing largely by volcanic activity, that replaced an original primitive atmosphere, which had been swept away by solar wind soon after the planets accreted 4.55 Ga ago. Thus, even if the three planets started with similar concentrations of carbon dioxide, much of the atmosphere on Mars was lost to space because of Mar's low gravity, and the absence of a biosphere on Venus prevented the deposition of carbon dioxide as calcium carbonate (unlike on Earth), and allowed its progressive increase in the atmosphere. This led to the beginning of an intense greenhouse effect on Venus: a runaway greenhouse if coexisting water became steam (Venus is too hot to permit water condensation), or a moist greenhouse if the water once had condensed to form an ancient former ocean. With either the dry or moist greenhouse concept the water that reached the upper atmosphere would have been dissociated by the Sun's component of ultraviolet light, followed by the hydrogen escaping into space and the oxygen combining with reduced iron minerals at the ground surface (the source of the red color of the dust that gives the planet its distinctive appearance in the night sky). The eventual loss of water, including a water ocean, would have prevented the deposition of carbon dioxide as calcium carbonate and led to its further increase in Venus' atmosphere and an even more intense greenhouse effect. Essentially, the differences in the distances of these three planets from the Sun meant that most water on Venus would exist as steam, in contrast with the liquid water on Earth and water ice on Mars, with the resulting production of three different exogenic morphologies on the planets.

Earth's primitive atmosphere, like that of the other rocky planets, was lost to the solar wind soon after accretion 4.55 Ga ago. It was replaced by a different atmosphere, one composed of carbon dioxide, nitrogen, and trace amounts of other gases, all produced by degassing of Earth. This degassing was vastly increased by the effects of the impact of a Mars-size planetesimal on Earth within 100 Ma after the accretion of Earth. The impact removed much of the Earth's upper mantle to form the Moon, exposing the lower mantle and allowing degassing of mantle rocks to add considerably to the volume and composition of the new atmosphere beyond the gases provided by only crustal or lithospheric degassing on Venus and Mars. During the early history of the Earth solar radiation may have been low enough to have allowed average temperatures to be below freezing on the Earth's surface, so that water may have existed there mainly as ice (as now on Mars). However, the carbon dioxide in the Earth's atmosphere apparently was sufficiently concentrated to produce a mild greenhouse effect that allowed water to exist as a liquid ocean instead of only as ice. This early greenhouse effect lasted until the Sun's luminosity increased, and an ice-laden Earth was avoided. A runaway greenhouse effect from the continuous buildup of carbon dioxide during early Archean time when extensive continents were not yet developed, was prevented by leaching of cations (calcium, magnesium, and iron) from the ocean floor by hydrothermal waters. Release of these cations led to the consumption of bicarbonate, a consumption that was enhanced by the decrease in solubility of carbonate minerals with increasing temperature. Biological activity associated with the hydrothermal systems on the komatiitic mid-ocean ridges also aided in maintaining the carbon dioxide content at a reasonable level. As the continents grew, subaerial erosion of silicate rocks further removed the atmospheric carbon dioxide through widespread deposition of carbonates in the marine realm. By 3.8 Ga ago marine life had evolved, and plants began to convert carbon dioxide, water, nitrate, and phosphate into tissue and free oxygen using chemosynthesis and photosynthesis processes. The widespread presence of this free oxygen set the scene for the formation of an ozone layer and for the development and evolution of animal life on Earth.

Essentially, the differences in the atmospheres of Venus, Earth, and Mars, their favorability for the development of life, the planets' surface morphology, and even their subsurface composition and structure are effects of the different distances of these planets from the Sun. These differences are heavily influenced by the varied concentrations of carbon dioxide in the different planetary atmospheres, also largely a function of distance from the Sun.

10.7 Life and Its Role in Structure and Morphology

Knowledge of the mutual effects of environment and life is necessarily biased by the fact that life appears to exist on only one of the planets of the Solar System: Earth. This exclusiveness is based upon direct observation on Earth contrasted with information for the other planets coming mostly from telescopic and satellite observations (absence of seasonal changes in plant distribution patterns), measurements of temperature (no life below the freezing point and above the boiling point of water), measurements of atmospheric compositions (no life in the absence of organic building materials, of oxygen for higher forms, and of sufficient water for habitation), and inconclusive measurements on martian soils at the two Viking lander sites. All the planets have had the same overall length of time, 4.55 Ga, for modification of their environments and for evolution of life.

The atmosphere that was formed by degassing of the Earth's mantle was devoid of free oxygen and was chemically reducing. Sediments deposited between 3.8 and 2.4 Ga ago included siliceous banded iron-mineral deposits whose lithified remains are widespread iron ores such as taconites and others. The banding was an alternation of reduced ferrous iron (black to blue) and oxidized ferric iron (white to red) minerals. Also deposited were layers of uraninite and pyrite that similarly required a reducing oxygen-free environment. The alternate reduced and oxidized layers of iron minerals indicate the occurrence of many widespread but sudden changes from reducing to oxidizing conditions and vice versa, probably originating because of episodic influxes of dissolved reduced iron alternating with rapid growth and oxygen production by early life forms. These forms must have been the earliest bacteria-like organisms that produced organic matter and free oxygen in anaerobic environments, using photosynthesis in shallow water and chemosynthesis in water too deep to be reached by sunlight. Chemosynthesis is the energy process still used by similar organisms in the polysulfide hot springs that are common along the plate-divergence belts of the present mid-ocean ridges. Dissolved hydrogen sulfide in the springs served as an oxygen sink, just like the role of the reduced dissolved iron, thereby excluding the originally poisonous dissolved oxygen. Such springs may well have been one of the sites of the initial evolution of life on the Earth and they would have been protected by deep overlying water from the physical destruction wrought by the early planetesimal bombardment.

The long-lasting reducing and oxygen-free (except for brief episodes) stage of Earth's history, dominated by procaryotic organisms came to an end about 2.4 Ga ago, when oxygen produced by these primitive organisms eventually became more abundant in the hydrosphere and atmosphere. By about 2.8 Ga ago came the first widespread life forms able to live in the presence of free oxygen, and they mainly used photosynthsis to produce organic matter from carbon dioxide and nutrients, liberating still more oxygen in the process. The former reduced iron-mineral deposits largely ceased being deposited, because weathering on land slowed in the presence of oxygen, and iron minerals then tended to remain as residual deposits rather than being transported to the ocean. Instead, widespread oxidized iron deposits (red beds) began to be formed. By about 1.5 Ga ago eucaryotic cells appeared and they became abundant and diversified by 1.0 Ga ago, about the time that the ozone screen formed and life was protected from most ultraviolet radiation. Eucaryotic cells differ from the earlier evolved, but still living (as blue-green algae), procaryotic cells in their sexual reproduction and their content of oxygenaze enzymes. These enzymes require oxygen pressures greater than 0.001% of the present level of atmospheric pressure, the same level at which procaryotes switch from anaerobic to aerobic metabolism. The earliest fossil remains of eucaryotic cells also have been found in rocks of about this age. These facts imply that the Earth's atmosphere had reached that oxygen pressure about 1.4 Ga ago. Subsequently, the oxygen content of the hydrosphere and atmosphere rapidly increased, passing about 7% of the present pressure about 570 Ma ago, and reaching a plateau of 21% about 400 Ma ago – after land plants had developed and supplemented the production of oxygen by water-living plants. As a result of the increased available oxygen, metazoan animals evolved and used oxygen to oxidize organic materials from ingested plants. The first marine animals had no hard skeletons; consequently, their earliest remains are poorly preserved and their fossils are not evident until during the Cambrian (570 Ma ago), as though they had just then invented chitinous and calcareous exoskeletons, and later internal skeletal elements. The first vertebrates (marine fish) evolved about 450 Ma ago, and with further evolution vertebrates moved onto land as amphibians, reptiles, and eventually mammals, in progressive adjustment to the presence of many new forms of plants to serve as basic food supplies.

The effects of plant life on the Earth and its habitability are manifold as a chain of effects. Most important and primary was the function of plants in converting the outgassed anaerobic carbon-dioxide-rich atmosphere into an aerobic atmosphere. This change prevented a severe greenhouse effect (as happened on Venus) that could have led to very high surface and atmospheric temperatures. The high temperatures, in turn, would have driven water vapor to the upper atmosphere where it would have been dissociated by ultraviolet light, so that the hydrogen would have been lost to space. A resulting absence of water on the surface of Earth would have meant no oceans, slow and different weathering of rocks, no erosion by streams, glaciers, groundwater, waves, or currents, and no transportation and deposition of particulate and dissolved sediments by these geological agents. In short, Earth's morphology would have lacked almost all the physiographic features familiar to us as scenery and textbook illustrations – all exogenic origin. Instead, the morphology would have consisted only of exotic and endogenic features and there would have been a dominance of impact craters, volcanic craters and flows, and patterns of faults and folds in igneous rocks (as on the Moon, Mercury, and Venus).

Even the endogenic morphologies would have been different from those of the present Earth if oceans had not been present, because plate tectonics requires the presence of an ocean in several ways. Most evident is the function of aqueous erosion in reducing the height and breadth of continental masses accreted by plate convergence, with the debris being deposited along continental margins as thick aprons. The presence of water also reduces the viscosity of mantle and magma, making for greater mobility and easier plate movements. Consequently, plate tectonism would have been absent or minimal in the absence of an ocean, and, in fact, the morphology produced by plate tectonism has not been recognized on the other terrestrial planets and satellites which have no liquid water. The absence of water also suggests that the magnitude of pyroclastic activity on the other rocky planets is minor, although such volcanism may have been more extensive on Mars when the planet was wetter than it is now. In the absence of plate tectonics, volcanic structures can attain immense dimensions as the magma chamber remains fixed to the structure above. Under these conditions a volcano can grow as long as the lithostatic pressure of the magma chamber is greater than the pressure exerted by the column of magma from the chamber to the summit of the volcano. Accumulation on some of the volcanic structures of Venus and Mars may be billions of years in duration. These features differ from those of Earth in having proportionally larger and deeper summit calderas.

Plant life on Earth not only allowed the retention of surface water but also, with the production of oxygen, led to the development of animal life and its evolutionary changes in enormous numbers of forms, from phyla to subspecies. Some of these forms have produced directly large exogenic morphological features such as coral reefs; other forms have produced indirect features or added to non-biological morphologies. The latter function is illustrated by enhanced weathering of rocks, leading to increased erosion, transportation, and deposition of sediments in stream valleys, along shores, and on ocean floors. Some of the filling in basins consists of nonbiological sediment as well as skeletal parts constructed by plants and animals from minerals dissolved in water.

The animal race that currently is most active in forming unique exogenic morphologies on Earth is the human one. These morphologies are both erosional and depositional in nature. Examples of erosional activities are large holes left from mining operations, and enhanced erosion by increased stream erosion that results from unwise mining, farming, lumbering, and land-clearing practices. Perhaps more widespread are features of deposition related to human activities, either direct or indirect in origin. Cities, highways, river dams, and waste dumps are examples of direct deposition. Morphologies produced by indirect deposition through human activities are illustrated by increased rates of filling of stream valleys and growth of river deltas. Most erosional and depositional activities of humans are directed toward obtaining perceived better facilities: food, clothing, shelter, and protection from other humans seeking the same facilities. The desire for protection so far has had few morphological consequences other than local features designed for defence or produced by explosives used in offense. However, the rapid rate of increase of the human populations (population pollution) and waste pollution may reverse the past role of biology in adapting to planetary environments, ending with the destruction of *Homo sapiens* and of many other life forms as well. Humans seem to have become a law unto themselves and not answerable to anything or anyone else. Only by cultivating a harmony with nature will this trend be reversed. It is our consciousness that has given meaning to our surroundings, and if we cease to exist then our knowledge of the Universe also will cease. However, a continuation on the present course, including the existence of nuclear bombs and contamination, is unlikely to make major direct changes in the Earth's morphology, whether exotic, endogenic, or exogenic. Indirect changes are more likely and may include the possible elimination of many life forms, including our own, and the cessation of their production of morphological features.

10.8 The Improbability of an Earth

It is interesting to note the many favorable events that had to happen for the morphology and life forms of the Earth to develop to their present stage, a stage that may exist nowhere else in the Solar System. Perhaps first among these events is that our solar nebula formed late in the history of our galaxy as a descendant or associate of one or more supernovas whose high temperatures produced the heavy elements that were retained by the Sun and now are concentrated in the Sun's family of planets and other rocky bodies. Part of the solar nebula condensed into smaller bodies that became the planets, moons, asteroids, comets, and meteorites that orbit the Sun in approximately the same plane and move in the same general direction. When the Sun condensed from the nebula about 4.55 Ga ago it was a small red star of the sort that can be expected to last about 10 Ga. It is considered to be too small to eventually become a supernova that would destroy its family. An appreciably larger and hotter star would have lasted only a short time before becoming a supernova, probably too short a time for life to have evolved. A smaller and cooler star may not have been able to provide the materials and thermal energy needed to construct and maintain a family of planets.

Earth was intermediate in size – large enough to hold a moderately thick atmosphere, unlike Mercury, Mars, and the Moon. It orbits the Sun at a moderate distance, far enough from the Sun to avoid the searing insolation experienced by Mercury and Venus, and near enough to avoid the very low insolation of the icy outer planets. The soon-to-follow solar wind removed the primitive atmosphere with its gases that would have prevented many existing exogenic geological processes as well as the development of life as we know it on Earth. The heavy infall of debris from the solar nebula that had not already been collected by the planets, impacted Earth, churning and melting part of the proto-crust and leaving many craters. The largest impact, by a Mars-size planetesimal, removed part of the lithosphere and mantle, thereby forming our Moon and increasing the surface area and depth of Earth susceptible to degassing, which in turn provided a new and different kind of atmosphere and a hydrosphere. Impacts of large planetesimals continue episodically even to the present, but oceans containing deep water derived from degassing provided shelter from impacts for early procaryotic forms of life; such shelter was probably within hot springs associated with the earliest plate tectonics.

The initiation of life about 3.8 Ga ago marked the beginning of an evolution of the Earth in ways that are nearly independent of events in space and allowed the morphology of Earth to become almost completely different from that of the other planets and their moons. Some of the procaryotic organisms and the later eucaroytic ones (beginning about 2.8 Ga ago) were able to produce organic matter from the carbon dioxide and nitrogen compounds of the atmosphere and hydrosphere, and from compounds of nitrogen, phosphorus, sulfur, and silica in sediments previously weathered from rocks. Even more importantly, these organisms liberated free oxygen in the process. This change from a carbon-dioxide-dominated atmosphere prevented a runaway greenhouse effect akin to the one on Venus and which causes its atmosphere and surface to have temperatures greatly exceeding the boiling point of water, an effective block to the development of life there. The free oxygen in the atmosphere and hydrosphere of Earth set the scene for the evolution of metazoans and eventually of humans. It also led to the presence in the upper atmosphere of an ozone layer that shields the organisms from the destructive effects of ultraviolet radiation from the Sun. To the simple procaryotic and eucaryotic cell structures we owe the liquid water which with its organic materials and chemical derivatives, has permitted weathering of all kinds of rocks. Most of these weathered products are carried by water (provided by life and largely kept liquid by life) in suspension and solution, and while part is deposited en route most eventually reaches the ocean. In the ocean these materials serve as components for the tissues of plants and animals, and for sedimentary deposits on the margins and floors of the basins, eventually being added to the continents through plate convergence. The plate tectonics made possible through the water released and kept liquid by life forms later also caused deformation of the continental crust, as well as its growth and its fragmentation, thereby forcing continents and ocean basins to constantly shift position and, in turn altering the migration paths and environments of both marine and continental plants and animals. These migrations and the resulting evolution of species are among the causes of paleontological changes that serve to date many of the geological events on Earth.

In an overall summary, it is interesting to speculate whether Earth is unique as a site of habitation by thinking animals or is only one of many such sites. The results obtained during the past few decades' study of the Solar System from Earth and by artificial satellites sent from Earth rule out all bodies of this family as past or potential future sites of long-term habitation but this does not necessarily exclude the possibility of using them as sites for short- or long-term observations or scientific study, or perhaps as sites for industrial establishments wishing to make use of lower gravities than that on Earth. Knowledge about planets orbiting other stars is still non-existent. In view of the estimates of 10^{11} stars in our galaxy and of 10^{11} galaxies in the Universe, one might suppose that there must be many such habitable planets. Direct verification of their presence and degree of similarity to

Earth, however, seems impossible because of the vast distance that separates most of them from Earth, a distance of at least 10^{23} km (10 Ga for light from them to reach the Earth). Among the 31 stars nearest the Earth (within 15 light years), 15 have some evidence of being binary or multiple systems. The two nearest stars (Alpha Centauri and Barnard's star, 4.3 and 6.0 light-years distant from the Sun, respectively) may have two companions each, and 70% of all stars in our galaxy have been estimated to have two or more companions – stars or planets. The creation of the Universe 10 to 20 Ga ago means that all galaxies and their stars have moved radially outward from the site of the big bang for that length of time, a span that exceeds the lifetime of most stars, including our Sun, which is a second- or third-generation star probably having had ancestors that perished as super-novas. Other stars with ages similar to that of the Sun must be dotted throughout our galaxy and other galaxies. Only during the past few decades have we, on our Earth (that formed with the Sun 4.55 Ga ago), reached the stage of seriously wondering about the presence of thinking life forms elsewhere in the Universe. If life evolved elsewhere at about the same rate as on Earth, we might assume that the best prospects are for life on planets associated with other stars that formed 4.5 to perhaps 6.0 Ga ago. Most systems that are similar to ours but older should have vanished as supernovas, and younger ones may not have had enough time for life to have evolved to a thinking stage.

Conceivably, the possible inhabitants of star systems that were about to become supernovas might have decided to move and colonize another planet not then inhabited by thinking beings or by displacing those who may already have inhabited such a planet. Such has been suggested by science-fiction writers for colonization of the Earth, but the paleontological record of fossils on Earth and their pattern of evolution precludes such a sudden colonization by humans, and no records or artifacts of such a colonization have been identified on Earth. Of course, the actual colonization of a foreign planet would have required identification and confirmation of a suitable environment, probably involving several exploratory visits. Up to the present time, humans have made six landings on the Moon (only 1.3 light-seconds from Earth), and these journeys required about 2.5 days for one-way travel, exclusive of take-off and landing orbits. The escape velocity from Earth is 11 km/s. No humans have visited other planets, although environmental conditions on Mercury and Mars are more like those on the Moon than are conditions on other planets and their moons. Unmanned exploration satellite trips to Mars and to near Jupiter have taken 157 and 650 days, respectively. In view of the much slower speed of actual space travel than of travel of light and radio waves, the likelihood of interstellar travel appears to be unlikely to have occurred during the past and only a slight possibility for the future. The great distances could involve generations aboard a spacecraft traveling to the nearest suitable stellar-planetary system – a really confining experience!

Even without space travel, it is possible that thinking life on other planets in our galaxy and beyond was established and evolved during the past. Because of this possibility several programs have been designed on Earth to receive and send intelligible messages by radio transmission from Earth, and other recorded messages have been incorporated within satellites destined to travel far beyond the Solar System after their primary exploratory missions have been completed; all this has happened since the advent of Sputnik in October 1957. Radio and light transmissions provide the best means for seeking communication because of their speed. However, the distances involved to reach most of the stars are so great that the questions asked by one generation, even via such speedy transmissions, are likely to be lost and forgotten before an answer is received and identified by a later generation. However, the insatiable curiosity of humans and the prompting by science-fiction writers probably guarantee that efforts toward interstellar communication will continue until Earth itself is destroyed by the coming eventual decay of our Sun, or, perhaps more likely, until Earth's life is destroyed much sooner by conflict between unwisely proliferating members of the human population of Earth.

In the meantime, the wide expanses of hydrosphere and biosphere give Earth its unique appearance from outer space – the appearance of a planet whose surface is colored with wide areas of blues, browns, and greens, and with wisps of white in its atmosphere. Previously, our geological studies have focused on the nature of the floor of Earth's ocean, but with this book we have expanded our geological studies to the Earth's solar companions. It is then fitting that we close our voyage to the Solar System with the words of Rhysling, the blind singer of the spaceways in the short story by Robert A. Henlein, *The Green Hills of Earth*:

'Let me rest my eyes on the fleecy skies
and the cool green hills of Earth'

References

Abbott D, Menke W (1990) Length of the global plate boundary at 2.4 Ga. Geology 18:58–61

Abelson PH (1966) Chemical events on the primitive Earth. Proc Natl Acad Sci USA 55:1365–1372

Agar WM, Flint RF, Longwell CR (1929) Geology from original sources. Holt, New York, 527 pp

Ahrens TJ (1990) Earth accretion. In: Newson HE, Jones JH (eds) Origin of the Earth. Oxford University Press, New York; Lunar and Planetary Institute, Houston, pp 211–227

Ahrens TJ, O'Keefe JD (1989) Formation of atmospheres during accretion of the terrestrial planets. In: Atreya SK, Pollack JB, Matthews MS (eds) Origin and evolution of planetary and satellite atmospheres. University of Arizona Press, Tucson, pp 328–385

Albee AL, Arvidson RE, Palluconi FD (1992) Mars Observer mission. J Geophys Res 97:7665–7680

Alfven, H (1978a) Origin of the Solar System. In: Dermott SF (ed) The origin of the Solar System. Wiley, New York, pp 19–40

Alfven H (1978b) The band-structure of the Solar System. In: Dermott SF (ed) The origin of the Solar System. Wiley, New York, pp 41–48

Allen CC (1979) Volcano-ice interactions on Mars. J Geophys Res 84:8049–8059

Alvarez LW, Alvarez W, Asaro F, Michel HV (1980) Extra-terrestrial cause for the Cretaceous-Tertiary extinction. Science 208:1095–1108

Alvarez W, Asaro F (1990) An extraterrestrial impact. Sci Am 263(4):76–84

Alvarez W, Muller RA (1984) Evidence from crater ages for periodic impact on the Earth. Nature 308:718–720

Anderson DL (1980) Tectonics and composition of Venus. Geophys Res Lett 7:101–105

Anderson DL (1981) Plate tectonics on Venus. Geophys Res Lett 8:309–311

Anderson DL, Miller WF, Latham GV, Nakamura Y, Toksöz MN, Dainty AM, Duennebier FK, Lazarewicz AR, Kovach RL, Knight TCD (1977) Seismology on Mars. J Geophys Res 82:4524–4546

Anderson DL, Tanimoto T, Zhang Y-S (1992) Plate tectonics and hot spots; the third dimension. Science 256:1645–1651

Anonymous (1989) Maps of part of the northern hemisphere of Venus, scale 1:15 000 000. US Geological Survey, Miscel Invest Ser, Map I-2041, 3 sheets

Arculus RJ, Holmes RD, Powell R, Righter K (1990) Metal-silicate equilibria and core formation. In: Newson HE, Jones JH (eds) Origin of the Moon. Oxford University Press, New York; Lunar and Planetary Institute, Houston, pp 251–271

Arrhenius G (1978) Chemical aspects of the formation of the Solar System. In: Dermott SF (ed) The origin of the Solar System. Wiley, New York, pp 521–582

Arvidson RM, Coradini M, Carusi C, Coradini A, Fulchignoni M, Federico C, Funiciello R, Salomone M (1976) Latitudinal variation of wind erosion of crater ejecta deposits on Mars. Icarus 27:503–516

Arvidson RE, Guinnese E, Lee S (1978) The surface of Mars. Sci Am 238(3):76–89

Arvidson RE, Goettel EA, Hehenberg CM (1980) A post-Viking view of the geologic evolution of Mars. Rev Geophys Space Phys 18:565–603

Arvidson RE, Binder AB, Jones KL (1983) The surface of Mars. In: The planets. Readings from Scientific American. Freeman, New York, pp 48–59

Arvidson RE, Baker VR, Elachi C, Saunders RE, Wood JA (1991) Magellan; initial analysis of Venus surface modifications. Science 252:270–275

Arvidson RE, Greeley R, Malin MC, Saunders RS, Izenberg N, Plaut JL, Stofan ER, Shepard MK (1992) Surface modifications of Venus as inferred from Magellan observations on plains. J Geophys Res 97:13 303–13 317

Arvidson RE, Phillips RJ, Izenberg N (1992) Global view of Venus from Magellan. EOS, American Geophysical Union, Trans 73:161, 168–169

Aubele JA, Slyuta EN (1990) Small domes on Venus; characteristics and origin. Earth, Moon, and Planets 50/51:493–532

Aubele JC, Crumpler LS, Head JW (1992) Venus shield fields; characteristics and implications. American Geophysical Union, Spring Meet, Montreal, May 12–16, 1992, Program and Abstr, p 178

Bailey ME, Clube SVM, Napier WM (1990) Origin of comets. Pergamon, New York, 577 pp

Baird AK, Clark BC (1981) On the original sources of martian fines. Icarus 45:113–123

Baker VR (1973) Paleohydrology and sedimentology of Lake Missoula flooding in eastern Washington. Geol Soc Am Spec Pap 144: 79 pp

Baker VR (1979) Erosional processes in channelized water flows on Mars. J Geophys Res 84:7985–7993

Baker VR (1982) The channels of Mars. University of Texas Press, Austin, 198 pp

Baker VR, Kochel RC (1979) Martian channel morphology; Maja and Kasei Valles. J Geophys Res 84:7961–7983

Baker VR, Komatsu G, Parker TJ, Gulick VC, Kargel JS, Lewis JS (1992) Channels and valleys on Venus; preliminary analysis of Magellan data. J Geophys Res 97:13 421–13 444

Baksi AK (1990) Timing and duration of Mesozoic-Tertiary flood-basalt volcanism. EOS, American Geophysical Union, Trans 71:1835–1836, 1840

Baldwin RB (1963) The measure of the Moon. University of Chicago Press, Chicago, 488 pp

Baldwin RB, Wilhelms DE (1992) Historical review of a long-overlooked paper by R. A. Daly concerning the origin and early history of the Moon. J Geophys Res 97:3832–3843

Baneredt WB, Grimm CG (1992) Small-scale fracture patterns on the volcanic plains of Venus. J Geophys Res 97:16 149–16 166

Barlow NG (1990) Constraints on early events in martian history as derived from cratering record. J Geophys Res 95:14 191–14 201

Barrell J (1927) On continental fragmentation and the geologic bearing of the Moon's surficial features. Am J Sci, 5th series. 13:283–314

Barrow JD (1991) Theories of everything. The quest for ultimate explanation. Clarendon Press, Oxford, 223 pp

Barsukov VL, Basilevsky AT, Burba GA, Bobinna MN, Kryuchkov VP, Kuzmin RP, Nikolaeva OV, Pronin AA, Ronca LB, Chernaya IM, Shashkina VP, Kushky AV, Markov MS, Kotelnikov VA, Rzhiga ON, Petrov GM, Alexandrov YuN, Sidorenko AI, Bogomolov AF, Skrypnik GI, Bergman MYu, Kudrin LV, Bokshtein IM, Kronkod MA, Chochia PA,Tuflin YuS, Kadnichansky SA, Akim EL (1986) The geology and geomorphology of the Venus surface as revealed by radar images obtained by Veneras 15 and 16. 16th Lunar and Planetary Science Conf, Pt 2, Proc. J Geophys Res 91:D378–D398

Barton WH Jr (1939) The birth of the Earth. The Sky 3:24–27

Basilevsky AT, Ivanov BA (1990) Cleopatra Crater on Venus; Venera 15/16 data and impact/volcanic origin controversy. Geophys Res Lett 17:175–178

Basilevsky AT, Pronin AA, Ronca LB, Kryuchkov VP, Sukhanov AL (1986) Styles of tectonic deformations on Venus; analysis of Venera 15 and 16 data. 16th Lunar and Planetary Science Conf, Pt 2, Proc J Geophys Res 91:D399–D411

Basilevsky AT, Ivanov BA, Burba GA, Chernaya IH, Kuyuchkov VP, Nikolaeva OV, Campbell DB, Ronca LB (1987) Impact craters of Venus; a continuation of the analysis of data from the Venera 15 and 16 spacecraft. J Geophys Res 43:12 869–12 901

Basilevsky AT, Nikolaeva OV, Weitz CM (1992) Geology of Venera 8 landing site from Magellan data; morphological and geochemical considerations. J Geophys Res 97:16 315–16 336

Batson RM, Bridge PM, Inge SL (1979) Atlas of Mars; the 1:5 000 000 map series. National Aeronautics and Space Administration, Washington, DC, 146 pp

Becker RH, Pepin RO (1984) The case for a martian origin of the shergottites – nitrogen and noble gases in EETA 79001. Earth and Planetary Science Lett 69:225–242

Bell JF, Cruishank DP, Gaffey MJ (1985) The composition and origin of the Iapetus dark material. Icarus 61:192–207

Bell JF III, McCord TB, Owensby PD (1990) Observational evidence of crystalline iron oxides on Mars. J Geophys Res 95:14 447–14 461

Belton, M (1991) Galileo reveals new lunar features. American Geophysical Union, Trans 72:1, 4

Belton MJS, Head JW III, Pieters CM, Greeley R, McEwen AS, Neukum G, Klassen PK, Anger CD, Carr MH, Chapman CR, Davies ME, Fanale FP, Gierasch PJ, Greenberg R, Ingersoll AP, Johnson T, Paczkowiski B, Pilcher CB, and Veverka J (1992) Lunar impact basins, and crustal heterogeneity; new western limb and far side data from Galileo. Science 255:570–576

Belton MJS, Veverka J, Thomas P, Helfenstein P, Simonelli D, Chapman C, Davies ME, Greeley R, Greenberg R, Head J, Murchie S, Klaasen K, Johnson TV, McEwen A, Morrison D, Neukum G, Fanale F, Anger C, Carr M, Pilcher C (1992) Galileo encounter with 951 Gaspra; first pictures of an asteroid. Science 257:1547–1652

Benz W, Cameron AGW (1990) Terrestrial effects of the giant impact. In Newson HE, Jones JH (eds) Origin of the Earth. Oxford University Press, New York; Lunar and Planetary Institute, Houston, pp 61–67

Benz W, Cameron AGW, Melosh HJ (1989) The origin of the Moon and the single-impact hypothesis III. Icarus 81:113–131

Bercovici D, Schubert G, Glatzmaier GA (1989) Three-dimensional spherical models of convection in the Earth's mantle. Science 244:950–955

Bergstralh JT, Miner ED (1991) The uranian system; an overview. In: Bergstralh JT, Miner ED, Matthews MS (eds) Uranus. University of Arizona Press, Tucson, pp 3–28

Bertrand JML, Caby R (1978) Geodynamic evolution of the Pan-African orogenic belt; a new interpretation of the Hoggar Shield. Geol Rundsch 67:357–388

Biemann K, Orom J, Toulmin P III, Orgel LE, Nier AO, Anderson DM, Simmonds PG, Flory D, Diaz AV, Rushneck DR, Biller JE, Lafleur AL (1977) The search for organic substances and inorganic volatile compounds in the surface of Mars. J Geophys Res 82:4641–4658

Bills BG (1990) The rigid obliquity history of Mars. J Geophys Res 95:14 137–14 153

Bills BG, Ferrari AJ (1977) A harmonic analysis of lunar topography. Icarus 31:244–259

Bills BG, Ferrari AJ (1978) Mars topography harmonics and geophysical implications. J Geophys Res 83:3497–3508

Binder AB (1974) On the origin of the Moon by rotational fission. Moon 11:53–76

Binder AB (1975) On the petrology and structure of a gravitationally differentiated Moon of fission origin. Moon 13:431–473

Binder AB (1980) On the origin of lunar pristine crustal rocks. Conference on the Lunar Highlands Crust, Houston, November 14–16, 1979, Proc Geochim Cosmochim Acta Suppl 12:71–79

Binder AB, Arvidson RE, Guinness EA, Jones KL, Morris EC, Mutch TA, Pieri DC, Sagan C (1977) The geology of the Viking Lander 1 site. J Geophys Res 82:4439–4451

Bindschadler DL, Schubert G, Kaula WM (1990) Mantle flow tectonics and the origin of Ishtar Terra, Venus. Geophys Res Lett 17:1341–1348

Bindschadler DL, Schubert G, Kaula WM (1992a) Coldspots and hotspots; global tectonics and mantle dynamics of Venus. J Geophys Res 97:13 495–13 532

Bindschadler DL, Decharon A, Beratan KK, Smrekar SE, Head JW (1992b) Magellan observations of Alpha Regio; implications for formation of complex ridged terrains on Venus. J Geophys Res 97:13 563–13 577

Binzel RP, Van Flandern TC (1979) Minor planets – the discovery of minor satellites. Science 203:903–905

Binzel RP, Young EF (1992) Albedo distribution and surface evolution on Pluto. American Geophysical Union, Spring Meet May 12–16, 1992, Montreal, Canada, Program and Abstr p 190

Binzel RP, Barucci MA, Fulchignoni M (1991) The origin of the asteroids. Sci Am 265(4):88–94

Black DC (1978) Isotopic anomalies in Solar System material. In: Dermott SF (ed) The origin of the Solar System. Wiley, New York, pp 583–596

Black LP, Williams IS, Compson W (1986) Four zircon ages from one rock; the history of a 3930 Ma-old granulite from Mt. Sones, Enderby Land, Antarctica. Contrib Mineral Petrol 94:427–437

Blair RW (1986) Karst landforms and lakes. In: Short MN, Blair RW Jr (eds) Geomorphology from space. National Aeronautics and Space Administration, Washington, DC, Spec Publ 486, pp 407–446

Blake D, Allamandola L, Sandford S, Hudgins D, Freund F (1991) Clathrate hydrate formation in amorphous cometary ice analogs in vacuo. Science 254:548–551

Blasius KR, Cutts JA (1976) Shield volcanism and lithospheric structure beneath the Tharsus Plateau. 7th Lunar and Planetary Science Conf, Proc, pp 3561–3573

Blasius KR, Cutts JA, Guest JE, Masursky H (1977) Geology of the Valles Marineris; first analysis of imaging from Viking 1 orbiter primary mission. J Geophys Res 82:4067–4091

Bloom AL (1986) Coastal landforms. In: Short MN, Blair RW (eds) Geomorphology from space. National Aeronautics and Space Administration, Washington, DC, SP-486, pp 353–406

Bogard DD, Johnson P (1983) Martian gases in an Antarctic meteorite? Science 221:651–654

Bogard DD, Nyquist LE, Johnson P (1984) Noble gas content of shergottites and implications for the martian origin of SNC meteorites. Geochim Cosmochim Acta 48:1723–1739

Boon JD, Albritton CC Jr (1936) Meteorite craters and their possible relationship to 'cryptovolcanic structures'. Field Lab. 5:1–9

Borgia A, Burr J, Montero W, Morales LD, Alvarado GE (1990) Fault propagation folds induced by gravitational failure and slumping of the Central Costa Rica Volcanic Range; implications for large terrestrial and martian volcanic edifices. J Geophys Res 95:14 357–14 382

Boss AP (1990) 3D solar nebula models; implications for Earth origin. In: Newson HE, Jones JH (eds) Origin of the Earth. Oxford University Press, New York, Earth and Planetary Institute, Houston, pp 3–15

Boss AP, Peale SJ (1986) Dynamical constraints on the origin of the Moon. In: Hartmann WK, Phillips RJ, Taylor GJ (eds) Origin of the Moon. Lunar and Planetary Institute, Houston, pp 59–101

Bowin C (1983) Gravity, topography, and crustal evolution of Venus. Icarus 56:345–371

Bowin C (1985) Gravity field of Venus at constant altitude and comparison with Earth. 15th Lunar and Planetary Science Conf, Pt 2, Proc. J Geophys Res 90, Suppl: C757–C770

Bowin C, Simon B, Wollenhaupt WR (1975) Mascons; a two-body solution. J Geophys Res 80:4947–4955

Bowker DE, Hughes JK (1971) Lunar Orbiter Photograph Atlas of the Moon. National Aeronautics and Space Administration, Washington, DC, SP-206, 675 plates

Bowring SA, Housh TB, Isachsen CE (1990) The Acasta gneisses; remnant of Earth's early crust. In: Newson HE, Jones JH (eds) Origin of the Earth. Oxford University Press, New York; Lunar and Planetary Institute, Houston, pp 319–344

Brakenridge GR (1990) The origin of the fluvial valleys and early geologic history, Aeolis quadrangle, Mars. J Geophys Res 95:17 289–17 308

Brass GW, Harrison GCA (1982) On the possibility of plate tectonics on Venus. Icarus 49:86–96

Breed CS, Grolier MJ, McCauley JF (1979) Morphology and distribution of common 'sand' dunes on Mars. J Geophys Res 84:8183–8204

Brennan K, Wood JA (1991) Mineralogy of low-emissivity areas of Venus; Aphrodite Terra; Gula and Maxwell Montes. American Geophysical Union Spring Meet, Baltimore, May 28–31, 1991, Program and Abstr, p 173

Bretz JH (1923) The channeled scablands of the Columbia Plateau. J Geol 31:617–649

Brown GM (1978) Chemical evidence for the origin, melting, and differentiation of the Moon. In: Dermott SF (ed) The Origin of the Solar System. Wiley, New York, pp 597–610

Brown RH, Kirk RL, Johnson TV, Soderblom LA (1990) Energy sources for Triton's geyser-like plumes. Science 250:431–435

Brown RH, Johnson TV, Synnott S, Anderson JD, Jacobson RA, Dermott SF, Thomas PC (1991) Physical properties of the uranian satellites. In: Bergstralh JT, Miner ED, Matthews MS (eds) Uranus: University of Arizona Press, Tucson, pp 513–527

Brownlee D, Rajan R, Tomandl D (1977) A chemical and textural comparison between carbonaceous chondrites and interplanetary dust. In: Delsemme A (ed) Comets, asteroids, meteorites. University of Toledo Press, Toledo, pp 137–142

Brush SG (1986) Early history of selenology. In: Hartmann WK, Phillips RJ, Taylor GJ (eds) Origin of the Moon. Lunar and Planetary Institute, Houston, pp 3–15

Bryan WB (1986) Tectonic controls of initial rifting and the evolution of young ocean basins – a planetary perspective. Tectonophysics 132:103–115

Buddhue JD (1950) Meteoritic dust. University of New Mexico Press, Albuquerque, Publications in Meteoritics, no 2, 102 pp

Bunch TE, Chang S (1980) Carbonaceous chondrites – II. Carbonaceous chondrite phyllosilicates and light element geochemistry as indicators of parent body processes and surface conditions. Geochim Cosmochim Acta 44:1543–1577

Burbidge G (1992) Why only one big bang. Sci Am 266(2):120

Burns JA (1986a) Some background about satellites. In: Burns JA, Matthews MS (eds) Satellites. University of Arizona Press, Tucson, pp 1–38

Burns JA (1986b) Evolution of satellite orbits. In: Burns JA, Matthews MS (eds) Satellites. University of Arizona Press, Tucson, pp 117–158

Cameron AGW (1978) The primitive solar accretion and formation of the planets. In: Dermott SF (ed) The origin of the Solar System. Wiley, New York, pp 49–74

Cameron AGW, Truran JW (1977) The supernova trigger for formation of the Solar System. Icarus 30:447–461

Cameron AGW, Fegley B Jr., Benz W, Slattery WL (1988) The strange density of Mercury; theoretical considerations. In: Vilas F, Chapman CR, Matthews MS (eds) Mercury. University of Arizona Press, Tucson, pp 692–708

Campbell BA, Campbell DB (1990) Western Eistla Regio, Venus; radar properties of volcanic deposits. Geophys Res Lett 17:1353–1356

Campbell DB, Head JW, Harmon JK, Hine AA (1983) Venus; identification of banded terrain in the mountains of Isthar Terra. Science 221:644–646

Campbell DB, Head JW, Harmon JK, Hine AA (1989) Identifications of banded terrain in the mountains of Ishtar Terra. Science 221:644–647

Campbell DB, Stacy NJS, Newman WI, Arvidson RE, Jones EM, Musser GS, Roper AY, Schaller C (1992) Magellan observations of extended impact crater related features on the surface of Venus. J Geophys Res 97:16 249–16 278

Cande SC, LaBrecque JL, Larson RL, Pitman WC III, Golovchenko X, Haxby WF (1989) Magnetic lineations of the world's ocean basins; scale 1:27 400 000 at equator. American Association of Petroleum Geologists, Tulsa, 1 sheet

Cannizzo JK, Kaitchuck RH (1992) Accretion disks in interacting binary stars. Sci Am 266(1):92–99

Carlson RW (1991) The mechanism and chronology of early lunar differentiation. American Geophysical Union Spring Meet, Baltimore, May 28–31, 1991, Program and Abstr, p 183

Carr MH (1973) Volcanism on Mars. J Geophys Res 78:4049–4062

Carr MH (1974) Tectonism and volcanism of the Tharsis regions of Mars. J Geophys Res 79:3943–3949

Carr MH (1979) Formation of martian flood features by release of water from confined aquifer. J Geophys Res 84:2995–3007

Carr MH (1983) The geology of the terrestrial planets. US National Report, American Geophysical Union, Washington, DC, pp 160–172

Carr MH (1984) Mars. In Carr MH (ed) The geology of the terrestrial planets. National Aeronautics and Space Adminstration, Washington, DC, SP-469, pp 207–269

Carr MH, and Schaber GG (1977) Martian permafrost features. J Geophys Res 82:4039–4054

Carr MH, Masursky H, Saunders RS (1973) A generalized geologic map of Mars. J Geophys Res 78:4031–4036

Carr MH, Crumpler LS, Cutts JA, Greeley R, Guest JE, Masursky H (1977) Martian impact craters and emplacement of ejecta by surface flow. J Geophys Res 82:4055–4065

Cassen P, Peale SJ, Reynolds RT (1980) On the comparative evolution of Ganymede and Callisto. Icarus 41:232–239

Cassen PM, Peale J, Reynolds RT (1982) Structure and thermal evolution of the Galilean satellites. In: Morrison D (ed) Satellites of Jupiter. University of Arizona Press, Tucson, pp 93–128

Chamberlin RT (1945) The Moon's lack of folded ranges. J Geol 53:361–373

Chamberlin TC, Salisbury RS (1930) College text-book of geology. Pt II. Historical geology. Holt, New York (rewritten and revised by Chamberlin RT and MacClintock P), 878 pp

Chapman CR (1974) Cratering on Mars. I. Cratering and obliteration history. Icarus 22:272–291

Chapman CR (1977) The evolution of asteroids as meteorite parent-bodies. In: Delsemme A (ed) Comets, asteroids, meteorites. University of Toledo Press, Toledo, pp 265–276

Chapman CR (1983) Asteroids and comets. Rev Geophys Space Phys 2:196–206

Chapman CR (1988) Mercury; introduction to an end-member planet. In: Vilas F, Chapman CR, Matthews MS (eds) Mercury. University of Arizona Press, Tucson, pp 1–23

Chapman CR, Davis DR (1975) Asteroid collisional evolution – evidence for a much larger early population. Science 190:553–555

Chapman DR, Larson HK (1963) On the lunar origin of tektites. J Geophys Res 68:4305–4358

Christensen EJ (1975) Martian topography derived from occultation, radar, spectral, and optical measurements. J Geophys Res 80:2909–2913

Christensen EJ, Balmino G (1979) Development and analysis of a twelfth degree and order gravity model for Mars. J Geophys Res 84:7943–7953

Cintala MJ (1981) The Mercurian regolith; an evaluation of impact glass production by micrometeoritic impact. Abstracts of Papers Submitted to the 12th Lunar and Planetary Science Conf, Houston, pp 141–143

Cintala MJ (1992) Impact-induced thermal effects in the lunar and mercurian regoliths. J Geophys Res 97:947–973

Cisowoski SM (1986) Magnetic studies on Shergotty and other SNC meteorites. Geochim Cosmochim Acta 50:1043–1048

Claeys P, Casier J-C, Margolis SV (1992) Microtektites and mass extinctions; evidence for a Late Devonian asteroid impact. Nature 257:1102–1107

Clark BC, Baird AK, Rose HJ Jr., Toulmin P III, Christian RP, Kelliher WC, Castro AJ, Rowe CD, Keil K, Huss GR (1977) The Viking X ray fluorescence experiment; analytical methods and early results. J Geophys Res 82:4577–4594

Clark DL (1981) Geology and geophysics of the Amerasian Basin. In: Nairn AEM, Churkin M Jr., Stehli FG (eds) The ocean basins and margins, vol 5, The Arctic Ocean. Plenum, New York, pp 599–634

Clark PE, Leake MA, Jurgens RF (1988) Goldstone radar observations of Mercury. In: Vilas F, Chapman CR, Matthews MS (eds) Mercury. University of Arizona Press, Tucson, pp 77–100

Clark RN, Swayze GA, Singer RB, Pollack JB (1990) High-resolution reflectance spectra of Mars in the 2.3-μ region; evidence for the mineral scapolite. J Geophys Res 95:14 463–14 480

Clayton DD (1977) Solar isotopic anomalies – supernova neighbor or presolar carriers. Icarus 32:255–269

Clayton DD (1978) Precondensed matter – key to the early solar system. Moon Planets 19:107–137

Cloud P (1978) Cosmos, Earth, and man; a short history of the Universe. Yale University Press, New Haven, 372 pp

Coey JMD, Mørup S, Madsen MB, Knudsen JM (1990) Titanomagnetite in magnetic soils on Earth and Mars. J Geophys Res 95:14 423–14 425

Cole S (1991) Magellan 'venusquake' evidence reconsidered. EOS, American Geophysical Union, Trans 72:426–427

Collinson DW (1986) Magnetic properties of Antarctic Shergottite meteorites EETA 79001 and ALHA 77005 – possible relevance to a martian magnetic field. Earth Planet Sci Lett 77:159–164

Comins N (1991) The Earth without the Moon. Astronomy 19(2):48–53

Condie KC (1989) Plate tectonics & crustal evolution. Pergamon, New York, 476 pp

Connerney JE, Ness NF (1988) Mercury's magnetic field and interior. In: Vilas F, Chapman CR, Matthews MS (eds) Mercury. University of Arizona Press, Tucson, pp 494–513

Cook AF, Franklin FA (1970) An explanation of the light curve of Iapetus. Icarus 13:282–291

Coradini M, Fulchignoni M, Visicchio F (1980) The hypsometric curve of Mars. Moon Planets 22:201–210

Counselman CC III, Gourevitch SA, King RW, Loriot GB (1980) Zonal and meridional circulation of the lower atmosphere of Venus determined by radio interferometry. J Geophys Res 85:8026–8030

Courtillot VE (1990) A volcanic eruption. Sci Am 263(4):85–92

Courvoisier J-L, Robson EI (1991) The Quasar 3C 273. Sci Am 264(6):50–57

Craddock RA, Maxwell TA (1990) Resurfacing of the martian highlands in the Amenthese and Tyrrhena region. J Geophys Res 95:14 265–14 278

Craddock RA, Greeley R, Christensen PR (1990) Evidence for an ancient impact in Daedalia Planum, Mars. J Geophys Res 95:10 729–10 741

Croft SK (1983) A proposed origin for palimpsests and anomalous pit craters on Ganymede and Callisto. 14th Lunar and Planetary Science Conference, Pt 1, Proc. J Geophys Res 88, Suppl: B71–B89

Croft SK, Soderblom LA (1991) Geology of the uranian satellites. In: Bergstralh JT, Miner ED, Matthews MS (eds) Uranus. University of Arizona Press, Tucson, pp 561–628

Croft SK, Kieffer SW, Ahrens TJ (1979) Low-velocity impact craters in ice and ice-saturated sand with implications for martian crater count ages. J Geophys Res 84:8023–8032

Cronin JR, Pizzarello S, Cruikshank DP (1988) Organic matter in carbonaceous chondrites, planetary satellites, asteroids and comets. In: Kerridge JF, Matthews MS (eds) Meteorites and the early Solar System. University of Arizona Press, Tucson, pp 819–857

Crough ST (1983) Hotspot swells. Earth Planet Sci, Annu Rev 11:165–193

Crowley T, Baum SK (1991) Toward reconciliation of Late Ordovician (∼ 440 Ma) glaciation with very high CO₂ levels. J Geophys Res 96:22 597–22 610

Cruikshank DP, Brown RH (1986) Satellites of Uranus and Neptune, and the Pluto-Charon system. In: Burns JA, Shapley MS (eds) Satellites. University of Arizona Press, Tucson, pp 836–874

Cruikshank DP, Degewiji J, Zellner BT (1982) The outer satellites of Jupiter. In: Morrison D (ed) Satellites of Jupiter. University of Arizona Press, Tucson, pp 129–146

Cruikshank DP, Allamandola LJ, Hartmann WK, Tholen DJ, Brown RH, Matthews CN, Bell JF (1991) Solid C=N bearing material on outer solar system bodies. Icarus 94:345–353

Crumpler LS (1990) Eastern Aphrodite Terra on Venus; characteristics, structure, and mode of origin. Earth, Moon Planets 50/51:343–388

Crumpler LS, Head JW III, Harmon JK (1987) Regional linear cross-strike discontinuities in western Aphrodite Terra, Venus. Geophys Res Lett 14:607–610

Crumpler LS, Aubele JC, Head JW (1992) The global distribution of volcanism on Venus; results from Magellan. American Geophysical Union, Spring Meet, Montreal, May 12–16, 1992, Program and Abstr, p 177

Cutts JA (1973a) Wind erosion in the martian polar regions. J Geophys Res 78:4211–4221

Cutts JA (1973b) Nature and origin of layered deposits of the martian polar regions. J Geophys Res 78:4231–4249

Cutts JA, Blasius KR (1981) Origin of martian outflow channels; the eolian hypothesis. J Geophys Res 86:5075–5102

Cutts JA, Lewis BH (1982) Models of climate cycles recorded in martian polar layered deposits. Icarus 50:216–244

Cutts JA, Smith RSU (1973) Eolian deposits and dunes on Mars. J Geophys Res 78:4139–4154

Dalrymple CB, Ryder G (1992) ⁴⁰Ar/³⁹Ar age spectra of Apollo 15 impact melt rocks using a continuous laser system. American Geophysical Union Spring Meet Montreal, May 12–16, 1992, Program and Abstr, p 362

Daly RA (1946) Origin of the Moon and its topography. Proc Am Philos Soc 90:104–119

Dalziel IWD (1991) Pacific margins of Laurentia and east Antarctica-Australia as a conjugate rift pair; evidence and implications for an Eocambrian supercontinent. Geology 19:598–601

Dana JD (1846) On the volcanoes of the Moon. Am J Sci, 2nd series. 2:335–353

Darwin GH (1881) On the tidal friction of a planet surrounded by several satellites and on the evolution of the solar system. Philos Trans R Soc Lond 172:491–535

Davies DW (1979) The relative humidity of Mars' atmosphere. J Geophys Res 84:8335–8340

Davies GF (1980) Review of oceanic and global heat flow estimates. Rev Geophys Space Phys 18:718–722

Davies GF (1990) Heat and mass transport in the early Earth. In: Newson HE, Jones JH (eds) Origin of the Earth. Oxford University Press, New York; Lunar and Planetary Institute, Houston, pp 175–194

Davies ME, Gault DE, Devornik SE, Strom RG (1978) Atlas of Mercury. National Aeronautics and Space Adminstration, Washington, DC, SP-423, 128 pp

Davis DR, Chapman C, Greenberg R, Weidenschilling S, Harris A (1979) Collisional evolution of asteroids; populations, rotations, and velocities. In: Gehrels T (ed) Asteroids. University of Arizona Press, Tucson, pp 528–557

Davis M, Hut P, Muller RA (1984) Extinction of species by periodic comet showers. Nature 308:715–717

Davis PA, Golombek MP (1990) Discontinuities in the shallow martian crust at Lunae, Syria, and Sinai plana. J Geophys Res 95:14 231–14 248

Delsemme AH, (1978) Comets and the origin of the Solar System. In: Dermott SF (ed) The origin of the Solar System. Wiley, New York, pp 463–468

de Wit MJR, Hart A, Hart RJ (1987) The Jamestown ophiolitic complex, Barberton mountain belt. A section through 3.5 Ga oceanic crust. J Afr Earth Sci 6:681–730

Dietz RS (1946) The meteoritic impact origin of the Moon's surface features J Geol 54:359–375

Dietz RS (1947) Meteorite impact suggested by the orientation of shatter-cones at the Kentland, Indiana disturbance. Science 105:76

Dietz RS (1959) Shatter cones in cryptoexplosion structures (meteorite impact?). J Geol 67:496–505

Dietz RS (1964) Sudbury structure as an astrobleme. J Geol 72:412–434

Dietz RS (1972) Sudbury astrobleme, splash emplaced sub-layer and possible cosmogenic ores. Geol Assoc Canada Spec Pap 10:29–40

Dietz RS (1977) Elgygytgyn Crater, Siberia; probable source of Australian tektite field. Meteoritics 12:145–157

Dietz RS, Holden JC (1965) Earth and Moon; tectonically contrasting realms. Ann NY Acad Sci 123:631–640

Dietz RS, McHone J (1974) Kaaba Stone; not a meteorite, probably an agate. Meteoritics 9:173–179

Dionne J-C (1992) Ring structures made by shore ice in a muddy tidal flat, St. Lawrence estuary, Canada. Sediment Geol 76:285–292

Dolginov SS, Yeroshenko YG, Zhugarov LN (1973) Magnetic field in the very close neighborhood of Mars according to data from Mars 2 and Mars 3 space craft. J Geophys Res 78:4779–4786

Donahue TM, Hoffman JH, Hodges RR Jr, Watson AJ (1982) Venus was wet; a measurement of the ratio of deuterium to hydrogen. Science 216:630–633

Duffield WA (1972) A naturally occurring model of global plate tectonics. J Geophys Res 77:2543–2555

Duncan RA, (1991) Ocean drilling and the volcanic record of hot spots. GSA Today l(10):214–216, 219

Duxbury TC,Veverka J (1977) Viking imaging of Phobos and Deimos; an overview of the primary mission. J Geophys Res 82:4203–4212

Dzurisin D (1978) The tectonic and volcanic history of Mercury as inferred from studies of scarps, ridges, troughs and other lineaments. J Geophys Res 83:4883–4906

Eberli GP, Ginsburg RN (1987) Segmentation and coalescence of Cenozoic carbonate platforms, southwest Great Bahama Bank. Geology 15:75–79

Edgett KS, Christensen PR (1992) The particle size of martian aeolian dunes. J Geophys Res 96:22 765–22 776

EEZ-SCAN 87, Scientific Staff (1991) Atlas of the US Exclusive Economic Atlantic Continental Margins. US Geological Survey, Miscel Invest Ser, I-2054, 174 pp

El-Baz F, Breed CS, Grolier MJ, McCauley JF (1979) Eolian features in the Western Desert of Egypt and some applications to Mars. J Geophys Res 84:8205–8221

Emery KO, Aubrey DG (1991) Sea levels, land levels, and tide gauges. Springer, Berlin, Heidelberg, New York, 237 pp

Emery KO, Uchupi E (1984) The geology of the Atlantic Ocean. Springer, Berlin, Heidelberg, New York, 1050 pp

Everett JR, Morisawa M, Short NM (1986) Tectonic landforms. In: Short NM, Blair RW Jr (eds) Geomorphology from space. National Aeronautics and Space Administration, Washington, DC, SP-486, pp 27–184

Everhart E (1979) Chaotic orbits in the Solar System. In: Gehrels T (ed) Asteroids. University of Arizona Press, Tucson, pp 283–288

Fanale FP (1976) Martian volatiles; their degassing history and geochemical fates. Icarus 28:179–202

Fanale FP, Baznerdt WB, Elson LS, Johnson TV, Zurek RW (1982) Io's surface; its phase composition and influence on Io's atmosphere and Jupiter's magnetosphere. In: Morrison D (ed) Satellites of Jupiter. University of Arizona Press, Tucson, pp 756–806

Feierberg MA, Drake MJ (1980) The meteorite-asteroid connection – the infrared spectra of eucrites, shergottites, and Vesta. Science 209:805–807

Feierberg MA, Larson HP, Fink U, Smith HA (1980) Spectroscopic evidence for two achondritic parent bodies; asteroids 349 Dembowska and 4 Vesta. Geochim Cosmochim Acta 44:513–524

Feierberg MA, Lebofsky LA, Larson HP (1981) Spectroscopic evidence for aqueous alteration products on the surface of low-albedo asteroids. Geochim Cosmochim Acta 45:971–981

Feierberg MA, Lebofsky LA, Tholen DJ (1985) The nature of C-class asteroids from 3 micron spectrophotometry. Icarus 63:183–191

Fielder G, Fielder F (1971) Lava flows and the origin of small craters in Mare Imbrium. In: Fielder G (ed) Geology and physics of the Moon. Elsevier, New York, pp 15–26

Fink JH, Park SO, Greeley R (1983) Cooling and deformation of sulfur flows. Icarus 56:38–50

Fink U, Smith BA, Benner DC, Johnson JR, Reitsema HJ (1980) Detection of a CH_4 atmosphere on Pluto. Icarus 44:62–71

Finnerty AA, Ransford GA, Pieri DC, Collerson KD (1980) Is Europa's surface cracking due to thermal evolution? Nature 289:24–27

Fisher AG, Arthur MA (1977) Secular variations in the pelagic realm. In: Cook HE, Enos P (eds) Deep-water carbonate environments. Soc Econ Paleontol Mineral, Spec Publ 25:19–50

Fisher C (1943) The story of the Moon. Doubleday, Doran, Garden City, 301 pp

Florenskiy KP, Bazilevskiy AT, Burba GA, Nokolaveya OV, Pronin AS, Selivanov AS, Narayeva MK, Panfilov AS, Chemadanov VP (1983) Panorama of Venera 9 and 10 landing sites. In: Hunten DM, Colin L, Donahue TM, Moroz VI (eds) Venus. University of Arizona Press, Tucson, pp 137–153

Flynn GF, McKay DS (1990) An assessment of the meteoritic contribution to the martian soil. J Geophys Research 95:14 497–14 509

Ford PG, Pettengill GH (1992) Venus topography and kilometer-scale slopes. J Geophys Res 97:13 103–13 114

Ford PG, Senske DA (1990) The radar scattering characteristics of Venus landforms. Geophys Res Lett 17:1361–1364

Francis PW, Wade G (1983) Olympus Mons aureole; formation by gravitational spreading. J Geophys Res 88:8333–8344

Francis PW, Wood CA (1982) The absence of silicic volcanism on Mars; implications for crustal composition and volatile abundance. J Geophys Res 87:9881–9889

Frank SL, Head JW (1990) Ridge belts on Venus; morphology and origin. Earth, Moon Planets 50/51:421–470

Freeman, JW (1993) Space physics and undergraduate education: sins of omission. EOS, American Geophysical Union, Trans 74:52–53

French RG, Nicholson PD, Porco CC, Marouf EA (1991) Dynamics and structure of uranian rings. In: Bergstralh JT, Miner ED, Matthews MS (eds) Uranus. University of Arizona Press, Tucson, pp 327–409

Frey HV, Grant TD (1990) Resurfacing history of Tempe Terre and surroundings. J Geophys Res 95:14 249–14 264

Frey H, Schultz RA (1988) Large impact basins and the mega-impact origin for the crustal dichotomy on Mars. Geophys Res Lett 15:229–232

Frey HV, Schultz RA (1990) Speculation on the origin and evolution of the Utopia–Elysium lowlands of Mars. J Geophys Res 95:14 203–14 213

Frey H, Lowry BL, Chase SA (1979) Pseudocraters on Mars. J Geophys Res 84:8075–8086

Froude DO, Ireland TR, Kinny PO, Williams IS, Compston W (1983) Ion microprobe identification of 4100–4200 Ma-old terrestrial zircons. Nature 304:616–618

Fryer RJ (1971) Origin of the lunar craters. In: Fielder G (ed) Geology and physics of the Moon. Elsevier, New York, pp 105–114

Fuentes C (1992) The buried mirror; reflections on Spain and the New World. Houghton Mifflin, Boston, 399 pp

Gabrilov IV (1972) Gipsograficheskaya krivaya Luny. In: Martynov DI (ed) Fizika Luny i Planet. Izdatel'stvo Nauka, Moscow, pp 144–148

Gaffey MJ (1990) Thermal history of the Asteroid Belt; implications for accretion of the terrestrial planets. In: Newson HE, Jones JH (eds) Origin of the Earth. Oxford University Press, New York; Lunar and Planetary Institute, Houston, pp 17–28

Gaffey MJ, McCord TB (1978) Asteroids surface materials – mineralogical characterizations from reflectance spectra. Space Sci Rev 21:555–628

Garvin JB (1990) Global budget of impact-derived sediment on Venus. Earth, Moon Planets 50/51:175–190

Garvin JB, Williams RS (1990) Small domes on Venus; probable analogs of Icelandic lava shields. Geophys Res Lett 17:1381–1384

Gaskell RW, Sinnott SP, McEwen AS, Schaber GG (1988) Large-scale topography of Io; implications for internal structure and heat transfer. Geophys Res Lett 15:581–584

Gault DE, Guest JE, Murray JB, Dzurisin D, Malin MC (1975) Some comparisons of impact craters on Mercury and the Moon. J Geophys Res 80:2444–2460

Geissler PE, Singler RB, Lucchitta BL (1990) Dark materials in Valles Marineris; indications of the style of volcanism and magmatism on Mars. J Geophys Res 95:14 399–14 413

Gilbert GK (1893) The Moon's face; a study of the origin of its features. Bull Philos Soc Wash 12:241–292

Glass BP (1990) Tektites and microtektites; key facts and inferences. Tectonophysics 171:393–404

Goettel KA (1988) Present bounds of the bulk composition for planetary formation processes. In: Vilas F, Chapman CR, Matthews MS (eds) Mercury. University of Arizona Press, Tucson, pp 613–621

Goettel KA, Barshay SS (1978) The chemical equilibrium model for condensation in the Solar System nebula – assumptions, implications, and limitations. In: Dermott SF (ed) The origin of the Solar System. Wiley, New York, pp 611–628

Goettel KA, Shields JA, Decker DA (1981) Density constraints on the composition of Venus. 11th Lunar and Planetary Science Conf, Proc 12B:1501–1516

Golombeck MP (1979) Structural analysis of lunar grabens and the shallow structure of the Moon. J Geophys Res 84:4657–4666

Golombeck MP, Banerdt WB, Tanaka KL, Tralli DM (1992) A prediction of mare seismicity from surface faulting. Science 258:979–981

Goudas CL, Katsiaris GA, Halioulas AA (1978) Dynamical considerations of the origin of the Solar System. In: Dermott SF (ed) The Origin of the Solar System. Wiley, New York, pp 259–266

Gould SJ (1989) Wonderful life. Noreton, New York, 347 pp

Gradie J, Tedesco E (1982) Compositional structure of the Asteroid Belt. Science 216:1405–1407

Gradie J, Thomas P, Veverka J (1980) Surface composition of Amalthea. Icarus 44:373–387

Grant M (1990) The visible past. Scribner Sons, New York, 252 pp

Greeley R (1985) Planetary landscapes. Allen & Unwin, London, 265 pp

Greeley R (1987) Planetary landscapes (revised paperback edition). Allen and Unwin, Boston, 275 pp

Greeley R, Crown DA (1990) Volcanic geology of Tyrrhena Patera, Mars. J Geophys Res 95:7133–7149

Greeley R, Schneid BD (1991) Magma generation on Mars; amounts, rates, and comparisons with Earth, Moon, and Venus. Science 254:996–998

Greeley R, Spudis P (1978) Volcanism in the cratered terrain hemisphere of Mars. Geophys Res Lett 5:453–455

Greeley R, Theilig E, Guest JE, Carr MH, Masursky H, Cutts JA (1977) Geology of Chryse Planitia. J Geophys Res 82:4093–4109

Greeley R, Fink JH, Gault DE, Guest JE (1982) Experimental simulation of impact cratering on icy satellites. In: Morrison D (ed) Satellites of Jupiter. University of Arizona Press, Tucson, pp 340–378

Greeley R, Belton M, Head JW, McEwen A, Neukum G (1991) Galileo SSI results; lunar maria. American Geophysical Union, Spring Meet, Baltimore, May 28–31, 1991, Program and Abstr, p 182

Greeley R, Arvidson RE, Elachi C, Geringer ME, Plaut JJ, Saunders RS, Schubert G, Stofan ER, Thouvenot EJP, Wall SD, Weitz CM (1992) Aeolian features on Venus; preliminary Magellan results. J Geophys Res 97:13 319–13 345

Greenberg R (1989) Planetary accretion. In: Atreya SK, Pollack JB, Matthews MS (eds) Origin and evolution of planetary atmospheres. University of Arizona Press, Tucson, pp 137–164

Greenberg R, Wacker JF, Hartmann WK, Chapman CR (1978a) Planetesimals to planets – numerical simulation of collision evolution. Icarus 35:1–26

Greenberg R, Hartmann WK, Chapman CR, Wacker J (1978b) The accretion of planets from planetesimals. In: Gehrels T, Matthews M (eds) Protostars and planets. University of Arizona Press, Tucson, pp 599–622

Greenberg R, Croft SK, Janes DM, Kargel JS, Lebofsky LA, Lunine JI, Marcialis RL, Melosh HJ, Ojakangas GW, Strom RG (1991) Miranda. In: Bergstralh JT, Miner ED, Matthews MS (eds) Uranus. University of Arizona Press, Tucson, pp 693–735

Grieve RAF (1982) The record of impact on Earth; implications for a major Cretaceous/Tertiary impact event. In: Silver LT, Schultz PH (eds) Geological implications of impacts of large asteroids and comets on the Earth. Geol Soc Am Spec Pap 190:25–37

Grieve RAF (1987) Terrestrial impact structures. Earth Planet Sci, Annu Rev 15:245–270

Grieve RAF (1990) Impact cratering on the Earth. Sci Am 262(4):66–73

Grieve RA, Stoffler D, Deutsch A (1991) The Sudbury structure; controversial or misunderstood. J Geophys Res 96:22 753–22 764

Grimm RE, Phillips RJ (1990) Tectonics of the Lakshmi Planum, Venus; tests for Magellan: Geophys Res Lett 17:1349–1352

Grimm RE, Phillips RJ (1992) Anatomy of a venusian hot spot: geology, gravity, and mantle dynamics of Eistla Regio. J Geophys Res 97:16 035–16 054

Grimm RE, Solomon SC (1989) Tests of crustal divergence models of Aphrodite Terra, Venus. J Geophys Res 94:12 103–12 131

Grinspoon DH, Lewis JS (1988) Cometary water on Venus; implications of stochastic impacts. Icarus 74:21–35

Grossman L, Olsen E, Lattimer JM (1979) Silicon in carbonaceous chondrite metal – relic of high-temperature condensation. Science 206:449–451

Guest JE (1971a) Centres of igneous activity in the maria. In: Fielder G (ed) Geology and physics of the Moon. Elsevier, New York, pp 41–54

Guest JE (1971b) Geology of the farside crater Tsiolkovsky. In: Fielder G (ed) Geology and physics of the Moon. Elsevier, New York, pp 93–104

Guest JE, Bulmer MH (1992) Volcanic edifices and volcanism in the plains of Venus. American Geophysical Union, Spring Meet, Montreal, May 12–16, 1992, Program and Abstr, p 178

Guest JE, Greeley R (1983) Geologic map of the Shakespeare quadrangle of Mercury. US Geological Survey, Miscel Invest Ser, Map I-1408, 1 sheet

Guest JE, O'Donnell WP (1977) Surface history of Mercury; a review. Vistas Astron 20:273–300

Guest JE, Butterworth PS, Greeley R (1977) Geological observation in the Cydonia region of Mars from Viking. J Geophys Res 82:4111–4120

Guest JE, Bulmer MH, Aubele J, Beratan K, Greeley R, Head JW, Michaels G, Waitz C, Wiles C (1992) Small volcanic edifices and volcanism in the plains of Venus. J Geophys Res 97:15 949–15 966

Guinness EA, Arvidson RE, Gehret DC, Bolef LK (1979) Color changes at the Viking sites over the course of a Mars year. J Geophys Res 84:8355–8364

Gulick VC, Baker VC (1990) Origin and evolution of valleys on martian volcanoes. J Geophys Res 95:14 325–14 344

Gurnis M (1988) Large-scale mantle convection and the aggregation and dispersal of supercontinents. Nature 332:695–699

Gurnis M (1990) Plate-mantle coupling and continental flooding. Geophys Res Lett 17:623–626

Haff PK, Watson CC (1979) Ion erosion on the Galilean satellites of Jupiter. 10th Lunar and Planetary Science Conf, Proc, pp 1685–1699

Haggerty SE, Sautter V (1990) Ultradeep (greater than 300 kilometers) ultramafic upper mantle xenoliths. Science 248:993–996

Hale W (1980) Orientation of central peaks in lunar craters: implications for regional structural trends. Conf on the Lunar Highlands Crust, Houston, November 14–16, 1979, Proc Geochim Cosmochim Acta Suppl 12:197–209

Hall JL, Solomon SC, Head JW (1981) Lunar floor-fractured craters: evidence for viscous relaxation of crater topography. J Geophys Res 86:9537–9552

Halliwell JJ (1991) Quantum cosmology and the creation of the Universe. Sci Am 265(6):76–85

Han T-M, Runnegar B (1992) Megascopic eukaryotic algae from the 2.1 billion-year-old Negaunee iron-formation, Michigan. Science 257:232–235

Hansen CJ, McEwen AS, Ingerson AP, Terrile RJ (1990) Surface airborne evidence for plumes and wind in Triton. Science 250:421–424

Hargraves RB, Cullicott CE, Deffeyes KS, Hougen S, Christiansen PP, Fiske PS (1990) Shatter cones and shocked rocks in southwestern Montana; the Beaverhead impact structure. Geology 18:832–834

Harmon JK, Campbell DB (1988) Radar observations of Mercury. In: Vilas F, Chapman CR, Matthews MS (eds) Mercury. University of Arizona Press, Tucson, pp 101–117

Harmon JK, Campbell DB, Head JW, Shapireom II (1986) Radar altimetry of Mercury; a preliminary analysis. J Geophys Res 91:385–401

Harris AW (1977) An analytical theory of planetary rotation rates. Icarus 31:168–174

Harris AW (1978) Dynamics of planetesimal formation and planetary accretion. In: Dermott SF (ed) The origin of the Solar System: Wiley, New York, pp 469–492

Harrison TM, Copeland P, Kidd WSF, Yin A (1992) Raising Tibet. Science 255:1663–1670

Hartmann WK (1973) Martian cratering, 4, Mariner 9 initial analysis of cratering chronology. J Geophys Res 78:4096–4116

Hartmann WK (1974) Geological observations of Martian arroyos. J Geophys Res 79:3951–3957

Hartmann WK (1977) Relative crater production rates on planets. Icarus 31:260–276

Hartmann WK (1979) Diverse puzzling asteroids and a possible unified explanation. In: Gehrels T (ed) Asteroids. University of Arizona Press, Tucson, pp 466–479

Hartmann WK (1980) Dropping stones in magma oceans; effects of early lunar cratering. Conference on the Lunar Highlands Crust, Houston, November 14–16, 1979, Proc Geochim Cosmochim Acta, Suppl 12:155–171

Hartmann WK (1983) Moons and planets, 2nd ed. Wadsworth, Belmont, 509 pp

Hartmann WK, Davis DR (1975) Satellite-sized planetesimals and lunar origin. Icarus 24:504–515

Hartmann WK, Hartmann AC (1968) Asteroid collision and evolution of asteroidal mass distribution and meteoritic flux. Icarus 8:361–381

Hartmann WK, Strom RG, Weidenschilling SJ, Blasius KR, Woronow W, Dence MR, Grieve RAF, Diaz J, Chapman CR, Shoemaker EM, Jones KL (1981) Chronology of planetary volcanism by comparative studies of planetary cratering. In: Basaltic volcanism on the terrestrial planets. Basaltic Volcanism Study Project. Pergamon, New York, pp 1048–1127

Hartmann WK, Cruikshank DP, Degewij DP (1982) Remote comets and related bodies – colorimetry and surface materials. Icarus 52:377–408

Hatfield CB, Camp MJ (1970) Mass extinction correlated with periodic galactic events. Geol Soc Am Bull 81:911–914

Hawke BR, Lucey PG, Taylor GJ, Bell JF, Peterson CA, Blewett DT, Horton K, Spudis PD (1991) Remote sensing studies of the Orientale basin region of the Moon. American Geophysical Union, Spring Meet, Baltimore, May 28–31, 1991, Program and Abstr, p. 183

Hawking SW (1988) A brief history of time – from the big bang to black holes. Bantam Books, New York, 198 pp

Hays, JD, Imbrie J, Shackleton NJ (1976) Variations in the Earth's orbit; pacemaker of the ice ages. Science 194:1121–1132.

Head JW (1976) Lunar volcanism in space and time. Rev Geophys Space Phys 14:265–300

Head JW (1990a) Venus trough-and-ridge tessera; analog to Earth oceanic crust formed at fast spreading centers? J Geophys Res 95:7119–7132

Head JW (1990b) Assemblages of geologic/geomorphic units in the northern hemisphere of Venus. Earth, Moon Planets 50/51:391–408

Head JW (1990c) Processes of crustal formation on Venus; an analysis of topography, hypsometry, and crustal thickness variation. Earth, Moon Planets 50/51:25–55

Head JW, Campbell DB (1982) Identification of banded terrain in the mountains of Ishtar Terra, Venus. National Aeronautics and Space Administration, Washington, DC, TM-85127, pp 80–82

Head JW III, Crumpler LS (1987) Evidence for divergent plate-boundary characteristics and crustal spreading on Venus. Science 238:1380–1385

Head JW, Crumpler LS (1990) Venus geology and tectonics; hot spot and crustal spreading models and questions for the Magellan mission. Nature 346:525–533

Head JW, Gifford A (1980) Lunar mare domes; classification and modes of origin. Moon Planets 22:235–258

Head JW, Solomon SC (1981) Tectonic evolution of the terrestrial planets. Science 213:62–76

Head JW, Wilson L (1986) Volcanic processes and landforms on Venus; theory, predictions, and observations. J Geophys Res 91:9407–9446

Head JW, Vorder Bruegge RW, Crumpler LS (1990) Venus orogenic belt environment; architecture and origin. Geophys Res Lett 17:1337–1340

Head JW, Campbell DB, Elachi C, Guest JE, McKenzie DP, Saunders RS, Schaber GG, Schubert G (1991a) Venus volcanism; initial analysis from Magellan data. Science 252:276–288

Head J, Guest J, Bulmer M, Wiles C, Lancaster M, Schaber G, Campbell DB, Schubert GG, Saunders S, DeJong E, Parker T, Roberts K, Senske D, Keddie S, Pavri B, Basilesvsky A, Klose B (1991b) Volcanic styles on Venus; recent Magellan results. American Geophysical Union, Spring Meet, Baltimore, May 28–31, 1991, Program and Abstr, p 171

Head JW, Carr M, Chapman C, Davies M, Fanale F, Robinson M, Pieters C, Murchie S, Sunshine J, Fisher E, Pluchak J, Greeley R, Sullivan R, Pilcher C, Greenberg R, Veverka J, Helfstein R, Neukum G, Hoffman H, Juamann R, Klaasenm K, Johnson T, McEwen A, Becker T (1991c) Orientale and South-Aitken basins: Galileo imaging results. American Geophysical Union, Spring Meet, Baltimore, May 28–31, 1991, Program and Abstr, p 182

Head JW, Crumpler LS, Aubele JC (1992) Venus global volcanism; implications for resurfacing style and rates. American Geophysical Union, Spring Meet, Montreal, May 12–16, 1992, Program and Abstr, p 17

Head JW, Crumpler LS, Aubele JC, Guest JE, Saunders RS (1992) Venus volcanism; classification of volcanic features and structures, and global distribution from Magellan data. J Geophys Res 97:13 153–13 197

Hédervári P (1960) A hold hipszografikusgörbéje: álap a hold általános morfológiájához. Int Lunar Soc J 1:154–161

Helfenstein P, Parmentier EM (1980) Fractures on Europa; possible response of an ice crust to tidal deformation. 11th Lunar and Planetary Science Conf, Proc, pp 1987–1998

Helfenstein P, Veverka J, McCarthy P, Lee P, Hiller J (1992) Large quasi-circular features beneath frost on Triton. Science 255:824–826

Henderson JS (1981) The world of the ancient Mayas. Cornell University Press, Ithaca, 271 pp

Heppenheimer TA (1978) On the interaction between nebular drag, planetary perturbations, and accretion. In: Dermott SF (ed) The origin of the Solar System. Wiley, New York, pp 511–520

Herkenhoff KE, Murray BC (1990) High-resolution and albedo of the south polar layered deposits on Mars. J Geophys Res 95:14 511–14 529

Herndron JM (1978) Chondrites and chondrules – macroscopic chemical aspects for their origin. In: Dermott SF (ed) The origin of the Solar System. Wiley, New York, pp 629–634

Herrick RR, Grimm RE (1991) Comments on 'Terrestrial spreading centers under Venus conditions; evaluation of a crustal spreading model for western Aphrodite Terra' by Senske DA, Head JW, Parmentier EM. Earth Planet Sci Lett 104:114–115

Herrick RR, Phillips RJ (1992) Geological correlation with the interior density structure of Venus. J Geophys Res 97:16 017–16 034

Herrick RR, Bills BG, Hall SA (1989) Variations in effective compensation depth across Aphrodite Terra, Venus. Geophys Res Lett 16:543–546

Herschel W (1787, XX) An account of three volcanoes in the Moon. Philos Trans 77:229–232

Higgins CG (1984) Piping and sapping; development of land forms by groundwater outflow. In: LaFleur PG (ed) Ground water as a geomorphic agent. Allen & Unwin, Boston, pp 18–58

Hildebrand AR, Boynton WV (1990) Proximal Cretaceous-Tertiary boundary impact deposits in the Caribbean. Science 248:843–846

Hildebrand AR, Penfield GT (1990) A buried 180 km-diameter impact crater on the Yucatan Peninsula, Mexico. EOS, American Geophysical Union, Trans 71:1425

Hill RI, Campbell IH, Davies GF, Griffiths RW (1992) Mantle plumes and continental tectonics. Science 256:186–193

Hills HK, Freeman W (1991) Evidence for a lunar water vapor event revisited. American Geophysical Union, Spring Meet, Baltimore, May 28–31, 1991, Program and Abstr, p 178

Hirschmann M (1991) Phase equilibria constraints on the Ni content of the lunar core. American Geophysical Union, Spring Meet, Baltimore, May 28–31, 1991, Program and Abstr, p 179

Hodges CA, Moore HJ (1979) The subglacial birth of Olympus Mons and its aureoles. J Geophys Res 84:8061–8074

Hoffman JH, Oyama VI, von Zahn U (1980) Measurements of the Venus lower atmosphere composition; a comparison of results. J Geophys Res 85:7871–7881

Hoffman PF (1991) Did the breakout of Laurentia turn Gondwanaland inside-out? Science 252:1409–1412

Hooke R (1665) Micrographia or some physiological descriptions of minute bodies made by magnifying glasses with observations and inquiries thereupon. Royal Soc London, 270 pp (reprinted in facsimile 1961 by Dover Publications, New York)

Horanyi M (1991) Lunar dust, erosion and deposition. American Geophysical Union, Spring Meet, Baltimore, May 28–31, 1991, Program and Abstr, p 178

Horgan J (1991) In the beginning. Sci Am 264(2):117–125

Horowitz NH, Hobby GL, Hubbard JS (1977) Viking on Mars; the carbon assimilation experiments. J Geophys Res 82:4659–4662

Hsui A, Toksöz M (1977) Thermal evolution of planetary bodies. 8th Lunar Science Conf, Proc, 1: pp 447–461

Hughes GH, App FM, McGetchin TR (1977) Global seismic effects of basin-forming impacts. Phys Earth Planet Interiors 15:251–262

Hunten DM (1979) Capture of Phobos and Deimos by protoatmospheric drag. Icarus 37:113–123

Hunten DM, Watson AJ (1982) Stability of Pluto's atmosphere. Icarus 51:665–677

Hunten DM, Morgan TH, Shemansky DE (1988) The Mercury atmosphere. In: Vilas F, Chapman CR, Matthews MS (eds) Mercury. University of Arizona Press, Tucson, pp 562–612

Ingersoll AP, Tryka KA (1990) Triton's plumes; the dust devil hypothesis. Science 250:435–437

Intriligator DS, Smith EJ (1979) Mars in the solar wind. J Geophys Res 84:8427–8435

Ivanov BA, Nemchinov IW, Svetsov VA, Provalov AA, Khazins RM, Phillips RJ (1992) Impact cratering on Venus: physical and mechanical models. J Geophys Res 97:16 167–16 182

Izenberg NR, Arvidson RE, Phillips RJ (1992) Venus resurfacing; building the global view. American Geophysical Union, Spring Meet, Montreal, May 12–16, 1992, Program and Abstr, p 179

Izett GA (1991) Tektites in Cretaceous-Tertiary boundary rocks on Haiti and their bearing on the Alvarez impact extinction hypothesis. J Geophys Res 96:20 897–20 905

Jakosky BM, Farmer CB (1982) The seasonal global behavior of water vapor in the Mars atmosphere; complete global results of the Viking atmospheric water detector experiment. J Geophys Res 87:2999–3019

James TGH (1988) Ancient Egypt; the land and its legacy. University of Texas Press, Austin, 223 pp

Janes DM, Squyres SW (1991) Mantle diapirism on Venus; constraints on diapir size from coronae. American Geophysical Union, Spring Meet, Baltimore, May 28–31, 1991, Program and Abstr, p 174

Janes DM, Squyres SW, Bindschaler DL, Baer G, Schubert G, Sharpton VL, Stofan ER (1992) Geophysical models for formation and evolution of coronae on Venus. J Geophys Res 97:16 055–16 068

Jansa LF, Pe-Piper G, Robertson PB, Friedenreich O (1989) Montagnais; a submarine impact structure on the Scotian shelf, eastern Canada. Geol Soc Am Bull 101:450–463

Jeans JH (1916) The part played by rotation in cosmic evolution. R Astronom Soc Mon Not 77:186–199

Jeffreys H (1916) On certain possible distributions of meteoritic bodies in the Solar System. R Astronom Soc Mon Not 77:84–112

Jeffreys H (1918) On the early history of the solar system. R Astronom Soc Mon Not 78: pp 424–442

Jeffreys H (1929) The Earth; its origin, history and physical constitution, 2nd edn. Cambridge University Press, Cambridge, 346 pp

Jewitt DC (1982) The rings of Jupiter. In: Morrison D (ed) Satellites of Jupiter. University of Arizona Press, Tucson, pp 44–64

Jewitt DC, Danielson GE, Synnott SP (1979) Discovery of a new Jupiter satellite. Science 206:951

Johnson CL, Sandwell DT (1992) Joints in venusian flows. J Geophys Res 97:13 601–13 610

Johnson DH, Toksöz MN (1977) Internal structure and properties of Mars. Icarus 32:73–84

Johnson TV (1991) Galileo's lunar observation. American Geophysical Union, Spring Meet, Baltimore, May 28–31, 1991, Program and Abstr, p 182

Johnson TV, McGetchin TR (1973) Topography on satellite surfaces and shape of asteroids. Icarus 18:612–620

Johnson TV, Soderblom LA (1982) Volcanic eruptions on Io; implications for surface evolution and mass loss. In: Morrison D (ed) Satellites of Jupiter. University of Arizona Press, Tucson, pp 634–646

Joksch HC (1957) Die hypsometrische Kurve des Mondes. Z Geophys 23:250–255

Jones JH, Hood LL (1990) Does the Moon have the same chemical composition as the Earth's upper mantle? In: Newson HE, Jones JH (eds) Origin of the Earth. Oxford University Press, New York; Lunar and Planetary Institute, Houston, pp 85–100

Jones KL (1974) Evidence for episode of crater obliteration intermediate in martian history. J Geophys Res 79:3917–3931

Kahn AR (1938) On the meteoritic origin of the Black Stone of the Ka'bah. Popul Astron 46:403–407

Kargel JS, Strom RG (1992) Ancient glaciation on Mars. Geology 20:3–7

Kasting JF (1987) Theoretical constraints on oxygen and carbon dioxide concentrations in the Precambrian atmosphere. Precambrian Res 34:563–576

Kasting JF (1992) Paradox lost and paradox found. Nature 355:676–677

Kasting JF, Pollack JB (1983) Loss of water from Venus, 1. Hydrodynamic escape of hydrogen. Icarus 53:479–508

Kasting JF, Toon OB (1989) Climate evolution on the terrestrial planets. In: Atreya SK, Pollack JB, Matthews MS (eds) Origin and evolution of planetary and satellite atmospheres. University of Arizona Press, Tucson, pp 423–449

Kasting JF, Pollack JB, Ackerman TP (1984) Response of Earth's atmosphere to increases in solar flux and implications for loss of water from Venus. Icarus 57:335–355

Kaula WM (1990a) Venus; a contrast in evolution to Earth. Science 247:1191–1196

Kaula WM (1990b) Mantle convection and crustal evolution on Venus. Geophys Res Lett 17:1401–1403

Kaula WM, Muradian LM (1982) Could plate tectonics on Venus be concealed by volcanic deposits? Geophys Res Lett 9:1021–1024

Kaula WM, Phillips RJ (1981) Quantitative tests for plate tectonics on Venus. Geophys Res Lett 8:1187–1190

Kaula WM, Bindschadler DL, Grimm RE, Hansen VL, Roberts KM, Smrekar SE (1992) Styles of deformation in Ishtar Terra and their implications. J Geophys Res 97:16 085–16 120

Keil K (1983) Meteorites from Moon, Mars. Geotimes 28:25–27

Kerr RA (1983) A lunar meteorite and maybe some from Mars. Science 220:288–289

Kerr RA (1990) Commotion over Caribbean impacts. Science 250:1081

Kerr RA (1991a) Do plumes stir Earth's entire mantle? Science 252:1078–1079

Kerr RA (1991b) A core-mantle link. Science 252:1617–1618

Kerr RA (1991c) Venus caught in a geologic act? Science 253:1208

Kerr RA (1991d) Magellan finds a flock of venusian volcanoes. Science 254:803

Kerr RA (1992) Geophysicists take a tour around the Solar System. Science 256: 1634–1635

Kerridge JF (1978) Aspects of accretion in the early solar system. In: Dermott SF (ed) The origin of the Solar System. Wiley, New York, pp 493–510

Kerridge JF, Bunch TE (1979) Aqueous activity on asteroids – evidence from carbonaceous meteorites. In: Gehrels T (ed) Asteroids. University of Arizona Press, Tucson, pp 745–764

Kerridge JF, Mackay AL, Boynton WV (1979) Magnetite in CI carbonaceous meteorites – origin by igneous activity on a planetesimal surface. Science 205:395–397

Kieffer HH, Chase SC Jr, Martin TZ, Miner ED, Palluconi FD (1976) Martian north pole summer temperatures; dirty water ice. Science 194:1341–1344

Kieffer SW (1982) Dynamics and thermodynamics of volcanic eruptions; implications for plumes on Io. In: Morrison D (ed) Satellites of Jupiter. University of Arizona Press, Tucson, pp 647–723

Kieffer SW (1990) A reexamination of the spreading center hypothesis for Ovda and Tethis regions, Venus. Geophys Res Lett 17:1373–1376

Kieffer SW, Hagler BH (1992) A mantle plume model for the equatorial highlands of Venus. J Geophys Res 96:20947–20966

King EA (1977) The origin of tektites; a brief review. Am Sci 65:212–218

King JS, Scott DH (1990) Geologic map of the Beethoven Quadrangle of Mercury. US Geological Survey, Miscel Invest Ser, Map I-2048 (H-7), 1 sheet

Kirby SH, Durham WB, Stern LA (1991) Mantle phase changes and deep-earthquake faulting in subducting lithosphere. Science 252:216–225

Kirk RL, Brown RH, Soderblom LA (1990) Subsurface energy storage of transport for solar-powered geysers on Triton. Science 250:424–429

Kirova OA (1964) Scattered matter from the area of fall of the Tunguska cometary meteorite. In: Cassidy WA (ed) Cosmic dust. Ann NY Acad Sci 119:235–242

Kirschner CE, Granz A, Mullen MW (1992) Impact origin of the Avak structure, Arctic Alaska, and genesis of the Barrow Gas Field. Am Assoc Pet Geol Bull 76:651–679

Kirsten T (1978) Time and the Solar System. In: Dermott SF (ed) The origin of the Solar System. Wiley, New York, pp 267–346

Klepeis JE, Schafer KJ, Barbee TW III, Ross M (1991) Hydrogen-helium mixtures at megabar pressures; implications for Jupiter and Saturn. Science 254:986–989

Klootwijk CT, Gee JS, Pierce JW, Smith GM, McFadden PL (1992) An early India-Asia contact; paleomagnetic constraints from Ninetyeast Ridge, ODP Leg 121. Geology 20:395–398

Klose B, Wood JA (1991) Mineralogy of low-emissivity areas on Venus; Aphrodite Terra, Gula and Maxwell Montes. American Geophysical Union, Spring Meet, Baltimore, May 28–31, 1991, Program and Abstr, p 173

Klose KB, Wood JA, Hashimoto A (1992) Mineral equilibria and high radar reflectivity of Venus mountaintops. J Geophys Res 97:16353–16370

Knoll AH (1991) End of the Proterozoic Eon. Sci Am 265(4):64–73

Kochel RG, Peake RT (1984) Quantification of waste morphology in martian fretted terrain. J Geophys Res 89 (Suppl): C336–C350

Kominz MA, Bond CC (1991) Unusually large subsidence and sea-level events during middle Paleozoic; new evidence supporting mantle convection models for supercontinent assembly. Geology 19:56–60

Kossinna E (1921) Die Tiefen des Weltmeeres. Institut fr Meereskunde, Berlin University, Veröff Geograph-Naturwiss 9:70 pp

Kowal CT (1988) Asteroids; their nature and utilization. Wiley, New York, 152 pp

Kuhi LV (1966) T Tauri stars; a short review. J Astronom Soc Can 60:1–14

Kuhi LV (1978) Rotation of pre-main-sequence stars in NGC2264 and the solar angular momentum problem. Moon Planets 19:199–202

Kuhn WR, Atreyha SK, Postawko SE (1979) The influence of ozone on the martian atmospheric temperature. J Geophys Res 84:8341–8342

Kuiper GP (1955) On the origin of binary stars. Astron Soc Pac Publ 67:387–396

Kuiper GP (1962) Infrared spectra of stars and planets. I. Photometry of the infrared spectrum of Venus, 1–2.5 microns. Lunar and Planetary Laboratory, University of Arizona, Commun 1:83–117

Kumar S, Hunten DM (1982) The atmosphere of Io and other satellites. In: Morrison D (ed) Satellites of Jupiter. University of Arizona Press, Tucson, pp 782–806

Lambeck K (1979) Comments on the gravity and topography of Mars. J Geophys Res 84:6241–6247

Lambeck K (1980) The Earth's rotation – geophysical causes and consequences. Cambridge University Press, Cambridge, 449 pp

Lanzerotti LJ, Brown WL, Poate JM, Augustyniak WM (1978) On the contribution of water products from Galilean satellites to the Jovian magnetosphere. Geophys Res Lett 5:154–157

Larimer JW (1978) Meteorites – relics from the early solar system. In: Dermott SF (ed) The origin of the Solar System. Wiley, New York, pp 347–394

Larson RB (1978) Collapse dynamics and collapse models. In Dermott SF (ed) The origin of the Solar System. Wiley, New York, pp 237–254

Larson RL (1991) Latest pulse of Earth; evidence for a mid-Cretaceous superplume. Geology 19:547–550

Leake MA (1982) The intercrater plains of Mercury and the Moon; their nature, origin, and role in terrestrial planet evolution. In: Advances in planetary geology. National Aeronautics and Space Administration, Washington, DC, TM-8894, 535 pp

Lebofsky LA (1978) Asteroid 1 Ceres—evidence for water of hydration. R Astronom Soc Mon Not 182:17–21

Lebofsky LA (1980) Infrared reflectance spectra of asteroids – a search for water hydration. Astron J 85:573–585

Lebofsky LA, Feierberg MA, Tokunafa AT, Larson HP, Johnson JR (1981) The 1.7 to 4.2 micron spectrum of Asteroid 1 Ceres-evidence for structural water in clay minerals. Icarus 48:453–459

Lee SW, Thomas PC, Veverka J (1982) Wind streaks in Tharsis and Elysium; implications for sediment transport by slope winds. J Geophys Res 87:10025–10041

Leighton RB, Murray BC (1966) Behavior of carbon dioxide and other volatiles on Mars. Science 153:136–144

Leinert C, Richter I, Pitz E, Planks B (1981) The zodiacal light from 1.0 to 3.0 AU as observed by Helios space probe. Astron Astrophys 103:177–188

Leith AC, McKinnon WB (1991) Terrace width variations in complex mercurian craters and the transient strength of crater mercurian and lunar crust. J Geophys Res 96:20923–20932

Lenardie A, Kaula WM, Bindschadler DL (1991) The tectonic evolution of western Isthar Terra, Venus. Geophys Res Lett 12:2207–2212

Leovy CB (1983) The atmosphere on Mars. In: The planets. Readings from Scientific American. Freeman, New York, pp 38–47

Levin GV, Straat PA (1977) Recent results from the Viking labeled release experiment on Mars. J Geophys Res 82:4663–4667

Lewis JS (1988) Origin and composition of Mercury. In: Vilas F, Chapman CR, Matthews MS (eds) Mercury. University of Arizona Press, Tucson, pp 651–666

Lewis JS, Kreimendahl FA (1980) Oxidation state of the atmosphere and crust of Venus from Pioneer–Venus results. Icarus 42:330–337

Linkletter E (1966) The conquest of England. Dorset, New York, 318 pp

Lipskii YuN, Kazimirov DA, Rodionova ZhF, Shuvayeva VA (1975) O volnoobraznom kharaktere gipsograficheskai krivoi i nekotorykh gipsometrichekikh osobennoctyakh poverkhosti Marsa. Dokl Akad Nauk SSSR, Moscow, 224:58–61

Lissauer JJ (1985) Can cometary bombardment disrupt synchronous rotation of planetary satellites? J Geophys Res 90:11289–11293

Liu L (1988) Water in the terrestrial planets and the Moon. Icarus 74:98–107

Lopes RMC, Kilburn CRJ (1990) Emplacement of lava flow field; application of terrestrial studies to Alba Patera, Mars. J Geophys Res 95:14383–14397

Lorell J, Born GH, Christensen EJ, Jordon JF, Laing PA, Martin WA, Sjogren VL, Shapiro II, Reasenberg RD, Slater GL (1972) Mariner 9 celestial mechanics experiment. Science 175:317–320

Lovelock JE (1979) Gaia; a new look at life on Earth. Oxford University Press, Oxford, 157 pp

Lowman PD Jr (1989) Comparative planetology and origin of continental crust. Precambrian Res 44:171–195

Lowman PD (1992) The Sudbury structure as a terrestrial mare basin. Rev Geophys 30:227–243

Lucchitta BK (1978) Geologic map of the north side of the Moon. US Geological Survey, Miscel Invest Ser, Map I-1062, 1 sheet

Lucchitta BK (1979) Landslides in Valles Marineris, Mars. J Geophys Res 84:8097–8113

Lucchitta BK, Ferguson HM (1992) A martian landslide caught in the act? Abstracts of Papers Submitted to 23rd Conf, Lunar and Planetary Science, Houston, p 2 HI-PE, pp 815–816

Lucchitta BK, Soderblom LA (1982) The geology of Europa. In: Morrison D (ed) Satellites of Jupiter. University of Arizona Press, Tucson, pp 521–555

Lucchitta BK, Anderson DM, Shoji H (1981) Did ice streams carve martian outflow channels? Nature 290:759–763

Lunine JI, Stevenson DJ, Yung YL (1983) Ethane ocean on Titan. Science 222:1229–1230
Lunine JI, Atreya SK, Pollack JB (1989) Present state and chemical evolution of the atmospheres of Titan, Triton, and Pluto. In: Atreya SK, Pollack JB, Matthews MS (eds) Origin and evolution of planetary and satellite atmospheres. University of Arizona Press, Tucson, pp 605–665
Lyell C (1850) Principles of geology, 8th edn. Murray, London, 811 pp (lst edn, 1830)
Lyttleton RA (1940) On the origin of the Solar System. R Astronom Soc Mon Not 100:546–553
Macdonald KC, Fox PJ (1990) The mid-ocean ridge. Sci Am 262(6):72–79
Macdonald KC, Fox PJ, Perram LJ, Eisen MF, Haydom RM, Miller SP, Carbotte SM, Cormier M-H, Shor AN (1988) A new view of the mid-ocean ridge from the behaviour of ridge-axis discontinuities. Nature 335:217–225
MacKinnon DJ, Tanaka KL (1989) The impact martian crust; structure, hydrology, and some geologic implications. J Geophys Res 94:17 359–17 370
Malin MC (1976) Age of martial channels. J Geophys Res 81:4825–4845
Malin MC (1992) Mass movements on Venus; preliminary results from Magellan Cycle 1 observations. J Geophys Res 97:16 337–16 352
Malin MC, Dzurisin D (1977) Landform degradation on Mercury, the Moon, and Mars; evidence from crater depth/diameter relationships. J Geophys Res 82:376–388
Marcialis RL, Rieke GH, Lebofsky LA (1987) The surface composition of Charon; tentative identification of water ice. Science 237:1349–1351
Margolis SV, Claeys P, Kyte FT (1991) Microtektites, microkrystites, and spinels from a Late Pliocene asteroid impact in the Southern Ocean. Science 251:1594–1597
Marshall JR, Greeley R (1992) An experimental study of aeolian structures on Venus. J Geophys Res 97:1007–1016
Marshall JR, Fogleman G, Greeley R, Hixon R, Tucker D (1991) Adhesion and abrasion of surface materials in the venusian aeolian environment. J Geophys Res 96:1931–1947
Masursky H, Boyce J M, Dial AL, Schaber GG, Strobell ME (1977) Classification and time of formation of martian channels based on Viking data. J Geophys Res 82:4016–4038
Masursky H, Schaber GG, Soderblom LA, Strom RG (1979) Preliminary geological mapping of Io. Nature 280:725–729
Masursky H, Eliason E, Ford PG, McGill GE, Pettengill GH, Schaber GG, Schubert G (1980) Pioneer Venus radar results; geology from images and altimetry. J Geophys Res 85:8232–8260
Mather KF, Mason SL (1939) A source book in geology. McGraw-Hill, New York, 702 pp
Matsui T, Abe Y (1986) Impact-induced atmosphere and oceans on Earth and Venus. Nature 322:526–528
Maury MF (1855) The physical geography of the sea (3rd edn). Harper & Brothers, New York, 287 pp (1st edn 1854)
McCauley JF (1973) Mariner 9 evidence for wind erosion in the equatorial and mid-latitude regions of Mars. J Geophys Res 78:4123–4137
McCauley JF, Carr MH, Cutts JA, Hartmann WK, Masursky H, Milton DJ, Sharp RP, Wilhelms DE (1972) Preliminary Mariner 9 report on the geology of Mars. Icarus 17:289–327
McCauley JF, Breed CS, El-Baz F, Whitney MI, Grolier MJ, Ward AW (1979) Pitted and fluted rocks in the Western Desert of Egypt, Viking comparisons. J Geophys Res 84:8222–8232
McCord TB, Clark RN (1979) The Mercury soil; presence of Fe^{2+}. J Geophys Res 84:7664–7668
McCrea WH (1978) The formation of the Solar System; a protoplanet theory. In: Dermott SF (ed) The origin of the Solar System: Wiley, New York, pp 75–110
McFarlane EA, and Drake MJ (1990) Element partitioning and the early thermal history of the Earth. In: Newson HE, Jones JH (eds) Origin of the Earth. Oxford University Press, New York; Lunar and Planetary Institute, Houston pp. 135–150
McGill GE (1991) Structural evolution of venusian plains. American Geophysical Union, Spring Meet, Baltimore, May 28–31, 1991, Program and Abstr, p 173
McGill GE, Dimitriou AM (1990) Origin of the martian global dichotomy by crustal thinning in the late Noachian or early Hesperian. J Geophys Res 95:12 595–12 605
McGill GE, Hills LS (1992) Origin of giant martian polygons. J Geophys Res 97:2633–2647
McGill GE, Warner JL, Malin MC, Arvidson RE, Eliason E, Nozette S, Reasenberg RD (1983)

Topography, surface properties, and tectonic evolution. In: Hunten DM, Colin L, Donahue TM, Moroz VI (eds) Venus. University of Arizona Press, Tucson, pp 69–130

McGovern PJ, Solomon S (1992) Patterns of deformation and volcanic flows associated with lithospheric loading by large volcanoes on Venus. American Geophysical Union, Spring Meet, Montreal, May 12–16, 1992, Program and Abstr, p. 178

McKay CP, Toon OB, Kasting F (1991) Making Mars habitable. Nature 352:489–496

McKenzie D, McKenzie JM, Saunders RM (1992) A dike emplacement on Venus and Earth. J Geophys Res 97:15 977–15 990

McKenzie D, Ford PG, Liu F, Pettengill GH (1992a) Pancake domes of Venus. J Geophys Res. 97:15 967–15 976

McKenzie D, Ford PG, Johnson C, Sandwell D, Parsons B, Saunders S, Solomon SC (1992b) Features on Venus generated by plate boundary processes. American Geophysical Union, Spring Meet, Montreal, May 12–16, 1992, Program and Abstr, p. 304

McKenzie D, Ford PG, Johnson C, Parsons B, Sandwell D, Saunders S, Solomon SC (1992c) Features on Venus generated by plate boundary processes. J Geophys Res 97:13 533–13 544

McKinnon WB (1983) Origin of the E-Ring; condensation of impact vapor or boiling of impact melt? Abstracts of Papers Submitted to the 14th Lunar and Planetary Science Conf, Houston, pp 487–488

McKinnon WB, Melosh HJ (1980) Evolution of planetary lithosphere; evidence from multiringed structures on Ganymede and Callisto. Icarus 44:454–471

McKinnon WB, Parmentier EM (1986) Ganymede and Callisto. In: Burns JA, Matthews MS (eds) Satellites. University of Arizona Press, Tucson, pp 718–763

McKinnon WB, Tanaka KL (1989) The impacted martian crust: structure, hydrology, and some geological implications. J Geophys Res 94:17 359–17 370

McKinnon WB, Chapman CR, Housen KR (1991) Cratering of the uranian satellites. In: Bergstralh JT, Miner ED, Matthews MS (eds) Uranus. University of Arizona Press, Tucson, pp 629–692

McSween HY (1979) Are carbonaceous chondrites primitive or processed? A review. Rev Geophys Space Phys 17:1059–1075

McSween HY (1987) Meteorites and their parent planets. Cambridge University Press, New York, 237 pp

Meissner R, Mooney WD (1991) Speculations on continental crustal formation. EOS, American Geophysical Union, Trans 72:585, 590

Melosh HJ (1977) Global tectonics of a despun planet. Icarus 31:221–243

Melosh HJ (1982) A simple mechanical model of Valhalla Basin, Callisto. J Geophys Res 87:1880–1890

Melosh HJ (1989) Impact cratering – a geologic process. Oxford University Press, New York; Lunar and Planetary Institute, Houston, 245 pp

Melosh HJ (1990) Giant impacts and the thermal state of the early Earth. In: Newson HE, Jones JH (eds) Origin of the Earth. Oxford University Press, New York; Lunar and Planetary Institute, Houston, pp 69–83

Melosh HJ, Dzurisin D (1978a) Tectonic implications for the gravity structure of Caloris basin, Mercury. Icarus 33:141–144

Melosh HJ, Dzurisin D (1978b) Mercurian global tectonics; a consequence of tidal despinning. Icarus 35:227–236

Melosh HJ, McKinnon WB (1988) The tectonics of Mercury. In: Vilas F, Chapman CR, Matthews MS (eds) Mercury. University of Arizona Press, Tucson, pp 374–400

Menard HW, Smith SM (1966) Hypsometry of ocean basin provinces. J Geophys Res 71:4305–4325

Merrill RT, McElhinny MW (1983) The Earth's magnetic field. Its history, origin, and planetary perspective. Academic Press, New York, 401 pp

Merritt A (1932) Dwellers in the mirage. Dover Books, New York, 222 pp

Milton DJ (1973) Water and processes of degradation in the martian landscape. J Geophys Res 78:4037–4045

Molnar P, England E (1990) Late Cenozoic uplift of mountain ranges and global climate change; chicken or egg? Nature 346:29–34

Molnar P, Tapponnier R (1975) Cenozoic tectonics of Asia; effects of a continental collision. Science 189:419–426

Moorbath SPN, Taylor PN, Jones NW (1986) Dating oldest terrestrial rocks – fact and fiction. Chem Geol 57:63–86

Moore HJ, Schenk P (1992) Thick lava flows on Venus; distribution, morphology and terrestrial comparison. American Geophysical Union, Spring Meet, Montreal, May 12–16, 1992, Program and Abstr, p 179

Moore HJ, Spitzer CR, Bradford KZ, Cates PM, Hutton RE, Shorthill RW (1979) Sample fields of the Viking landers, physical properties, and aeolian processes. J Geophys Res 84:8365–8377

Moore HJ, Plaut JJ, Chenk PM, Head JW (1992) An unusual volcano on Venus. J Geophys Res 97:13 479–13 493

Moore JM (1984) The tectonic and volcanic history of Dione. Icarus 59:205–220

Moore JM (1990) Nature of the mantling deposited in the heavily cratered terrain of northeastern Arabia, Mars. J Geophys Res 95:14 279–14 289

Moore JM, Ahern JL (1983) The geology of Tethys. 13th Lunar and Planetary Science Conference, Pt 2, Proc, J Geophys Res 88 (Suppl): A577–A584

Moore JM, Horner VM (1984) The geomorphologic features of Rhea. Abstracts of Papers Submitted to the 15th Lunar and Planetary Science Conf, Pt 2, Houston, pp 560–561

Moore JM, Horner VM, Greeley R (1985) Geomorphology of Rhea; implications for geologic history and surface processes. 15th Lunar and Planetary Science Conference, Pt 2, Proc J Geophys Res 90 (Suppl):C785–C796

Moore P (1988) The planet Neptune. Halsted, New York, 144 pp

Moore P (1990) Mission to the planets. Norton, New York, 128 pp

Moore RC (1933) Historical geology. McGraw-Hill, New York, 673 pp

Moores EM (1991) The southwest US – east Antarctica (SWEAT) connection; a hypothesis. Geology 19:425–428

Morel P, Irving E (1978) Tentative paleocontinental maps of the early Phanerozoic. J Geol 86:535–561

Morgan P, Phillips RJ (1983) Hot spot heat transfer; its application to Venus and implications to Venus and Earth. J Geophys Res 88:8305–8317

Moroz VI (1983) Summary of preliminary results of the Venera 13 and Venera 14 missions. In: Hunten DM, Colin L, Donahue TM, Moroz VI (eds) Venus. University of Arizona Press, Tucson, pp 45–68

Morrison D (1982) Introduction to the satellites of Jupiter. In: Morrison D, Matthews MS (eds) Satellites of Jupiter. University of Arizona Press, Tucson, pp 3–43

Morrison D, Owen T, Soderblom LA (1986) The satellites of Saturn. In: Burns JA, Matthews MS (eds) Satellites. University of Arizona Press, Tucson, pp 764–801

Mouginis-Mark P (1979) Martian fluidized crater morphology; variations with crater size, latitude, altitude, and target material. J Geophys Res 84:8011–8022

Mouginis-Mark PJ, McCoy TJ, Taylor GJ, Keil K (1992) Martian parent craters for SNC meteorites. J Geophys Res 97:10 213–10 225

Muhleman DO, Butler BJ, Grossman AW, Slade MA (1991) Radar images of Mars. Science 253:1508–1513

Murphy JB, Nance RD (1992) Mountain belts and the supercontinent cycle. Sci Am 266(4):84–91

Murray BC, Malin MC (1973) Polar volatiles – theory versus observation. Science 182:437–443

Murray BC, Soderblom LA, Cutts JA, Sharp RP, Milton DJ, Leighton RB (1972) A geological framework for the south polar region of Mars. Icarus 17:328–345

Murray BC, Belton MJS, Danielson GE, Davies ME, Gault DE, Hapke B, O'Leary B, Strom RGV, Suomi V, Trask N (1974) Mercury's surface; preliminary description and interpretation from Mariner 10 pictures. Science 185:169–179

Murray BC, Strom RG, Trask NJ, Gault DE (1975) Surface history of Mercury; implications for terrestrial planets. J Geophys Res 80:2508–2514

Murray J (1888) On the height of the land and the depth of the ocean. Scott Geogr Mag 4:1–41

Murray J, Hjort J (1912) The depths of the ocean. Macmillan, London, 821 pp

Murray J, Renard AF (1891) Deep–sea deposits based on the specimens collected during the voyage of H.M.S. Challenger in the years 1872 to 1876. Her Majesty's Stationery Office, London, 583 pp

Murray JB (1971) Sinuous rilles. In: Fielder G (ed) Geology and geophysics of the Moon; a study of some fundamental problems. Elsevier, Amsterdam, pp 27–39

Murray JB, Dolfus A, Smith B (1972) Cartography of the surface markings of Mercury. Icarus 17:576–584

Mutch TA (1972) Geology of the Moon; a stratigraphic view. Princeton University Press, Princeton, 391 pp

Mutch TA, Arvidson RE, Head JW III, Jones KL, Saunders RS (1976) The geology of Mars. Princeton University Press, Princeton, 400 pp

Mutch TA, Arvidson RE, Binder AB, Guinness EA, Morris EC (1977) The geology of the Viking Lander 2 site. J Geophys Res 82:4452–4467

Nance RD, Worsley TR, Moody JB (1986) Post-Archean biogeochemical cycles and long-term episodicity in tectonic processes. Geology 14:514–518

Needham J (1959) Science and civilisation in China, mathematics and the sciences of the heavens and the Earth, vol III. Cambridge University Press, Cambridge, 877 pp

Neiman VB (1964) A comparative description of hypsographic data of some planets. In: Fedynskii VV (ed) (1968) Zemlya vo vselennoi, Moscow MYSL' (The Earth in the Universe, translated from Russian): Jerusalem, Israel Program for Scientific Translations, pp 164–270

Nesbit EG (1987) The young Earth. Allen & Unwin, Boston, 402 pp

Ness MF, Acuña MH, Lepping RP, Burlaga LF, Bohannon KW, Neubauer FM (1979a) Magnetic field studies at Jupiter by Voyager 1; preliminary results. Science 204:982–987

Ness MF, Acuña MH, Lepping RP, Burlaga LF, Bohannon KW, Neubauer FM (1979b) Magnetic studies at Jupiter by Voyager 2; preliminary results. Science 206:966–971

Ness MF, Acuña MH, Behannon KW, Burlaga, LF, Connerney JEP, Lepping RP, Neubauer FM (1982) Magnetic field studies by Voyager; preliminary results of Saturn. Science 215:558–563

Ness MF, Acuña MH, Behannon KW, Burlaga LF, Connerney JEP, Lepping RP, Neubauer FM (1986) Magnetic field of Uranus. Science 233:85–97

Ness MF, Acuña MH, Burlaga LF, Connerney JEP, Lepping RP, Neubauer FM (1989) Magnetic fields of Neptune. Science 246:1473–1478

Neukum G, Hiller K (1981) Martian ages. J Geophys Res 86:3097–3132

Neukum G, and Wise DU (1976) Mars; a standard crater curve and possible new time scale. Science 194:1381–1387

Newson HE (1980) Hydrothermal alteration of impact sites with implications for Mars. Icarus 44:207–216

Newson HE (1990) Accretion and core formation in the Earth; evidence from siderophile elements. In: Newson HE, Jones JH (eds) Origin of the Earth. Oxford University Press, New York; Lunar and Planetary Institute, Houston, pp 273–288

Newson HE, Sims KWW (1991) Core formation during early accretion of the Earth. Science 252:926–933

Nikishin AM (1990) Tectonics of Venus; a review. Earth, Moon Planets 50/51, p. 101–125.

Nikolaveya OV (1990) Geochemistry of the Venera 8 material demonstrates the presence of continental crust on Venus. Earth, Moon Planets 50/51:329–341

Nyquist LE (1983) Do oblique impacts produce martian meteorites? 13th Lunar and Planetary Science Conference, Proc. J Geophys Res 88 (Suppl): A785–A798

Nyquist LE (1984) The oblique impact hypothesis and relative probabilities of lunar and martian meteorites. 14th Lunar and Planetary Science Conference, Proc, Pt 2. J Geophys Res 89 (Suppl):B631–B640

Oberbeck VR, Quaide WL, Arvidson RE, Aggarwal HR (1977) Comparative studies of lunar, martian, and mercurian craters and plains. J Geophys Res 82:1681–1698

O'Connor JE, Baker VR (1992) Magnitude and implications of peak discharge from glacial Lake Missoula. Geol Soc Am Bull 104:267–279

Officer CB, Drake CL, Pindell JL, Meyerhoff AE (1992) Cretaceous-Tertiary events and the Caribbean caper. Today 2:69–70, 73–74

O'Keefe JA (1976) Tektites and their origin. Elsevier, New York, 254 pp

O'Keefe JA (1976) Tektites and their origin. Elsevier, New York, 254 pp

O'Keefe JA (1980) The terminal Eocene event; formation of a ring around the Earth? Nature 285:309–311

O'Keefe JD, Ahrens TJ (1986) Oblique impact – a process for obtaining meteorite samples from other planets. Science 234:346–349

Oort JH (1950) The structure of the cloud of comets surrounding the Solar System and a hypothesis concerning its origin. Bull Astron Inst Neth 11:91–110

Oparin AI (1953) The origin of life. Dover, New York, 157 pp

Owen T, Biemann K, Rushneck DR, Biller JE, Howarth DW, Lafleur AL (1977) The composition of the atmosphere at the surface of Mars. J Geophys Res 82:4635–4639

Oyama VI, Berdahl BJ (1977) The Viking gas exchange experiment results from Chryse and Utopia surface samples. J Geophys Res 82:4669–4676

Padolak M, Hubbard WB, Stevenson DJ (1991) Models of Uranus' interior and magnetic field. In: Bergstralth JT, Miner ED, Matthews MS (eds) Uranus. University of Arizona Press, Tucson, pp 29–61

Page T, Page LW (eds) (1966) The origin of the Solar System. Genesis of the Sun and planets, and life on other worlds. Macmillan, New York, 336 pp

Paige DA, Herkenhoff KE, Murray BC (1990) Mariner 9 observations of the south polar cap of Mars; evidence for residual CO_2 frost. J Geophys Res 95:1319–1335

Parker TJ, Saunders RS, Schneeberger DM (1989) Transitional morphology in West Deuteronilus Mensae, Mars: implications for modification of the lowland/upland boundary. Icarus 82:111–145

Parmentier EM, Squyres SW, Head JW, Allison ML (1982) The tectonics of Ganymede. Nature 295:290–293

Passey QR (1983) Viscosity of the lithosphere of Enceladus. Icarus 53:105–120

Passey QR, Shoemaker EM (1982) Craters and basins of Ganymede and Callisto. In: Morrison D (ed) Satellites of Jupiter. University of Arizona Press, Tucson, pp 379–434

Patterson C (1956) Age of meteorites and of the Earth. Geochim Cosmochim Acta 10:230–237

Pavri B, Head JW III, Klose KB, Wilson L (1992) Steep-sided domes on Venus; characteristics, geologic setting, and eruption conditions from Magellan data. J Geophys Res 97:13 445–13 478

Pearl JC, Sinton WM (1982) Hot spots of Io. In: Morrison D (ed) Satellites of Jupiter: University of Arizona Press, Tucson, pp 724–755

Pechmann JC (1980) The origin of polygonal troughs on the north plains of Mars. Icarus 42:185–210

Phillips RJ (1990) Convection-driven tectonics on Venus. J Geophys Res 95:1301–1316

Phillips RJ, Lambeck K (1980) Gravity fields of the terrestrial planets; long-wavelength anomalies and tectonics. Rev Geophys and Space Phys 18:27–76

Phillips RJ, Malin MC (1983) The interior of Venus and tectonic implications. In: Hunten DM, Colin L, Donahue TM, Moroz VI (eds) Venus. University of Arizona Press, Tucson, pp 159–214

Phillips RJ, Saunders RS (1975) The isostatic state of martian topography. J Geophys Res 80:2893–2898

Phillips RJ, Kaula WM, McGill GE, Malin MC (1981) Tectonic evolution of Venus. Science 212:879–887

Phillips RJ, Grimm RE, Malin MC (1991) Hot-spot evolution of the global tectonics of Venus. Science 252:651–658

Phillips RJ, Arvidson RE, Boyce JM, Campbell DB, Guest JE, Schaber GG, Soderblom LA (1991) Impact craters on Venus. Science 252:288–297

Phillips RJ, Raubertas RE, Arvidson RE, Sarkar IC, Herrick RR, Izenberg N, Grimm RE (1992) Impact craters and Venus resurfacing history. J Geophys Res 97:15 923–15 948

Pickersgill AO, Hunt GE (1979) The formation of martian lee waves generated by a crater. J Geophys Res 84:8317–8331

Pieri DC (1979) Geomorphology of martian valleys. PhD Thesis, Cornell University, Ithaca, 280 pp

Pieri DC (1980a) Martian valleys; morphology, distribution, age, and origin. Science 210:895–897

Pieri DC (1980b) Lineament and polygonal patterns on Europa. Nature 289:17–21

Pieters CM, Belton M, Backer T, Davies M, Fanale F, Fischer E, Gaddis L, Head JW, Helfenstein P, Hoffmann H, Johnson TV, Klassen K, McEwen A, Murchie S, Neukum G, Oberst J, Plutchak J,

Robinson M, Sullivan R, Sunshine J (1991) Compositional heterogeneity and crustal stratigraphy of the Moon from Galileo SSI multi-spectral images. American Geophysical Union, Spring Meet, Baltimore, May 28–31, 1991, Program and Abstr, pp 182–183

Pike RJ (1988) Geomorphology of impact craters on Mercury. In: Vilas F, Chapman CR, Matthews MS (eds) Mercury. University of Arizona Press, Tucson, pp 165–273

Pinet P, Chevrel S (1990) Spectral identification of geologic units on the surface of Mars related to the presence of silicates from Earth-based near-infrared telescopic charge-couple device imaging. J Geophys Res 95:14 435–14 446

Plaut JF (1992) Results from the Magellan mission to Venus. EOS, American Geophysical Union, Trans 73:22

Plaut JJ, Arvidson RE, Jurgens RF (1990) Radar characteristics of the equatorial plains of Venus from Goldstone observations; implications for interpretation of Magellan data. Geophys Res Lett 17:1357–1360

Plescia JB (1983) The geology of Dione. Icarus 56:255–277

Plescia JB (1991) Graben and extension in northern Tharsis, Mars. J Geophys Res 96:18 883–18 985

Plescia JB, Boyce JM (1982) Crater densities and the geological history of Rhea, Dione, Mimas and Tethys. Nature 295:285–290

Plescia JB, Boyce JM (1983) Crater numbers and the geological histories of Iapetus, Enceladus, Tethys and Hyperion. Nature 301:666–670

Plescia JB, Saunders RS (1982) Tectonic history of the Tharsis region, Mars. J Geophys Res 87:9775–9791

Poag CW, Low D (1987) Unconformable sequence boundaries at Deep Sea Drilling Project site 612, New Jersey transect; their characteristics and stratigraphic significance. Initial Reports of the Deep Sea Drilling Project, US Government Printing Office, Washington, DC, vol 95, pp 453–498

Poag CW, Powars DS, Poppe LJ, Mixon RS, Edwards LE, Folger DW, Bruce S (1992) DSDP Site 612 bolide event: new evidence of a Late Eocene impact-wave deposit and a possible impact site, U. S. east coast. Geology 20:771–774

Poirier JP (1982) Rheology of ices; a key to the tectonics of the ice moons of Jupiter and Saturn. Nature 299:683–688

Pollack JB, Fanale F (1982) Origin and evolution of the Jupiter satellite system. In: Morrison D (ed) Satellites of Jupiter. University of Arizona Press, Tucson, pp 872–910

Pollack JB, Colburn D, Kahn R, Hunter J, Van Camp W, Carlston CE, Wolf MR (1977) Properties of aerosols in the Martian atmosphere, as inferred from Viking Lander imaging data. J Geophys Res 82:4479–4496

Pollack JB, Toon OB, Boese R (1980) Greenhouse models of Venus high surface temperature, as constrained by Pioneer Venus measurements. J Geophys Res 85:8223–8231

Pollack JB, Lunine JI, Tittlemore WC (1991) Origin of the uranian satellites. In: Bergstralh JT, Miner ED, Matthews MS (eds) Uranus. University of Arizona Press, Tucson, pp 469–512

Pope KO, Ocampo AC, Duller CE (1991)Mexican site for K/T impact crater? Nature 351:105.

Porada H (1979) The Damara-Reibeira orogen of the Pan-African-Brasiliana cycle in Namibia (southwest Africa) and Brazil as interpreted in terms of continental collision. Tectonophys 57:237–265

Porco CC (1991) An explanation for Neptune's ring arcs. Science 253:995–1000

Powell CS (1991) Peering inward. Sci Am 264(6):101–111

Powell CS (1992) The golden age of cosmology. Sci Am 267(1):17–22

Prentice AJR (1978) Towards a modern Laplacian theory for the formation of the Solar System. In Dermott SF (ed) The origin of the Solar System. John Wiley, New York, pp 111–162

Prinn RG, Fegley B (1987) The atmospheres of Venus, Earth, and Mars; critical comparison. Earth Planet Sci Annu Rev 15:171–212

Purves NG, Pilcher CB (1980) Thermal migration of water on the Galilean satellites. Icarus 43:51–55

Quirke S, Spencer J (eds) (1992) The British Museum book of ancient Egypt. Thames and Hudson, London, 240 pp

Rampino MR, Stothers RB (1984) Terrestrial mass extinctions, cometary impacts and the Sun's motion perpendicular to the galactic plane. Nature 308:709–712

Ransford GA, Finnerty AA, Collerson KD (1980) Europa's petrological thermal history. Nature 289:21–24

Raup DM, Serkoski JJ (1984) Periodicity of extinctions in the geologic past. Proc Natl Acad Sci USA 81:801–805

Reeves H (1978a) The origin of the Solar System. In: Dermott SF (ed) The origin of the Solar System. Wiley, New York, pp 1–18

Reeves H (1978b) The collapse of interstellar matter in stars. In: Dermott SF (ed) The origin of the Solar System. Wiley, New York, pp 255–258

Reimers CE, Komar PD (1979) Evidence for explosive volcanic density currents on certain martian volcanoes. Icarus 39:88–110

Richardson RM (1992) Ridge forces, absolute plate motions, and the intraplate stress field. J Geophys Res 97:11 739–11 748

Ringwood AE (1970) Origin of the Moon – the precipitation hypothesis. Earth Planet Sci Lett 8:131–140

Ringwood AE (1977) Structure and composition of the Earth. Encyclopaedia Britannica, vol 6. Encyclopaedia Britannica Inc., Chicago, pp 48–57

Ringwood AE (1990) Earliest history of the Earth-Moon system. In: Newson HE, Jones JH (eds) Origin of the Earth. Oxford University Press, New York; Lunar and Planetary Institute, Houston, pp 101–133

Riordan M, Schramm DN (1991 The shadows of creation – dark matter and the structure of the Universe. Freeman, New York, 277 pp

Roberts KM, Head JW (1990a) Western Ishtar Terra and Lakshmi Planum, Venus; models of formation and evolution. Geophys Res Lett 17:1341–1344

Roberts KM, Head JW (1990b) Lakshmi Planum; characteristics and models of origin. Earth, Moon Planets 50/51:193–249

Roberts KM, Head JW III, Guest JE (1991) Mylitta Fluctus, Venus; characteristics of a complex lava flow field. American Geophysical Union, Spring Meet, Baltimore, May 28–31, 1991, Program and Abstr, p 174

Roberts KM, Guest JE, Head JW, Lancaster MG (1992) Mylitta Fluctus, Venus; rift-related, centralized volcanism and emplacement of large-volume flow units. J Geophys Res 97:15 991–16 016

Ross MN, Schubert G, Spohn T, Gaskell RW (1990) Internal structure of Io and the global distribution of its topography. Icarus 85:309–325

Rossbacher LA, Judson S (1981) Ground ice on Mars; inventory, distribution, and resulting landforms. Icarus 45:39–59

Rubey WW (1951) Geologic history of sea water. Geol Soc Am Bull 62:1111–1147

Rubincam DP (1992) Mars secular obliquity change due to seasonal polar caps. J Geophys Res 97:2629–2632

Ruddiman WF, Kutzbach JE (1990) Late Cenozoic uplift and climate change. Trans R Soc Edinb Earth Sci 81:301–314

Russell CT (1991) Lunar magnetism. American Geophysical Union, Spring Meet, Baltimore, May 28–31, 1991, Program and Abstr, p 179

Ryder G (1990) Lunar samples, lunar accretion and the early bombardment of the Moon. EOS, American Geophysical Union, Trans 71:313, 322–323

Sachs A (1955) Late Babylonian astronomical and related texts. Brown University Press, Providence, 271 pp

Sagan C (1973) Sandstorms and eolian erosion on Mars. J Geophys Res 78:4155–4161

Sagan C, Dermott SF (1982) The tide in the seas of Titan. Nature 300:731–733

Sagan C, Veverka J, Fox P, Dubisch R, French R, Gierasch P, Quam L, Lederberg L, Levinthal E, Tucker R, Eross B (1973) Variable features on Mars, 2, Mariner 9 global results. J Geophys Res 78:4163–4196

Sandwell DT, Schubert G (1991) Lithospheric flexure due to thermal subsidence of coronae. American Geophysical Union, Spring Meet, Baltimore, May 28–31, 1991, Program and Abstr, p 174

Sandwell DT, Schubert G (1992a) Evidence for retrograde lithospheric subduction on Venus. Science 257:766–770

Sandwell DT, Schubert G (1992b) Flexural ridges, trenches, and outer rises around coronae on Venus. J Geophys Res 97:16 069–16 084

Sasaki S (1990) The primary solar-type atmosphere surrounding the accreting Earth: H_2O inducted high surface temperature. In: Dawson HE, Jones JH (eds) Origin of the Earth. Oxford University Press, New York; Lunar and Planetary Institute, Houston, pp 195–209

Saunders RS, Carr MH (1984) Venus. In: Carr MH (ed) The geology of the terrestrial planets. National Aeronautics and Space Administration, Washington, DC, SP-469, pp 57–78

Saunders RS, Pettengill GH (1991) Magellan; mission summary. Science 252:247–249

Saunders RS, Dobovolskis AR, Greeley R, Wall SD (1990) Large-scale patterns of eolian sediment transport on Venus; predictions for Magellan. Geophys Res Lett 17:1365–1368

Sautter V, Haggerty SE, Field S (1991) Ultradeep (> 300 kilometers) xenoliths; petrologic evidence from the transition zone. Science 252:827–830

Schaber GG (1973) Lava flows of Mare Imbrium; geologic evaluation from Apollo orbital photography. 4th Lunar Science Conf, Proc, pp 73–92

Schaber GG (1980) The surface of Io; geologic units, morphology and tectonics. Icarus 43:302–333

Schaber GG (1982) The geology of Io. In: Morrison D (ed) Satellites of Jupiter. University of Arizona Press, Tucson, pp 556–597

Schaber GG (1991) Impact cratering on Venus. American Geophysical Union, Spring Meet, Baltimore, May 28–31, 1991, Program and Abstr, p 171

Schaber GG, McCauley JF (1980) Geological map of the Tostoj Quadrangle of Mercury (H-8), US Geological Survey, Miscel Invest Ser, Map I-1199, 1 sheet

Schaber GG, Hortsmann KC, Dial AL (1978) Lava flow materials in the Tharsis region of Mars: 9th Lunar and Planetary Science Conf, Proc, pp 3433–3548

Schaber GG, Strom RG, Moore HJ, Soderblom LA, Kirk RL, Chadwick DJ, Dawson DD, Gaddis LR, Boyce JM, Russell J (1992) Geology and distribution of impact craters on Venus; what are they telling us? J Geophys Res 97:13 327–13 301

Schaefer MW (1990) Geochemical evolution of the northern plains of Mars; early hydrosphere, carbonate development, and present morphology. J Geophys Res 95:14 291–14 300

Schenk PM (1991) Fluid volcanism on Miranda and Ariel; flow morphology and composition: J Geophys Res 96:1887–1906

Schenk PM, Seyfert CK (1980) Fault offsets and proposed plate motions for Europa. EOS, American Geophysical Union, Trans 61:286

Schenk P, Head JW III, Guest JE (1991) Viscous rhyolite-like volcanic flow complex on Venus. American Geophysical Union, Spring Meet, Baltimore, May 28–31, 1991, Program and Abstr, p 174

Schidlowski M, Eichmann R, Junge CE (1975) Precambrian sedimentary carbonates; carbon and oxygen geochemistry and implications for terrestrial oxygen budget. Precambrian Res 2:1–69

Schneeberger DM, Pieri DC (1991) Geomorphology and stratigraphy of Alba Patera, Mars. J Geophys Res 96:1907–1930

Schneider N, Spencer J (1992) Io and Jupiter; the volcano-magnetosphere connection. EOS, American Geophysical Union, Trans 73:55

Schopf JW (1976) Evidence of Archean life. In: Windley BF (ed) The early history of the Earth. John Wiley, New York, pp 589–593

Schopf JW, Hayes JM, Walter MR (1983) Evolution of the Earth's earliest ecosystem; recent progress and unsolved problems. In: Schopf JW (ed) The Earth's earliest biosphere; its origin and evolution. Princeton University Press, Princeton, pp 361–384

Schubert G, Spohn T, Reynolds RT (1986) Thermal histories, composition, and internal structures of the moons of the Solar System. In: Burns JA, Matthews MS (eds) Satellites. University of Arizona Press, Tucson, pp 224–292

Schubert G, Ross MN, Stevenson DJ, Spohn T (1988) Mercury's thermal history and the generation of its magnetic field. In: Vilas F, Chapman CR, Matthews MS (eds) Mercury. University of Arizona Press, Tucson, pp 429–460

Schubert G, Bercovici D, Glatzmaier GA (1990) Mantle dynamics in Mars and Venus; influence of an immobile lithosphere on three-dimensional mantle convection. J Geophys Res 95:14 105–14 129

Schultz PH (1988) Cratering on Mercury. In: Vilas F, Chapman CR, Matthews MS (eds) Mercury. University of Arizona Press, Tucson, pp 274–335

Schultz PH (1992a) Atmospheric effects on cratering efficiency. J Geophys Res 97:975–1005

Schultz PH (1992b) Atmospheric effects on ejecta emplacement and crater formation on Venus from Magellan. J Geophys Res 97:16 183–16 248

Schultz PH, Gault DE (1975) Seismic effects from major basin formation on the Moon and Mercury. Moon 12:159–177

Schultz PH, Glicken H (1979) Impact crater and basin control of igneous processes on Mars. J Geophys Res 84:8033–8047

Schultz PH, Liaza RE (1992) Recent grazing impacts recorded in the Rio Cuart crater field, Argentina. Nature 355:234–237

Schultz PH, Spudis PP (1983) Beginning and end of lunar mare volcanism. Nature 302:233–236

Schultz PH, Srnka LJ (1980) Cometary collisions on the Moon and Mercury. Nature 284:22–26

Schultz RA (1992) Development of Coprates Chasma and western Ophir Planum, Valles Marineris, Mars. J Geophys Res 96:22 777–22 792

Schultz RA, Frey HV (1990) A new survey of multiring impact basins on Mars. J Geophys Res 95:14 175–14 189

Schumm SA (1974) Structural origin of large martian channels. Icarus 22:371–384

Schwab WC, Danforth WW, Scanlon KM, Masson DG (1991) A giant submarine slope failure on the northern insular slope of Puerto Rico. Mar Geol 96:237–246

Schwartz K, Schubart G (1969) The early despinning of the Sun. Astrophys Space Sci 5:444–447

Schwartz RD, James PB (1984) Periodic mass extinctions and the Sun's oscillation about the galactic plane. Nature 308:712–713

Sclater JC, Jauport G, Galson D (1980) The heat through oceanic and continental crust and the heat loss of the Earth. Rev Geophys Space Phys 189:269–311

Scott DH, Carr MH (1978) Geologic map of Mars, scale: 1:25 000 000. US Geological Survey, Miscel Invest Ser, Map I-1083, 1 sheet

Scott DH, Tanaka KL (1986) Geologic map of the western equatorial regions of Mars; scale 1:15 000 000, 67°S–57°N, 0°–180°W. US Geological Survey, Miscel Invest Ser, Map I-1802A, 1 sheet

Scott DH, McCauley JF, West MN (1977) Geologic map of the west side of the Moon. US Geological Survey, Miscel Invest Ser, Map I-1034, 1 sheet

Sears DWG, Dodd RT (1988) Overview and classification of meteorites. In: Kerridge JF, Matthews MS (eds) Meteorites and the early Solar System. University of Arizona Press, Tucson, pp 3–31

Seiff A, Kirk DB (1977) Structure of the atmosphere of Mars in summer at mid-latitudes. J Geophys Res 82:4364–4378

Senske DA (1990) Geology of equatorial region from Pioneer Venus radar images. Earth, Moon Planets 50/51:305–327

Senske DA, Stofan ER (1991) Geology of western Eistla Regio, Venus; results from Magellan radar data. American Geophysical Union, Spring Meet, Baltimore, May 28–31, 1991, Program and Abstr, p 174

Senske DA, Schaber GG, Stofan ER (1992) Regional topographic rises on Venus; geology of western Eistla Regio and comparison to Beta Regio and Atla Regio. J Geophys Res 97:13 395–13 420

Shaler NS (1903) A comparison of the features of the Earth and the Moon. Smithson Contrib Knowledge 34(1438):130 pp

Shapley H (1937) Conspicuous astronomical advances. The Sky 1(3):4–5, 24

Sharp RP (1973a) Mars; troughed terrain. J Geophys Res 78:4063–4072

Sharp RP (1973b) Mars; fretted and chaotic terrains. J Geophys Res 78:4073–4083

Sharp RP (1973c) Mars; south polar pits and etched terrains. J Geophys Res 78:4222–4230

Sharp RP, Malin MC (1975) Channels on Mars. Geol Soc Am Bull 86:4073–4083

Sharpton VL, Head JW III (1986) A comparison of the regional slope characteristics of Venus and Earth; implications for geologic processes on Venus. J Geophys Res 91:7545–7554

Shirley K (1992) Overlooked 'hole' found. Am Assoc Pet Geol, Tulsa, Explorer May Issue, pp 1, 12–13

Shoemaker EM (1960) Penetration mechanics of high velocity meteorites, illustrated by Meteor Crater, Arizona. Int Geol Congr, 21st session, Copenhagen, vol 18, pp 418–434

Shoemaker EM (1962) Exploration of the Moon's surface. Am Sci 50:99–130

Shoemaker EM (1964) The geology of the Moon. Sci Am 211(6):38–47

Shoemaker EM, Hackman RJ (1962) Stratigraphic basis for a lunar time scale. In: Kopal Z, Mikhailov ZK (eds) The Moon. Academic Press, London, pp 289–300

Shoemaker EM, Wolfe RF (1982) Cratering time scales for the Galilean satellites. In: Morrison D (ed) Satellites of Jupiter. University of Arizona Press, Tucson, pp 277–339

Shoemaker EM, Luccitta BK, Squyres SW, Wilhelms DE (1982) The geology of Ganymede. In: Morrison D (ed) Satellites of Jupiter. University of Arizona Press, Tucson, pp 435–520

Short NM (1975) Planetary geology. Prentice-Hall, Englewood Cliffs, 361 pp

Sill GT, Clark RN (1982) Composition of the surfaces of the Galilean satellites. In:Morrison D (ed) Satellites of Jupiter. University of Arizona Press, Tucson, pp 174–212

Simarski LT (1991) Galileo captures asteroid image. EOS, American Geophysical Union, Trans 72:537–538

Simow PD, Wood JA (1992) Fluid outflows from Venus impact craters; analysis from Magellan data. J Geophys Res 97:13 643–13 665

Singer SF (1986) Origin of the Moon by capture. In:Hartmann WH, Phillips RJ, Taylor GJ (eds) Origin of the Moon. Lunar and Planetary Institute, Houston, pp 471–486

Siscoe GL (1991) Aristotle on the magnetosphere. EOS, American Geophysical Union, Trans 72:69–70

Sjogren WL, Phillips RJ, Birkeland PW, Wimberly RM (1980) Gravity anomalies on Venus. J Geophys Res 85:8295–8302

Slade M, Muhleman D, Butler B, Jurgens R (1991) An ice cap on the hottest planet? Science 254:935

Smit J, Klaver G (1981) Sanidine spherules at the Cretaceous-Tertiary boundary – cometary material? Nature 292:47–49

Smit J, Montanari A, Swinburne NHM, Alvarez W, Hildebrand AR, Margolis SV, Claeys P, Lowrie W, Asaro F (1992) Tektite-bearing deep-water clastic unit at the Cretaceous-Tertiary boundary in northeastern Mexico. Geology 20:99–103

Smith BA, Soderblom LA, Beebe R, Boyce J, Briggs G, Carr M, Collins SA, Cook AF II, Danielson GE, Davies ME, Hunt GE, Ingersoll A, Johnson TV, Masursky H, McCauley J, Morrison D, Owen T, Sagan C, Shoemaker EM, Strom R, Suomi VE, Veverka J (1979a) The Galilean satellies and Jupiter; Voyager 2 imaging science results. Science 206:927–950

Smith BA, Soderblom LA, Johnson TV, Ingersoll AP, Collins SA, Shoemaker EM, Hunt GE, Masursky H, Carr MH, Davies ME, Cook AF II, Boyce J, Danielsen GE, Owen T, Sagan C, Beebe RF, Veverka J, Strom RG, McCauley JF, Morrison D, Briggs GA, Suomi VE (1979b) The Jupiter system through the eyes of Voyager 1. Science 204:951–972

Smith BA, Soderblom L, Beebe R, Boyce J, Briggs G, Bunker A, Collins SA, Hansen CJ, Johnson TV, Mitchell JL, Terrile RJ, Carr M, Cook AFII, Cuzzi J, Pollack JB, Danielson GE, Ingersoll A, Davies ME, Hunt GE, Masursky E, Shoemaker E, Morrison D (1981) Encounter with Saturn; Voyager 1 and imaging science results. Science 212:163–191

Smith BA, Soderblom L, Batson L, Bridges P, Inge J, Masursky H, Shoemaker E, Beebe T, Boyce J, Briggs G, Bunker A, Collins SA, Hansen CJ, Johnson TV, Mitchell JL, Terrile RJ, Cook AFII, Cuzzi J, Pollack JB, Danielson GE, Morrison D, Owen T, Sagan C, Veverka J, Strom R, Suomi VE (1982) A new look at the Saturn System; the Voyager 2 images. Science 251:504–537

Smith BA, Soderblom LA, Beebe R, Bliss D, Boyce JM, Brahic A, Briggs GA, Brown RH, Collins SA, Cook AFII, Croft SK, Cuzzi JN, Danielson GE, Davies ME, Dowling TE, Godfrey D, Hansen CJ, Harris C, Hunt GE, Ingersoll AP, Johnson TV, Krauss RJ, Masursky H, Morrison D, Owen T, Plescia JB, Pollack JB, Porco CC, Rages K, Sagan C, Shoemaker EM, Sromvsky LA, Stoker C, Strom RG, Suomi VE, Synnott SP, Terrile RJ, Thomas P, Thompson WR, Veverka J (1986) Voyager 2 in the uranian system; imaging science results. Science 233:43–64

Smith BA, Soderblom LA, Banfield D, Barnet C, Basilevsky AI, Beebe RF, Bollinger K, Boyce JM, Branic A, Briggs GA, Brown RH, Chyba C, Collins SA, Colvin T, Cook AFII, Crisp D, Croft SK,

Cruikshank D, Cuzzi JN, Danielson GE, Davies ME, De Jong E, Dones L, Godfrey D, Goguen J, Grenier I, Haemmerle VR, Hammel H, Hansen CJ, Helfenstein CP, Howell C, Hunt GE, Ingersoll AP, Johnson TV, Kargel J, Kirk R, Kuehn DI, Limaye S, Masursky H, McEwen A, Morrison D, Owen T, Owen W, Pollack JB, Porgo CC, Rages K, Rogers P, Ruby D, Sagan C, Schwartz J, Shoemaker EM, Showalter M, Sicardy B, Simonelli D, Spencer J, Stromovsky LA, Stoker C, Strom RG, Suomi VE, Synott SP, Terrile RJ, Thomas P, Thompson WR, Verbischer A, Veverka J (1989) Voyager 2 at Neptune; imaging science results. Science 246:1422–1449

Smith DE, Lerch FJ, Chan JC, Chinn DS, Iz HB, Mallama A, Patel CB (1990) Mars gravity field error analysis from simulated radio tracking of Mars Observer. J Geophys Res 95:14 155–14 167

Smith DK, Cann JR (1990) Hundreds of small volcanoes on the median valley floor of the Mid-Atlantic Ridge at 24°–30°N. Nature 348:152–155

Smith DK, Cann JR (1992) The role of seamount volcanism in crustal construction at the Mid-Atlantic Ridge (24°–30°N). J Geophys Res 97:1645–1658

Smith EJ, David L, Colerman PJ, Jones DE (1965) Mariner IV measurements near Mars – initial results; magnetic field measurements near Mars. Science 149:1241–1242

Smoluchowski R (1968) Mars; retention of ice. Science 159:1348–1350

Smrekar SE (1991) A comparison of gravitational spreading of high terrain on Venus and on Earth. American Geophysical Union, Spring Meet, Baltimore, May 28–31, 1991, Program and Abstr, p 174

Smrekar SE, Solomon SC (1992) Gravitational spreading of high terrain in Ishtar Terra, Venus. J Geophys Res 97:16 121–16 148

Snyder CW (1979) The planet Mars as seen at the end of the Viking mission. J Geophys Res 84:8487–8519

Soderblom LA, Johnson TV (1983) The moons of Saturn. In: Murray B (ed) The planets. Freeman, San Francisco, pp 95–107

Soderblom LA, Kreidler TJ, Masursky H (1973a) Latitudinal distribution of a debris mantle on the martian surface. J Geophys Res 78:4117–4122

Soderblom LA, Malin MC, Cutts JA, Murray BC (1973b) Mariner 9 observations of the surface of Mars in the north polar region. J Geophys Res 78:4197–4210

Soderblom LA, West RA, Herman BM, Kreidler TJ, Conduit CD (1974) Martian planetwide crater distribution; implications for geologic history and surface processes. Icarus 22:239–263

Soderblom LA, Kieffer SW, Becker TL, Brown RH, Cook AFII, Hansen CJ, Johnson TV, Kirk RL, Shoemaker EM (1990) Triton's geyser-like plumes; discovery and basic characterizations. Science 250:410–415

Solomon SC (1977) The relationship between crustal tectonics and internal evolution in the Moon and Mercury. Phys Earth Planet Interiors 15:135–145

Solomon SC (1978) On volcanism and thermal histories on one-plate planets. Geophys Res Lett 5:461–464

Solomon SC (1991) Tectonic processes on Venus; comparison and contrasts with Earth. American Geophysical Union, Spring Meet, Baltimore, May 28–31, 1991, Program and Abstr, p 171

Solomon SC, Head JW (1982a) Mechanisms for lithospheric heat transport on Venus; implications for tectonic style and volcanism. J Geophys Res 87:9236–9242

Solomon SC, Head JW (1982b) Evolution of the Tharsis province of Mars; the importance of heterogenous lithospheric thickness and volcanic construction. J Geophys Res 87:9755–9774

Solomon SC, Head JW (1990) Lithospheric flexure beneath the Freyka Montes foredeep, Venus; constraints on lithospheric thermal gradient and heat flow. Geophys Res Lett 17:1393–1396

Solomon SC, Head JW (1991) Fundamental issues in the geology and geophysics of Venus. Science 252:252–259

Solomon SC, Head JW, Kaula WM, McKenzie D, Parsons B, Phillips RJ, Schubert G, Talwani M (1991) Venus; tectonics; initial analysis from Magellan. Science 252:297–312

Solomon SC, Smrekar SE, Bindschadler DL, Grimm RE, Kaula WM, McGill GE, Phillips RJ, Saunders RS, Schubert G, Squyres SW, Stofan ER (1992) Venus tectonics; an overview of Magellan observations. J Geophys Res 97:13 199–13 255

Sotin C, Senske DAS, Head JW, Parmentier EM (1989) Terrestrial spreading centers under Venus conditions; evaluation of a crustal model for western Aphrodite Terra. Earth Planet Sci Lett 95:321–333

Spencer JR (1984) A tectonic geomorphological classification of the walls of Valles Marineris. Reports of the Planetary Geology Program – 1983. National Aeronautics and Space Administration, TM-86246, pp 243–245

Spencer JR, Fanale FP (1990) New models for the origin of Valles Marineris closed depressions. J Geophys Res 95:14 301–14 313

Spergel DN, Turok NG (1992) Textures and cosmic structures. Sci Am 266(3):52–59

Spitzer L (1939) The dissipation of planetary filaments. Astrophys J 90:675–688

Spitzer L (1941) The encounter theory falls. Sky 5:6–7

Spudis PD, Guest JE (1988) Stratigraphy and geologic history of Mercury. In: Vilas F, Chapman CR, Matthews MS (eds) Mercury. University of Arizona Press, Tucson, pp 118–164

Spurr JE (1944) Geology applied to selenology. I. The Imbrian Plain region of the Moon, pp 1–112. II. The features of the Moon, pp 1–318. The Science Press, Lancaster

Squyres SW (1979) The distribution of lobate debris aprons and similar flows on Mars. J Geophys Res 84:8087–8096

Squyres SW (1980) Volume changes in Ganymede and Callisto and the origin of grooved terrain. Geophys Res Lett 7:793–796

Squyres SW (1984) The history of water on Mars. Annu Rev Earth Planet Sci 12:83–106

Squyres SW, Croft SK (1986) The tectonics of icy satellites. In: Burns JA, Matthews MS (eds) Satellites. University of Arizona Press, Tucson, pp 293–341

Squyres SW, Reynolds RT, Cassen PM, Peale SJ (1983a) Liquid water and active resurfacing on Europa. Nature 301:225–226

Squyres SW, Reynolds RT, Cassen PM, Peale SJ (1983b) The evolution of Enceladus. Icarus 53:319–331

Squyres SW, Jakowski DG, Simons M, Solomon SC, Hager BH, McGill GE (1992a) Plains tectonism on Venus; the deformation belts of Lavinia Planitia. J Geophys Res 97:13 579–13 599

Squyres SW, Janes DM, Baer G, Bindschadler DL, Schubert G, Sharpton VL, Stofan ER (1992b) The morphology and evolution of coronae on Venus. J Geophys Res 97:13 611–13 634

Stern A (1993) The Pluto reconnaissance flyby mission. EOS, American Geophysical Union, Trans 74:73, 76–78

Stern PC, Young OR, Druckman D (eds) (1992) Global environmental change; understanding the human dimensions:National Academy Press, Washington, DC, 308 pp

Stern SA (1991) On the number of planets in the outer Solar System; evidence of substantial population of 1000-km bodies. Icarus 90:271–281

Stevenson DJ (1978) The outer planets and their satellites. In: Dermott SF (ed) The origin of the Solar System. Wiley, New York, pp 395–432

Stevenson DJ (1982) Volcanism and igneous processes in small icy satellites. Nature 298:142–144

Stevenson DJ (1990) Fluid dynamics of core formation. In: Newson HE, Jones JH (eds) Origin of the Earth: Oxford University Press, New York; Lunar and Planetary Institute, Houston, pp 231–249

Stevenson DJ, Harris AW, Lunine JI (1986) Origins of satellites. In: Burns JA, Matthews MS (eds) Satellites. University of Arizona Press, Tucson, pp 39–88

Stofan ER (1992) Coronae on Venus; geologic characteristics of hotspots on Venus. American Geophysical Union, Spring Meet, Montreal, May 12–16, 1992, Program and Abstr, pp 177–178

Stofan ER, Saunders RS (1990) Geologic evidence of hotspot activity on Venus; predictions for Magellan. Geophys Res Lett 17:1377–1380

Stofan ER, Bindschadler DL, Head JW, Marcparmentier E (1991) Corona structures on Venus; models of origin. J Geophys Res 96:20 933–20 946

Stofan ER, Sharpton VL, Schubert G, Baer C, Bundschadler DL, Jones DM, Squyres SW (1992) Global distribution and characteristics of coronae and related features on Venus; implications for origin and relation to mantle processes. J Geophys Res 97:13 347–13 378

Stone EC, Miner ED (1981) Voyager 1 encounter with the saturnian system. Science 212:159–162

Stone EC, Miner ED (1986) The Voyager 2 encounter with the uraniam system. Science 233:39–43

Stone EC, Miner ED (1989) The Voyager 2 encounter with the neptunian system. Science 246:1417–1421

Strom RG (1977) Origin and relative age of lunar and mercurian intercrater plains. Phys Earth Planet Interiors 15:156–172

Strom RG (1984) Mercury. In: Carr MH (ed) The geology of the terrestrial planets. National Aeronautics and Space Administration, SP-469, pp 13–55

Strom RG, Fielder G (1971) Multiphase eruptions associated with craters Tycho and Aristarchus. In: Fielder G (ed) Geology and physics of the Moon. Elsevier, New York, pp 55–92

Strom RG, Neukum G (1988) The cratering record on Mercury and the origin of impacting objects. In: Vilas F, Chapman CR, Matthews MS (eds) Mercury. University of Arizona Press, Tucson, pp 336–373

Strom RG, Schneider NM (1982) Volcanic eruption plumes on Io. In: Morrison D (ed) Satellites of Jupiter. University of Arizona Press, Tucson, pp 598–633

Strom RG, Trask NJ, Guest JE (1975) Tectonism and volcanism on Mercury. J Geophys Res 80:2478–2507

Strom RG, Croft SK, Boyce JM (1990a) The impact cratering record on Triton. Science 250:437–439

Strom RG, Malin MC, Leake MA (1990b) Geologic map of the Bach region of Mercury. US Geological Survey, Miscel Invest Ser, Map I-2015d (H-15), 1 sheet

Strouhal E (1992) Life of the ancient Egyptians. University of Oklahoma Press, Norman, 279 pp

Struve O (1949) New trends in cosmogony. Sky Telescope 8 (12):302–305

Stuart-Alexander DE (1978) Geologic map of the central far side of the Moon. US Geological Survey, Miscel Invest Ser, Map I-1047, 1 sheet

Stump E (1992) The Ross Orogen of the Transantarctic Mountains in light of the Laurentia-Gondwana split. GSA Today 2:25–27, 30–31

Sukhanov AL, Pronin AA, Burba GA, Nikishin AM, Kryuchkov VP, Basilevsky AT, Markov MS, Kuzmin RO, Bobina NN, Shashkina VP, Slyuta EE, Chernaya IM (1989) Geomorphic/geologic map of part of the northern hemisphere of Venus, scale 1:15 000 000. US Geological Survey, Miscel Invest Ser, Map I-2059, 1 sheet

Surkov Yu A (1983) Studies of Venus rocks by Venera 8, 9 and 10. In: Hunten DM, Colin C, Donahue TM, Moroz VI (eds) Venus. University of Arizona Press, Tucson, pp 154–158

Swisher CCIII, Grajales-Nishimura JM, Montanari A, Margolis SV, Claeys P, Alvarez W, Renne P, Cedillo-Pardo E, Maurrasse FJ-MR, Curtis GH, Smit J, McWilliams MO (1992) Coeval ^{40}Ar/^{39}Ar ages of 65.0 million years ago from Chicxulub Crater melt rock and Cretaceous-Tertiary boundary tektites. Science 257:954–958

Synnott SP (1980) The discovery of a previously unknown jovian satellite. Science 210:786

Synnott SP (1981, 1979) J3; discovery of a previously unknown satellite of Jupiter. Science 212:1392

Tajika E, Matsui T (1990) The evolution of the terrestrial environment. In: Newson HE, Jones JH (eds) Origin of the Earth. Oxford University Press, New York; Lunar and Planetary Institute, Houston, pp 347–370

Tanaka KL, Chapman MG (1990) The relation of catastrophic flooding of Mangala Valles, Mars, to faulting of Memnonia Fossae and Tharsis volcanism. J Geophys Res 95:14 315–14 323

Tanaka KL, Golombek MP, Baneredt WB (1991) Reconciliation of stress and structural histories of the Tharsis region of Mars. J Geophys Res 96:15 617–15 633

Taylor EGR, Richey MW (1969) Le marin géométrique. Éditions Maritimes et d'Outre-Mer, Paris, 160 pp

Taylor GJ (1991) Diversity of basaltic impact melts on the Moon. American Geophysical Union, Spring Meet, Baltimore, May 28–31, 1991, Program and Abstr, p 179

Taylor SR (1975) Lunar science – a post apollo view; scientific results and insights from the lunar samples. Pergamon, New York, 372 pp

Taylor SR (1982) Planetary science – a lunar perspective. Lunar and Planetary Institute, Houston, 481 pp

Taylor SR, Norman MD (1990) Accretion of differentiated planetesimals to the Earth. In: Newson HE, Jones JH (eds) Origin of the Earth. Oxford University Press, New York; Lunar and Planetary Institute, Houston, pp 29–43

Theilig E, Greeley R (1979) Plains and channels in the Lunae Planum-Chryse Planitia region of Mars. J Geophys Res 84:7994–8010

Thein J (1987) A tektite layer in upper Eocene sediments of the New Jersey continental slope (Site 612, Leg 95). Initial Reports of the Deep Sea Drilling Project, US Government Printing Office, Washington, DC, vol 95, pp 565–574

Thomas P, Veverka J (1979) Seasonal and secular variation of wind streaks on Mars; an analysis of Mariner 9 and Viking data. J Geophys Res 84:8131–8146

Thomas P, Veverka J (1982) Amalthea. In: Morrison D (ed) Satellites of Jupiter. University of Arizona Press, Tucson, pp 147–173

Thomas P, Veverka J, Dermott S (1986) Small satellites. In: Burns JA, Matthews MS (eds) Satellites. University of Arizona Press, Tucson, pp 802–835

Thomas PG, Masson P, Fleitout L (1988) Tectonic history of Mercury. In: Vilas F, Chapman CR, Matthews MS (eds) Mercury. University of Arizona Press, Tucson, pp 401–428

Thomas P, Squyres SW, Carr MH (1990) Flank tectonics of martian volcanoes. J Geophys Res 95:14 345–14 355

Thompson WR, Sagan C (1990) Color and chemistry on Triton. Science 250:415–418

Tittman BR (1979) Brief note for consideration of active seismic exploration on Mars. J Geophys Res 84:7940–7942

Todesco E, Drummond J, Candy M, Birch P, Nikoloff I, Zellner B (1978) 1580 Betulia – an unusual asteroid with an extraordinary light curve. Icarus 35:344–359

Tonks WB, Melosh HJ (1990) The physics of crystal setting and suspension in a turbulent magma ocean. In: Newson HE, Jones JH (eds) Origin of the Earth. Oxford University Press, New York; Lunar and Planetary Institute, Houston, pp 151–174

Toulmin P III, Baird AK, Clark BC, Keil K, Rose HJ Jr, Christian RP, Evens PH, Kelliher WC (1977) Geochemical and mineralogical interpretation of the Viking inorganic chemical results. J Geophys Res 82:4625–4634

Trask NJ, Guest JE (1975) Preliminary geologic terrain map of Mercury. J Geophys Res 80:2461–2477

Tsoar H, Greeley R, Peterfreud AR (1979) Mars; the north polar sand sea and related wind patterns. J Geophys Res 84:8167–8180

Tucholke BE (1992) Massive submarine rockslide in the rift-valley wall of the Mid-Atlantic Ridge. Geology 20:129–132

Turco RP, Toon OB, Park C, Whitten RC, Pollack JB, Noerdlinger P (1982) An analysis of the physical, chemical, optical, and historical impacts of the 1908 Tungska meteor fall. Icarus 50:1–52

Turcotte DL (1991) Implications of Magellan images for rates of crustal production and recycling on Venus. American Geophysical Union, Spring Meet, Baltimore, May 28–31, 1991, Program and Abstr, p 173

Turcotte DL, Schubert G (1988) Tectonic implications and radiogenic noble gases in planetary atmospheres. Icarus 75:36–46

Twichell DC, Parson LM, Paull CK (1990) Variations in the styles of erosion along the Florida Escarpment, eastern Gulf of Mexico. Mar Pet Geol 7:253–266

Uchupi E, Emery KO (1991a) Genetic global geomorphology. In: Osborne RH (ed) From shoreline to abyss: contributions in marine geology honoring Francis Parker Shepard. Soc Econ Paleontol Mineral Spec Publ 46:273–290

Uchupi E, Emery KO (1991b) Pangaean divergent margins; historical perspective. Mar Geol 102:1–28

US Geological Survey (1976) Topographic map of Mars. US Geological Survey, Miscel Invest Ser, Map I-961, 1 sheet

US Geological Survey (1979) Shaded relief map of Mercury. US Geological Survey, Miscel Invest Ser, Map I-1149, 1 sheet

US Geological Survey (1985) Controlled photomosaic of the Ismenius Lacus south-central quadrangle of Mars (revised). US Geological Survey, Miscel Invest Ser, Map M2M39/326CH; MC-5S-C; Atlas of Mars, scale 2 000 000, Topographic Ser, 1 sheet

US Geological Survey (1988) The southern hemispheres of the uranian satellites. US Geological Survey, Miscel Invest Ser, Map I-1920, 3 sheets

US Geological Survey (1989) Topographic maps of the western, eastern equatorial and polar regions of Mars. US Geological Survey, Miscel Invest Ser, Map I-2030, 3 sheets

Veverka J, Duxbury TC (1977) Viking observations of Phobos and Deimos. J Geophys Res 82:4213–4224

Veverka J, Thomas P, Davies D, Morrison D (1981) Amalthea; Voyager imaging results. J Geophys Res 86:8675–8692

Veverka J, Brown RH, Bell JF (1991) Uranus satellites; surface properties. In: Bergstrahl JT, Miner ED, Matthews MS (eds) Uranus. University of Arizona Press, Tucson, pp 528–560

Veverka J, Belton M, Chapman C, and the Galileo Imaging Team (1993) 951 Gaspra: first pictures of an asteroid. EOS, American Geophysical Union, Trans 74:70

Vilas F (1988) Surface composition of Mercury from reflectance spectrophotometry. In: Vilas F, Chapman CR, Matthews MS (eds) Mercury. University of Arizona Press, Tucson, pp 59–76

Vogt PR, Perry RK, Feden RH, Fleming HS, Cherkis NZ (1981) The Greenland – Norwegian Sea and Iceland environment; geology and geophysics. In: Nairn AEM, Churkin M Jr, Stehli FG (eds) The ocean basins and margins 5. The Arctic Ocean. Plenum Press, New York, pp 493–598

Vorder Bruegge RW, Head JW (1990) Tectonic evolution of eastern Ishtar Terra, Venus. Earth, Moon Planets 50/51:251–304

Voyager Imaging Science Team (1988) The southern hemispheres of the uranian satellites. US Geological Survey, Miscel Invest Ser, Map I-1920, 3 sheets

Walker D (1983) Lunar terrestrial crust formation. J Geophys Res 88 (Suppl):B17–B25

Walker JCD (1985) Carbon dioxide on the early Earth. Origins Life 16:117–127

Walker JCG (1990a) Precambrian evolution of the climate system. Palaeogeogr, Palaeoclimatol, Paleoecol 82:261–289

Walker JCG (1990b) Origin of an inhabited planet. In: Newson HE, Jones JH (eds) Origin of the Earth. Oxford University Press, New York; Lunar and Planetary Institute, Houston, pp 371–375

Wallace D, Sagan C (1979) Evaporation of ice in planetary atmospheres; ice–covered rivers on Mars. Icarus 39:385–400

Wang K (1992) Glassy microspheres (microtektites) from an Upper Devonian limestone. Science 256:1547–1550

Wang S, Yuezhen W, Mengxian B, Liwu D, Sufang W (1981) A possible satellite of 9 Metis. Icarus 46:285–287

Ward AW (1979) Yardangs on Mars; evidence of recent wind erosion. J Geophys Res 84:8147–8165

Ward WR (1973) Large–scale variations in the obliquity of Mars. Science 181:260–262

Ward WR (1979) Present obliquity oscillations of Mars; fourth–order accuracy in orbital e and I. J Geophys Res 84:237–241

Warner JL, Morrison DA (1978) Planetary tectonics. I. The role of water. Abstracts of Papers submitted to the 9th Lunar and Planetary Science Conf Houston, vol 9, p 1217

Warren PH (1985) The magma ocean concept and lunar evolution. Earth Planet Sci Annu Rev 13:201–240

Warren PH (1987) Mars regolith versus SNC meteorites; evidence for abundant crustal carbonates? Icarus 70:153–161

Wasson JT (1988) The building stones of the planets. In: Vilas F, Chapman CR, Matthews MS (eds) Mercury. University of Arizona Press, Tucson, pp 622–650

Watson AJ, Donahue TM, Kuhn WR (1984) Temperatures in a runaway greenhouse on the evolving Venus. Earth Planet Sci Lett 68:1–6

Watters TR (1991) Origin of periodically spaced wrinkle ridges on the Tharsis Plateau of Mars. J Geophys Res 96; E1:15 599–15 616

Wegener A (1921) Die Entstehung der Mondkrater, vol 55. Sammlung Vieweg, Braunschweig, 48 pp (translated into English by A.M. Celâl Sengör, 1975, The origin of lunar craters. Moon 14:211–236)

Weidenschilling SJ (1980) Hector – nature and origin of a binary asteroid. Icarus 44:807–809

Weidenschilling SJ (1991) A plurality of worlds. Science 352:190–192

Weissman PR (1984) Cometary showers and unseen solar companions. Nature 380:380

Weissman PR (1989) The impact history of the Solar System; implications for the origin of atmospheres. In: Atreya SK, Pollack JB, Matthews MS (eds) Origin and evolution of planetary atmospheres. University of Arizona Press, Tucson, pp 230–267

Weissman PR (1990) Are periodic bombardments real? Sky Telescope 79:266–271

Wells HG (1898) War of the worlds. Harper, New York, 173 pp

Wetherill GW (1976) The role of large bodies in the formation of the Earth. Proc 7th Lunar Science Conf, pp 3245–3257

Wetherill GW (1985) Occurrence of giant impacts during the growth of the terrestrial planets. Science 228:877–879

Wetherill GW (1988) Accumulation of Mercury from planetesimals. In: Vilas F, Chapman CR, Matthews MS (eds) Mercury. University of Arizona Press, Tucson, pp 670–791

Wetherill GW, Chapman CR (1988) Asteroids and meteorites. In: Kerridge JF, Matthews MS (eds) Meteorites and the early Solar System. University of Arizona Press, Tucson, pp 35–67

Whitmore DP, Jackson AAIV (1984) Are periodic mass extinctions driven by a distant solar companion? Nature 308:713–715

Whitmore DP, Matese JJ (1985) Periodic comet showers and planet X. Nature 313:36–37

Wichman RW, Schultz PH (1989) Sequence and mechanisms of deformation around the Hellas and Isidis impact basins on Mars. J Geophys Res 94:17 333–17 357

Wiens RC, Beckert RH, Pepin RO (1986) The case for a martian origin of the shergottites, II Trapped and indigenous gas components in the EETA 79001 glass. Earth Planetary Sci Lett 77:149–158

Wilford JN (1990) Mars beckons – the mysteries, the challenges, the expectations of our next great adventure in space. Knopf, New York, 244 pp

Wilhelms DE (1974) Comparison of martian and lunar geologic provinces. J Geophys Res 79:3933–3941

Wilhelms DE (1984) Moon. In: Carr MH (ed) The geology of the terrestrial planets. National Aeronautics and Space Administration, Washington, DC, SP-469, pp 107–206

Wilhelms DE (1987) The geologic history of the Moon. US Geological Survey, Professional Paper 1348, 302 pp

Wilhelms DE, El-Baz F (1977) Geologic map of the east side of the Moon. US Geological Survey, Miscel Invest Ser, Map I-948, 1 sheet

Wilhelms DE, McCauley JF (1971) Geologic map of the near side of the Moon. US Geological Survey, Miscel Invest Ser, Map I-703, 1 sheet

Wilhelms DE, Squyres SW (1984) The martian hemisphere dichotomy may be due to a giant impact. Nature 309:138–140

Wilkening LL (1978) Carbonaceous chondritic material in the Solar System. Naturwissenschaften 65:73–79

Willey GR (1966) An introduction to American archeology, vol 1, North and Middle America:Prentice-Hall, Englewood Cliffs, 530 pp

Williams DR, Gaddis L (1991) Stress analysis of Tellus Regio, Venus, based on gravity and topography; comparison with Venera 15/16 radar images. J Geophys Res 96:18 841–18 859

Williams DR, Pan V (1990) Consequences of a hydrous mantle for Venus tectonics; predictions for Magellan. Geophys Res Lett 17:1397–1400

Williams IP (1978) Chemical segregation in protoplanetary theories. In:Dermott S F (ed) The origin of the Solar System. Wiley, New York, pp 517–520

Wilson KM, Hay WW, Wold CN (1991) Mesozoic evolution of exotic terranes and marginal seas, western North America. Mar Geol 102:311–361

Wilson PW (1937) The romance of the calendar. Norton, New York, 351 pp

Wise DU (1966) Origin of the Moon by fission. In: Marsden B, Cameron A (eds) The Earth-Moon system. Plenum Press, New York, pp 213–223

Wise DU, Golombek MP, McGill GE (1979) Tectonic evolution of Mars. J Geophys Res 84:7934–7939

Wolfteich C (1988) SNC meteorites and the case for a Martian origin. University of Houston, Unpubl Rep, 19 pp

Wolfteich C (1989) CI and CM chondrites – source regions and distribution in the Solar System. University of Houston, unpubl report, 19 pp

Wood CA, Ashwal LD (1981) SNC meteorites; igneous rocks from Mars. 12th Lunar and Planetary Science Conf, Proc, pp 1359–1375

Wood CA, Head JW, Cintala MJ (1978) Interior morphology of fresh martian craters; the effects of target characteristics. 9th Lunar and Planetary Science Conf, Proc, pp 3691–3709

Wood JA (1986) Moon over Mauna Loa; review of hypotheses of formation of the Moon. In: Hartmann WK, Phillips RJ, Taylor GJ (eds) Origin of the Moon. Lunar Planetary Institute, Houston, pp 17–55

Woolfson MM (1978) The evolution of the Solar System. In: Dermott SF (ed) The origin of the Solar System. Wiley, New York, pp 199–218

Woronow A (1977) A size-frequency study of large martian craters. J Geophys Res 82:5807–5820

Woronow A, Strom RG, Gurnis G (1982) Interpreting the cratering record; Mercury to Ganymede and Callisto. In: Morrison D (ed) Satellites of Jupiter. University of Arizona Press, Tucson, pp 237–276

Worsley TR, Nance RD, Moody JB (1986) Tectonic cycles and the history of the Earth's biogeochemical and paleoceanographic record. Paleoceanogr 1:233–263

Wright FE (1935) The surface features of the Moon. Sci Mon 40:101–115

Wright JD (1993) Deglaciation triggered by the resumption of North Atlantic Deep Water. EOS, American Geophysical Union, Trans 74:24

Wu SSC, Schafer FJ, Nakata GM, Jordan R, Blasius KR (1973) Photogrammetric evaluation of Mariner 9 photography. J Geophys Res 78:4405–4410

Yelle RV (1992) The effect of surface roughness on Triton's volatile distribution. Science 255:1553–1555

Zahnle KJ (1992) Airburst origin of dark shadows on Venus. J Geophys Res 97:10 243–10 255

Zimbelman JR, Edgett KS (1992) The Tharsis Montes, Mars: comparison of volcanic and modified landforms. 23rd Lunar and Planetary Science Conf, Proc vol 22, pp 31–44

Zoback ML (1992) First- and second-order patterns of stress in the lithosphere; the world stress map project. J Geophys Res 97:11 703–11 728

Zolensky M, McSween HY (1988) Aqueous alteration. In: Kerridge JF, Matthews MS (eds) Meteorites and the early Solar System. University of Arizona Press, Tucson, pp 114–143

Zuber MT (1990) Ridge belts; evidence for regional- and local-scale deformation on the surface of Venus. Geophys Res Lett 17:1369–1372

Zuber MT, Aist, LL (1990) The shallow structure of the martian lithosphere in the vicinity of the ridged plains. J Geophys Res 95:14 215–14 230

Index

Springer-Verlag
and the Environment

We at Springer-Verlag firmly believe that an international science publisher has a special obligation to the environment, and our corporate policies consistently reflect this conviction.

We also expect our business partners – paper mills, printers, packaging manufacturers, etc. – to commit themselves to using environmentally friendly materials and production processes.

The paper in this book is made from low- or no-chlorine pulp and is acid free, in conformance with international standards for paper permanency.